T0310052

MICROWAVE AND MILLIMETER WAVE CIRCUITS AND SYSTEMS

MICROWAVE AND MILLIMETER WAVE CIRCUITS AND SYSTEMS

EMERGING DESIGN, TECHNOLOGIES, AND APPLICATIONS

Editors

Apostolos Georgiadis
CTTC, Spain

Hendrik Rogier
Ghent University, Belgium

Luca Roselli
University of Perugia, Italy

Paolo Arcioni
University of Pavia, Italy

A John Wiley & Sons, Ltd., Publication

This edition first published 2013
© 2013 John Wiley & Sons, Ltd

Registered office
John Wiley & Sons, Ltd, The Atrium, Southern Gate, Chichester, West Sussex, PO19 8SQ, United Kingdom

For details of our global editorial offices, for customer services and for information about how to apply for permission to reuse the copyright material in this book please see our website at www.wiley.com.

The right of the author to be identified as the author of this work has been asserted in accordance with the Copyright, Designs and Patents Act 1988.

All rights reserved. No part of this publication may be reproduced, stored in a retrieval system, or transmitted, in any form or by any means, electronic, mechanical, photocopying, recording or otherwise, except as permitted by the UK Copyright, Designs and Patents Act 1988, without the prior permission of the publisher.

COST logos have been reproduced by permission of © 2012 COST Office. The EU flag is reproduced courtesy of © European Union, 1995-2012 (http://europa.eu/geninfo/legal_notices_en.html).

Wiley also publishes its books in a variety of electronic formats. Some content that appears in print may not be available in electronic books.

Designations used by companies to distinguish their products are often claimed as trademarks. All brand names and product names used in this book are trade names, service marks, trademarks or registered trademarks of their respective owners. The publisher is not associated with any product or vendor mentioned in this book. This publication is designed to provide accurate and authoritative information in regard to the subject matter covered. It is sold on the understanding that the publisher is not engaged in rendering professional services. If professional advice or other expert assistance is required, the services of a competent professional should be sought.

Library of Congress Cataloging-in-Publication Data

Microwave and millimeter wave circuits and systems : emerging design, technologies, and applications / editors, Apostolos Georgiadis ... [et al.].
    p. cm.
  Includes bibliographical references and index.
  ISBN 978-1-119-94494-2 (cloth)
    1. Microwave circuits–European Union countries–Design and construction. 2. Electronic circuit design 3. Millimeter waves–Research–European Union countries. I. Georgiadis, Apostolos.
  TK7876.M5225 2012
  621.381'32–dc23

                                                                                    2012015786

A catalogue record for this book is available from the British Library.

ISBN: 9781119944942

Set in 10/12pt Times by Thomson Digital, Noida, India.
Printed and bound in Singapore by Markono Print Media Pte Ltd.

Contents

About the Editors   xiii

About the Authors   xvii

Preface   xxxi

List of Abbreviations   xli

List of Symbols   xlv

Part I   DESIGN AND MODELING TRENDS

1   Low Coefficient Accurate Nonlinear Microwave and Millimeter Wave
     Nonlinear Transmitter Power Amplifier Behavioural Models   3
     1.1   Introduction   3
            1.1.1   Chapter Structure   4
            1.1.2   LDMOS PA Measurements   4
            1.1.3   BF Model   7
            1.1.4   Modified BF Model (MBF) – Derivation   8
            1.1.5   MBF Models of an LDMOS PA   13
            1.1.6   MBF Model – Accuracy and Performance Comparisons   15
            1.1.7   MBF Model – the Memoryless PA Behavioural Model of Choice 22
            Acknowledgements   24
            References   24

2   Artificial Neural Network in Microwave Cavity Filter Tuning   27
     2.1   Introduction   27
     2.2   Artificial Neural Networks Filter Tuning   28
            2.2.1   The Inverse Model of the Filter   29
            2.2.2   Sequential Method   30
            2.2.3   Parallel Method   31
            2.2.4   Discussion on the ANN's Input Data   33
     2.3   Practical Implementation – Tuning Experiments   36
            2.3.1   Sequential Method   36
            2.3.2   Parallel Method   41

2.4   Influence of the Filter Characteristic Domain on Algorithm Efficiency   43
2.5   Robots in the Microwave Filter Tuning   47
2.6   Conclusions   49
       Acknowledgement   49
       References   49

3   Wideband Directive Antennas with High Impedance Surfaces   51
3.1   Introduction   51
3.2   High Impedance Surfaces (HIS) Used as an Artificial Magnetic Conductor
       (AMC) for Antenna Applications   52
       3.2.1   AMC Characterization   52
       3.2.2   Antenna over AMC: Principle   55
       3.2.3   AMC's Wideband Issues   55
3.3   Wideband Directive Antenna Using AMC with a Lumped Element   57
       3.3.1   Bow-Tie Antenna in Free Space   57
       3.3.2   AMC Reflector Design   59
       3.3.3   Performances of the Bow-Tie Antenna over AMC   60
       3.3.4   AMC Optimization   61
3.4   Wideband Directive Antenna Using a Hybrid AMC   64
       3.4.1   Performances of a Diamond Dipole Antenna over the AMC   65
       3.4.2   Beam Splitting Identification and Cancellation Method   69
       3.4.3   Performances with the Hybrid AMC   73
3.5   Conclusion   78
       Acknowledgments   80
       References   80

**4   Characterization of Software-Defined and Cognitive Radio Front-Ends for
       Multimode Operation   83**
4.1   Introduction   83
4.2   Multiband Multimode Receiver Architectures   84
4.3   Wideband Nonlinear Behavioral Modeling   87
       4.3.1   Details of the BPSR Architecture   87
       4.3.2   Proposed Wideband Behavioral Model   89
       4.3.3   Parameter Extraction Procedure   92
4.4   Model Validation with a QPSK Signal   95
       4.4.1   Frequency Domain Results   95
       4.4.2   Symbol Evaluation Results   98
       References   99

**5   Impact and Digital Suppression of Oscillator Phase Noise in Radio
       Communications   103**
5.1   Introduction   103
5.2   Phase Noise Modelling   104
       5.2.1   Free-Running Oscillator   104
       5.2.2   Phase-Locked Loop Oscillator   105
       5.2.3   Generalized Oscillator   107

5.3   OFDM Radio Link Modelling and Performance under Phase Noise   109
        5.3.1   Effect of Phase Noise in Direct-Conversion Receivers   110
        5.3.2   Effect of Phase Noise and the Signal Model on OFDM   110
        5.3.3   OFDM Link SINR Analysis under Phase Noise   113
        5.3.4   OFDM Link Capacity Analysis under Phase Noise   114
5.4   Digital Phase Noise Suppression   118
        5.4.1   State of the Art in Phase Noise Estimation and Mitigation 119
        5.4.2   Recent Contributions to Phase Noise Estimation and
                Mitigation   122
        5.4.3   Performance of the Algorithms   128
5.5   Conclusions   129
        Acknowledgements   131
        References   131

6   **A Pragmatic Approach to Cooperative Positioning in Wireless
    Sensor Networks**   **135**
    6.1   Introduction   135
    6.2   Localization in Wireless Sensor Networks   136
            6.2.1   Range-Free Methods   136
            6.2.2   Range-Based Methods   139
            6.2.3   Cooperative versus Noncooperative   142
    6.3   Cooperative Positioning   142
            6.3.1   Centralized Algorithms   143
            6.3.2   Distributed Algorithms   144
    6.4   RSS-Based Cooperative Positioning   147
            6.4.1   Measurement Phase   147
            6.4.2   Location Update Phase   148
    6.5   Node Selection   150
            6.5.1   Energy Consumption Model   152
            6.5.2   Node Selection Mechanisms   153
            6.5.3   Joint Node Selection and Path Loss Exponent Estimation 156
    6.6   Numerical Results   160
            6.6.1   OLPL-NS-LS Performance   164
            6.6.2   Comparison with Existing Methods   164
    6.7   Experimental Results   166
            6.7.1   Scenario 1   166
            6.7.2   Scenario 2   169
    6.8   Conclusions   169
            References   170

7   **Modelling of Substrate Noise and Mitigation Schemes for UWB Systems** **173**
    7.1   Introduction   173
            7.1.1   Ultra Wideband Systems – Developments and Challenges 174
            7.1.2   Switching Noise – Origin and Coupling Mechanisms   175
    7.2   Impact Evaluation of Substrate Noise   176
            7.2.1   Experimental Impact Evaluation on a UWB LNA   177

	7.2.2	Results and Discussion	178
	7.2.3	Conclusion	181
7.3	Analytical Modelling of Switching Noise in Lightly Doped Substrate		182
	7.3.1	Introduction	182
	7.3.2	The GAP Model	185
	7.3.3	The Statistic Model	192
	7.3.4	Conclusion	195
7.4	Substrate Noise Suppression and Isolation for UWB Systems		195
	7.4.1	Introduction	195
	7.4.2	Active Suppression of Switching Noise in Mixed-Signal Integrated Circuits	196
7.5	Summary		204
	References		205

Part II   APPLICATIONS

8   Short-Range Tracking of Moving Targets by a Handheld UWB Radar System   **209**
8.1	Introduction	209
8.2	Handheld UWB Radar System	210
8.3	UWB Radar Signal Processing	210
	8.3.1 Raw Radar Data Preprocessing	211
	8.3.2 Background Subtraction	212
	8.3.3 Weak Signal Enhancement	213
	8.3.4 Target Detection	214
	8.3.5 Time-of-Arrival Estimation	215
	8.3.6 Target Localization	217
	8.3.7 Target Tracking	217
8.4	Short-Range Tracking Illustration	218
8.5	Conclusions	223
	Acknowledgement	224
	References	224

9   Advances in the Theory and Implementation of GNSS Antenna Array Receivers   **227**
9.1	Introduction	227
9.2	GNSS: Satellite-Based Navigation Systems	228
9.3	Challenges in the Acquisition and Tracking of GNSS Signals	230
	9.3.1 Interferences	232
	9.3.2 Multipath Propagation	232
9.4	Design of Antenna Arrays for GNSS	233
	9.4.1 Hardware Components Design	234
	9.4.2 Array Signal Processing in the Digital Domain	239
9.5	Receiver Implementation Trade-Offs	244
	9.5.1 Computational Resources Required	244

	9.5.2	*Clock Domain Crossing in FPGAs/Synchronization Issues*	247
9.6	Practical Examples of Experimentation Systems		248
	9.6.1	*L1 Array Receiver of CTTC, Spain*	248
	9.6.2	*GALANT, a Multifrequency GPS/Galileo Array Receiver of DLR, Germany*	253
	References		272

10 Multiband RF Front-Ends for Radar and Communications Applications   275

10.1	Introduction		275
	10.1.1	*Standard Approaches for RF Front-Ends*	275
	10.1.2	*Acquisition of Multiband Signals*	276
	10.1.3	*The Direct-Sampling Architecture*	277
10.2	Minimum Sub-Nyquist Sampling		278
	10.2.1	*Mathematical Approach*	278
	10.2.2	*Acquisition of Dual-Band Signals*	279
	10.2.3	*Acquisition of Evenly Spaced Equal-Bandwidth Multiband Signals*	282
10.3	Simulation Results		284
	10.3.1	*Symmetrical and Asymmetrical Cases*	284
	10.3.2	*Verification of the Mathematical Framework*	285
10.4	Design of Signal-Interference Multiband Bandpass Filters		287
	10.4.1	*Evenly Spaced Equal-Bandwidth Multiband Bandpass Filters*	288
	10.4.2	*Stepped-Impedance Line Asymmetrical Multiband Bandpass Filters*	289
10.5	Building and Testing of Direct-Sampling RF Front-Ends		290
	10.5.1	*Quad-Band Bandpass Filter*	290
	10.5.2	*Asymmetrical Dual-Band Bandpass Filter*	291
10.6	Conclusions		293
	References		294

11 Mm-Wave Broadband Wireless Systems and Enabling MMIC Technologies   295

11.1	Introduction		295
11.2	V-Band Standards and Applications		297
	11.2.1	*IEEE 802.15.3c Standard*	297
	11.2.2	*ECMA-387 Standard*	299
	11.2.3	*WirelessHD*	300
	11.2.4	*WiGig Standard*	301
11.3	V-Band System Architectures		302
	11.3.1	*Super-Heterodyne Architecture*	302
	11.3.2	*Direct Conversion Architecture*	303
	11.3.3	*Bits to RF and RF to Bits Radio Architectures*	305
11.4	SiGeV-Band MMIC		306
	11.4.1	*Voltage Controlled Oscillator*	307
	11.4.2	*Active Receive Balun*	310
	11.4.3	*On-Chip Butler Matrix*	313
	11.4.4	*High GBPsSiGeV-Band SPST Switch Design Considerations*	317

11.5   Outlook   320
       References   322

12   Reconfigurable RF Circuits and RF-MEMS   **325**
12.1   Introduction   325
12.2   Reconfigurable RF Circuits – Transistor-Based Solutions   326
      12.2.1   *Programmable Microwave Function Arrays*   326
      12.2.2   *PROMFA Concept*   327
      12.2.3   *Design Example: Tunable Band Passfilter*   331
      12.2.4   *Design Examples: Beamforming Network, LNA and VCO*   333
12.3   Reconfigurable RF Circuits Using RF-MEMS   335
      12.3.1   *Integration of RF-MEMS and Active RF Devices*   336
      12.3.2   *Monolithic Integration of RF-MEMS in GaAs/GaN*
                MMIC Processes   337
      12.3.3   *Monolithic Integration of RF-MEMS in SiGeBiCMOS Process*   342
      12.3.4   *Design Example: RF-MEMS Reconfigurable LNA*   344
      12.3.5   *RF-MEMS-Based Phase Shifters for Electronic Beam Steering*   348
12.4   Conclusions   353
       References   353

13   MIOS: Millimeter Wave Radiometers for the Space-Based Observation
of the Sun   **357**
13.1   Introduction   357
13.2   Scientific Background   358
13.3   Quiet-Sun Spectral Flux Density   359
13.4   Radiation Mechanism in Flares   361
13.5   Open Problems   361
13.6   Solar Flares Spectral Flux Density   363
13.7   Solar Flares Peak Flux Distribution   364
13.8   Atmospheric Variability   365
13.9   Ionospheric Variability   366
13.10   Antenna Design   369
13.11   Antenna Noise Temperature   371
13.12   Antenna Pointing and Radiometric Background   373
13.13   Instrument Resolution   373
13.14   System Overview   374
13.15   System Design   376
13.16   Calibration Circuitry   378
13.17   Retrieval Equations   381
13.18   Periodicity of the Calibrations   381
13.19   Conclusions   384
       References   384

14   Active Antennas in Substrate Integrated Waveguide (SIW) Technology   **387**
14.1   Introduction   387
14.2   Substrate Integrated Waveguide Technology   388

14.3 Passive SIW Cavity-Backed Antennas   388
    14.3.1 Passive SIW Patch Cavity-Backed Antenna   389
    14.3.2 Passive SIW Slot Cavity-Backed Antenna   391
14.4 SIW Cavity-Backed Antenna Oscillators   395
    14.4.1 SIW Cavity-Backed Patch Antenna Oscillator   395
    14.4.2 SIW Cavity-Backed Slot Antenna Oscillator with
    *   Frequency Tuning*   397
    14.4.3 Compact SIW Patch Antenna Oscillator with
    *   Frequency Tuning*   401
14.5 SIW-Based Coupled Oscillator Arrays   406
    14.5.1 Design of Coupled Oscillator Systems for Power Combining   407
    14.5.2 Coupled Oscillator Array with Beam-Scanning Capabilities   412
14.6 Conclusions   414
    References   415

15 Active Wearable Antenna Modules   **417**
15.1 Introduction   417
15.2 Electromagnetic Characterization of Fabrics and Flexible Foam Materials   419
    15.2.1 Electromagnetic Property Considerations for Wearable Antenna
    *   Materials*   419
    15.2.2 Characterization Techniques Applied to Wearable Antenna
    *   Materials*   419
    15.2.3 Matrix-Pencil Two-Line Method   420
    15.2.4 Small-Band Inverse Planar Antenna Resonator Method   427
15.3 Active Antenna Modules for Wearable Textile Systems   436
    15.3.1 Active Wearable Antenna with Optimized Noise Characteristics   436
    15.3.2 Solar Cell Integration with Wearable Textile Antennas   445
15.4 Conclusions   451
    References   452

16 Novel Wearable Sensors for Body Area Network Applications   **455**
16.1 Body Area Networks   455
    16.1.1 Potential Sheet-Shaped Communication Surface Configurations   456
    16.1.2 Wireless Body Area Network   460
    16.1.3 Chapter Flow Summary   460
16.2 Design of a 2-D Array Free Access Mat   460
    16.2.1 Coupling of External Antennas   462
    16.2.2 2-D Array Performance Characterization by Measurement   464
    16.2.3 Accessible Range of External Antennas on the 2-D Array   467
16.3 Textile-Based Free Access Mat: Flexible Interface for Body-Centric
    Wireless Communications   467
    16.3.1 Wearable Waveguide   470
    16.3.2 Summary on the Proposed Wearable Waveguide   475
16.4 Proposed WBAN Application   476
    16.4.1 Concept   476

16.5 Summary 478
Acknowledgment 478
References 478

17 Wideband Antennas for Wireless Technologies: Trends and Applications **481**
17.1 Introduction 481
  17.1.1 Antenna Concept 482
17.2 Wideband Antennas 483
  17.2.1 Travelling Wave Antennas 483
  17.2.2 Frequency Independent Antennas 484
  17.2.3 Self-Complementary Antennas 485
  17.2.4 Applications 486
  17.2.5 Ultra Wideband (UWB) Arrays: Vivaldi Antenna Arrays 489
  17.2.6 Wideband Microstrip Antennas: Stacked Patch Antennas 495
17.3 Antenna Measurements 496
17.4 Antenna Trends and Applications 498
  17.4.1 Phase Arrays and Smart Antennas 499
  17.4.2 Wearable Antennas 502
  17.4.3 Capsule Antennas for Medical Monitoring 503
  17.4.4 RF Hyperthermia 503
  17.4.5 Wireless Energy Transfer 503
  17.4.6 Implantable Antennas 503
Acknowledgements 504
References 504

18 Concluding Remarks **509**

Index **511**

About the Editors

Apostolos Georgiadis was born in Thessaloniki, Greece. He received his BS degree in Physics and MS degree in Telecommunications from the Aristotle University of Thessaloniki, Greece, in 1993 and 1996, respectively. He received his PhD degree in Electrical Engineering from the University of Massachusetts at Amherst, in 2002. Since 2007, he has been a Senior Research Associate at Centre Tecnològic de Telecomunicacions de Catalunya (CTTC), Barcelona, Spain, in the area of communications subsystems, where he is involved in active antennas and antenna arrays and more recently with radio frequency identification (RFID) technology and energy harvesting. Dr. Georgiadis is an IEEE Senior Member. He was the recipient of a 1996 Fulbright Scholarship for graduate studies with the University of Massachusetts at Amherst, the 1997 and 1998 Outstanding Teaching Assistant Award presented by the University of Massachusetts at Amherst, the 1999, 2000 Eugene M. Isenberg Award presented by the Isenberg School of Management, University of Massachusetts at Amherst, and the 2004 Juan de la Cierva Fellowship presented by the Spanish Ministry of Education and Science. He is involved in a number of technical program committees and serves as a reviewer for several journals, including *IEEE Transactions on Antennas and Propagation* and *IEEE Transactions on Microwave Theory and Techniques*. He is the Chairman of COST Action IC0803, RF/Microwave Communication Subsystems for Emerging Wireless Technologies (RFCSET), and the Coordinator of the Marie Curie Industry–Academia Pathways and Partnerships Project on the Symbiotic Wireless Autonomous Powered (SWAP) System. He is a Member of Technical Committees MTT-24 – RFID Technologies – and MTT-26 – Wireless Power Transfer. He was the Chair of the 2012 IEEE RFID-TA Conference and 2012 IEEE MTT-S IMWS Workshop on Millimeter Wave Integration Technologies.

Hendrik Rogier was born in 1971. He received Electrical Engineering and PhD degrees from Ghent University, Gent, Belgium, in 1994 and in 1999, respectively. He is currently Associate Professor with the Department of Information Technology. From October 2003 to April 2004, he was a Visiting Scientist at the Mobile Communications Group of Vienna University of Technology. He authored and co-authored about 75 papers in international journals and about 100 contributions in conference proceedings. He is serving as a member of the Editorial Board of IET Science, Measurement Technology and acts as the URSI Commission B representative for Belgium. His current research interests are the analysis of electromagnetic waveguides, electromagnetic simulation techniques applied to electromagnetic compatibility (EMC), and signal integrity (SI) problems, as well as to indoor propagation and antenna design, and in smart antenna systems for wireless networks. Dr. Rogier was twice awarded the URSI Young Scientist Award, at the 2001 URSI Symposium on Electromagnetic Theory and at the 2002 URSI General Assembly. He is a Senior Member of the IEEE.

Luca Roselli received the Laurea degree in Electronic Engineering from the University of Florence, Florence, Italy, in 1988. From 1988 to 1991 he worked at the University of Florence on SAW devices. In November 1991, he joined the Institute of Electronics at the University of Perugia, Perugia, Italy, as a Research Assistant. Since 1992 he has been an IEEE Member. Since 1994 he has been teaching Electronic Devices at the same University. In the same year he was the Director of Works for the realization of a 35 square meters clean room for a 100 class at the University of Perugia. Since 1996 he has been in the reviewer list of the *IEEE Microwave and Guided Wave Letters* (MGWL), now MWCL. Since 1998 he has been in the reviewer list of the *IEEE Transactions on Microwave Theory and Techniques* (MTT). Since that same year he has been a member of the technical program committee of the Electrosoft Conference and of the MTT–International Microwave Symposium. Since 2000 he has been Associate Professor at the University of Perugia where he teaches Microwave Electronics, High Frequency Components and Applied Electronics. Since that time he has been coordinating the research activity of the High Frequency Electronics (HFE) Lab. In the same year he founded the WiS (Wireless Solutions) Srl, a spin-off company operating in the field of microwave electronic systems with which he is currently cooperating as a consultant. In 2001 He was nominated an IEEE Senior Member. Since 2003 he has been in the steering committee of CEM-TD (Computational ElectroMagnetic-Time Domain). In the same year he was a member of the spin-off committee of the University of Perugia and in 2005 he founded a second spin-off company: DiES (Digital Electronic Solutions) Srl. In 2007 he was Chairman of the VII Computational Electromagnetic in the Time Domain Workshop; he was the guest editor of the special issue of the *International Journal of Numerical Modelling* on the VII CEM-TD. Again in 2008 he was the Co-PI (Principal Investigator) of the Project ADAHELI (a project founded by the Italian Space Agency – ASI – for the solar observation from the satellite. He is a Chairman of Technical Committee MTT-24 – RFID Technologies – and Member of MTT-25 – RF Nanotechnologies – and MTT-26 – Wireless Power Transfer. He is Member of the list of experts of the Italian Ministry of Research and University (MIUR), Director of the Science and Technology Committee of the Research Centre 'il Pischiello', and Coordinator of the High Frequency Electronic Lab (HFE-Lab) at the University of Perugia. His research interests mainly focus on the design of high-frequency electronic circuits and systems, including the development of numerical methods for electronic circuit analysis with special attention to RFID-NFC systems, new material electronics, and near-field wireless power transfer. In these fields he has published almost 200 contributions to international reviews and peer-reviewed conferences, the interest in which is testified by an HF index of 18 (PoP font) and more than 900 citations on international reviews and peer-reviewed conferences.

Paolo Arcioni is Full professor at the Department of Electronics of the University of Pavia. His research activity started in 1973 and first concerned the design and the characterization of compensated structures for linear accelerators, the development of microwave equipment for EPR, and the investigation of ferrite tuning of power magnetrons. Subsequently, his scientific interest concentrated on the study of a novel class of numerical methods, based on a hybrid representation of the electromagnetic field (Boundary Integral – Resonant Mode Expansion, BI-RME). These methods have been applied successfully to the development of efficient codes for the design and the optimization of waveguide components and to the modeling of active and passive quasi-optical components for millimeter and submillimeter wavelengths. More recently, the same approach has been extended to the modeling of microstrip circuits on multilayered substrates. At present, he is involved in the electromagnetic modeling of integrate passive components for millimeter-wave circuits realized in standard CMOS technology. He collaborated with many research groups in Italy and all over the world, as well as in the framework of a project funded by the European Space Agency, by the European Commission (TMR and RTN Networks), and by the Italian Ministry of Education (PRIN and FIRB Projects). He also carried out many applied research activities commissioned by the Department by important industries and small-medium enterprises active in the microwave field. He authored many scientific papers published in the most significant journals in the field, two patents, and three book chapters. In particular, he co-authored the chapter 'Wideband modelling by the BI-RME method' in the book *Advanced Modal Analysis* by G. Conciauro, M. Guglielmi, and R. Sorrentino, published by John Wiley & Sons, Ltd (2000). He acts as a reviewer for the *IEEE Transaction on Microwave Theory and Techniques* and for the *European Microwave Week*. Paolo Arcioni is Senior Member of the IEEE (MTT Society) and Member of the European Microwave Association (EuMA) and of the Società Italiana di Elettromagnetismo (SIEM).

About the Authors

Chapter 1

Máirtín O'Droma, BE (NUI), PhD (NUI), CEng, FIET, SMIEEE is Director, Telecommunications Research Centre and Senior Lecturer, University of Limerick, Ireland. Recent positions and activities include IEEE Subject Matter Expert and an ITU-T Invited Expert; founding partner and steering committee member of the European Union Networks of Excellence TARGET (Top Amplifier Research Groups in a European Team); and ANWIRE (Academic Network for Wireless Internet Research in Europe); Ireland's delegate to four European Science Foundation (ESF) Cooperation in the Field of Science and Technology (COST) Research Actions: Modelling and Simulation Tools for Research in Emerging Multi-service Telecommunications (285); Traffic and QoS Management in Wireless Multimedia Networks (290); Wireless Networking for Moving Objects, WiNeMO (IC0906); and RF/Microwave Communication Subsystems for Emerging wWireless Technologies, RFCSET (IC0803). His research interests include: behavioural modeling linearization and efficiency techniques in multimode, multiband, multicarrier broadband nonlinear RF power amplifiers; complex wireless telecommunication systems simulation; wireless network and protocol infrastructural innovations for next generation networks (NGNs), the ubiquitous consumer wireless world (UCWW), and for cognitive networks. He has authored and co-authored over 180 research publications in learned journals, international conferences, and book chapters.

Yiming Lei, BSc (Peking University, 2001), PhD (University of Limerick, 2011) is currently a post-doctoral research fellow in the School of Electronics Engineering and Computer Science, Peking University, China. He is also an external post-doctoral research fellow in the Telecommunications Research Centre, University of Limerick. His research interests include mechanism analysis and linearization of nonlinear wireless communications systems and the application of information technology in medicine.

Chapter 2

Jerzy Julian Michalski was born in 1974 in Poland. He received MS and PhD degrees in Electrical Engineering in 1998 and 2002, respectively. His research has so far concerned the omega pseudochiral materials and ferrite materials longitudinally magnetized in applications for microwave devices. Currently, he works at TeleMobile Electronics Ltd. His main research interest concerns new efficient methods based on artificial intelligence algorithms applied in the automatic tuning of filters and diplexers. He is the author of numerous software tools for GSM, DCS, and EDGE. Since 2007 he has been an IEEE member.

Jacek Gulgowski was born in Gdansk, Poland, in 1972. He received an MS degree in Informatics from Technical University of Gdansk in 1996, MS degree in Mathematics from University of Gdansk in 1997, and PhD degree in Mathematics from University of Gdansk in 2001. From 1997 he has been a lecturer in Faculty of Mathematics in the University of Gdansk with scientific interests directed towards nonlinear analysis. In 2010 he joined Tele-Mobile Electronics Ltd in Gdynia, Poland, to work on the microwave filter tuning methods and algorithms.

Tomasz Kacmajor was born in 1986 in Poland. He received an MS degree in Electrical Engineering in 2010. Currently he works at TeleMobile Electronics Ltd. His main research concerns are computational intelligence in the application for microwave technique, microwave filters, and optimization methods. Since 2010 he has been an IEEE member.

Mateusz Mazur was born in Gdynia, Poland, in 1975. He received an MSc in Communication Engineering from Gdańsk University of Technology (GUT) in 1999 and PhD in Electrical Engineering from Telecommunication Research Institute in Warsaw (PIT) in 2007. Since 2000 he has been with PIT; his main areas of interest are antennas, passive devices, and radar systems. In 2010 he joined TeleMobile Electronics Ltd, where he participates in the COST project: 'New optimization methods and their investigation for the application to physical microwave devices that require tuning'. In 2011 he received a grant from NCBiR to realize his project: 'Implementation and correction of measurements methods of antenna pattern in near field' at GUT.

Chapter 3

Anne Claire Lepage received an MS degree in fundamental physics in 1999 from Henri Poincaré University, Nancy, France, and an MS degree in Electronics and Telecommunications in 2001 from Pierre and Marie Curie University, Paris. She received a PhD degree from Telecom ParisTech, Paris, in 2005. She joined the Radiofrequency and Microwave Group of Telecom ParisTech as Associate Professor in 2007. Her research activities encompass compact wideband bipolarized antennas and metamaterials for wideband low profile antennas and are related to various applications fields like cognitive radio, LTE, airborne and space systems.

Julien Sarrazin received his MS and PhD degrees from the University of Nantes in France, in 2005 and 2008, respectively. Until the end of 2008, he was working on antennas for MIMO systems in the IREENA Research Institute in Nantes. In 2009 and 2010, he worked in the BK Birla Institute of Technology (BKBIET) of Pilani in India, where he was in charge of the telecommunication and electromagnetism teaching. He is now working as a Research Engineer in Telecom ParisTech, Paris, France. His current research interests focus on meta-materials, including high impedance surfaces, dedicated to antenna design.

Xavier Begaud was born on December 11, 1968 in Chateaudun, France. He received an MS degree in optics, optoelectronics, and microwaves, from Institut National Polytechnique de Grenoble (INPG) in 1992. He received a PhD degree from the University of Rennes in 1996 and the Habilitation degree from Pierre and Marie Curie University (Paris 6) in 2007. He joined Telecom ParisTech (formerly Ecole Nationale des Telecommunications) in 1998, where he is presently Professor in the Communications and Electronics Department.

He works with the Radiofrequency and Microwave Group of CNRS, the French National Scientific Research Centre. His research topics include theory, conception, modeling, and characterization of wideband, bipolarized, and 3D antennas with special emphasis on numerical methods. Currently, research activities include design of metamaterials, channel sounding, and mutual coupling analysis in the framework of ultra-wideband and software radio. He has published over 120 journal papers, patents, book chapters, and conference articles. He is a member of the editorial board of the journal *Advanced Electromagnetics* and has organized two international conferences as the general chairman and edited one book.

Chapter 4

Nuno Borges Carvalho received the diploma and doctoral degrees in Electronics and Telecommunications Engineering from the Universidade de Aveiro, Aveiro, Portugal, in 1995 and 2000 respectively. From 1997 to 2000 he was an Assistant Lecturer at the same University and a Professor since 2000. Currently he is an Associate Professor with 'Agregacao' at the same University and a Senior Research Scientist at the Instituto de Telecomunicações. He has worked as a scientist researcher at the Instituto de Telecomunicações and was engaged to different projects on nonlinear characterization, circuits and systems design for software defined radio and wireless power transmission. His main research interests include nonlinear circuits/systems characterization, nonlinear distortion analysis in microwave/wireless circuits, and systems and measurement of nonlinear phenomena. Recently he has also been involved in design of dedicated radios and systems for newly emerging wireless technologies. In 2007 he was a visitor researcher at the North Carolina State University and at the National Institute of Standards and Technology, NIST. He was the recipient of the 2000 IEE Measurement Prize. He is also the co-inventor of five registered patents. He is a reviewer for several magazines including the *IEEE Transactions on Microwave Theory and Techniques* and an associate editor for the *IEEE Microwave Magazine*. He is the chair of IEEE MTT-11 Technical Committee and the chair of the URSI-Portugal Metrology Group. Dr. Borges Carvalho is co-author of the book *Intermodulation in Microwave and Wireless Circuits* from Artech House, 2003.

Pedro Miguel Cruz was born in Ovar, Portugal, in 1982. He received a five year degree in Electronics and Telecommunications Engineering in July 2006 and MSc degree in July 2008, both from Universidade de Aveiro, Aveiro, Portugal. He is now a PhD student of Electrical Engineering at the same university. He has worked at Portugal Telecom Inovação, Aveiro, Portugal, as a trainee in a project of localization systems based on wireless devices from September 2006 to April 2007. Currently, he is a researcher with Instituto de Telecomunicações, Aveiro, Portugal, enrolled in the characterization and modeling of software-defined radio front-ends. His main research interests are connected to software defined radio and cognitive radio fields, also with great attention to high-speed wideband data converters (A/D and D/A). Mr. Cruz was a finalist in the student paper competition of the IEEE MTT-S International Microwave Symposium 2008. He was recognized with third place in the GAAS Association PhD Student Fellowship for EuMIC 2009.

Chapter 5

Mikko Valkama was born in Pirkkala, Finland, on November 27, 1975. He received MSc and PhD degrees (both with honors) in Electrical Engineering (EE) from Tampere University of Technology (TUT), Finland, in 2000 and 2001, respectively. In 2002 he received the Best PhD Thesis award by the Finnish Academy of Science and Letters for his dissertation entitled 'Advanced I/Q signal processing for wideband receivers: models and algorithms'. In 2003, he was working as a Post-Doc Research Fellow with the Communications Systems and Signal Processing Institute at SDSU, San Diego, CA. Currently, he is a Full Professor and Department Head at the Department of Communications Engineering at TUT, Finland. He has been involved in organizing conferences, like the IEEE SPAWC'07 (Publications Chair) held in Helsinki, Finland. He has published more than 130 refereed scientific articles in international journals and conference proceedings, including also highly cited individual articles, with beyond 100 ISI citations. His general research interests include communications signal processing, estimation and detection techniques, signal processing algorithms for software-defined flexible radios, radio transceiver architectures, cognitive radio technologies, digital transmission techniques, such as different variants of multicarrier modulation methods and OFDM, and radio resource management for ad hoc and mobile networks.

Ville Syrjälä was born in Lapua, Finland. He received an MSc degree (with honors) in Communications Engineering (CS/EE) from Tampere University of Technology (TUT), Finland, in 2007. Currently, he is working as a researcher with the Department of Communications Engineering at TUT, Finland. His general research interests are in communications signal processing and signal processing algorithms for flexible radios.

Risto Wichman received his MSc and DSc(Tech) degrees in Digital Signal Processing from Tampere University of Technology, Tampere, Finland, in 1990 and 1995, respectively. From 1995 to 2001, he worked at Nokia Research Centre as a senior research engineer. In 2002, he joined the Department of Signal Processing and Acoustics, Aalto University School of Electrical Engineering, where he has been a Professor since 2003. His major research interests include digital signal processing for applications in wireless communications.

Pramod Mathecken received a BE degree in Electronics and Communication from Coimbatore Institute of Technology, an autonomous college affiliated to Anna University, India, in 2005. From April 2006 to July 2008, he worked at the Analogue-devices DSP Learning Centre in IIT Madras, Chennai, India. In March 2011, he received an MSc degree in Communications Engineering from Aalto University School of Electrical Engineering, Helsinki, Finland. He is currently pursuing his DSc degree at Aalto University School of Science and Technology. His current research interests are toward analysis and compensation of RF-impairments and synchronization methods in wireless communications.

Chapter 6

Albert Bel Pereira was born in Sabadell, Spain, in 1983. He received the degree in telecommunication engineering in 2007 from the Universitat Autònoma de Barcelona (UAB). From March to July 2007 he stayed in the Università degli Studi di Siena doing his degree project. Since September 2007 he has been a PhD candidate at the Department of Telecommunications and Systems Engineering (UAB), being a member of the Signal Processing for Communications and Navigation (SPCOMNAV) Group. His research interests are positioning and tracking systems for wireless sensor networks.

José López Vicario was born in Blanes, Spain, in 1979. He received both the degree in Electrical Engineering and the PhD degree from the Universitat Politècnica de Catalunya (UPC), Barcelona, in 2002 and 2006, respectively. During 2002, he served as a DSP programmer (UPC, 2002) and participated in the Spanish government R&D Project TIC99-0849. From October 2002 to September 2006, he was a PhD candidate at UPC's Signal Theory and Communications Department and, from January 2003, he pursued his thesis at CTTC, where he was involved in several European R&D projects (IST ACE and IST NEWCOM). Since September 2006, he has been an Assistant Professor at the Universitat Autònoma de Barcelona, teaching courses in digital communications and signal processing. He has published 15 papers in recognized international journals and around 40 papers in conferences. Dr. Vicario is a member of the IEEE Signal Processing Society and received the 2005/06 best PhD prize in Information Technologies and Communications by the UPC.

Gonzalo Seco-Granados received MSc and PhD degrees in Telecommunication Engineering in 1996 and 2000, respectively, from Universitat Politècnica de Catalunya (UPC). He also received an MBA degree from IESE-University of Navarra, Barcelona, in 2002. From 2002 to 2005, he was a member of technical staff at the European Space Research and Technology Centre (ESTEC), ESA, Noordwijk, The Netherlands, involved in the Galileo Project and leading the activities concerning indoor GNSS. In 2006, he joined the Department of Telecommunications and Systems Engineering, Universitat Autònoma de Barcelona (UAB) as an Associate Professor and Director of the Signal Processing for the Communications and Navigation (SPCOMNAV) Group. His research interests are statistical signal processing, optimization, and the physical layer of wireless communications and positioning systems. Since 2009, he has been the director of one of Chairs of Knowledge and Technology Transfer 'UAB Research Park–Santander'. He is a Senior Member of the IEEE.

Chapter 7

Ming Shen received MSc and PhD degrees in Electronic Engineering from the Graduate University of the Chinese Academy of Sciences in 2005 and Aalborg University in 2010, respectively. He is currently working as a post-doc with the Department of Electronic Systems, Aalborg University. His research interests include on-chip substrate noise characterization, RF integrated circuit design for biomedical applications, and circuit design for energy harvesting.

Jan Hvolgaard Mikkelsen received MSc and PhD degrees in Electrical Engineering from Aalborg University, Denmark, in 1995 and 2005, respectively. Since 2001 he has been working as an IC design manager for the large-scale RF-IC design effort at Aalborg University, where he currently holds a position as Associate Professor. He has authored or co-authored over 50 peer-reviewed journal and conference papers. His main research interests include analog circuit design, RF/LF CMOS circuit design for various applications – communication, biomedical, and energy harvesting – as well as system level analysis of transceiver architectures. He serves as a reviewer for the IEE and IEEE.

Torben Larsen received MSc and DrTechn degrees from Aalborg University, Aalborg, Denmark, in 1988 and 1998, respectively. Since 2001, he has been a Full Professor with Aalborg University. He has industrial experience with Bosch Telecom and Siemens Mobile Phones. In 2005, he was appointed a member of the Research Council for Technology and

Production by the Minister of Science, Technology, and Innovation. He has authored or co-authored over 100 peer-reviewed journal and conference papers. His research interests include RF system design, integrated circuit design, wireless communications, and transceiver design.

Chapter 8

Dušan Kocur was born in 1961 in Košice, Slovakia. He received his Ing (MSc) and CSc (PhD) degrees in Radioelectronics from the Faculty of Electrical Engineering, Technical University of Košice, in 1985 and 1990, respectively. Now he is the Full Professor at the Department of Electronics and Multimedia Communications of his Alma Mater. His research interests are radar signal processing, UWB technologies, and physical layer of wireless communication systems.

Jana Rovňáková was born in 1983 in Michalovce, Slovakia. She received her MSc degree in Mathematics from the Faculty of Science, Pavol Jozef Šafárik University in Košice and her PhD degree in Electronics from the Faculty of Electrical Engineering, Technical University of Košice, in 2006 and 2009, respectively. Since then, she has been a researcher at the Department of Electronics and Multimedia Communications of the Technical University of Košice. Her research interests are focused on UWB radar signal processing.

Chapter 9

Javier Arribas is a Research Engineer and a PhD Candidate at CTTC since 2008. He received BSc and MSc degrees in Telecommunication Engineering in 2002 and 2004, respectively, at La Salle School of Engineering, Ramon Llull University, in Barcelona, Spain. In 2004 he joined to the Signal Theory and Communications Department at La Salle, where he was Teaching Assistant in the field of Electronic Measurements. He currently collaborates with La Salle as MSc and BSc theses advisor, where he is involved in the SalleSat Cubesat picosatellite development program as technical director. His primary areas of interest include statistical signal processing, GNSS synchronization, detection and estimation theory, and the design of RF front-ends. He has contrasted experience in GNSS software defined receivers and FPGA-based real-time systems.

Pau Closas received MSc and PhD (*cum laude*) degrees in Electrical Engineering from the Universitat Politècnica de Catalunya (UPC) in 2003 and 2009, respectively. During 2008 he was Research Visitor at the Stony Brook University (SBU), New York, USA. In September 2009 he joined the CTTC, where he currently holds a position as a Research Associate in the Communications Subsystems Area and coordinates the Positioning Systems research line. He has many years of experience in projects funded by the European Commission, Spanish and Catalan Governments, as well as the European Space Agency (ESA) in both technical and managerial duties. He has numerous contributions in his primary areas of interest, including statistical and array signal processing, estimation and detection theory, Bayesian filtering, robustness analysis, and game theory, with applications to positioning systems, wireless sensor networks, and mathematical biology.

Carles Fernández-Prades (Member, IEEE) received MSc and PhD (*cum laude*) degrees in Electrical Engineering from the Universitat Politècnica de Catalunya (UPC), Barcelona, Spain, in 2001 and 2006, respectively. In May 2006, he joined the Centre Tecnòlogic de

Telecomunicacions de Catalunya (CTTC), where he currently holds the position of a Research Associate and the Coordinator of the Communications Subsystems Area. His primary areas of interest include statistical signal processing, estimation theory, GNSS synchronization, digital communications, and design of radio frequency (RF) front-ends.

Manuel Cuntz received the diploma in Electrical Engineering in 2005 from the Technical University of Kaiserslautern. He joined the Institute of Communications and Navigation of the German Aerospace Centre (DLR) in 2006 as Research Assistant. Since then Manuel Cuntz has contributed to several national and international projects in the field of interference robust satellite navigation receivers. His fields of research are the development of analog front-end hardware and digital signal processing algorithms for interference detection and mitigation in satellite navigation receivers for safety critical applications.

Michael Meurer received the diploma and PhD degree from the University of Kaiserslautern, Germany. After graduation, he joined the Research Group for Radio Communications at the Technical University of Kaiserslautern, Germany, as a Senior Key Researcher, where he was involved in various international and national projects in the field of communications and navigation, both as project coordinator and as technical contributor. From 2003 till 2005, Dr. Meurer was active as a senior lecturer. Since 2005 he has been an Associate Professor (PD) at the same university. Additionally, since 2006 Dr. Meurer has been with the German Aerospace Centre (DLR), Institute for Communications and Navigation, where he is the Director of the Department of Navigation. His current research interests include GNSS signals, GNSS receivers, interference mitigation, and navigation for safety-critical applications.

Andriy Konovaltsev received his engineer diploma and PhD degree in Electrical Engineering from Kharkov State Technical University of Radio Electronics, Ukraine, in 1993 and 1996, respectively. He joined the Institute of Communications and Navigation of German Aerospace Centre (DLR) in 2001 as Research Assistant. Since then he was involved in various national and international projects in the field of satellite-based navigation, especially aimed at development of the European GALILEO navigation system. His research interests are primarily in array processing for satellite navigation systems and signal processing algorithms for robust navigation receivers, including synchronization, multipath, and radio interference mitigation techniques.

Chapter 10

Roberto Gómez-García received MSc and PhD degrees in Electrical Engineering from the Technical University of Madrid, in 2001 and 2006, respectively. Currently, he is with the Department of Signal Theory and Communications of the University of Alcalá, Madrid. His research interests are in the pursuit of new concepts to design advanced fixed-/reconfigurable-frequency RF filters and multiplexers in planar, hybrid, and MMIC technologies, multifunction circuits, and novel software-defined radio and radar architectures.

José-María Muñoz-Ferreras received MSc and PhD degrees in Electrical Engineering from the Technical University of Madrid in 2004 and 2008, respectively. Currently, he is with the Department of Signal Theory and Communications of the University of Alcalá, Madrid. His research activities are in the area of radar signal processing and advanced radar/communications systems and concepts. Specifically, the focusing of high-resolution inverse synthetic aperture radar images and the design and validation of radar/communications systems with the software-defined paradigm are issues of his interest.

Manuel Sánchez-Renedo received an MSc degree in Electrical Engineering from the Technical University of Madrid in 2005. He is currently working towards a PhD at the Department of Signal Theory and Communications of the University of Alcalá, Madrid. His current research interests are focused on the development of novel fixed/tunable, compact, and multiband microwave filters in planar and hybrid technologies.

Chapter 11

Jian Zhang was born in Anqing city, China. He received a BEng degree in Automation Engineering from Southeast University, Nanjing, China, in 1999 and a PhD degree in Micro-electrics and Solid State Electronics Engineering from Shanghai Institute of Microsystem and Information Technology of Chinese Academy of Sciences (CAS), Shanghai, China, in 2008. From 1999 to 2002, he was a research engineer with Wuhan Research Institute of Post and Telecommunication (WRI), Wuhan, China, during which time he was involved with the circuit design for an ATM-PON access network application. In the summer of 2008, he was a design engineer with RFMD Shanghai Design Centre, where he was engaged in power amplifier design and measurement system programming. From October 2008 to 2010, he has been a post-doctoral researcher with Microwave Electronics Laboratory, MC2, Chalmers University of Technology, Gothenburg, Sweden, where he has been engaged in monolithic microwave integrated circuit (MMIC) development using compound semiconductors such as GaAs pHEMT, GaInP HBT, and InP DHBT process technology. Since October 2010, he has been a Marie Curie experienced researcher at the High Frequency Electronic Circuits Division, Institute of Electronics and Communications and Information Technology (ECIT), Queens University of Belfast, United Kingdom, where he is currently engaged in the development of RF front-end circuits and communication systems for wireless personal area network (WPAN) applications using SiGe process technology.

Mury Thian obtained a BSc degree from Atma Jaya Catholic University of Indonesia, Jakarta, Indonesia, an MSc degree from Delft University of Technology, the Netherlands, and a PhD degree from Queens University of Belfast, United Kingdom, all in Electronics Engineering. He was with Philips Semiconductors (now NXP Semiconductors), Nijmegen, the Netherlands, and then with the University of Birmingham, United Kingdom, before returning to Queens University of Belfast in 2009 to work on radio technologies for short-range gigabit wireless communications. He is presently on a two-year secondment to Infineon Technologies, Villach, Austria, as a Marie Curie Fellow.

Guochi Huang received BEng and MEng degrees from Harbin Institute of Technology, China, in 2003 and 2005, respectively, and a PhD degree in Electrical and Computer Engineering from Sungkyunkwan University, South Korea, in 2009. From June 2009 to April 2010, he was a Research Assistant with ECIT at Queens University of Belfast, working on 60GHz VCOs and amplifiers using SiGe technology. He is currently with the RF Group of Analogue Devices Inc. in Limerick, Ireland. His research interests include RF/MMW frequency generated IC and transceiver design.

George Goussetis graduated from the Electrical and Computer Engineering School in the National Technical University of Athens, Greece, in 1998, and received a PhD degree from the University of Westminster, London, UK. In 2002 he also graduated with a BSc in Physics (first class) from University College London (UCL), UK. In 1998, he joined Space Engineering, Rome, Italy, as RF Engineer and in 1999 the Wireless Communications Research Group, University of Westminster, UK, as a Research Assistant. Between 2002 and 2006 he was a Senior Research Fellow at Loughborough University, UK. Between 2006 and 2009 he was a Lecturer (Assistant Professor) at the School of Engineering and Physical Sciences, Heriot-Watt University, Edinburgh, UK. He joined the Institute of Electronics Communications and Information Technology at Queens University Belfast, UK, in September 2009 as a Reader (Associate Professor). He has authored or co-authored over 150 peer-reviewed papers, three book chapters, and two patents. His research interests include the modeling and design of microwave filters, frequency-selective surfaces and periodic structures, leaky wave antennas, microwave sensing, and curing, as well as numerical techniques for electromagnetics. Dr. Goussetis received the Onassis Foundation Scholarship in 2001. In October 2006 he was awarded a five-year research fellowship by the Royal Academy of Engineering, UK. In 2010 he was visiting Professor in UPCT, Spain.

Vincent F. Fusco holds a Personal Chair in High Frequency Electronic Engineering. His research interests include active antenna and front-end MMIC techniques. Professor Fusco has published over 350 scientific papers in major journals and in referred international conferences. He has authored two textbooks, holds patents related to self-tracking antennas, and has contributed invited papers and book chapters. He serves on the Technical Programme Committee for various international conferences including the European Microwave Conference. He is a Fellow of the Royal Academy of Engineering, Royal Irish Academy, the IET, and the IEEE.

Chapter 12

Robert Malmqvist received his PhD degree from Linköping University in 2001. Since 2000 he has been with FOI (Swedish Defence Research Agency) where he currently holds a position as Senior Scientist at the Department of Radar Systems. He is currently also Research Scientist within the Microwave Group at Uppsala University, Sweden. He has authored or co-authored more than 50 scientific papers of which six are in peer-review journals and received two Best Paper awards at EuRAD2004 and IEEE CAS2009, respectively. He is currently engaged in research related to MMIC/RF-MEMS circuit design for reconfigurable microwave/mm-wave systems.

Aziz Ouacha received his PhD degree in Electronics in 1995 from Linköping University, Sweden. He is currently Adjunct Professor at the Department of Electrical Engineering (ISY), Linköping University, Sweden, and Research Director at the Division of Sensors and EW systems at the Swedish Defence Research Agency (FOI). His current research interests are in broadband active phased array antennas and in particular multifunction RF systems. He has published over 80 research papers in international journal and conference proceedings. His challenging goal is a fully reconfigurable and cognitive single RF front-end for multifunction phased array systems. He is a member of the IEEE and Association of Old Crows.

Mehmet Kaynak received his BS degree from the Electronics and Communication Engineering Department of Istanbul Technical University (ITU) in 2004, and received an MS degree from the Microelectronic Program of Sabanci University, Istanbul, Turkey, in 2006. He joined the technology group of IHP Microelectronics, Frankfurt (Oder), Germany, in 2008. He is currently working on development and integration of embedded MEMS technologies and leading the MEMS group in IHP.

Naveed Ahsan received his BS degree from UET Lahore, Pakistan Electronics in 1998. He was awarded an MSc degree from NUST, Rawalpindi, Pakistan, in 2002 and received his PhD degree from Linköping University, Sweden, in 2009. Dr. Naveed Ahsan has over ten years of research experience in the field of antennas and RF/microwave engineering. He has published more than 20 papers including two journal papers. His research interests include MMIC-based transceiver design and solid-state power amplifier design for radar and communication systems.

Joachim Oberhammer received his PhD degree from KTH, Royal Institute of Technology, Stockholm, Sweden, in 2004, where he is currently Associate Professor heading a research group in RF and microwave MEMS. By 2012, he had published over 90 research papers in this field and has served as a TPRC member of IEEE Transducers, IEEE IMS, and IEEE MEMS conferences since 2009. His work on RF MEMS phase-shifters and tunable capacitors received Best Paper awards at IEEE EuMIC in 2009 and IEEE APMC in 2010.

Chapter 13

Federico Alimenti was born in Foligno, Italy, in 1968. He received the Laurea degree (*magna cum laude*) and PhD from the University of Perugia, Italy, in 1993 and 1997, respectively, both in Electronic Engineering. In 1993 he held an internship at Daimler-Benz Aerospace, Ulm, Germany. In 1996 he was recipient of the URSI Young Scientist Award and Visiting Scientist at the Technical University of Munich, Germany. Since 2001 he has been with the Department of Electronic and Information Engineering at the University of Perugia as a Research Associate, teaching the classes of 'Telecommunication Electronics' and 'Microwave Electronics'. In 2006 he has been elected as a member of the Academic Senate at the same university. His scientific interests concern the design and characterization of microwave integrated circuits in CMOS and SiGe BiCMOS technologies. He has authored more than 120 papers in referred journals, conference proceedings and book chapters. Federico Alimenti is a Senior IEEE Member.

Andrea Battistini was born in Terni, Italy, in 1981. He received the Laura degree (*magna cum laude*) and PhD from the University of Perugia, Italy, in 2006 and 2008, respectively, both in Electronic Engineering. His research interests include analog design of microwave sensors (both radiometers and radar) for civil and space application. Since 2006 he has also been an R&D engineer at ART S.r.l., Passignano (PG), Italy.

Valeria Palazzari received the Laurea degree in Electronic Engineering and PhD degree in Information and Electronic Engineering from the University of Perugia, Italy, in 2000 and 2003, respectively. Since 2000 she joined the High Frequency Electronics Laboratory (HFE-lab) at the Department of Electronic and Information Engineering at the same university. In summer 2003 she joined the ATHENA Research Group at the Georgia Institute of

Technology, Atlanta, USA. She is currently working as a post-doc assistant, teaching the classes of 'Telecommunication Electronics'. Her research goes from the modeling, design, and realization of microwave integrated circuits in SiGe BiCMOS technology to the packaging technologies for RF and wireless systems.

Luca Roselli (see About the Editors)

Stephen M. White is a solar and stellar radio astronomer who has worked on solar radio imaging techniques across a wide range of wavelengths. His scientific interests include solar flares, solar magnetic fields, the solar atmosphere, and particle acceleration. He has studied the nature of radio emission on stars other than the Sun and the ways in which it differs from the solar case. After working for many years in the Department of Astronomy at the University of Maryland, he has moved to the Space Weather Group at the Air Force Research Laboratory.

Chapter 14

Francesco Giuppi was born in Pavia in 1983. He achieved a MS degree (110/110 *cum laude*) in Electronic Engineering at the University of Pavia in April 2008. Afterwards, he joined the Department of Electronics of the University of Pavia as a PhD student in Microwave Electronics. During his PhD he worked on the development of active antennas in substrate integrated waveguide (SIW) technology. Francesco was a visiting PhD student at the Catalonian Technologic Telecommunication Centre (CTTC) from September 2010 to September 2011. He was the recipient of four prizes at international conferences and workshops, including the Best Student Paper at the European Conference on Antennas and Propagation (EuCAP 2010) and the Best Student Paper at the IEEE International Microwave Workshop Series (IMWS 2011).

Apostolos Georgiadis (see About the Editors)

Ana Collado received MSc and PhD degrees in Telecommunications Engineering from the University of Cantabria, Spain, in 2002 and 2007, respectively. Since 2007 she has been a Research Associate at the Technological Telecommunications Centre of Catalonia (CTTC), Barcelona, Spain, where she performs her professional activities. Her professional interests include active antennas, substrate integrated waveguide structures, nonlinear circuit design, and energy harvesting solutions for self-sustainable and energy efficient systems. She is a Management Committee Member of the EU COST Action IC0803 RF/Microwave Communications Subsystems for Emerging Wireless Technologies, where she also serves as the Grant Holder Representative supervising the scientific activities, dissemination activities, and financial reporting. Among the mentioned activities she has collaborated in the organization of several international workshops in different countries of the European Union and also a Training School for PhD students. She is a Marie Curie Fellow of the FP7 Project Symbiotic Wireless Autonomous Powered (SWAP) system. She was also finalist in the 2007 IEEE International Microwave Symposium and the 2012 IEEE Radio Wireless Week Symposium Student Contests. She was the Technical Program Chair of the 2011 IEEE RFID Technologies and Applications Conference and the Technical Program Chair of the 2011 IEEE MTTS International Microwave Workshop Series on 'Millimetre Wave Integration Technologies', as well as part of the Local Organizing Team in both events. She serves on the Editorial Board of the 2011–2012 *Radioengineering Journal* and assists as reviewer of several IEEE publications.

Maurizio Bozzi was born in Voghera, Italy, in 1971. He received the 'Laurea degree in Electronic Engineering and PhD in Electronics and Computer Science from the University of Pavia, Italy, in 1996 and 2000, respectively. In 2002 he joined the Department of Electronics of the University of Pavia as an Assistant Professor in electromagnetics. He currently teaches the courses of 'Numerical Techniques for Electromagnetics' and of 'Computational Electromagnetics and Photonics'. He has held research positions in various universities worldwide, including the Technical University of Darmstadt, Germany, the University of Valencia, Spain, the Polytechnical University of Montreal, Canada, and the Centre Tecnologic de Telecomunicacions de Catalunya (CTTC), Spain. His research activities concern the development of numerical methods for the electromagnetic modeling of microwave and millimeter-wave components. Professor Bozzi received the Best Young Scientist Paper Award at the XXVII General Assembly of URSI in 2002 and the MECSA Prize for the best paper presented by a young researcher at the Italian Conference on Electromagnetics in 2000.

Luca Perregrini was born in Sondrio, Italy, in 1964. He received the 'Laurea' degree in Electronic Engineering and PhD in Electronics and Computer Science from the University of Pavia, Pavia, Italy, in 1989 and 1993, respectively. In 1992, he joined the Department of Electronics of the University of Pavia, where he is now an Associate Professor in Electromagnetics. His main research interests are in numerical methods for the analysis and optimization of waveguide circuits, frequency selective surfaces, reflectarrays, and printed microwave circuits. He co-authored the textbook *Fondamenti di Onde Elettromagnetiche* (McGraw-Hill Italia, Milano, Italy, 2003). Professor Perregrini was an Invited Professor at the Polytechnical University of Montreal, Montreal, Quebec, Canada, in 2001, 2002, and 2004. He was a consultant of the European Space Agency and of some European telecommunication companies.

Chapter 15

Frederick Declercq was born in 1983. He received an MSc degree in Electronic Engineering at Howest, University College West Flanders, Kortrijk, Belgium. He received a PhD degree in Electrical Engineering at the Faculty of Engineering at Ghent University in 2011. Since September 2005, he has been with the Electromagnetics Group, Department of Information Technology (INTEC) at Ghent University. He is currently conducting his post-doctoral research at Ghent University. His research interests are electromagnetic characterization of textile and flexible foam materials and the design of wearable active textile antennas.

Hendrik Rogier (see About the Editors)

Apostolos Georgiadis (see About the Editors)

Ana Collado (see Chapter 14)

Chapter 16

Chomora Mikeka was born in Malawi in 1978. He holds a PhD degree from the Division of Physics, Electrical and Computer Engineering at Yokohama National University, Japan. His PhD research was about power autonomous sensor radio based on cellular and digital TV RF energy harvesting. In 2010, as a Visiting Researcher at the Communications Subsystems Lab, CTTC in Barcelona (Spain), he collaborated in the design, simulations, and fabrication of an ultra low power DC–DC buck boost converter with regulated output for less than a milliwatt RF energy harvesting. He won the 2009 European Microwave Association (EuMW2009) Student Challenge Prize in Rome (Italy). He won the 2011 IEEE RWW, Biomedical Radio and Wireless Technologies, Network and Sensing Systems Second Best Paper Award in Phoenix, Arizona (USA). He won the IEEE RFID-TA 2011, Third Best Paper Award in Sitges, Barcelona (Spain). He won the 2010 Yokohama National University International Science Exchange Encouragement Award. His biography is included in the 27th edition of Marquis *Who's Who in the World*. He is the Laureate of IBC Top 100 Engineers in 2010. He is now a Lecturer in the Physics Department at Chancellor College of the University of Malawi, Africa.

Hiroyuki Arai received a BE degree in Electrical and Electronic Engineering and ME and DE degrees in Physical Electronics from Tokyo Institute of Technology in 1982, 1984 and 1987, respectively. After being a Research Associate in Tokyo Institute of Technology, he joined Yokohama National University as a Lecturer in 1989. Now he is a Professor in the Division of Electrical and Computer Engineering, Yokohama National University. He was a Visiting Scholar at University of California, Los Angeles, in 1997 and was Visiting Professor at Yonsei University, Seoul, in 2005.

Chapter 17

Bahattin Turetken has received MSc and PhD degrees from Istanbul Technical University, Istanbul, Turkey, in 1998 and 2002, respectively. He received an Associate Professor title in 2008. He has worked as Chief Researcher at TUBITAK-UEKAE (National Research Institute of Electronics and Cryptology) EMC and TEMPEST Test Centre. Now, he manages the Electromagnetic and Research Group (EMARG) and the project at the Antenna Test and Research Centre. He was awarded the 'Young Scientist Award' by URSI in 1999 and 'Young Scientist Paper' by MMET in 2000. His research topics are radar, antenna design and testing, computational electromagnetic, diffraction and scattering EM problems, and civilian and military EMC/EMI problems.

Umut Buluş has received a Bachelor degree in Electronics and Communication Engineering in Istanbul Technical University, İstanbul,Turkey, and a Master degree in Duisburg-Essen University, Duisburg, Germany. He worked in RF&Antenna Departments of IMST Gmbh and Fraunhofer Institute in Germany. He has been working as a Researcher in the Electromagnetics and Antennas Research Group at TUBITAK, Turkey, since 2008. His scientific interests concern design and measurement of antennas and microwave structures.

Erkul Başaran was born in Kocaeli, Turkey. He received a BS degree in Electrical and Electronics Engineering from Selcuk University, Konya, Turkey, in 2000, and MS and PhD degrees in Electronics Engineering from the Gebze Institute of Technology (GIT), Kocaeli,

Turkey, in 2002 and 2008, respectively. He is currently working at TUBITAK (The Scientific and Technological Research Council of Turkey) as a senior researcher. He is interested in the numerical analysis and simulation of the electromagnetic and underwater acoustic wave phenomena, the finite difference time domain method, radar cross-section prediction and measurement, antenna design, and testing.

Eren Akkaya has received a BS degree in Electronics and Communication Engineering from Yildiz Technical University, Istanbul, Turkey, in 2009. He has been working as a Researcher at TUBITAK-UEKAE (National Research Institute of Electronics and Cryptology) Electromagnetic and Antenna Research Group. His research topics are computational electromagnetics, UWB antenna design, and testing.

Koray Sürmeli has received a BSc degree in Electronics and Communication Engineering from Yildiz Technical University, Istanbul, Turkey, in 2009. He has been working as a Researcher at TUBITAK-UEKAE (National Research Institute of Electronics and Cryptology) Electromagnetic and Antenna Research Group. His research topics are radar, antenna design and testing, computational electromagnetic, diffraction, and scattering EM problems, and civilian and military EMC/EMI problems.

Hüseyin Aniktar has received BSc, MSc, and PhD degrees from Istanbul Technical University, Middle East Technical University, and Aalborg University in 1998, 2002, and 2007, respectively. All degrees are in the Electronics Engineering field. He worked at TES Electronic Solutions GmbH, Stuttgart, as Analogue and RF IC Design Engineer. He also worked in Aselsan Inc. as RF Design Engineer. Currently he is working as RF Design Engineer at TUBITAK-UEKAE. His research topics are RF IC design, frequency synthesizers, and power amplifiers.

Preface

Novel applications for microwave and millimeter wave subsystems emerge in fields such as monitoring, logistics, health, and security, leading to stricter performance requirements for the underlying microwave and millimeter wave front-ends. As a result, there is a necessity for novel circuit and system architectures and topologies, for fast, efficient, and accurate modeling and optimization techniques of circuit components and systems as well as for collaboration between the signal processing and microwave electronics communities.

The book addresses a number of topics that are of fundamental importance in the field of microwave and millimeter wave systems and are expected to play a leading role in the next 5 years. Microwave systems as well as millimeter wave systems in a lesser extent have enjoyed great evolution and application in recent years. Wireless systems, smart-phones, WiFi, satellite systems, and RFID technologies have become part of everyday life. Although the book is concerned with and strongly related to current state-of-the-art in microwave and millimeter wave systems, the authors specifically chose to discuss topics that have the potential of having a long lasting impact in the longer term. The motivation behind the book has been to provide a reference point for the readers working in the field of microwave and millimeter wave circuits and systems. Undoubtedly there are many different technical issues and problems within this field, ranging from a circuit perspective to a system perspective, and they are ultimately interconnected under the unifying umbrella of the various applications they address. The book intends to provide a starting point, a key reference useful in every designer's library. The aim of the book is to address a selected number of challenging emerging problems and provide, on the one hand, state-of-the-art information and, on the other hand, a perspective on promising new technologies such as textile electronics, substrate integrated waveguide technology, and selected architectures such as software-defined radio, digital transceivers, and ultrawideband (UWB) radar, all of which are expected to lead to new breakthroughs in terms of system performance in the next decade. Depending on the chapter, prior knowledge at the level of a Master of Science in Electrical Engineering with knowledge in electromagnetic fields, antennas, microwave engineering, and signal processing is a prerequisite.

This book aims to highlight selected research and technology trends that emerged as topics of the European Union Cooperation in Science and Technology (COST) Action IC0803 RF/Microwave Communication Subsystems from Emerging Wireless Technologies, a collaboration project between leading European and cooperating institutions worldwide, focusing on the design of novel microwave and millimeter circuits and systems.

COST is the oldest and widest European intergovernmental network for cooperation in research (www.cost.eu). Established by the Ministerial Conference in November 1971, COST is presently used by the scientific communities of 35 European countries to cooperate in common research projects supported by national funds. The funds provided by COST – less than 1% of the total value of the projects – support the COST cooperation networks (COST Actions) through which, with EUR 30 million per year, more than 30 000 European scientists are involved in research having a total value that exceeds EUR 2 billion per year. This is the financial worth of the European added value that COST achieves. A 'bottom up approach' (the initiative of launching a COST Action comes from the European scientists themselves), 'à la carte participation' (only countries interested in the Action participate), 'equality of access' (participation is open also to the scientific communities of countries not belonging to the European Union), and 'flexible structure' (easy implementation and light management of the research initiatives) are the main characteristics of COST. As precursor of advanced multidisciplinary research COST has a very important role for the realization of the European Research Area (ERA) anticipating and complementing the activities of the Framework Programmes, constituting a 'bridge' towards the scientific communities of emerging countries, increasing the mobility of researchers across Europe, and fostering the establishment of 'Networks of Excellence' in many key scientific domains such as: Biomedicine and Molecular Biosciences; Food and Agriculture; Forests, their Products and Services; Materials, Physical and Nanosciences; Chemistry and Molecular Sciences and Technologies; Earth System Science and Environmental Management; Information and Communication Technologies; Transport and Urban Development; Individuals, Societies, Cultures and Health. It covers basic and more applied research and also addresses issues of pre-normative nature or of societal importance.

Action IC0803 was established in 2008 for a period of four years and forms part of the Information and Communication Technologies (ICT) COST Domain. More than one hundred researchers from 57 entities have participated in the Action. The work program of the Action is divided into three Working Groups, on ultra-low power and power efficient technologies, smart and reconfigurable radio transceivers, and finally design and optimization methods towards highly integrated terminals and efficient communication systems. The first two Working Groups are driven by applications whereas the last one is focused on simulation and optimization techniques that provide for accurate and efficient tools for the design of systems addressed by the other two working groups.

The first Working Group targets energy efficient systems. Energy efficiency is considered both in terms of reducing the overall power dissipation as well as improving the efficiency of high power systems. There are many research efforts and technologies targeting these goals. In terms of low power systems one can distinguish design efforts towards extremely low power sensor networks and RFIDs. In terms of power efficient systems there are research efforts towards architectures that reduce the nonlinear distortion present in power amplifiers in order to accommodate nonconstant envelope modulation techniques as well as highly efficient architectures that inherently do not pose strict linearity requirements in the power amplifiers due to the use of constant envelope modulated signals. In both cases the resulting efficiency of the amplifiers is maximized. In all these types of systems there is a need for both circuit-oriented, signal-processing, and system-oriented advancements. A promising technology that is explored concerns MEMS-based devices, as they have displayed excellent RF/microwave characteristics in terms of loss, bandwidth, and power consumption and, consequently, have found numerous applications in radar and communication system applications. It should be noted that low power and power efficient designs, and ultimately battery-less circuits, lead to more environmental friendly solutions, which is one of the critical challenges of present and future subsystems.

Working Group 2 of the COST IC0803 Action dealt with Smart and reconfigurable RF radio transceivers, covering all hardware and software aspects of state-of-the art RF radio transceivers. The most important development of the last few years is the convergence of hardware and software related research. With the availability of large processing power at reasonable cost and energy consumption, typical tasks that were previously purely implemented in hardware are now also available in the digital domain. The transition of hardware to software implementations comes with its own new challenges. The most typical example is software-defined radio, requiring suitable algorithms to limit interference and to perform decent estimation and synchronization. Also adaptations are required in terms of hardware, such as the development of wideband adaptive antennas and transmitters as well as sparse sampling in efficient radio receivers. By means of DSPs, hardware impairments may be mitigated in wireless communication systems. Novel emerging communication techniques, such as MIMO, UWB, and body-centric systems, as well as cognitive radio and wireless sensor networks, also present new challenges in terms of software and hardware, and open novel applications and terms of localization, monitoring, and communication.

Finally, Working Group 3 focuses on developing new and efficient computer-aided design (CAD) techniques for the design of novel compact components and efficient systems. Efficient simulation methods are required in order to reduce the design time and, more importantly, allow for fast optimization methodologies that include multiple constraints based on the properties of the signals being transmitted. Such techniques will be applied in the design of new compact components that lead to more integrated designs. As an example, substrate integrated waveguide (SIW) resonators provide a low cost, high performance solution for filter design. SIW components combine advantages of rectangular waveguides such as high Q-factor and low losses, with compact size and ease of integration with traditionally used microstrip implementations as they share the same dielectric substrate. Another very promising technology that is explored is components based on meta-materials. In addition, the design of nonlinear circuits presents additional challenges associated with their rich dynamical behavior. Although linear simulation and optimization methods are very well developed, there is still a lot of work to be done in terms of nonlinear techniques applied in the accurate and efficient design nonlinear circuits that are used for the generation and conversion of frequencies (oscillators, frequency dividers, and mixers). Their nonlinear response and their potential instability make their experimental behavior very difficult to predict. In all of these topics it is essential to combine circuit and system theory with stability theory and nonlinear dynamics. The presence and the effects of modulated signals in the stability of such circuits also needs to be accurately analysed using efficient simulation methods. Such topics are addressed in this Working Group and the results are ultimately utilized in the various applications considered within the other Working Groups.

The applications, research problems, and challenges addressed within the Action Working Groups have led to the realization of this book. The book is divided into two parts, addressing design and modeling trends, on the one hand, and highlighting important applications, on the other. Furthermore, the material is ordered in such a way that it progresses from circuit challenges to systems and finally to applications. Part one of the book contains seven chapters, whereas part two includes ten chapters.

The first chapter, *Low coefficient accurate nonlinear microwave and millimeter-wave nonlinear transmitter power amplifier behavioural models,* by Máirtín O'Droma, from the Telecommunications Research Centre of the University of Limerick, Ireland, and Lei Yiming, from the State Key Laboratory of Advanced Optical Communication Systems and Networks, School of EECS, Peking University, China, provides comprehensive coverage of the new RF power amplifier (PA) modified Bessel–Fourier (MBF) behavioral model. It is shown to be most suitable for large signal PA behavioral modeling and superior to all other established low-order models used by the microwave and millimeter wave research and engineering design community, such as power series models, the Saleh model, the modified Saleh model, and the original Bessel–Fourier model, from which the MBF takes its origin.

The second chapter, *Artificial neural network in microwave cavity filter tuning,* by Jerzy Michalski, Jacek Gulgowski, Tomasz Kacmajor, and Mateusz Mazur from TeleMobile Electronics Ltd., Gdynia, Poland, is related to filter optimization. Presently, microwave filter tuning is a necessary step in the production process. This step typically consists of manual work performed by a trained operator and usually requires a considerable amount of time. Hence, there is great expectation among microwave filter production companies to automate the process. Automated methods of filter tuning based on artificial neural networks are suggested and different approaches to the problem are described with a series of experiments supporting the presented ideas.

The third chapter, *Wideband directive antennas with high impedance surfaces,* by Anne Claire Lepage, Julien Sarrazin, and Xavier Begaud from Telecom ParisTech, Paris, France, demonstrates that it is possible to design low profile wideband directive antennas with high impedance surfaces. The artificial magnetic conductor (AMC) behavior of such surfaces is utilized in order to improve performances. Limitations of AMC narrow bandwidth characteristics are overcome with two different approaches. A first technique leads to an optimized antenna using a lumped element based AMC whereas a second one leads to a hybrid AMC that enhances the antenna's performances over a wide band.

The fourth chapter, *Characterization of software-defined and cognitive radio front-ends for multimode operation,* by Pedro Miguel Cruz and Nuno Borges Carvalho from the Instituto de Telecomunicações – Departamento de Electrónica, Telecomunicações e Informática from Universidade de Aveiro, Portugal, addresses software radios. Software-defined radios (SDRs) are now being accepted as the most probable solution for resolving the need for integration between actual and future wireless communication standards. SDRs take advantage of the processing power of modern digital processor technology to replicate the behavior of a radio circuit. Such a solution allows inexpensive, efficient interoperability between the available standards and frequency bands, because these devices can be improved and updated and given new capabilities by a simple change in software algorithms. This SDR concept is also the basis for cognitive radio (CR) approaches, in which the underneath concept imposes strong changes in terms of both complexity and flexibility of operation due to its potential adaptation to the air interface. A promising application for this CR technology is to implement a clever management of spectrum occupancy by use opportunistic radios, where the radio will adapt and employ spectrum strategies in order to occupy spectra that are not being used by other radio systems at a given moment. Nevertheless, for the correct operation of these radios a correct behavioral model is fundamental, and thus this chapter is devoted to discuss these problems.

The fifth chapter, deals with the *Impact and digital suppression of oscillator phase noise in radio communications*, and is authored by Mikko Valkama and Ville Syrjälä, from Tampere University of Technology, Finland, and Risto Wichman and Pramod Mathecken from Aalto University, Finland. The design of compact and low-cost radios with high performance and reconfigurable capabilities is a challenging task due to the contradictory nature of these requirements, and is further hindered by the various imperfections and impairments of the analog electronics involved in the radio transceiver. Such imperfections are: mirror-frequency interference due to I/Q imbalance, nonlinear distortion due to mixer and amplifier nonlinearities, timing jitter and nonlinearities in sampling and analog-to-digital (A/D) converter circuits, and oscillator phase noise. These impairments can easily become a limiting factor in the performance of the radio device, especially when complex high-order modulated waveforms such as orthogonal frequency division multiplexing (OFDM) are being deployed. This chapter concentrates on phase noise, which has a complicated character and a large impact on the performance of multicarrier OFDM systems. A model of time-varying phase noise for free running and phase locked loop oscillators is presented, followed by a detailed description of the effects of phase noise in OFDM systems and time-varying channels. Finally, different algorithms for the compensation of phase noise in the digital base band are explored.

The sixth chapter, *A pragmatic approach to cooperative positioning in wireless sensor networks*, contributed by Albert Bel Pereira, Jose Lopez Vicario, and Gonzalo Seco Granados from the Universitat Autònoma de Barcelona – UAB, Spain, is devoted to the theoretical framework behind localization techniques in wireless sensor networks. In recent years, location estimation in wireless sensor networks (WSNs) has raised a lot of interest from researchers. In particular, much attention has been recently paid to cooperative positioning techniques, as accurate positioning estimates can be obtained in networks with low complexity and low-cost terminals. The objective of this chapter is to provide a review on cooperative schemes for WSN. Among all of them, special emphasis is given to receive signal strength (RSS) based techniques as they provide suitable solutions for practical implementation. Since the accuracy of RSS methods depends on the suitability of the propagation models, cooperative localization algorithms that dynamically estimate the path loss exponent are also described. Practical examples based on real WSN deployments are also presented.

Finally, the last chapter of this first part of the book, Chapter 7, is devoted to *Modelling of substrate noise and mitigation schemes for ultra-wideband (UWB) systems*, and is written by Ming Shen, Jan H. Mikkelsen, and Torben Larsen from Aalborg University, Denmark. In highly integrated mixed-mode designs, digital switching noise is an ever-present problem that needs to be taken into consideration. This is of particular importance when low-cost implementation technologies, for example lightly doped substrates, are aimed for. For traditional narrow-band designs much of the issue can be mitigated using tuned elements in the signal paths. However, for UWB designs this is not a viable option and other means are therefore required. Moreover, owing to the ultra-wideband nature and low power spectral density of the signal, UWB mixed-signal integrated circuits are more sensitive to substrate noise compared with narrow-band circuits. This chapter presents a study on the modeling and mitigation of substrate noise in mixed-signal integrated circuits (ICs), focusing on UWB system/circuit designs. Experimental impact evaluation of substrate noise on UWB circuits is presented. It shows how a wideband circuit can be affected by substrate noise. This chapter also presents a new analytical model for the estimation of the spectral content of the switching noise. In addition, a novel active noise mitigation scheme based on spectral information is presented.

The second part of the book begins with Chapter 8, *Short-range tracking of moving targets by a handheld UWB radar system* by Dušan Kocur and Jana Rovňáková from Technical University of Košice, Slovak Republic. The chapter gives a description of the signal processing methods for a handheld UWB radar system applied for a short-range detection and tracking of moving persons. It explains the importance of the particular phases and provides an overview of methods applied within them. The performance of the selected methods is illustrated by the results of processing of radar signals obtained by the M-sequence UWB radar system applied for tracking of people moving behind walls.

The ninth chapter, *Advances in the theory and implementation of GNSS antenna array receivers*, by Javier Arribas, Pau Closas, and Carles Fernández-Prades from CTTC, Spain, and Manuel Cuntz, Michael Meurer, and Andriy Konovaltsev from DLR (German Aerospace Centre), Germany, focuses on recent developments in the field of antenna arrays applied to the design of robust, high-performance global navigation satellite system (GNSS) receivers. This work mainly focuses on the architecture of a GNSS receiver with an adaptive antenna array. The specifics of the design of such receivers are discussed, covering both analog and digital signal processing blocks. Special attention is devoted to the design of critical components such as the antenna array, radio frequency (RF) front-ends, and analog-to-digital (A/D) converters. Array signal processing techniques are also addressed describing two main strategies based on (i) spatial filtering by means of digital beamforming and (ii) signal parameter estimation by statistical array processing. Finally, two operational prototypes of GNSS receivers with adaptive antennas are proposed, including their latest test results.

Chapter 10, by Roberto Gómez-García, José-María Muñoz-Ferreras, and Manuel Sánchez-Renedo from the University of Alcalá, Spain, is devoted to *Multiband RF front-ends for radar and communications applications*. New concepts and implementations of radio frequency (RF) front-ends are essential for emerging radar and wireless technologies. In this context, multiband approaches are interesting for acquiring multistandard services. This chapter focuses on deriving a mathematical framework for band allocation so that the minimum sub-Nyquist sampling frequency can be employed without aliasing in a digital multiband acquisition process. This justifies the use of direct-sampling architectures for the associated receiver, where the input RF multiband bandpass filter becomes the key element in terms of hardware. Design rules and experimental prototypes for a novel class of multiband bandpass filter based on signal-interference principles and suitable for these systems are also described.

Chapter 11, on *Mm-wave broadband wireless systems and enabling MMIC technologies*, is contributed by Jian Zhang, Mury Thian, Guochi Huang, George Goussetis, and Vincent F. Fusco, from Queen's University Belfast, UK. Millimeter wave bands provide large available bandwidths for high data rate wireless communication systems, which are envisaged to shift data throughput well in the GBps range. This capability has over the past few years driven rapid developments in the technology underpinning broadband wireless systems as well as in the standardization activity from various nongovernmental consortia and the band allocation from spectrum regulators globally. This chapter provides an overview of the recent developments on V-band broadband wireless systems with the emphasis placed on enabling MMIC technologies. An overview of the key applications and available standards is presented. System-level architectures for broadband wireless applications are being reviewed.

Examples of analysis, design, and testing on MMIC components in SiGe BiCMOS are presented and the outlook of the technology is discussed.

The twelfth chapter, *Reconfigurable RF circuits and RF-MEMS*, was contributed by Robert Malmqvist from Swedish Defence Research Agency (FOI) and Uppsala University, Sweden, Aziz Ouacha from FOI, Sweden, Mehmet Kaynak from IHP GmbH, Frankfurt (Oder), Germany, Naveed Ahsan from Linköping University, Sweden, and Joachim Oberhammer from KTH Royal Institute of Technology, Stockholm, Sweden. While most of today's RF circuits are designed for a specific (fixed) function and frequency range, a much higher degree of flexibility would be possible using highly reconfigurable circuit implementations and front-end architectures. This chapter presents examples of reconfigurable RF circuits that have been realized using either fully transistor based solutions or by employing RF microelectromechanical systems (RF-MEMSs). First a novel approach for implementing reconfigurable circuitry based on the concept of programmable microwave function arrays (PROMFA) is presented. Various reconfigurable circuit designs based on the emergence of high performance RF-MEMS switches being developed in GaAs, GaN, and SiGe RFIC/ MMIC process technologies are then reviewed. In the final section, an overview of state-of-the-art RF-MEMS based phase shifter designs intended for electronic beam-steering antennas and phased array systems is presented.

Chapter 13, authored by Federico Alimenti, Andrea Battistini, Valeria Palazzari, and Luca Roselli from the University of Perugia, Italy, and Stephen M. White from the University of Maryland, College Park, USA, is titled *MIOS: millimeter-wave radiometers for the space-based observation of the Sun*. Millimeter-wave observations of the Sun have never been carried out from a space-based platform. This chapter presents a feasibility study for a full-disk 90 GHz radiometer designed to detect the radio emission of solar flares. First, flare radiation mechanisms are introduced, showing that millimeter-waves are very sensitive probes of the highest energy electrons accelerated in solar flares. Then the fluctuation of the Sun to satellite radio path attenuation is studied by modeling the ionosphere as charged plasma. Finally, the science requirements and the system design are described.

The fourteenth chapter, *Active antennas in substrate integrated waveguide (SIW) technology*, by Francesco Giuppi, Apostolos Georgiadis and Ana Collado from CTTC, Spain, and Maurizio Bozzi and Luca Perregrini from the University of Pavia, Italy, presents the modeling and implementation of active cavity-backed antennas in substrate integrated waveguide (SIW) technology. The cavity-backed topology helps to suppress undesired surface-wave modes and provides improved antenna oscillator phase noise performance. The use of SIW technology allows for a compact and cost-effective implementation of the structure. SIW active antennas open new perspectives in the field of microwave and mm-wave low-cost radio systems and wireless sensors.

The fifteenth chapter, *Active wearable antenna modules*, by Frederick Declercq and Hendrik Rogier from Ghent University, Belgium, and Apostolos Georgiadis and Ana Collado from CTTC, Spain, introduces the novel concept of active wearable antenna modules, constructed entirely from breathable textiles and integrated flexible electronics, for body-centric communication systems. First, a broadband transmission line and an inverse small-band antenna characterization technique are presented. Both methods allow characterizing the electromagnetic properties of the materials as used in the final antenna design. The latter technique is especially developed for a fast and accurate environmental characterization of textile materials allowing predicting antenna performance as a function of relative humidity. Second, an active wearable antenna module is designed relying on computer-aided full-wave co-optimization techniques. Measurements prove that the proposed design technique together with an accurate material characterization provides an excellent performance prediction. Also, energy scavenging by means of integration of flexible solar cells on to wearable antennas is discussed by means of an example antenna design.

Chapter 16, *Novel wearable sensors for body area network applications*, by Chomora Mikeka and Hiroyuki Arai from Yokohama National University, Japan, describes a novel wearable waveguide: a flexible interface for body-centric wireless communications, capable of concentrating wireless communication within the so-called smart suit, and made of flexible, lightweight, conductive fabric: the SC8100 textile. A typical sensing and wireless data transmission design example is also presented.

Finally, Chapter 17 focuses on *Wideband antennas for wireless technologies: trends and applications*. It has been authored by Bahattin Türetken, Umut Bulus, Erkul Başaran, Eren Akkaya, Koray Sürmeli, and Hüseyin Aniktar, from TUBITAK-UEKAE, Turkey. Antennas form a critical part of every wireless system and wideband antennas form an important subset of antennas and arrays with many applications.

The last chapter of the book provides an overview of the development history, design methodologies, and applications of wideband antennas, and concludes with a review on emerging trends.

Monitoring, imaging, security, radar, Gbit communications, body area networks, RFID, and wireless sensor networks are fields of significant interest and primary importance worldwide, and represent examples where the use of microwave and millimeter wave technology will be central in the next decade. Some of the requirements these applications pose to the underlying front-ends are conformal topology, large bandwidths, linearity, miniaturization, MIMO, and diversity techniques relying on multiple antennas, as well as adaptive and reconfigurable transmission/reception capabilities including robust performance against interference. The Editors have made an effort to address selected distinct challenges in the design and modeling of microwave and millimeter wave circuits and systems, identify important applications, and focus on new technologies such as substrate-integrated waveguide and textile electronics.

The Editors would like to acknowledge the chapter authors and all members of EU COST Action IC0803 RF/Microwave Communication Subsystems for Emerging Wireless Technologies for their active participation, contributions, and friendship. The many fruitful scientific and other discussions with all the COST friends over these four years were highly appreciated. The work of Dr. Georgiadis was additionally supported by Project TEC2008-02685/TEC on Novel Architectures for Reconfigurable Reflectarrays and Phased Array Antennas (NARRA) of the Ministry of Science and Innovation, Spain, and by EU Marie Curie Project FP7-PEOPLE-2009-IAPP 251557 on Symbiotic Wireless Autonomous Powered System (SWAP). The work of Prof. Rogier and his group was supported by several project grants of the Special Research Fund of Ghent University, the Flemish Research Fund (Fonds voor Wetenschappelijk Onderzoek Vlaanderen, FWOV), and the Flemish Agency for Innovation by Science and Technology (IWT), by the 'Protection e-Textiles: MicroNano-Structured Fibre Systems for Emergency-Disaster Wear – ProeTex', Sixth Framework Programme Integrated Project, and the 'Antenna and fRont-end MOdules for pUblic Regulated Service applications (ARMOURS)', and EC-FP7 Galileo.2011.3.1-2: Collaborative Project. This publication is supported by COST. Neither the COST Office nor any person acting on its behalf is responsible for the use that might be made of the information contained in this publication. The COST Office is not responsible for the external websites referred to in this publication.

Apostolos Georgiadis
Hendrik Rogier
Luca Roselli
Paolo Arcioni

List of Abbreviations

3G	third generation mobile telecommunications
AC	alternating current (electricity; physics)
ACEPR	adjacent channel error power ratio
A/D	analog-to-digital
ADC	analog-to-digital converter
ADS	advanced design system
AE	(mean) absolute error
AGC	automatic gain control
ALU	arithmetic logic unit
AM–AM	PA EM amplitude modulation to amplitude modulation g envelope characteristic
AM–PM	PA EM amplitude modulation to phase modulation Φ envelope characteristic
BAN	body area network
BF	Bessel–Fourier
BOC	binary offset carrier
BP	blood pressure (medical)
BPSK	binary phase shift keying
BTS	base transceiver station
BW	bandwidth
CAD	computer aided design
CBOC	composite binary offset carrier
CCDF	complementary cumulative distribution function
cf.	compare
CME	coronal mass ejection
CMOS	complementary metal oxide semiconductor
CN0	carrier-to-noise density ratio, $C/N0$
COTS	commercial off-the-shelf
CP	circular polarization
CRPA	controlled reception pattern antenna
CST	computer simulation technology
Cu	copper (Cuprum)
CW	continuous wave
D/A	digital to analog
DC	direct current

DIRECT	dividing rectangles
DME	distance metering equipment
EBG	electromagnetic bandgap
ECG	electrocardiogram
EEG	electroencephalogram
EM	electromagnetic(s)
EM	equivalent memoryless
EMI	electromagnetic interference
ENOB	effective number of bits
ENR	excess noise ratio
FET	field-effect transistor
FIFO	first in first out
FMC	FPGA mezzanine card
FOC	full operational capability
FOM	figure of merit
FoV	field of view
FPGA	field programmable gate array
FSL	fast simple link
FSR	full scale range
GEO	geosynchronous Earth orbit
GLRT	generalized likelihood ratio test
GND	ground
GNSS	global navigation satellite system
GPS	global positioning system
HAGR	high gain advanced GPS receiver
HF	high frequency
HP	high precision
Hz	Hertz (Hz, MHz)
IA	instrumentation amplifier
IBO	input backoff
IC	integrated circuit
ICD	interface control document
IEEE	Institute of Electrical and Electronics Engineers (technology advancement organization)
IF	intermediate frequency
IGSO	inclined geosynchronous Earth orbit
ILS	instrument landing system
IMP	intermodulation products
IOV	in-orbit validation
IP	intellectual property
ISM	industrial, scientific, and medical (radio spectrum)
IVTC	instantaneous voltage transfer characteristics
LAN	local area network
LCMV	linearly constrained minimum variance
LDMOS	laterally diffused metal oxide semiconductor
LF	low frequency

LHCP	left-hand circular polarization
LMS	least mean squares
LNA	low-noise amplifier
LO	local oscillator
LORAN	long-range navigation
LOS	line-of-sight
LPF	lowpass filter
LTE	long-term evolution
MBF	modified Bessel–Fourier
MCU	microcontroller unit
MEMS	microelectromechanical systems
MEMUB	memory to EM upper-bound
MEO	medium Earth orbit
MF	matched filter
MIOS	millimiter-wave instrument for theobservation of the Sun
MLE	maximum likelihood estimator
MMIC	monolithic microwave integrated circuit
MMSE	minimum mean squared error
mmw	millimeter wave
MR	moisture regain
Ms	millisecond
MS	modified Saleh
MSE	mean squared error
MVDR	minimum variance distortionless response
N	north
Ni	nickel
NMSE	normalized mean square error
OBO	output backoff
OFDM	orthogonal frequency division multiplex
PA	power amplifier
PC	personal computer (generic term)
PDHU	processing data handling unit
PHEMT	pseudomorphic high electron mobility transistor
PIFA	planar inverted F antenna
ppm	part per million
PRN	pseudorandom noise
PS	power series
PSU	power supply unit
PVT	position, velocity, and time
PWR	power
QHA	Quadrafilar helix antenna
RAIM	receiver autonomous integrity monitoring
RAM	random access memory
RF	radio frequency
RFI	radio frequency interference
RH	relative humidity

RHCP	right-hand circular polarization
RLS	recursive least squares
$|S_{11}|$	input reflection coefficient of 50 Ω terminated output
$|S_{21}|$	forward transmission coefficient of 50 Ω terminated output
SA	selective availability
SAW	surface acoustic wave
sfu	solar flux unit
SINR	signal-to-interference plus noise ratio
SMA	submulti-assembly
SoL	safety-of-life
SP	standard precision
SUMO	surrogate modeling
TACAN	tactical air navigation system
TEM	transversal electromagnetic
TEMP	temperature
UHF	ultra high frequency
VHDL	very high speed integrated circuit hardware description language
VHF	very high frequency
VOR	VHF omnidirectional radio range
VSWR	voltage standing wave ratio
W	with
WAAS	wide-area augmentation system
WBAN	wireless body area network (implanted sensors that can take direct measurements of body chemistry)
WCDMA	wideband code division multiple access
W/O	without
WPR	wind profile radar
WPT	wireless power transmission

List of Symbols

π	3.141 592 653 589 793:
P_n	absolute noise power density referred to 290 K
$\|x\|$	absolute value of x
$|\cdot|$	absolute value operator
f_M	actual (measured) PA complex envelope transfer characteristic
$\upsilon_k(m)$	additive noise and ICI due to an extra spectral component of the phase noise
$\mathbf{z}_m$	additive white Gaussian noise vector for the mth OFDM symbol
T_a	ambient room temperature (K)
G_A	amplifier gain
F_A	amplifier noise factor
A	amplitude of an oscillating signal
A_{ref}	amplitude of the reference tone
φ	angle at the polar coordinate
ω_c	angular centre or carrier frequency
ω	angular frequency
$\arg(\cdot)$	argument of the complex exponential
α	attenuation constant (m^{-1})
α_d	attenuation constant attributed to dielectric losses (m^{-1})
$R_{vv}(t, t - \tau)$	autocorrelation with timing offset τ
G_A	available amplifier gain
σ_h^2	average magnitude response of the channel
σ_z^2	average power of the additive noise
σ_x^2	average power of the sent subcarrier symbols
$P_{o,mean}$	average signal power over the N output signal samples
ϕ	azimuth angle (rad)
$b(t, \tau)$	background
$\hat{b}(t, \tau)$	background estimation
B_i	bandwidth of the ith channel, Hz
B_L	bandwidth of the lower channel, Hz
B_U	bandwidth of the upper channel, Hz
$s(t)$	baseband complex signal
J	Bessel functions of the first kind

BF($L; \gamma$)	BF behavioural model of a PA EM envelope characteristic, having L coefficients or terms and where the model to actual PA dynamic range ratio is γ
C	capacitance (F)
f_c	centre frequency of a channel, Hz
B	channel bandwidth of an equal-bandwidth multiband signal, Hz
$\mathbf{h}_m$	channel impulse-response vector for the mth OFDM symbol
Δ	channel separation of a dual-band signal, hertz
Δf	channel separation of an equal-spaced multiband signal, Hz
Z_0	characteristic impedance (Ω)
Z	characteristic impedance of a transmission line segment, Ω
$\mathbf{H}_m$	circular channel convolution matrix
K	coefficient number
$\mathbf{X}^*$	complex conjugate of matrix $\mathbf{X}$ (also applied to scalars)
e^{jx}	complex exponential function with argument x
$v_c(t)$	complex oscillator signal
γ	complex propagation constant (m^{-1})
σ	conductivity
$A(t)$	continuous envelope amplitude of an RF PA input signal at time t
$\phi(t)$	continuous envelope phase of an RF PA input signal at time t
S_i	cooperating node group
$\mathbf{x}_i$	coordinates of node i
C	cost function
α_i	coupling factor ($i = 1,2$)
β_i	coupling factor ($i = 1,2$)
J_i	current density
P_{1dB}	1 dB compression point
BW_{3dB}	3 dB bandwidth
β	3-dB bandwidth of the phase noise
β_{ang}	3-dB bandwidth of the phase noise (in angular frequency)
I	DC current (A)
P_{DC}	DC power dissipation
V	DC voltage (V)
$\Delta \tau$	delay between successive observation time instants (impulse responses)
Δt	delay between successive propagation time instants
δ	delta Dirac function.
$n(\sigma)$	density distribution function of the PA (device and model) time domain output signal samples over the power domain
f_0	design centre frequency, Hz
$h_d(t, \tau)$	detector output
$\Phi_k(m)$	DFT of the sampled phase noise
c	diffusion rate
d_l^k	distance among the transmitting antenna Tx, target T_l and receiving antenna Rx_k
d	distance between antennas
d	distance between nodes

δ_{ij}	distance estimate between node i and node j
V_{DD}	drain DC bias voltage
σ	effective conductivity (S/m)
$\varepsilon_{r,eff}$	effective relative permittivity
E	electric field
θ	electrical length of a transmission line segment, rad
θ	elevation angle (rad)
f_{EM}	EM PA complex envelope transfer characteristics
$u_{o,EM}(i)$	EM PA sampled outputs, corresponding to the $u_{in}(i)$
$\in$	energy consumption
μ_{Tx}/μ_{Rx}	energy consumption to transmit/receive
ψ_k	energy of the phase noise around the kth subcarrier
$h_e(t, \tau)$	enhanced radar signals
A	envelope amplitude of a single PA input RF tone
e_n^{jit}	error caused by sampling jitter to the nth sample in charge sampling
$\widehat{\mathbf{X}}$	estimation and true value of parameter $\mathbf{X}$
$\oplus$	exclusive–or operation (modulo–2 addition)
α	exponential weighting factor
α	first subcarrier index of the current channel
s_{31}	forward transmission gain
$\text{FS}(n; \gamma)$	Fourier series approximation to the periodic extension of the PA RF IVTC, having n coefficients or terms, and having a period γD; usually $n = L/2$
$S_1(f)$	Fourier transform of $s_1(t)$
$S_2(f)$	Fourier transform of $s_2(t)$
f	frequency (in Hertz)
γ_c	frequency at which the phase noise PSD deviates from the nominal $1/f$ slope
ω_{ref}	frequency of the reference tone
$\|\mathbf{X}\|_F$	Frobenius norm of matrix $\mathbf{X}$; if $\mathbf{X}$ is $N \times N$
V_{GS}	gate-source voltage
n	generic sample index
$\max_x \{f(x)\}$	global maximum value of $f(x)$ considering all the possible values of x
f_H	highest frequency component of a bandpass signal, Hz
H_0	hypothesis about the absence of $s(t, \tau)$ in the radar signal
H_1	hypothesis about the presence of $s(t, \tau)$ in the radar signal
$r(t)$	I/Q modulated real signal
$\text{LPF}\{\cdot\}$	ideal low-pass filter operator
j	imaginary unit
Z	impedance
L	inductance (H)
$s_I(t)$	in-phase component of the baseband complex signal
P_{in}	input power
s_{11}	input reflection coeficient
IIP_3	input third-order intercept point
$v_i(t)$	instantaneous complex RF input voltage signal

$v_o(t)$ instantaneous complex RF output voltage signal

n integer (0.12,3, . . .)

$FS(2L; \gamma)_{from}$ IVTC FS approximation derived from an MBF($L; \gamma$)

$(M)BF(L; \gamma)$ or an BF($L; \gamma$) envelope model

b_k kth BF or MBF model complex coefficient

$R_k(m)$ kth DFT sample of the received signal from the mth OFDM symbol

$H_k(m)$ kth frequency bin of the channel transfer function for the mth OFDM symbol

c_k kth FS approximation complex coefficient

J_k kth spectral component phase noise complex exponential

$X_k(m)$ kth subcarrier symbol at the mth OFDM symbol

κ last subcarrier index of the current channel

d_l^k length of the main half-axis of the ellipse created on the basis of d_l^k

C_L loading capacitance

$\tan \delta$ loss tangent

f_L lowest frequency component of a bandpass signal, Hz

f_{L1} lowest frequency component of the first channel, Hz

G_{ampl} magnitude part of G

π mathematical constant pi, 3.14159265358979323846264338327 9 . . .

$TOA_{D\,max}$ maximal difference between the TOA estimated from both receivers and belonging to the same single target

f_{max} maximum frequency of oscillation

AE mean absolute error

f_{mod} memoryless complex envelope behavioural model of f_{EM}

NF_{min} minimum achievable noise figure

NF_{min} minimum noise figure

$f_{s,min}$ minimum sub-Nyquist sampling rate, Hz

$MBF(n; \gamma)$ modified BF EM envelope behavioural model of a PA, of n coefficients or terms where the model to actual PA dynamic range ratio is γ

$s_2(t)$ multiband signal with asymmetry with respect to the centre frequency

$s_1(t)$ multiband signal with symmetry with respect to the centre frequency

e Napier's constant, 2.71828182845904523536028747135 2 . . .

F noise factor

NF noise figure

NF noise figure

f_n nominal frequency

$\mathcal{N}(\mu, \sigma^2)$ normal distribution with expected value μ and variance σ^2

B_{Nz} normalized bandwidth

f_{Nz} normalized frequency

N_a number of active subcarriers in an OFDM symbol

N_1 number of anchor nodes

N number of channels of a multiband signal

L number of complex coefficients (BF and MBF models); model order

κ number of iterations

N_2 number of nonlocated nodes

S	number of points, distributed over the PA dynamic range, used in calculating AE
G	number of samples in a guard interval/cyclic prefix
n_α/n_δ	number of selected nodes (path loss selection/distance selection)
u	number of significant spectral components of phase noise
N	number of subcarriers in an OFDM symbol
N	number of WCDMA signal samples in the PA model extraction and validation signals
f_{Nyq}	Nyquist sampling rate, Hz
τ	observation time
$AE_{zero-model}$	AE of an envelope PA model where all model coefficients are set to zero
m	OFDM symbol index
$\Delta\omega_f$	offset from the carrier at which the flicker noise effect is dominating
$\Delta\omega_w$	offset from the carrier at which the white noise effect is dominating
$L(\Delta\omega)$	one-sided PSD phase noise measurement at offset $\Delta\omega$ from the carrier
R_{on}	on-state resistance
$\mathrm{diag}(\cdot)$	operator that creates a diagonal matrix from an input vector
$PE\{\cdot\}$	operator that yields the principal eigenvector of a matrix
P_{out}	output power
Γ_{out}	output reflection coefficient
s_{33}	output reflection coeficient
$u_{o,mod}(i)$	output samples of the memoryless PA behavioural model f_{mod}
γ	overall signal-to-interference plus noise ratio
P1	PA 1 dB compression point
D	PA actual or measured envelope dynamic range
$-D$ to D	PA actual RF IVTC dynamic range
g	PA EM AM–AM envelope characteristic, as a gain
Φ	PA EM AM–PM envelope characteristics
G	PA RF complex memoryless, or EM, IVTC
α	parameter inversely related to the model's dynamic range in the FS RF IVTC approximation and BF envelope model
G_p	passive antenna gain of an active antenna
α	path loss exponent
$G_{e;\gamma}$	periodic extension of G, with the period set to γD
μ_0	permeability of free space (H/m)
β	phase constant (m^{-1})
$\phi(t)$	phase noise at time t
ϕ_n	phase noise at time nT_s
G_{phase}	phase part of G
Φ_i	phase shift ($i = 1,2,3$)
P_{Rx}	power received
P_0	power received at a reference distance
$|S_{11}|$	power reflection S-parameter, dB
$S_v(\omega)$	power spectral density of $v(t)$ at frequency ω
$S_f(\omega)$	power spectral density of the flicker noise at frequency ω
$|S_{21}|$	power transmission S-parameter, dB

$h_p(t,\tau)$	preprocessed radar signals
k_0	propagation constant in free space (m^{-1})
t	propagation time
Q	Q-factor
$s_Q(t)$	quadrature component of the baseband complex signal
Q	quality factor
$h_b(t,\tau)$	radar signals with subtracted background
r	radius
τ	ratio of each dipole antenna's length
γ	ratio of the PA model's dynamic range to actual PA's dynamic range
$h(t,\tau)$	raw radar signals
$v(t)$	real oscillator signal
ρ	received signal-to-noise ratio
Z_0	reference impedance, Ω
$\Delta\phi$	relative phase shift:
ε_r	relative permittivity
$n_{BS}(t,\tau)$	residual noise present after using background subtraction methods
R	resistance (Ω)
s_{13}	reverse transmission gain
i	sample number
A_i	sampled envelope amplitude of an RF PA input signal at sample point i
ϕ_i	sampled envelope phase of an RF PA input signal at sample point i
Y_{meas}	sampled output spectra of the measured PA output signal
Y_{mod}	sampled output spectra of the modelled PA output signal
$u_{in}(i),\ i=1,N$	sampled PA input WCDMA envelope signal
$u_{o,M}(i)$	sampled PA output WCDMA envelope signal
T_s	sampling interval $(1/F_s)$
$\delta_{n,1}$	sampling jitter at the beginning of the integration interval in charge sampling
$\delta_{n,2}$	sampling jitter at the end of the integration interval in charge sampling
ζ_n	sampling jitter realization at the nth sample in voltage sampling
F_s	sampling rate or sampling frequency $(1/T_s)$
f_s	sampling rate, Hz
$\mathbf{q}_m$	scaling vector
$\overline{\mathbf{S}}$	scattering parameter matrix
$\mathbf{R}^{M\times N},\mathbf{C}^{M\times N}$	set of $M\times N$ matrices with real- and complex-valued entries, respectively
υ	shadowing effects
$s(t,\tau)$	signal scattered by the target
γ_k	signal-to-intereference plus noise ratio for the kth subcarrier in the OFDM symbol
$\text{sinc}(x)$	$\text{sinc}(x)=\sin(\pi x)/(\pi x)$
$S_{v,ss}(\omega)$	single-sided power spectral density of $v(t)$ at frequency ω
$S_{a,ss,CO}(\omega)$	single-sided power spectral density of the CO at frequency ω
$S_{a,ss}(\omega)$	single-sided power spectral density of the VCO at frequency ω
λ_k	spectral masque of the phase noise at the kth subcarrier
Δ_n	spectral separation of the nth channel and the $(n+1)$th channel, Hz

$B(t)$	standard Brownian motion at time t
B_n	standard Brownian motion at time nT_s
σ	standard deviation
$E\{\cdot\}$	statistical expectation operator
γ	step length factor
k	subcarrier index in an OFDM symbol
R_s	surface resistivity $(\Omega/\text{sq.})$
$X(t,\tau)$	testing (decision) statistics for the purposes of detection
$\gamma(t,\tau)$	threshold for the purposes of detection
V_{TH}	threshold voltage
t	time (in seconds)
τ	time offset (in seconds)
t	time variable
τ	time-delay factor
$\mathbf{r}_m$	time-domain received signal vector from the mth OFDM symbol
t_0	time-zero
$TOA_l^k(\tau)$	TOA estimated in the observation time instant τ from the receiving antenna $Rx_k(k \in 1,2)$ for the lth target $(l \in L)$
G_{tot}	total gain active antenna
L	total number of targets
$\text{Tr}\{\mathbf{X}\}$	trace of matrix $\mathbf{X}$
g_m	transconductance
G_T	transducer amplifier gain
$\overline{\overline{\mathbf{T}}}$	transfer scattering parameter matrix
W_T	transistor width
f_T	transit frequency
$\cup$	union of two sets
$\mathbf{e}_n$	unit vector of length N and unity as its nth element
$\mathbf{e}_{ij}$	unit vector with the orientation between node i and node j
$\arg\max_x\{f(x)\}$	value of x that maximizes $f(x)$
$\arg\min_x\{f(x)\}$	value of x that minimizes $f(x)$
σ^2	variance
σ_ϕ^2	variance of the phase noise
σ_v^2	variance of the shadowing effects
$\boldsymbol{\phi}_m$	vector of combined transmitter and receiver phase noise samples
$\mathbf{P}_m$	vector of pilot subcarrier values at the mth OFDM symbol
$\boldsymbol{\phi}_{m,R}$	vector of receiver phase noise samples for the mth OFDM symbol
$\boldsymbol{\phi}_{m,T}$	vector of transmitter phase noise samples for the mth OFDM symbol
c	velocity of propagation in the air
ε	a very small number
λ	wavelength
λ_0	wavelength in free space (m)

Part One

Design and Modeling Trends

1

Low Coefficient Accurate Nonlinear Microwave and Millimeter Wave Nonlinear Transmitter Power Amplifier Behavioural Models

Máirtín O'Droma[1] and Yiming Lei[2]
[1]Telecommunications Research Centre, University of Limerick, Limerick, Ireland
[2]State Key Laboratory of Advanced Optical Communication Systems and Networks, School of EECS, Peking University, Beijing, China

1.1 Introduction

The new modified Bessel–Fourier (MBF) nonlinear RF power amplifier (PA) memoryless behavioural model is fully derived and its attributes explored and described in this chapter. Its performance is compared most favourably with the other main competing models. Effectively it is shown in this chapter to be the model of choice when it comes to microwave and millimetre wave memoryless PA behavioural modelling.

This new model originated from efforts to find low order models with better model accuracy than that attainable from Bessel–Fourier (BF), itself along with the modified Saleh (MS) model being among the best memoryless small to large signal PA behavioural models in use today [1–6]. Good low order models are desirable in certain situations, such as where model parameters need to be constantly recomputed as, for example, in adaptive predistortion linearizers of PAs with linear memory [7–11] or in simulations of large multicarrier and/or multiband subsystem simulations containing nonlinear PAs. For these latter situations – a

Microwave and Millimeter Wave Circuits and Systems: Emerging Design, Technologies, and Applications,
First Edition. Edited by Apostolos Georgiadis, Hendrik Rogier, Luca Roselli, and Paolo Arcioni.
© 2013 John Wiley & Sons, Ltd. Published 2013 by John Wiley & Sons, Ltd.

single orthogonal frequency division multiplex (OFDM) air interface would be a common modern-day example – PA models possessing accessible decomposability properties are desirable. BF models and now the new MBF models are such examples. Decomposability here means the capacity to individually generate, isolate, include or exclude each and all nonlinear PA harmonics and intermodulation products (IMP), multipath and adjacent channel interference signals.

1.1.1 Chapter Structure

Demonstrating the superiority of the new model naturally requires a comparative analysis with other models. Here this analysis is benchmarked against the same physical measurements. Hence the physical context for model extraction for this analysis is first described in Section 1.1.2. There details on an L-band laterally diffused metal oxide semiconductor (LDMOS) PA and the modern wideband code division multiple access (WCDMA) signal, which are used throughout this chapter for model extraction, validation and comparison, are presented. This is a typical modern solid state PA; results found for other PAs, not presented here, are quite similar. Then, in Section 1.1.3, the BF model, for which the MBF model was sought as an improvement, is summarized. This is done especially from the perspective of model accuracy, highlighting in particular the anomalous accuracy gaps of low order BF models. In Section 1.1.4 the MBF derivation is set out. This necessitates a more indepth exposition of aspects of the origin and composition of the BF model. The concept of deriving hypothetical RF instantaneous voltage transfer characteristics (IVTC) of the PA, and complex FS approximation of these, is introduced. Exploring the relationship between the two enabled the discovery of better accuracy low order models. From this it is shown how to derive the new MBF model by means of which such improved accuracy low order models may be directly extracted. Further benefits from this exposition are the new useful insights gained into the BF model. In Section 1.1.5, various MBF models of the LDMOS PA are extracted and analysed in the context of their IVTCs. Section 1.1.6 focuses mainly on showing how much better model accuracy and behaviour prediction performance of third order MBF models is compared with other established low order models. Section 1.1.7 addresses key conclusions.

1.1.2 LDMOS PA Measurements

For model extraction, validation and evaluation measurements, an L-band LDMOS nonlinear PA manifesting some memory effects is driven at 5 dB input backoff (IBO) by a standard WCDMA signal having a bandwidth of 3.84 MHz and channel spacing of 5 MHz. Sample input and output signal spectra are shown in Figure 1.1. IBO and output backoff (OBO) here denote the signal's input and output powers normalized to the values at the 1 dB compression point (P1), which is that point where an output power is compressed by 1 dB relative to that yielded by the ideal linear PA equivalent for the same input power. This normalization rule is applied to all powers and voltages in this chapter, unless otherwise stated.

The N input WCDMA envelope signal samples, $u_{in}(i), i = 1, N$, may be written

$$u_{in}(i) = \int A(t)e^{j\phi(t)}\delta(t - i\tau)dt = A_i e^{j\phi_i} \tag{1.1}$$

where $A(t), \phi(t), A_i$ and ϕ_i are the continuous and sampled input envelope amplitude and phase at time t and sample point i respectively, and δ is the delta Dirac function. The number

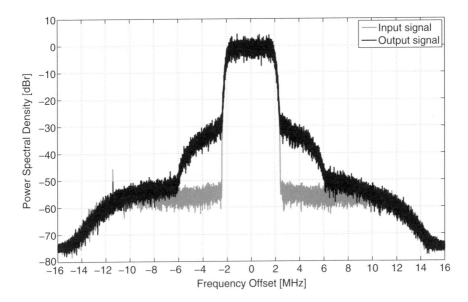

Figure 1.1 LDMOS PA 3G WCDMA input and output spectra, in dB relative (dBr) to the peak value, with the PA driven at 5 dB IBO.

of samples, N, is 10^5; at 32×10^6 samples/s, this corresponds to a 3.125 ms signal duration. The peak-to-average power ratio (PAPR) of the signals at the PA input and output under the operating conditions defined above are found to be 10.36 and 6.6 dB respectively. Hence the PA operation stretches deep into its large signal nonlinear region.

The corresponding N samples of the output envelope, $u_{o,M}(i)$ are as defined in Equation (1.2) below. These are graphed in Figure 1.2 (grey dots) in the form of RF envelope gain and phase versus IBO for each sample pair. It is clear from the spread of output samples at any input IBO point that the PA complex envelope transfer characteristic, f_M, manifests some memory effects. As shown also in Figure 1.2 (full lines), the gain (i.e. AM–AM, g) and phase (i.e. AM–PM, Φ) envelope characteristics of an EM PA, denoted f_{EM}, may be extracted from these by applying a moving average process over the sampled instantaneous input–output envelope responses. These EM characteristics are also graphed in Figure 1.3, but there the input and output amplitudes are voltages normalized to the corresponding voltages at the P1 point. The EM PA's sampled outputs, $u_{o,EM}(i)$, corresponding to the $u_{in}(i)$, are then read off these. The memoryless PA behavioural model f_{mod} is of f_{EM} and its output samples are denoted $u_{o,mod}(i)$.

The relationship between the f_M, f_{EM} and f_{mod} PA complex nonlinear envelope and the input–output sample sets may be expressed as

$$u_{o,M}(i) = \int f_M(A(t)e^{j\phi(t)})\delta(t - i\tau)\mathrm{d}t \tag{1.2}$$

$$\cong u_{o,EM}(i) = f_{EM}(A_i e^{j\phi_i}) = g(A_i)e^{j(\phi_i + \Phi(A_i))} \tag{1.3}$$

$$\cong u_{o,mod}(i) = f_{mod}(A_i e^{j\phi_i}) \tag{1.4}$$

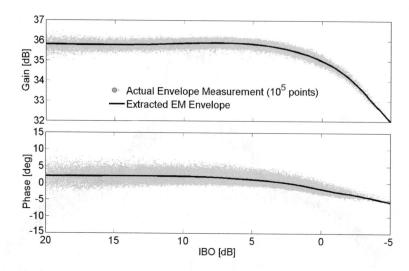

Figure 1.2 Measured samples of AM–AM (gain, g) and AM–PM (Φ) versus IBO responses of LDMOS PA driven by a 3G WCDMA signal (10^5 samples; grey dots), together with the extracted EM envelope characteristics f_{EM} (full lines).

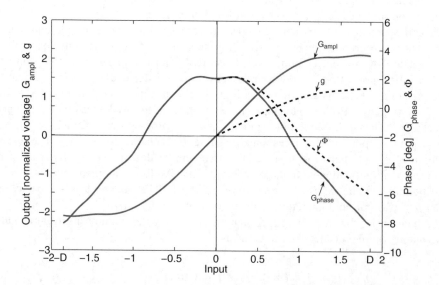

Figure 1.3 The magnitude and phase IVT characteristics of G over the PA's measured dynamic range $-D$ to D, which are denoted G_{ampl} and G_{phase} respectively, together with the extracted EM AM–AM, g, and AM–PM, Φ, envelope characteristics. The 'input' and 'output' are normalized to their respective voltages at P1.

The behavioural prediction performance figures of merit (FOMs), used here, for example in Equation (1.13) below, are based on the difference between output measurement samples $u_{o,M}(i)$ and the model's prediction of these, $u_{o,mod}(i)$. The difference between $u_{o,M}(i)$ and $u_{o,EM}(i)$ amounts to a memory to EM error, be that memory linear or nonlinear, or both [1, 7]. This clearly sets a performance upper-bound to the closeness the behavioural prediction of any memoryless model of this PA EM characteristics can come to the actual PA behaviour. Hence it is denoted 'memory to EM upper-bound' (MEMUB).

1.1.3 BF Model

A memoryless BF model of a PA, f_{mod} of f_{EM}, of order L may be written [3]

$$f_{mod}(A) = \sum_{k=1}^{L} b_k J_1(\alpha k A) \tag{1.5}$$

where J represents Bessel functions of the first kind, $b_k, k = 1, L \ldots, L$, are the model's L complex coefficients and A is the envelope amplitude of a single PA input RF tone. Parameter α is shown in O'Droma [3] to be inversely related to the model's dynamic range. As such it should be harmonized with the actual or measured dynamic range, D, of the PA being modelled, rather than be arbitrarily set as other researchers have done, for example Shimbo & Nguyen, [12]. Using a modelled-to-measured PA dynamic range ratio parameter γ (cf. Equation (1.9) below, where it is defined in relation to α, and associated explanations) the notation BF($L; \gamma$) is used to denote these models, i.e., as defined in Equation (1.5). While theoretically extensible to infinity, usually any $L \geq 7$ will yield excellent full range (small to large signal dynamic range) model accuracy of the f_{EM} envelope characteristics of most PAs.

Coefficients may be extracted through minimizing an error function such as the mean absolute error, AE, between the model's envelope characteristics and the device's EM envelope characteristics, that is

$$AE = \frac{1}{S} \sum_{s=1}^{S} \left| g(A_s) e^{j\Phi(A_s)} - \sum_{k=1}^{L} b_k J_1(\alpha k A_s) \right| \tag{1.6}$$

AE is taken over S points, distributed over the PA dynamic range. To reflect any internal minor deviations between the EM measurements and their model, a reasonably large value for S is advisable, for example more than 40 measurements. Here we use 81, but much smaller numbers yielded almost identical results. Below, in Section 1.1.6, AE is also employed as a model goodness FOM in model comparisons.

Graphs of AE versus α for model orders ranging from 2 to 10, 15 and 20, for BF models of the L-band LDMOS PA, are presented in Figure 1.4. A 'zero model' is included as a useful reference. It is that model where all model coefficients in Equation (1.6) are set to zero; hence its AE, denoted $AE_{zero-model}$, is equivalent to the average absolute EM envelope amplitude, normalized here of course. These graphs immediately convey why Shimbo and other authors of References [2, 12, 13], were successful with their 'arbitrary' choice of 0.6 for α in creating good tenth order BF models, but why Vuong and Moody, authors of Reference [14], where they strongly criticised the model, were quite unsuccessful because of their bad choice of 200.

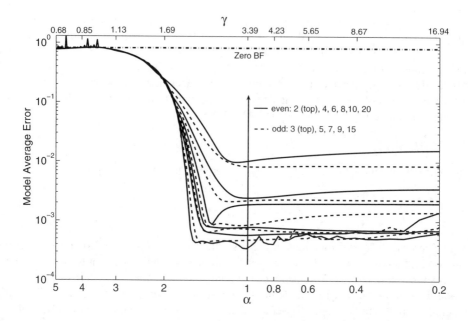

Figure 1.4 Model accuracy of the BF models of the LDMOS PA EM envelope characteristics with 2 to 10, 15 and 20 coefficients as a function of $\alpha\gamma$. Log scales used. Full line: even order models; dashed line: odd order models.

1.1.3.1 BF Model Accuracy Anomalies that Spurred the Development of the MBF Model

Values of $\alpha \leq 1.2$ yield viable and good BF models, with accuracy improving exponentially with model order, hitting an AE 'floor' for orders ≥ 8, as may be seen in Figure 1.4. The sixth and seventh order models manifest optima that almost reach this floor. The AE for the low order $BF[3; \gamma]$ model, at 0.0083, for the best we could extract via Equation (1.6), is more than 10 times worse than that for a $BF[10; \gamma]$ model, which at 0.0006 is excellent. However, what is more notable is that it is about five times worse than the quite good $BF[4; \gamma]$ and $BF[5; \gamma]$ models. A search for a better third order model to bridge this 'accuracy gap' is what led to the discovery of the modified BF models (MBF). Key to this is a deeper understanding of the relationship between AE and α; this is examined further in Section 1.1.4.

1.1.4 Modified BF model (MBF) – Derivation

The proposed new MBF model is derived by exploiting the link between the BF envelope model in Equation (1.5) and the complex Fourier series (FS) approximation of the periodic extension $G_{e;\gamma}$ of the PAs RF complex EM IVTC, denoted G. This latter may be expressed as

$$v_o(t) = G[v_i(t)] = G_{ampl}[v_i(t)]e^{G_{phase}[v_i(t)]}$$

$$= \sum_{k=-\infty}^{\infty} c_k e^{j\alpha k v_i(t)} - D \leq v_i(t) \leq D \tag{1.7}$$

where $v_i(t)$ and $v_o(t)$ are the instantaneous complex RF input and output voltage signals, G_{ampl} and G_{phase} are the magnitude and phase parts of G and c_k is the kth complex coefficient of the FS approximation of G.

In Reference [3], O'Droma has shown both that the relationship between the coefficients b_k in the envelope BF model, Equation (1.5), and the FS c_k coefficients here is

$$b_k = j(c_k - c_{-k}) \qquad (1.8)$$

and that parameter α is the same in both Equations (1.5) and (1.7). Implicit to the definition of the FS is that the period of the periodic extension $G_{e;\gamma}$ of G is $2\pi/\alpha$. This period is effectively the model's dynamic range and may be so defined. Hence linking it to D, the PA measured envelope dynamic range, which in turn is half that of the actual IVTC range, is important and may be achieved by defining a dynamic range ratio parameter γ such that

$$\frac{2\pi}{\alpha} = \gamma D \qquad (1.9)$$

In the LDMOS PA example here, the value of the normalized dynamic range of the PA envelope amplitude is $D = 1.855$. Just as the relationship between α and the period of the FS approximation of the periodic extension of the associated IVTC $G_{e;\gamma}$ has been missed in some seminal papers on the BF model, for example Shimbo, and Vuong and Moody in References [12, 14], so also the linking of the ratio of the modelled to measured dynamic ranges through γ has been missed. However, as will be seen below, all this plays an important role in both the BF models and in the new MBF models.

1.1.4.1 PA IVT Characteristics from BF Envelope Models

From the extracted coefficients b_k of a BF model of the PA EM envelope characteristics, Equation (1.5), an FS approximation of the periodic extension $G_{e,\gamma}$ of a hypothetical non unique memoryless IVTC of the PA, G, may be derived, by obtaining the c_k coefficients in Equation (1.7) via Equation (1.8). As the dynamic range of the PA EM envelope is 0 to D, so the dynamic range of G is $-D$ to D. While being hypothetical, in a mathematical sense, this IVTC model will be a good model if the originating BF envelope model is good. In fact, there is an unlimited number of such 'derived' IVTC models, depending on how one chooses to fix the relationship between c_{-k} and c_k in Equation (1.8). Presumably at least one of these will match the actual PAs IVTC, which is unknown here and may remain unknown without affecting the validity of the theory being set out below; cf. also Blachman's approach in Reference [15]. In Equation (1.8), setting

$$c_{-k} = -c_k \qquad (1.10)$$

yields IVTC models manifesting odd and even symmetry for the amplitude and phase components respectively. Deriving such an IVTC, G, for this LDMOS PA from an optimum 10th order $BF(10; 2.7)$ model yields an excellent match to the measured EM envelope characteristics over the dynamic range D. This derived G is shown in Figure 1.3, in its G_{ampl} and G_{phase} components, together with the PA EM AM–AM, g, and AM–PM, Φ, envelope characteristics.

1.1.4.2   Finding Good Low Order BF PA Envelope Models from a Derived PA IVTC

Treating this derived G as though it was a measured EM IVTC of the PA, new FS approx-imations (models) now may be extracted from it, via Equation (1.7). Different FS approximations may be found depending on how the periodic extension of G is achieved, quite apart from the order of the FS. These models in turn may be converted back into new BF envelope models by applying Equation (1.8) to the extracted FS coefficients. Such BF models obviously cannot be better than the originating excellent $BF(10; 2.71)$ used to find G but, as there is some freedom in choosing how to make the periodic extension $G_{e;\gamma}$ of G for the new FS models, following O'Droma and Lei [16], there is a potential to find, low order BF models that are more accurate than those directly extracted.

The first step in finding such improved models is to control the ratio of the period, γD, of the periodic extension $G_{e;\gamma}$ to the $2D$ dynamic range of G. This can be done directly via control of γ (or α, Equation (1.9)); hence the inclusion of the subscript γ in $G_{e;\gamma}$. Making $\gamma > 2$ opens a gap of $(\gamma - 2)D$ width between the recurring G sections in the middle of each period of the $G_{e;\gamma}$, creating the possibility, the second step, to control the shape of the curve in this gap, which connects these G sections. Different gap widths and different shapes of connecting curves will yield different FS approximations. A 'good' FS model extracted from $G_{e;\gamma}$, following Equation (1.7), is one that is a good match to the IVTC G that is in the '$-D$ to D' range only of $G_{e;\gamma}$.

That the dynamic range of the model, the period γD, should be greater than or at least equal to that of G, the dynamic range of the PA being modelled, which results when $\gamma \geq 2$, is a natural expectation. However, for a fuller understanding, it is useful to consider in more depth the modelling dynamics as a function of the actual values of γ ($>2, = 2$ and <2), as addressed in the following.

$\gamma = 2$: poor models with strong Gibbs effect
With $\gamma = 2$ the period of the FS model and the PAs IVTC dynamic range are equal, that is both are $2D$. Hence $G_{e;2}$ will consist simply of G sections repeated with a step discontinuity 'connecting' them at each period transition, as seen in Figure 1.5 for two periods of the amplitude component of $G_{e;2}$ and in Figure 1.6 for the corresponding phase component. An FS approximation (to $G_{e;2}$) will naturally manifest a Gibbs effect [17] at these discontinu-ities, resulting in an inherently poor IVTC model accuracy, especially in the region of the discontinuity, that is in the PAs large signal region. An $FS(20; 2)$ approximation of this, which was extracted but is not shown here, is indistinguishable from the graphs marked 'FS from $BF(10; 2)$' in these figures; the Gibbs effect is clearly visible in both.

This impact on model accuracy for values of γ in and closely around $\gamma = 2$ is also evident from the AE graphs in Figure 1.4, where $\gamma = 2$ falls in the transition area between the good AE models on the right and the poor models on the left.

$\gamma > 2$: Good models, especially around $\gamma = 4$
Depending on the size of this gap $(\gamma - 2)D$ between the G sections in each period of the $G_{e;\gamma}$ and the shape of the curve connecting the G sections within the gap, different models result. It is intuitive to seek a smooth transition between the end points thereby eliminating disconti-nuities and thus also the Gibbs effects in the FS approximations. Further, it makes sense to choose the shape of a 'transition' curve so that attributes present in the G sections are exploited. One such clear attribute is the resemblance of the typical shape of the large signal

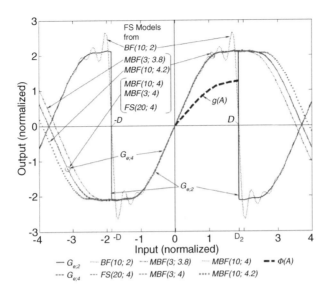

Figure 1.5 The periodic extensions, $G_{e;2}$ and $G_{e;4}$, of the PA EM IVT amplitude characteristic G (the full line from $-D$ to D; also Figure 1.3) and the amplitude parts of FS approximations of these shown over the normalized input range -4 to 4. The periodic extensions are with γ values of 2 and 4 respectively, the former with discontinuities at $\pm D$, $\pm 3D$ and so on, the latter without, as it is constructed following Equation (1.11). There are graphs of six FS approximations superimposed on the $G_{e;2}$ and $G_{e;4}$. One is of $G_{e;2}$, derived from $BF(10; 2)$. The Gibbs effect is visible and strong. The other five are of $G_{e;4}$: one, the $FS(20; 4)$ approximation, is directly extracted from $G_{e;4}$, Equation (1.7), and four are derived from optimized and nonoptimized 3 and 10 coefficient MBF envelope models, via Equations (1.7) to (1.9). The γ optimized ones are $MBF(3; 3.8)$ and $MBF(10; 4.2)$ and non-optimized are $MBF(3; 4)$ and $MBF(10; 4)$. All five models are visually indistinguishable from $G_{e;4}$ and $G_{e;2}$ over the relevant $-D$ to D range, although outside this range, the best ones, from the γ optimized $MBF(10; 4.2)$ and $MBF(3; 3.8)$ models, are clearly distinguishable. The extracted EM $g(A)$ characteristic is also shown. Input and output are normalized with respect to their respective voltages at P1.

nonlinear amplitude component of a PA IVTC to a half period $(-\pi/2, \pi/2)$ of a sinusoid. Hence if the period of the periodic extension is widened to $4D$, that is $\gamma = 4$, and thus the gap between the G sections in each period increased to $2D$, then making an even reflection of the IVTC about its end points, that is about the G_{ampl} end points at D and $-D$, will tend to complete the resemblance to a full sinusoid. Besides greatly reducing or eliminating the discontinuity in the amplitude component of the IVTC, this kind of manipulation of $G_{e;4}$ significantly increases the likelihood of finding reduced coefficient-count FS approximations with acceptable accuracy, as there is a good potential to achieve a significant matching of the dominant amplitude part of $G_{e;4}$ by the first term of the FS approximation. Such a periodic extension, with odd symmetry about zero for the amplitude characteristic, is yielded by

$$G_{e;4}(2nD + x) = (-1)^n G(x), \quad -D \le x \le D \quad \text{with } n = 0, \pm1, \pm2, \ldots \quad (1.11)$$

The graph of the $G_{e;4}$ periodic extension in accordance with Equation (1.11) for the LDMOS PA IVTC G is shown in Figure 1.5 (amplitude component). An $FS(20; 4)$ model,

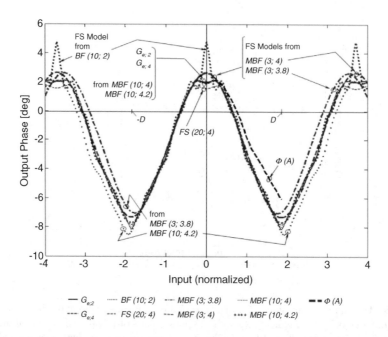

Figure 1.6 The $G_{e;2}$ and $G_{e;4}$ phase characteristics and their FS approximations corresponding to the Figure 1.5 graphs. Apart from the FS approximation derived from $BF(10; 2)$, in the relevant $-D$ to D range, the models match the $G_{e;4}$ and $G_{e;2}$ very well. For the most part they are visually indistinguishable from one another, especially those from $MBF(10; 4)$ and $MBF(10; 4.2)$. Only in a narrow region in the vicinity of zero and $\pm D$ may the directly extracted $FS(20; 4)$ approximation and the models from $MBF(3; 3.8)$ and $MBF(3; 4)$ be visually distinguished. As with the amplitude models from $MBF(10; 4.2)$ and $MBF(3; 3.8)$, here also these are clearly distinguishable outside the $-D$ to D range.

extracted from $G_{e;4}$ in Equation (1.7), is graphed there also. It is visually indistinguishable (without zooming) from $G_{e;4}$. Further, over the dynamic range $-D$ to D, the $G_{e;4}$ graph is indistinguishable from other superimposed FS models that are derived from the newly extracted MBF models – yet to be described (Section 1.1.5 below).

In Figure 1.6, the phase component graphs corresponding to those in Figure 1.5 are shown. There the $FS(20; 4)$ approximation of $G_{e;4}$ is visually distinguishable from $G_{e;4}$ (and $G_{e;2}$) only in the vicinity of the mild discontinuities at zero, $0, \pm D, \pm 2D$ and so on. The differences are really quite insignificant, being of the order of a fraction of a degree.

Given the nature of the periodic extension rule used, Equation (1.11), the even coefficients of the $FS(20; 4)$ approximation of $G_{e;4}$ will be zero. The spectrum of the odd coefficient magnitudes are found to decrease exponentially and rapidly towards zero with increasing coefficient number. This is as would be expected. It is almost identical in fact to the spectrum for the new $MBF(10; 4)$ model, shown later in Figure 1.12(b); the positive order coefficients only are shown here. This means of course that the significance for model accuracy of the higher order terms reduces exponentially with order value. While this attribute might in general be expected in a series-based approximation model, in fact it is not normally the case for the coefficient sets directly extracted for $BF(L; \gamma)$ models via Equation (1.6). However, here, following the algorithm described, the attribute does apply and hence good lower order FS models of the IVTC, and thus derived BF models of the envelope characteristics, may be found, either by just dropping coefficients or by simply extracting new low order FS models of $G_{e;4}$.

$\gamma < 2$ and the 'Zero Model' asymptote

Reducing γ below 2 means that the period of the periodic extension of the IVTC will be shorter than the $2D$ dynamic range of G and so the ends of each period will overlap. It is similar to the aliasing errors in DSP when undersampling happens. With this, it is impossible to reproduce the G IVTC in the overlap region. Moreover, considering that the fundamental FS component itself will be $2/\gamma$ sinusoid periods long in the $2D$ IVTC dynamic range and as such a poor match to the monotonic half-sinusoid form of the IVTCs G_{ampl} over this $2D$ range, then the coefficient extraction optimization process will tend to attenuate this first FS coefficient; the smaller the γ, the greater the attenuation. Hence it is intuitive to appreciate that the model extraction optimization process will quickly drive the FS coefficient values towards zero as γ decreases and the overlap increases. This behaviour of the FS, and thus of the BF, coefficients is reflected in Figure 1.4, shown on the left where all models tend towards the 'zero model' asymptotically.

1.1.4.3   The Modified BF Model

Having set out a process (above) that makes for a good low order model – viz. deriving a good IVTC G from an excellent high order BF model, finding good low order FS models of it and converting these back to good low-order BF models – a new MBF model may be adduced as a way to find these good low order models directly, that is bypassing the algorithm's intermediate steps. In can be achieved with two modifications of the BF model. The first is, following the logic of Equation (1.11), to impose the constraint of allowing only odd order coefficients. The second is to substitute α in the BF model with $2\pi/(\gamma D)$, in accordance with Equation (1.9), and thereby express the period of the underlying FS approximation of the IVTC in terms of the measured dynamic range D of the PA being modelled and have control of the ratio, γ, of both. For low order models the value of γ should be kept in the region of 4 so as to optimize the intuitive match of the fundamental component of the directly linked underlying FS approximation of that underlying hypothetical IVTC, which will have the sinusoidal-like form of the $G_{e;4}$ as described above and in Equation (1.11). Hence this modified BF (MBF) may be expressed as

$$g(A)e^{j\Phi(A)} = \sum_{k=1}^{L} b_{(2k-1)}J_1\left(\frac{2\pi}{\gamma D}(2k-1)A\right) \qquad (1.12)$$

The following section analyses MBF models extracted using this equation.

1.1.5   MBF Models of an LDMOS PA

Tenth and third order MBF models of the LDMOS PA above were directly extracted, using Equation (1.12), with $\gamma = 4$, Equation (1.11), and with values of γ that were optimized to yield minimum AE, those values found here being 4.2 and 3.8 respectively.

Table 1.1 LDMOS PA model performance comparisons: MBF vs classical models

Model	L^a	γ	Average error (AE)	NMSE (dB)	ACEPR (dB)
MBF (optimum)	3	3.8	0.0030	−33.3	−43.1
MBF	3	4	0.0034	−33.0	−42.8
MBF	10	4	0.0006	−33.5	−43.5
MBF (optimum)	10	4.2	0.0005	−33.5	−43.5
BF (optimum)	3	3.64	0.0083	−31.3	−40.5
BF (optimum)	10	2.71	0.0006	−33.5	−43.5
Saleh[b]	2	—	0.0514	−27.7	−36.6
MS	4	—	0.0063	−32.1	−43.1
PS(3)[c]	3	—	0.0085	−31.0	−39.8
MEMUB[d]	10–40	—	—	−33.5	−43.5

[a]L is the number of coefficients only; it does not include γ or α in MBF and BF, or the various optimizable exponent parameters in Saleh and MS.
[b]With AM–AM only, hence two parameters; for this PA, a Saleh AM–PM model is unextractable [6].
[c]Three complex coefficients: first, third and fifth.
[d]FOM's "memory to equivalent memory upper bound" for the LDMOS PA.

To check these models in the IVTC domain against the $G_{e;4}$ characteristic and the FS(20; 4) model extracted from $G_{e;4}$, they were translated into FS approximations using Equations (1.7), (1.6) and (1.10). These then are approximations of hypothetical periodically extended IVTCs of the PA. They are graphed in Figures 1.5 and 1.6 as the graphs marked 'FS models from *MBF(10; 4), MBF(10; 4.2), MBF(3; 4)* and *MBF(3; 3.8)*'. For the amplitude component, over the key G dynamic range $-D$ to D, they appear excellent and visually indistinguishable from $G_{e;4}$ and from the FS(20; 4) model. This is the range where model accuracy matters. Outside this region, $\leq -D$ and $\geq D$, the two FS approximations derived from the optimized MBF models can be distinguished, just as might be expected. A similar result may be noted for the phase component in the range $-D$ to D, except in the vicinity of the mild discontinuities present here at 0 and $\pm D$, although interestingly the two 10-term MBF models are also indistinguishable from the IVTC here, unlike the FS(20; 4) model. This latter is an indication that higher order MBF models are better than their equivalent BF models, even if only slightly. That the optimized MBF envelope models do yield better results may be seen in the accuracy and performance comparisons in the sequel in Section 1.1.6 (e.g. Table 1.1).

As might be intuited from the foregoing, for lower order models, especially third order ones, the actual value of γ for optimum results relates mainly to the kind of extension needed to migrate the underlying extended IVTC shape into a sinusoidal resemblance with period γD, and hence enable its periodical extension $G_{e;\gamma}$ to be largely approximated through the first coefficient of the FS approximation. Whether γ is > 4 or < 4, and by how much, depends inversely on how deeply the EM characteristics (from extracted measurements) penetrate into the PA saturation region. For high order models, there is little difference in the AE between the optimum and a model with $\gamma = 4$, as may be observed in Figure 1.7 for *MBF(10; γ)* and *MBF(7; γ)*. There AE is more determined by the interplay between the coefficient terms.

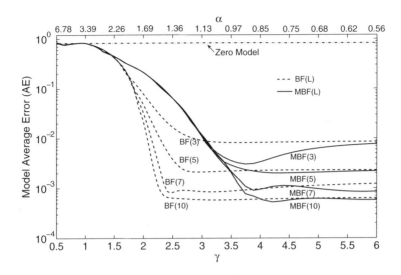

Figure 1.7 Accuracy of the BF and MBF envelope models, with 3, 5, 7 and 10 coefficients, of the LDMOS PA EM envelope characteristics as a function of γ and α.

The tendency in the typical PA AM–AM characteristic, $g(A)$, towards quarter wavelength sinusoidal resemblance means the model is particularly suited to large signal modelling, with measurements deep into the PA saturation region, for example extending up to -6 dB IBO.

1.1.6 MBF Model – Accuracy and Performance Comparisons

Comparisons in the following are focused on low order models for the most part – on third order MBF models being compared with third order BF and with other established low order power series (PS), Saleh and modified Saleh models [1, 5, 6]. All models are extracted from the EM LDMOS PA. Excellent tenth order MBF and BF models are also included. The behavioural prediction performance is set against the measurements of PA amplifying WCDMA signals, as described in Section 1.1.2, and generally follows traditional comparative analysis [8, 9]. These comparison results are summarized as follows:

1. MBF and BF model accuracy.
2. Gain characteristics – model accuracy.
3. Behaviour prediction performance – NMSE and ACEPR.
4. Regeneration of the envelope amplitude's CCDF.
5. Model large signal margin of reliability.
6. Model extensibility, MBF versus BF.

These are addressed in Table 1.1 under the following subsection headings.

1.1.6.1 MBF and BF Model Accuracy

Through the *AE* model accuracy FOM, that is Equation (1.6) for BF and an equivalent for MBF, models may be compared. Figure 1.7 shows these comparisons as a function of γ (and α, top axis).

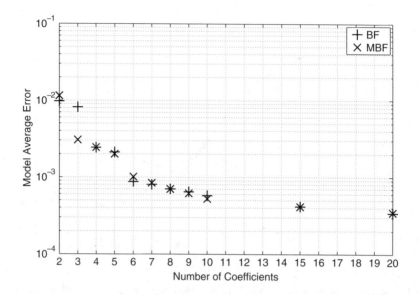

Figure 1.8 Accuracy of optimum BF and MBF models as function of model order, *L*.

The behaviour on the left (poor *AE*, asymptotically driven towards the zero model) and in the transition region follows the logic already discussed above for the BF model in Section 1.1.4.2. On the right, in the region with $\gamma \geq 4$, the *AE* improves (exponentially) with increasing model order; this is also seen in Figure 1.8. For each order all model coefficients are separately extracted.

MBF models of 2 and 4 coefficients (and others) are omitted for clarity; as with the 5; 7 and 10 coefficient MBF models, they provide no special *AE* improvement over their BF equivalents in the $\gamma \geq 4$ region. Orders ≥ 10 are omitted but these are found to yield negligible further improvement; this may already be observed for BF models in Figure 1.4.

Of particular note, and the main outcome of this comparison, is the greater sensitivity to γ of the third order MBF model; it manifests a pronounced optimum occurring at $\gamma = 3.8$. At this point *MBF(3; 3.8)* has an *AE* of more than 250% better than any third order BF model, that is, any *BF(3; γ)* model directly extracted, via Equations (1.5) and (1.6). Examination of other PA characteristics yield similar results (not shown here), with optimized γ always being found in a narrow region around 4. The *AE* comparisons with the other low order models may be seen in Table 1.1, where MBF performs better than twice as good as its nearest competitor, the MS model.

For higher order models the *AE* improvement of MBF over their BF equivalents is not significant. In a sense all higher order ($L \geq 7$) BF and MBF models are good. Optimization points for γ are so mild as to be nearly indefinable; hence setting $\gamma \geq 4$ (or $\alpha \leq 0.85$) will always yield good near-optimum models. Both MBF and BF models are better than equivalent power series models, and of course better than the modified Saleh (MS) and Saleh models [1, 6].

1.1.6.2 Gain Characteristics – Model Accuracy

Complementing the *AE* view (Table 1.1) are graphs shown in Figure 1.9 comparing a selection of gain characteristics of lower order models over the envelope dynamic range.

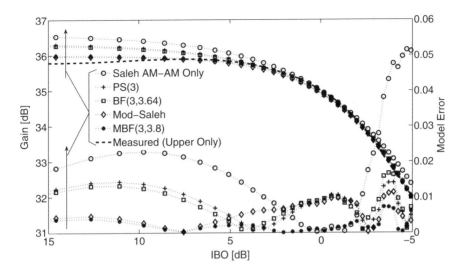

Figure 1.9 *MBF(3, 3.8), BF(3, 3.64)*, Saleh, MS and *PS(3)* AM–AM gain models (top part) and modelling errors with respect to EM 'measured' gain (bottom part) over the LDMOS PA input dynamic range.

Models included are the best *MBF(3; 3.8), BF(3; 3.64)*, MS, two-parameter AM–AM Saleh and three-term power series *PS(3)* with complex coefficients that could be extracted for the LDMOS PA used here. The MBF model is clearly quite superior over the input dynamic range, and especially through the important large signal region. The four-parameter MS is a close second.

IVTC Domain Accuracy

For completeness the associated underlying FS models derived from the *MBF(3; 3.8)* and *MBF(3; 4)*, following Equations (1.7), (1.8) and (1.10), have been included in the IVTC domain graphs in Figures 1.5 and 1.6. Addressing the amplitude characteristics, in the relevant $-D$ to D dynamic range of G (i.e. of the $G_{e,4}$ and/or $G_{e,2}$) the FS models cannot be distinguished from the $G_{e,4}$ and/or $G_{e,2}$; a zoom at saturation regions would show small differences (not shown here). Outside that range, *MBF(3; 3.8)* can be distinguished, as would be expected, but this is not relevant to model accuracy.

1.1.6.3   Behaviour Prediction Performance – NMSE and ACEPR

This is done here by comparing the output signal from the model to that of the actual LDMOS PA output when using a WCDMA validation signal, a different signal from that used for model extraction. That signal's input and output PAPR values were similar at 10 and 6.57 dB respectively. Two figures of merit, FOMs, are compared in Table 1.1: normalized mean square error (NMSE) and adjacent channel error power ratio (ACEPR[1, 18, 19]. The NMSE effectively evaluates the autocovariance of the difference between the measured and modelled PA outputs [19]. In ACEPR the power of the error between the measured and the modelled signal outputs within the adjacent channel is evaluated and compared to the power in the carrier channel [20]. As ACEPRs can be taken of the upper or lower adjacent

channels both are calculated and the worst case selected for presentation here. Both FOMs
are defined as follows [1]:

$$NMSE = 10 \times \log \left[\frac{\sum\limits_{i=1}^{N} |u_{o,M}(i) - u_{o,mod}(i)|^2}{\sum\limits_{i=1}^{N} |u_{o,M}(i)|^2} \right] \tag{1.13}$$

$$ACEPR = 10 \times \log \left[\frac{\int\limits_{f_{adj}} |Y_{meas}(f) - Y_{mod}(f)|^2 df}{\int\limits_{f_{ch}} |Y_{meas}(f)|^2 df} \right] \tag{1.14}$$

$$\cong 10 \times \log \left[\frac{\sum\limits_{i=n_2}^{n_3} |Y_{meas}(i) - Y_{mod}(i)|^2}{\sum\limits_{i=-n_1}^{n_1} |Y_{meas}(i)|^2} \right] \tag{1.15}$$

where the sampled output spectra Y_{meas} and Y_{mod} of the measured and modelled output signals are obtained by fast Fourier transforms of $u_{o,M}$ and $u_{o,mod}$ respectively. The values of n_1,
and of n_2, n_3, are chosen so as to correctly define the inband channel components over ± 1.92
MHz and the standard first upper or lower adjacent channel components over $\pm(5 \pm 1.92)$
MHz respectively relative to the 3G WCDMA centre frequency; cf. Figure 1.1.

PA memory effects are present in the actual PA output. With memory omitted from the
models, there is then a 'memory to equivalent memoryless' upper bound (MEMUB) on each
of the FOMs, that is an upper bound on the performance achievable by the EM models. This
can be estimated by calculating the signal output differences between a near 'perfect' (high
order) EM PA model and the actual PA device with memory. To set the MEMUB results for
the FOM bounds (row 'MEMUB' in Table 1.1), several excellent BF, MBF and higher order
piecewise linear models of the PA EM envelope characteristics were examined (covering the
model order range of 10 to 40 terms) and none could improve on the *MBF(10; 4)* and optimized *MBF(10; 4)* and *BF(10; 2.71)* models. This is not unexpected, as the *AE* improvement
with ever higher order BF or MBF models is negligible.

In Table 1.1 it is clear that, for these FOMs, *MBF(3; 3.8)* is significantly better than a
classical *BF(3)*, a *PS(3)* and the Saleh model, and is also better than the MS model, though
this latter compares well with it.

1.1.6.4 Regeneration of the Envelope Amplitude's CCDF

A comparison is made of the performance of the models to regenerate or preserve the
statistical distribution of the envelope amplitude. This is done here through the complementary cumulative distribution functions (CCDFs) of the output envelopes of the actual

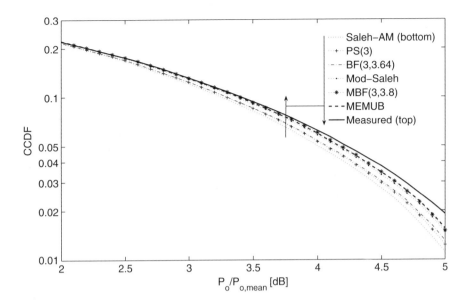

Figure 1.10 A comparison of the CCDF regeneration performance of the low order models of the LDMOS EM envelope characteristics set against the actual CCDF extracted from the measurements ('measured') and against the 'memory to EM upper bound' CCDF regeneration (MEMUB); the upper two graphs. The model graphs, in order going from bottom (worst case) to top (best case), are Saleh (AM–AM only), *PS(3)*, *BF(3; 3.64)*, MS and *MBF(3; 3.8)*.

LDMOS PA ('measured') and its models obtained using the validation WCDMA signal. For the N output samples,

$$CCDF(P_o) = \frac{1}{N} \int_{\sigma=P_o}^{\infty} n(\sigma)\,\mathrm{d}\sigma \qquad (1.16)$$

where $n(\sigma)$ is the density distribution function of the time domain output signal samples over the power domain. These are shown in Figure 1.10. There $P_{o,mean}$ is the average signal power over the N output signal samples. It is clear that here again the MBF performs best among the low order models, with the MS a close second. The CCDF result for the *MBF(3; 3.8)* model is indistinguishable from the MEMUB graph, that is the graph obtained using the near 'perfect' model, which is the regenerated CCDF upper bound achievable for this LDMOS when memory is omitted from the model.

1.1.6.5 Model Large Signal Margin of Reliability

A further advantage of the MBF model over its corresponding BF model, be these high or low order models, is its behaviour beyond the PA measured dynamic range used to extract the model. If the model extraction signal drove the PA deep into saturation, even into flat compression as for the LDMOS PA here, then the MBF envelope model and its corresponding IVTC model, both of which can be considered 'extended' when employing a value of γ

in the region of 4, through their even reflection about the upper limit of the measured dynamic range (i.e. about D for the envelope measurement and about both the D and $-D$ points for the IVTC model), effectively continue this compression for some (short) distance beyond the measured dynamic range. This may be observed in Figure 1.5. Such continuation of this compression is what would be expected to be seen in the actual behaviour of a PA. This can be regarded as providing a modelling 'margin of reliability'. It will even do this when falling short of the deep PA saturation. It is an attribute that can be important when a model is used in a context where the signals being amplified in a simulation or behavioural analysis experiment have PAPRs that are greater than the model extraction signal.

This is notable as normally models, and model extraction procedures, do not set any constraints on the behaviour in the region outside the measurement dynamic range (except sometimes by artificial extrapolation of the characteristics). It can happen that the behaviour of some models in this region is erratic. This is the case for BF models. Therefore, if one seeks a $BF(L; \gamma)$ model by a direct fitting process to the EM envelope characteristics of Equation (1.5) and using a γ that one judges to be good that is correctly following the thinking outlined above, then the resulting model will in all likelihood be good – excellent for high order L – within the measurement dynamic range D (i.e. for the envelope characteristics or from $-D$ to D for the IVTCs) but its behaviour outside this range will be erratic and not amenable to prediction. If not foreseen, it could result in significant behaviour prediction errors when the envelope dynamic range of the signal being analysed is even slightly greater than D.

An example of a good directly extracted $BF(10; 4)$ model which does not have this margin of reliability may be observed in Figure 1.11. There the amplitude characteristics of it and its

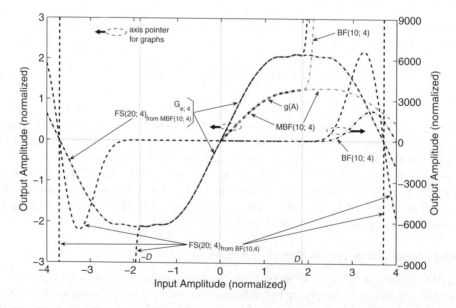

Figure 1.11 Amplitude parts of the actual (right axis) and zoomed (left axis) *BF(10; 4)* envelope model obtained by direct fitting to the LDMOS PA EM characteristics and its corresponding *FS(20; 4)* IVTC approximation set against the well-behaved *MBF(10; 4)* model and its *FS(20; 4)* approximation; the measured EM *g(A)* and IVTC $G_{e;4}$ are also included.

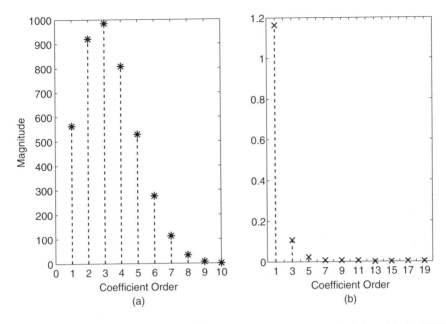

Figure 1.12 The coefficient spectra of (a) the *FS(20; 4)* IVTC model derived from (a) the *BF(10; 4)* model (one side) and (b) the *MBF (10; 4)* model.

corresponding *FS(20; 4)* IVTC model, derived from it using Equations (1.7), (1.8) and (1.10), are shown (actual and zoomed) and set against those of the *MBF(10; 4)* and its derived *FS(20; 4)*. All the models perform well within the dynamic range of the measurements: the derived amplitude IVTC FS approximations *FS(20; 4)ₜfromBF(10;4)* and *FS(20; 4)fromMBF(10;4)* within the −D to D range and the *BF(10; 4)* and *MBF(10; 4)* models within the envelope AM–AM dynamic range 0 − D. However, immediately outside this, the *FS(20; 4)BF(10;4)* and *BF(10; 4)* manifest erratic behaviour and radically 'take off' vertically. With a zoom-out scale of 18 000 (right-hand side axis), it is possible to show their full behaviour. The behaviour of the phase characteristic outside the measurement dynamic range is likewise just as erratic (not shown here).

This effect is also reflected in the magnitude values in the coefficient spectrum (Figure 1.12a). Besides, there being a growth in the coefficient magnitude until the third coefficient and only thereafter an exponential decline, there are several order-of-magnitude differences in the coefficient values compared to the *MBF(10; 4)* coefficients (Figure 1.12b). Numerical computational precision is a further concern when handling such large order-of-magnitude numbers mixed with other small numbers. The MBF coefficients notably have values in close proportion to the (normalized) PA measurements and have an exponential decline to zero with coefficient order.

1.1.6.6 Model Extensibility – MBF versus BF

Both BF and MBF models are extensible. However, MBF has an additional distinguishing extensibility property. For a given MBF model of a given order L, the model *AE* will deteriorate gradually and monotonically, as terms are dropped, starting with the highest

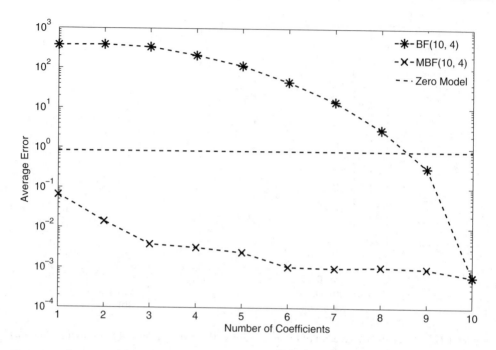

Figure 1.13 A comparison of model *AE* deterioration as a function of an orderly dropping of coefficients terms, starting with the highest order ones, for directly extracted *BF(10; 4)* and *MBF(10; 4)* models.

order coefficients. Such behaviour would be analogous to that expected in power series models. However, as is implicit in Figure 1.12(a), this is not normally the case with the BF model class. There, the integrity of the coefficient set generally must be maintained; dropping even one coefficient regardless of its order can cause radical deterioration in accuracy. An example of this is illustrated in Figure 1.13, which shows a comparison of *AE* deterioration for a *BF(10; 4)* versus an *MBF (10; 4)* model as a function of an orderly dropping of coefficients.

1.1.7 MBF Model – the Memoryless PA Behavioural Model of Choice

The evolution and derivation of the MBF PA behavioural model has been set out. Its most immediate and significant contribution is the superiority of its low order models over all existing established low order models, and this is especially so where large signal behaviour is being considered. This has been demonstrated here, where three-term MBF models are extracted of equivalent memory characteristics of an L-band LDMOS nonlinear PA that displays some memory effects and compared to other low order models of the PA, namely a three-term BF model, a three-term power series model, the modified Saleh model and the classical Saleh model. It yields the best model accuracy – more than twice as good as its nearest competitor – and behavioural performance prediction among all these low order models (with margins of several dB in some instances). The reason for its good low order modelling of large signal PA behaviour modelling capability is shown to be an inherent suitability to match the general quarter wavelength sinusoidal resemblance of the typical heavily saturated PA AM–AM characteristic, $g(A)$. The behavioural prediction performance analysis and comparisons are based on the PA amplifying a 3G WCDMA. In fact, for the performance FOMs considered – NMSE, ACEPR and the fidelity with which a model regenerates the CCDF of the signal's envelope – the MBF comes exceptionally close to the upper-bound values set by the memoryless approximation of the PA 'with memory'.

In evolving the MBF, useful concepts introduced include the 'equivalent memoryless' model of those PAs that manifest linear and nonlinear memory, and the use of this in setting FOM memory to equivalent memory upper-bounds (MEMUB) as well as the behavioural prediction performance achievable by any memoryless model of a nonlinear PA with memory effects and the hypothetical instantaneous voltage transfer characteristic (IVTC) domain, where one may find any number of hypothetical PA IVTCs that correspond to the PA EM envelope characteristic. This IVTC domain was most useful in showing the direct relationship of complex FS approximations of the PA IVTCs in this domain with MBF and BF approximations in the envelope characteristic domain. This linkage has been known but never really explored, or exploited to yield the MBF model, as done here. While an MBF model is directly extracted in the PA envelope characteristic domain, its principal novel modelling attribute is to implicitly exploit the possibility, in the corresponding hypothetical IVTC domain, of shaping the extension of the IVTC outside the PA measured dynamic range in such a way as to achieve a good (even optimum) alignment between a periodic extension of this and the first (fundamental) FS term. This is why MBF is so good as a low order model and also why it is particularly suited to large signal memoryless PA behavioural modelling applications with a signal dynamic range extending beyond saturation, while simultaneously maintaining good modelling accuracy in the small signal region. Since for best low order MBF models, parameter γ, the ratio of the PA model to measured dynamic ranges, should be optimized in a narrow region around a value of 4; then perhaps it is more correct to call the model an $(L + 1)$ parameter model. This optimization, however, while important for third order models, is unnecessary for other orders and particularly orders greater than 7, as the accuracy improvement is negligible.

Another benefit of evolving the MBF model has been the deeper insight it afforded into both BF and MBF model classes. How to always get accurate stable models of either type has been fully clarified. Importantly the new MBF model class inherits all attributes and capabilities of the BF model, such as its instantaneous RF and equivalent baseband or envelope behavioural modelling capability, and its accessible intermodulation product and harmonic decomposability property, which is so useful in multicarrier and multiband system analysis.

Further advantages it manifests over the BF model include augmented extensibility and an additional margin of reliability. It augments the BF model's series-extensibility properties, in that higher order coefficients monotonically and exponentially increment model accuracy by ever smaller amounts, which is not the case in the BF model, where high order coefficients can be very important in their contribution to accuracy and thus cannot be dropped with an expectation of a small change in accuracy. Unlike the BF model, the MBF has the potential to inherently introduce a margin of reliability into large signal model behaviour for some distance beyond the measured dynamic range. Thus, and also for higher order models, these subtle but important advantages and distinctions over the BF model would recommend it as the model of choice when selecting between the BF and MBF models.

As a final concluding remark, MBF is recommended as the model of choice among all models for small or large signal behavioural modelling of microwave and millimetre-wave nonlinear PAs using low or high order models.

Acknowledgements

The authors wish to gratefully acknowledge the support of: the Telecommunications Research Centre (TRC), University of Limerick (UL); the Irish Government's IRCSET/ EMBARK research postgraduate fund; the European Union's TARGET Network of Excellence IST-1-507893; and Technology University of Vienna, Austria (for the LDMOS PA measurements; The State Key Laboratory of Advanced Optical Communication Systems and Networks, School of EECS, Peking University, China; and the General Financial Grant from China Postdoctoral Science Foundation (Contract No. 2012M510272).

References

1. Schreurs, D., O'Droma, M., Goacher, A.A. and Gadringer, M. (2009) *RF Power Amplifier Behavioural Modeling. The Cambridge RF and Microwave Engineering Series*, Cambridge University Press, UK.
2. Shimbo, O. (1988) *Transmission Analysis in Communication Systems*, Computer Science Press, USA.
3. O'Droma, M. (1989) Dynamic range and other fundamentals of the complex Bessel function series approximation model for memoryless nonlinear devices. *IEEE Transactions on Communications*, 37 (4), April, 397–398.
4. O'Droma, M. and Mgebrishvili, N. (2005) Signal modeling classes for linearized OFDM SSPA behavioral analysis. *IEEE Communications Letters*, 9 (2), February, 127–129.
5. Saleh, A. A. M. (1981) Frequency-independent and frequency-dependent nonlinear models of TWT amplifiers. *IEEE Transactions on Communications*, 29 (11), November, 1715–1720.
6. O'Droma, M., Meza, S. and Lei, Y. (2009) New modified Saleh models for memoryless nonlinear power amplifier behavioural modelling. *IEEE Communications Letters*, 13 (6), June, 399–401.
7. Gilabert, P. L., Cesari, A., Montoro, G. *et al.* (2008) Multi look-up table FPGA implementation of an adaptive digital predistorter for linearizing RF power amplifiers with memory effects. *IEEE Transactions on Microwave Theory and Techniques*, 56 (2), Febuary, 372–384.
8. Hammi, O., Ghannouchi, F. M. and Vassilakis, B. (2008) A compact-memory polynomial for transmitters modeling with application to baseband and RF-digital predistortion. *IEEE Microwave and Wireless Components Letters*, 18 (5), May, 359–361.
9. Isaksson, M., Wisell, D. and Rönnow, D. (2006) A comparative analysis of behavioural models of RF power amplifiers. *IEEE Transactions on Microwave Theory and Techniques*, 54 (1), January, 348–359.
10. Ding, L., Zhou, G. T., Morgan, D. R. *et al.* (2004) A robust digital baseband predistorter constructed using memory polynomials. *IEEE Transactions on Communications*, 52 (1), January, 159–165.
11. He, Z., Ge, J., Geng, S. and Wang, G. (2006) An improved look-up table predistortion technique for HPA with memory effects in OFDM systems. *IEEE Transactions on Broadcasting*, 52 (3), March, 87–91.
12. Shimbo, O. and Nguyen, L. (1982) Further clarification on the use of the Bessel function expansion to approximate TWTA nonlinear characteristics. *IEEE Transactions on Communications*, 30 (2), February, 418–419.
13. O'Droma, M. and Mgebrishvili, N. (2005) On quantifying the benefits of SSPA linearization in UWC-136 systems. *IEEE Transactions on Signal Processing*, 53 (7), July, 2470–2476.
14. Vuong, X. T. and Moody, H. J. (1980) Comments on a general theory for intelligible crosstalk between frequency-division multiplexed angled modulated carriers. *IEEE Transactions on Communications*, 28 (11), November, 1939–1943.
15. Blachman, N. M. (1971) Detectors, bandpass nonlinearities, and their optimisation: inversion of the Chebyshev transform. *IEEE Transactions on Information Theory*, 17 (4), April, 398–404.

16. O'Droma, M. and Lei, Y. (2008) A novel optimization method for nonlinear Bessel–Fourier PA model using an adjusted instantaneous voltage transfer characteristic. In the Proceedings of the 38th European Microwave Conference, *"Bridging Gaps," EUMW'08, Amsterdam,* Netherlands, 27 Oct. – 1 Nov., 1269–1272.

17. Jeruchim, M. C., Balaban, P. and Shanmugan, K. Sam (2002) Section 3.3.5, in *Simulation of Communication Systems*, 2nd edn, Kluwer Academic Publishers, USA.

18. Zhou, D. and DeBrunner, V. E. (2007) Novel adaptive nonlinear predistorter based on the direct learning algorithm. *IEEE Transactions on Signal Processing,* **55** (1), January, 120–133.

19. Westwick, D. T. and Kearney, R. E. (2003) *Identification of Nonlinear Physiological Systems*, John Wiley & Sons, Inc., USA.

20. Silva, C. P., Clark, C. J., Moulthrop, A. A. and Muha, M. S. (2005) Survey of characterization techniques for nonlinear communication components and systems. In the Proceedings of the IEEE Aerospace Conference, vol. 1, 1–25.

2

Artificial Neural Network in Microwave Cavity Filter Tuning

Jerzy Julian Michalski, Jacek Gulgowski, Tomasz Kacmajor and Mateusz Mazur
TeleMobile Electronics Ltd, Pomeranian Science and Technology Park, Gdynia, Poland

2.1 Introduction

The growth and progress in the mobile telecommunication industry drives a big demand for microwave filters. Looking from the perspective of rapidly developing industry one must face the technological obstacles limiting its growth. In the case of microwave filters the production process is very difficult because of the requirements imposed on the final product and the precision that is implied by these requirements. It turns out that even a modern assembly line cannot guarantee the satisfactory precision that is repeatable in the industrial scale. This is the basic reason for the situation where the microwave filter manufacturing process is now only partially automated. When the device leaves the automatic assembly line there is no guarantee that it satisfies the requirements assumed in the design process. It may be said that the filter is *pre-tuned* and waits for the intervention of the skilled human operator – this phase is called the *tuning process*. The filter's construction allows for some tuning elements (practically screws) that influence the resonant frequencies of cavities and values of couplings between them.

In the literature many different attitudes to the filter tuning process are described. The basic idea is to find the relations between characteristics of the filter being tuned and the positions of the tuning screws. In a wider aspect this means finding the relation between measurable physical properties of the filter and the change to the tuning screws that must be applied to make these properties fulfil the technical requirements.

Here should be mentioned, especially, the methods that identify the tuning element that has to be changed, like those based on the time domain response [1, 2] or the coupling matrix

Microwave and Millimeter Wave Circuits and Systems: Emerging Design, Technologies, and Applications,
First Edition. Edited by Apostolos Georgiadis, Hendrik Rogier, Luca Roselli, and Paolo Arcioni.
© 2013 John Wiley & Sons, Ltd. Published 2013 by John Wiley & Sons, Ltd.

parameter extraction method [3, 4]. Only a few approaches to filter tuning are discussed here but the interested reader may be directed to specific descriptions [5–9] and to an overview of methods [10].

2.2 Artificial Neural Networks Filter Tuning

Looking at the filter tuning process from a more formal perspective, we have the device that is described by a filter characteristic function (or set of numbers). The filter may be treated as a function that converts the position of tuning elements into the filter characteristics. Therefore, when we denote the space of filter characteristics by C and we assume that there are R tuning elements, then we have the function $F : R^R \to C$. In this context there are many possible choices for the space C. It may be the space of real- or complex-valued functions, the space of coupling matrices, the space R^K for some K or many others, but what really matters is the inverse of this mapping, that is $F^{-1} : C \to R^R$. It may happen that such mapping does not exist as a single-valued one, but even if a given element of C may be achieved for multiple positions of tuning screws, it is still acceptable. It is enough to know how much the positions of tuning screws should be changed in order to achieve the desired filter characteristics.

It is hard to expect that we will be able to find the inverse mapping F^{-1} in an analytic form, as the numerical approximations are of huge complexity. That is why the artificial neural network (ANN) seems to be a good choice. The idea is to find the ANN representation of the unknown function F^{-1} and see if it may be used in the filter tuning process. This representation is called the *inverse model* of the filter. Such an inverse model, for the detuned characteristic of a filter S_d, generates the tuning element deviations (distance vector from properly tuned filter) Δz, which, after applying to the current filter screw positions z, makes the filter tuned (Figure 2.1).

The theoretical ideas behind ANN are not presented in this chapter – only basic intuitions related to the subject are briefly described. The ANN may be perceived as a certain extrapolation of a function defined by a discrete set of samples. The extrapolation is built in the so-called *learning process*. In this phase the arguments of the function and the expected values

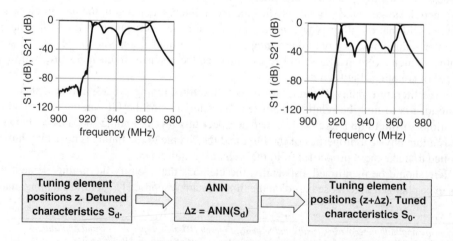

Figure 2.1   Tuning process with the use of an ANN approximator.

are specified. The larger the set of learning samples, the better it is – but only to some extent. If we supply too many or too few learning data, then the ANN loses its *generalization* properties. The generalization ability is the real value that we are chasing here – the network is expected to correctly 'guess' the values of the function for the argument outside the learning set. The power of this tool is hidden just here: to generate the correct values of the function not from the well-known formula but from the dependencies, patterns and similarities 'discovered' during the learning process.

The ANN cannot be treated as the universal tool that will solve all the problems and approximate all unknown functions. The examples showing that the ANN approximation of function F^{-1} may be useful in the filter tuning process are to be presented in this chapter. It will be shown how such a filter's inverse models may be built based on the ANN and used in the filter tuning procedures. In the more detailed descriptions the following notations will be used: N will denote the filter order, R will define the number of tuning elements of a filter, typically it will be the filter order plus number of tunable couplings and cross-couplings, while M will denote the number of points that represent the discretized filter characteristic.

2.2.1   The Inverse Model of the Filter

The inverse model of the filter is built as the artificial neural network of a certain architecture. In the described case the three-layer, feedforward multilayer-perceptron ANN [11] is used. Its input layer contains the fixed number of N_I neurons, the middle layer contains N_{ML} neurons and the output layer has N_O neurons. The number N_I depends on the measured characteristics of the filter – this issue will be discussed a little bit later, but one may consider it to represent a set of complex values of the reflection characteristics $S_{11}(f)$ measured at a certain number of frequency points. The value of N_O corresponds to the number of tuning elements of a filter. In the case of sequential tuning there is always $N_O = 1$, while in the case of parallel methods we have $N_O = R$, where R is the number of tuning elements. The number N_{ML} of middle (hidden) layer neurons is selected experimentally for the given filter.

The inverse model is always trained using training pairs (input and output vectors) of the form $\{s_n, \Delta z_n\}$, where n varies through some finite set of indexes. In this pair s_n is the input vector of dimension N_I while Δz_n is the expected output, being a vector of dimension N_O. Vectors s_n represent the detuned filter characteristics and Δz_n the corresponding deviations of tuning elements responsible for filter detuning. The training set may be selected according to different methodologies: the certain number of vector pairs may be selected randomly, where one may select points distributed uniformly in specified hypercube in the ΔZ space, but there is always a necessity to measure the filter's characteristics for the filter being detuned in a controlled way. This shows the most important drawback of the model based on the ANN: the measurements for different positions of tuning elements must be repeated many times. The size of the learning set appears to be an important problem. Collecting learning data takes a lot of time and moreover it may happen that the network loses its generalization capabilities.

Looking at the filter tuning problem one may follow one of two main paths: one is sequential tuning and the other is parallel tuning. Both methods treat the tuning process as a sequence of tuning steps but the first one assumes that we are adjusting all elements of the filter one by one – one at a time. In this method, before tuning, the filter has all tuning elements removed. In some situations, depending on the filter design, it is better to put the cavity tuning screws maximally down. The second method assumes that before tuning a filter is pre-tuned and during the tuning process all employed tuning elements are changed at the same time. The two methods imply different approaches towards the ANN architecture and learning process – and each of them has some advantages and disadvantages.

It must be mentioned here that whatever automated tuning method is chosen, the reduction in the number of steps necessary to have the filter tuned should be treated as the main goal. It means much more than saving time necessary to tune the filter. By minimizing the number of tuning element changes, it also reduces the problem of passive intermodulation (PIM), which can appear if a tuning element is changed many times. In this case small metal elements of the tuning screw can drop into a cavity and can be the source of PIM. One must remember that each time the position of tuning screws is changed, it influences the physical properties of the filter. This is not a problem when the number of tuning steps is limited, but it may appear when the tuning is performed by a human operator. The automated solutions generally lead to a smaller number of tuning steps.

2.2.2   Sequential Method

The description of the sequential tuning method will be presented here. Following the description given in Reference [12], in the sequential tuning method the general inverse model for a filter is not built but a set of inverse models for many subfilters of the given filter is created. How may this be achieved? At the beginning of collecting the data (the learning vectors) for algorithm preparation a filter must be correctly tuned. It is used as the inverse model of the entire filter. Then the R-th tuning element is removed (or put maximally down in the case of cavity),[1] which gives us a filter with simpler topology (we have one tuning element less). This changes the filter's properties but one may build the inverse model for such a subfilter. The procedure of removing (or inserting) tuning elements is repeated one by one, which results in the set of inverse models. Figure 2.2 presents the eighth subfilter of a sixth order filter consisting of 13 tuning elements (tuning elements: 2, 4, 6, 8, 10, 12 – cavities, tuning elements: 1, 3, 5, 7, 9, 11, 13 – couplings).

Now let us assume building the inverse model for the r-th subfilter. The set of samples $P_n = \{S_n^k, \Delta z_n^k\}$ is collected, where Δz_n^k goes through the set of tuning element positions in the form

$$\Delta z_n^k = \{-Ku, -Ku + u, -Ku + 2u, \ldots, 0, \ldots, Ku - 2u, Ku - u, Ku\}$$

which gives $(2K + 1)$ training points. For each Δz_n^k value the filter characteristic S_n^k is read from the vector network analyser (VNA). In this model u is the unit change of the tuning screw position, while Ku is the maximal change in the tuning element position. The values of u and K depend on the sensitivity of the tuning elements and for each filter type should be chosen experimentally. The important observation is that for a given subfilter the size of the training set is not too big and one may expect that for each subfilter it will be approximately the same. This means that the size of the training set for the entire filter (the set of all R inverse models) grows linearly with the number of tuning elements R. The r-th inverse model is presented in Figure 2.3.

---

[1] In practice, it is enough to change the tuning element to shift the resonance frequency outside the observed band.

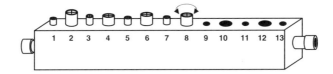

Figure 2.2   The eighth subfilter model of the six cavity filter, with $R = 13$ tuning elements. The tuning elements 1–7 are tuned, 9–13 are removed. Reproduced courtesy of The Electromagnetics Academy.

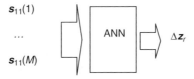

Figure 2.3   The inverse model in the case of the sequential filter tuning method.

When the r-th inverse model is trained the ANN learning ability is checked with the error function given by

$$L_r = \frac{\sum_{K=1}^{2K+1} \left| \Delta z_r^{k_0} - \Delta z_r^{k_x} \right|}{2K+1} [u] \tag{2.1}$$

where k is the index of the measurement, $\Delta z_r^{k_0}$ is the correct value of the tuning element deviation (known value) and $\Delta z_r^{k_x}$ is the tuning element deviation as a response of the ANN.

When the set of R inverse models is trained, one may use them in the process of microwave filter tuning whose path is inverted compared to the data collecting order. The tuning process starts from the filter input. Initially it is referred to the latest built inverse model. During tuning the tuning screw is put in an initial position and then, based on measured characteristics, the first inverse model generates the Δz_1 value. The value of Δz_1 must be applied to the tuning element to enable the subfilter to be tuned; in this case the ANN answers $\Delta z_1 = 0$. Then the second tuning element is processed by putting it in some initial position and looking for the response of the second inverse model to the measured filter's characteristics. The value Δz_2 is obtained and needs to be applied to the second tuning element. The procedure is repeated until $\Delta z_2 = 0$, which means that the second element has found its correct position. This scheme is repeated for all R inverse models and all R tuning elements.

2.2.3   Parallel Method

The sequential method described above requires rather simple ANN architecture and offers a tuning algorithm that is very straightforward. Later it is proved that this concept may be applied practically, but the user must be aware of its limitations. It should be especially mentioned that when the filter is not tuned properly in a certain step (for a certain subfilter) then the ANN responses in the following steps will not lead to a tuned filter. In other words, if there is a need to return to one of the elements that is already

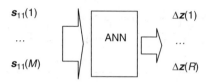

Figure 2.4 The inverse model in the case of parallel filter tuning. Reproduced courtesy of The Electromagnetics Academy.

tuned there will be no ability to do it. The model is not going to answer the question 'which element is incorrectly tuned'.

The parallel method allows one to look at the filter from a different perspective. In this case the inverse model is treated as the vector-valued function returning a vector of dimension R. So each response of the inverse model tells how much the positions of all tuning elements that require tuning should be changed. Moreover, one may possibly expect that the filter will be tuned in one step only (assuming that positions of all tuning screws may be changed in a single step – at the same time) and even if this theoretical expectation is not observed practically, one may expect that the number of tuning steps is very small. This sounds like a much more universal attitude than the sequential tuning, but starts numerous problems that were not observed previously. The inverse model used in the parallel method of filter tuning is presented in Figure 2.4.

The first problem that emerges now is the selection of the training set. In the case of sequential tuning when uniformly distributed positions of tuning elements were selected, the size of the learning set kept growing linearly with the order of the filter. Theoretically there is no problem now; if one wants to have tuning element positions uniformly distributed in some hypercube of R^R then one must use $(2K + 1)^R$ positions of tuning elements! In practice this is not acceptable because it increases the time of the learning process considerably. The example of distribution of samples in three-dimensional space is presented in Figure 2.5.

Therefore, some other methods of learning point selection should be chosen. One of the methods is to select the samples randomly from the set of acceptable tuning element variations $[-Ku, Ku]^R$. One may also select points by varying one tuning element only – this looks like a serious limitation of the space of possible settings of tuning elements, but, surprisingly, appears to work.

The number of points representing the discretized filter characteristic that should be selected is also an issue. The performed tests [13] show that it may be relatively small (of order $2R$), but it is good to increase it to have better learning and generalization properties. The exact dependences between the training vector number and network topology (number of weights) are very difficult to establish. The theoretical estimation can be done using the Vapnik–Chervonenkis dimension [14]. In practice this relation should be chosen experimentally. In general, if the number of elements of the training set is too low the network can be overtrained and will poorly generalize. If the number of elements of the training set is too high the training process of the network can be very hard and time consuming or even impossible to perform successfully. The training process for an ANN designed for parallel tuning method is based on vector pairs from one set $P_L = \{S_L, \Delta z_L\}$ (learning set). The ANN generalization ability, during the training process, is checked using the vector pairs

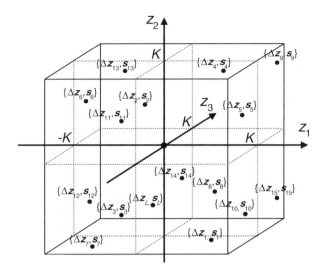

Figure 2.5 The example of sampling in three-dimensional space ($R = 3$). Learning vectors are marked as pairs $\{\Delta z_i, s_i\}$, $i = 1, 2, \ldots, L$, where L is the number of learning pairs. Reproduced courtesy of The Electromagnetics Academy.

from a different set $P_T = \{S_T, \Delta z_T\}$ (training set). The generalization error in the case of the parallel method is calculated as

$$GErr = 2K \sum_{a=1}^{TE} \sum_{b=1}^{R} \left| \Delta z_{0T}^a(b) - \Delta z_{xT}^a(b) \right| / (TE \times R) \tag{2.2}$$

where TE is the number of testing elements, R is the number of output neurons (i.e. tuning elements of the filter), K is as defined before, the maximal deviation of tuning the element in both directions, and $\Delta z_{0T}^a(b)$ and $\Delta z_{xT}^a(b)$ denote respectively the known and expected ANN response for the filter characteristic s_T from the testing set.

Now, assuming that the inverse model is ready (i.e. the ANN is trained), the tuning algorithm may be described following the much simpler method than in the case of sequential filter tuning – the same steps are repeated as many times as the filter's characteristics differ from the one of the correctly tuned filter. This single step is to put the measured filter's characteristics as an ANN input and read the response that indicates which screws should be changed and by how much. After the screw's positions are adjusted we look for the filter's characteristics again and repeat the step until $\Delta z(b) = 0, b = 1, 2, \ldots, R$.

2.2.4 Discussion on the ANN's Input Data

In general, the inverse modelling idea is based on the transformation of the filter's characteristics into the change of tuning element's position. However, there are many possible choices of what should be used to describe the filter, that is the filter characteristic. It appears that not all of them are useful as far as the filter tuning is concerned. The first idea that was used – and appeared to be a good choice – is to take the discrete sample of the reflection characteristic – as the complex-valued function. Theoretically for Chebyshev filters (without cross-couplings) the complex reflection characteristic is a full filter representation. All experiments presented below consider the reflection characteristics only. Alternatively, or additionally, the transmission S_{12} can be used. If we decide to use both the reflection and transmission characteristics the dimension of the input layer of the neural network must be doubled. What should be highlighted here is the fact that performing the tuning experiments on cross-coupled filters, when basing this on the reflection characteristics only, we obtained very good results.

Let us assume now that the certain number of frequency points f_i, where $i = 1, 2, \ldots, M$ are taken and the ANN input data are defined as the set of complex values of function $S_{11}(f_i)$. This set appears to carry the entire information on the reflection characteristic that may be extracted from the analyser – the question remains, how many elements should be taken? From the theoretical analysis it occurs that at least $2N$ data points are needed (to recreate the formula of $S_{11}(f)$, which is the rational function, being the ratio of two polynomials of degree N, with the coefficient of the highest degree of the denominator equal to 1).

As presented in Reference [15], in general, the transmission and reflection characteristics of a two-port filter network, composed of a series of N intercoupled resonators, can be defined as a ratio of two polynomials

$$S(\omega) = \frac{A(\omega)}{B(\omega)} = \frac{\sum_{i=0}^{T} a_i \omega^i}{\sum_{j=0}^{N} b_j \omega^j} \tag{2.3}$$

This equation may be transformed into

$$\sum_{i=0}^{T} a_i \omega^i = S(\omega) \sum_{j=0}^{N} b_j \omega^j \tag{2.4}$$

and in matrix form it can be written as

$$X_{M \times (T+1)} a_{(T+1) \times 1} - Y_{M \times (N+1)} b_{(N+1) \times 1} = 0 \tag{2.5}$$

and then

$$\left[X_{M \times (T+1)} - Y_{M \times (N+1)} \right] \begin{bmatrix} a_{(T+1) \times 1} \\ b_{(N+1) \times 1} \end{bmatrix} = 0 \tag{2.6}$$

which gives the final matrix form of the homogeneous linear equation as

$$X_{M \times (T+N+2)} d_{(T+N+2) \times 1} = 0 \tag{2.7}$$

This equation looks homogeneous but actually it is not because of necessity to impose some restrictions on coefficients – especially that $b_N = 1$. Therefore we have $T + N + 1$

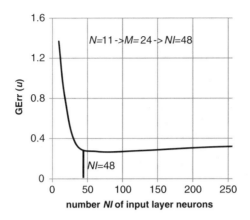

Figure 2.6   ANN error defined by Equation (2.2) for the TX filter from Figure 2.7.

unknown values. With $M < T + N + 1$ equations the system is expected to have multiple solutions. That is not acceptable because there is no clear criteria to select one of them. On the other hand, if $M > T + N + 1$ the system is overdetermined and is rather expected to be inconsistent, but one may still look for a least-square type solution, that is the unknown vector that gives the value closest to zero. Equation (2.7) may therefore be solved if the $S(\omega)$ characteristic of Equation (2.3) is determined in at least $M = T + N + 1$ frequency points. Considering the reflection characteristic S_{11} then $T = N$ [15]. Having such a discrete set of points $S(\omega_l), l = 1, 2, \ldots, M$, it is possible to restore the whole characteristic in an analytical form [13]. Considering this, it is assumed that for $M \geq 2N + 1$, unambiguous mapping between the reflection characteristic S_{11} determined at M frequency points to tuning element positions is possible. Each sampled complex point of the reflection characteristic requires two input neurons, one for the real and the second for the imaginary part. This gives relations between the filter order N and the number of input layer neurons $N_l \geq 2M$.

A series of tests were done showing that the number of points may be limited (also the number of input layer neurons) to a number close to $2N$, keeping the generalization error on nearly the same level [13]. Below Figure 2.6 is shown the generalization error defined by Equation (2.2) for the filter presented in Figure 2.7. In this experiment the learning and testing sets have 1000 and 100 elements respectively.

What about the other possible representations of the reflection characteristics? Practically, when the filter is tuned by the human operator the only information that is used is the modulus of reflection characteristics $|S_{11}(f)|$. It is therefore reasonable to expect that the ANN input data may be limited to a modulus $|S_{11}(f)|$ so decreasing the number of input layer neurons by half. Unfortunately, performed experiments were not successful – it turned out that the generalization error could not drop below a certain level. Apparently the modulus is not sufficient to recreate the inverse model in a reasonable way.

It is also worth mentioning that other partial representations of the complex-valued functions, such as phase, only real or only imaginary parts, did not lead to reasonable generalization properties of the inverse model. In general, the method can be customized with regard to any filter characteristics that fully describe a filter, constituting the filter tuning goal.

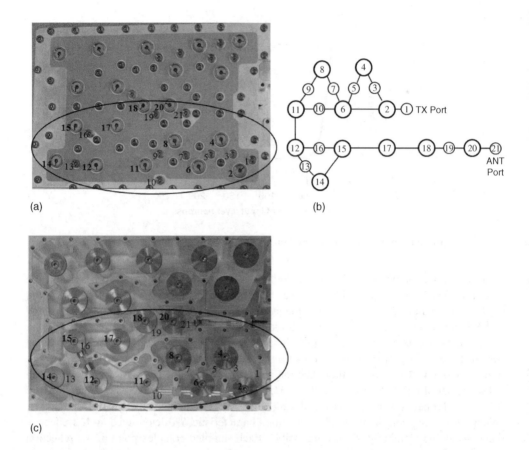

(a)   (b)

(c)

Figure 2.7 The layout (a, c) and the topology (b) of the filter used in the experiment. Small circles represent tunable couplings and cross-couplings. Bigger circles represent cavities. There are no coupling tuning elements between cavities 15–17 and 17–18. Fixed cross-coupling occurs between cavities 2–6. Reproduced courtesy of The Electromagnetics Academy.

2.3 Practical Implementation – Tuning Experiments

The tests were performed to check whether the procedures described above may be implemented in practice. The testing environment was completely automated and consisted of the programmed robot that was responsible for changing the positions of the tuning screws according to responses of the appropriate inverse model.

The multiple tests were performed for filters with different topologies. Below we present results of selected and representative tests of the most general case, that is filters where not only cavities but also couplings may be adjusted in order to have a filter tuned.

2.3.1 Sequential Method

Following the example given in Reference [12] this method represents two relatively complex filters – one of order 11 and the other of order 8. The first tuning experiment for the filter of order 11 is described (Figure 2.7), with the lower part of diplexer – a TX filter marked

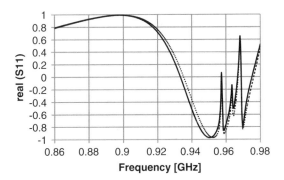

Figure 2.8 The real part of the reflection. All tuning elements, 1–21, are removed. Solid line – IMT characteristic, dotted line – TF characteristic. Reproduced courtesy of The Electromagnetics Academy.

with numbers – and the upper part of the diplexer – an RX filter not used in the tuning experiment. One of the most important features of the presented approach is that the filter is treated as a black box; no information on the filter topology and its technical details is necessary in the algorithm customization.

The inverse models were built based on the following tuning elements extraction path: 21, 20, 19, . . . , 2, 1. The training sets for each inverse model were prepared for 41 positions of the tuning screw, ranging from $-20u$ to $+20u$, where $u = 22.5°$. In the experiment, the inverse filter models were built based on the complex reflection characteristics collected from one filter, which can be described as an inverse model template (IMT). The tuning process was performed for another filter of the same type, which is defined as a tuned filter (TF). The reflection characteristics were represented by 256 complex points. The scattering characteristics obtained in the following steps of the tuning process are presented in the following pictures. Apart from the scattering characteristics in dB we present the real part of S_{11}, that is the characteristic that was used in the process of preparing ANN.[2]

The filter used in the present experiment should be considered as tuned if the reflection characteristic level is below -18 dB within the passband centred at $f_0 = 943.5$ MHz and with bandwidth $BW = 35$ MHz. Figures 2.8 to 2.13 present the reflection characteristic of the filter during the tuning process where successive tuning elements are set in the proper position. While observing the final tuning results in Figures 2.12 and 2.13 one may conclude that the filter is properly tuned. Experiments were performed for five different filters of the same type using the inverse models generated based on the inverse model template. All filters were tuned within the time of a couple of minutes, which is very short compared to the time needed by skilled technicians to do it manually.

Now let us proceed to the next example of the filter of order 8, which is an RX part of the 900 MHz GSM combiner (Figure 2.14). The inverse models were built on the following tuning element extraction path: 8, 7, 9, 6, 10, 5, 11, 4, 12, 3, 13, 2, 14, 1. In this case the training sets for each inverse model were prepared for 21 settings starting from $-10u$ up to $+10u$,

---

[2] In the process of preparing an ANN an imaginary part of S_{11} was also used. To avoid the mess in figures, authors decided to include only the real part.

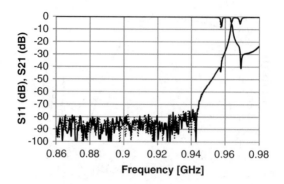

Figure 2.9 The transmission and the reflection (in dB). All tuning elements, 1–21, are removed. Solid line – IMT, dotted line – TF characteristics. Reproduced courtesy of The Electromagnetics Academy.

where $u = 45°$. The reflection characteristics were represented by a discrete set of 32 points. The steps of the tuning process are presented in Figures 2.15 to 2.20.

The last characteristics, depicted in Figures 2.19 and 2.20, show the filter condition when tuning is completed, with all elements set. The centre frequency for this filter is at $f_0 = 897.5$ MHz and the bandwidth equals $BW = 35$ MHz. The technical filter specification requires that within the passband we should have $|S_{11}|$ below the level of -16 dB, so that the filter is properly tuned. The tuning time for this filter is about 3 minutes and in this case it requires one tuning iteration for each tuning element, so each tuning element is positioned only once. The differences between the inverse model template characteristics (which we treat as the ideal reference) and the tuned filter may be explained by small physical discrepancies appearing as a result of the production process. Due to those discrepancies in some cases we may expect that in practice the filter may require fine tuning, which can be performed very easily even by a mildly experienced human operator.

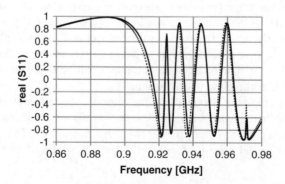

Figure 2.10 The real part of the reflection. Tuning elements, 1–11, are tuned. Solid line – IMT characteristic, dotted line – TF characteristic. Reproduced courtesy of The Electromagnetics Academy.

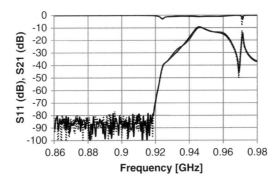

Figure 2.11   The transmission and the reflection (in dB). Tuning elements, 1–11, are tuned. Solid line – IMT characteristics, dotted line – TF characteristics. Reproduced courtesy of The Electromagnetics Academy.

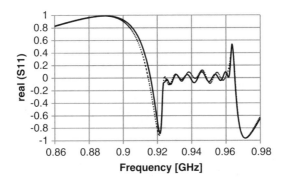

Figure 2.12   The real part of the reflection. All tuning elements, 1–21, are tuned. Solid line – IMT characteristic, dotted line – TF characteristic. Reproduced courtesy of The Electromagnetics Academy.

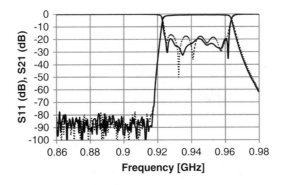

Figure 2.13   The transmission and the reflection (in dB). All tuning elements, 1–21, are tuned. Solid line – IMT, dotted line – TF characteristics. Reproduced courtesy of The Electromagnetics Academy.

(a)   (b)

Figure 2.14 The picture (a) and the topology (b) of the filter used in the experiment. Small circles represent tunable couplings and cross-couplings. Bigger circles represent cavities. There is no tunable coupling element between cavities 7–8. Fixed cross-couplings can be found between the cavities 1–7 and 8–14. Reproduced courtesy of The Electromagnetics Academy.

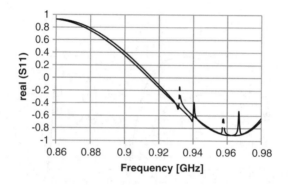

Figure 2.15 The real part of the reflection. All tuning elements, 1–14, are removed. Solid line – IMT characteristic, dotted line – TF characteristic. Reproduced courtesy of The Electromagnetics Academy.

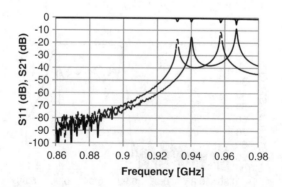

Figure 2.16 The transmission and the reflection (in dB). All tuning elements, 1–14, are removed. Solid line – IMT, dotted line – TF characteristics. Reproduced courtesy of The Electromagnetics Academy.

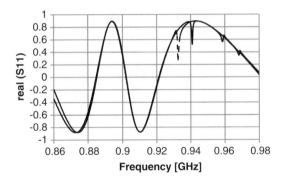

Figure 2.17 The real part of the reflection. Tuning elements, 1–5, 10–14 are tuned. Solid line – IMT characteristic, dotted line – TF characteristic. Reproduced courtesy of The Electromagnetics Academy.

2.3.2 Parallel Method

To test the parallel method the tuning experiments were performed for two filters: the first one of fourth order and the second one of fifth order. The topologies of these filters are presented accordingly in Figures 2.21 and 2.22. For both filters the relatively small learning sets are collected: for each tuning element the S_{11} characteristics were collected (512 points) for 11 positions of each tuning screw (from $-5u$ to $+5u$), where $u = 3.6°$ (for cavity) and $u = 216°$ (for coupling). This gave the learning sets consisting of 77 and 99 elements for fourth order and fifth order filters respectively.

Tables 2.1 and 2.2 show the positions of the tuning elements for the investigated filters before and after tuning. The corresponding reflection and transmission characteristics are depicted in Figures 2.23 and 2.24.

The performed filter tuning process, for both filters, required changes to all tuning elements – both cavities and couplings, which makes the experiment essentially different from the one described in Reference [16, 17]. The described change in the tuning element position was applied to all elements one by one and each element was touched only once. The fifth

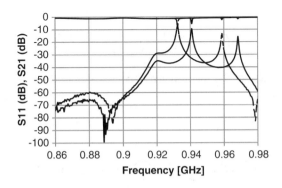

Figure 2.18 The transmission and the reflection in (dB). Tuning elements, 1–5, 10–14 are tuned. Solid line – IMT characteristics, dotted line – TF characteristics. Reproduced courtesy of The Electromagnetics Academy.

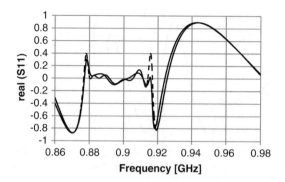

Figure 2.19 The real part of the reflection. All tuning elements, 1–14, are tuned. Solid line – IMT characteristic, dotted line – TF characteristic. Reproduced courtesy of The Electromagnetics Academy.

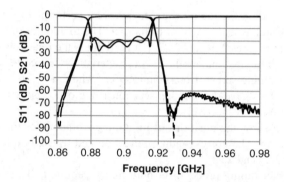

Figure 2.20 The transmission and the reflection (in dB). All tuning elements, 1–14, are tuned. Solid line – IMT, dotted line – TF characteristics. Reproduced courtesy of The Electromagnetics Academy.

Figure 2.21 The topology of the fourth order filter used in the experiment. Large circles denote cavities, smaller couplings.

Figure 2.22 The topology of the fifth order filter used in the experiment. Large circles denote cavities, smaller couplings.

Table 2.1 Deviations of tuning elements before and after the tuning process for the fourth order filter

Tuning element number	1	2	3	4	5	6	7
$\Delta z(m)$ before tuning (u)	5	−2	5	1	−5	−5	−3
$\Delta z(m)$ after tuning (u)	0	0	0	0	0	0	0

order filter's centre frequency is $f_0 = 1.95$ GHz and the bandwidth $BW = 20$ MHz. For the fourth order filter we have $f_0 = 2.14$ GHz and the bandwidth $BW = 20$ MHz. For both devices the requirement was to have the reflection characteristics level in the passband below -20 dB. Therefore one can see that both filters are tuned correctly. In practice the tuning process is not always completed in one step and the additional fine tuning procedure is required, but it does not require special experience from the human operator and is relatively quick and straightforward.

We refer the reader to the paper [16] to see test results for some other filters with more complex topology. The tuned device is a 17 cavity diplexer with a sixth order filter having a single cross-coupling between cavities 2 and 5 as the RX part, with the TX part being an 11 cavity filter with two cross-couplings (between cavities 2 and 5, and between cavities 5 and 8). As mentioned, the testing environment was different from the one described in the present chapter, because only cavities were used in the tuning process. Some additional factors were investigated including utilization of a number of filters in the testing process (in this work only one is used to collect learning vectors). Also the influence of the learning set size on the generalization error was discussed. The learning and generalization error curves for different learning sets, varying from 25 up to 1000 elements, were compared. What may be observed is the dependence of the learning set cardinality on the learning and generalization error – with an increased number of learning elements a defined level of learning/generalization error appears, but with the learning set large enough (over 200 elements) the error level that is achieved is basically the same in all cases. Thus learning may take longer but final learning and generalization effects are approximately the same.

2.4 Influence of the Filter Characteristic Domain on Algorithm Efficiency

As mentioned before, considerable reduction of the dimension of the input vector, keeping the generalization error (i.e. the inverse model accuracy) at a reasonable level, is possible. Still the issue of selection of the sampling frequency points appears to be very important. The lower the number of points, the more important it is to specify how they should be selected. Also additional problems start appearing – especially lower stability of the generalization process resulting from the measurement errors. It appears that a minor fluctuation of a single frequency point value starts to be meaningful.

Table 2.2 Deviations of tuning elements before and after the tuning process for the fifth order filter

Tuning element number	1	2	3	4	5	6	7	8	9
$\Delta z(m)$ before tuning (u)	3	1	−1	−4	−5	2	3	1	−5
$\Delta z(m)$ after tuning (u)	0	0	0	0	0	0	0	0	0

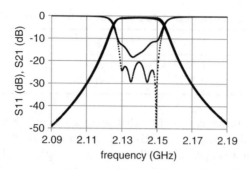

Figure 2.23 The scattering characteristics of the fourth order filter. Solid lines represent character-istics of the detuned filter. Dashed lines show results of the tuning process.

However, the problem of limiting the dimension of the input data vector remains very important. It is therefore worth mentioning that recently two different ideas on input data compression were reported [18, 19]. The first one is related to discrete wavelet transform compression and the other to principal component analysis (PCA). In both cases the basic idea is very similar: instead of passing the complete characteristic as the pattern that is required to be recognized, authors extract 'the most important data' from it, ignoring the 'meaningless' remainder. In the case of the wavelet transform D4 a discrete wavelet trans-form is used and experiments show that the relatively long tail of the transform (Figures 2.25 and 2.26) may be ignored, keeping the generalization error on the nearly unchanged level. On the other hand, the PCA method suggests the reduction of the input data vector dimen-sion by the selection of the appropriate basis. The experiments show that even if the dimension of the space spanned by the newly selected basis is much smaller than the original one, the generalization error is kept at the acceptable level.

First, let us focus on the Daubechies D4 wavelet transform as a method of filter reflection characteristics compression. The details about that transform and its application in the reflec-tion characteristics representation are described in References [19] to [21] respectively.

In Figures 2.25 and 2.26 the original signal and its D4 transform are presented. When looking at the Daubechies D4 transforms one may observe that the biggest value of the trans-formed signal occurs at its beginning. This observation raises a question: at which moment can the transform signal be truncated without significant influence to the inverse model? To

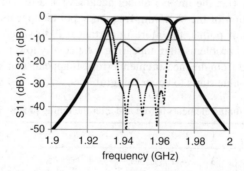

Figure 2.24 The scattering characteristics of the fifth order filter. Solid lines represent characteristics of the detuned filter. Dashed lines show results of the tuning process.

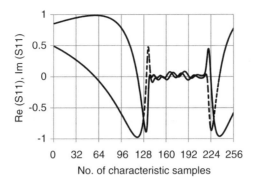

Figure 2.25 The reflection characteristic of a filter. Real part – solid line, imaginary part – dashed line. © 2011 European Microwave Association. Reprinted, with permission, from [19], Fig. 1.

answer this question, some numerical investigations have been performed. Starting from 512 complex-valued points different compression rates were taken into account – starting from the first eight transform values up to 512 values (meaning no compression). Figure 2.27 presents how the generalization error for the trained ANN depends on the number of learning epochs for different compression levels (denoted by different values of C). One may come to the conclusion that if the signal is highly compressed and fewer than 56 transform points are left, the ANN cannot be properly trained, causing high generalization errors. If 64 or more transform points are left, the ANN trains very well and practically no improvement is achieved with additional transform points.

The other compression idea is related to the principal component analysis (known also as Karhunen–Loeve transformation – for details see References [11] and [22]). The brief description of the method is to transform the signal into the so-called feature space and select only the most important features of the signal. This is achieved by

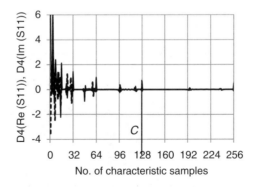

Figure 2.26 D4 transform of the reflection characteristic of a filter. Real part – solid line, imaginary part – dashed line. Point C defines the location from which the transform is completed with zeros or truncated in further considerations. © 2011 European Microwave Association. Reprinted, with permission, from [19], Fig 2.

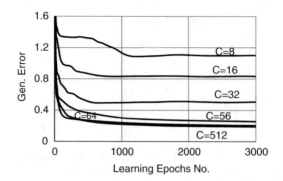

Figure 2.27 Generalization error of the ANN trained with D4 filter characteristics for different compression levels C, where C is the point at which the D4 transform is truncated. © 2011 European Microwave Association. Reprinted, with permission, from [19], Fig 3.

changing the appropriate coordinates, so the biggest variance is in the subspace generated by the first basis vector, then in the subspace spanned by the second basis vector and so on. Similarly, as in case of wavelet transform, we can truncate the least important basis vectors compressing the signal [18]. Figures 2.28 and 2.29 present the original and compressed signals.

Similarly, we can compare the generalization error depending on the number of learning epochs for different dimensions C of the space of the compressed signal (Figure 2.30). As one may notice, if there are less than 16 components the ANN may not be trained on a satisfactory level, but for more than 16 components further increasing the number of components does not affect the efficiency of the network.

The two compression techniques presented in this section show that the input data dimension may be reduced considerably. This results in a far less complex internal structure of the ANN and finally in a much faster learning process. The tests described above show that this might be used in practice based on real measurement data.

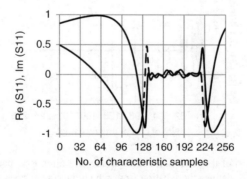

Figure 2.28 The reflection characteristic of a filter. Real part – solid line, imaginary part – dashed line. © 2011 IEEE. Reprinted, with permission, from [18], Fig. 1.

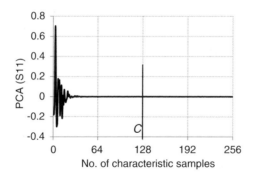

Figure 2.29 PCA transform of reflection characteristic of a filter. Point C defines the location from which the transform is completed with zeros or truncated in further considerations. © 2011 IEEE. Reprinted, with permission, from [18], Fig. 2.

2.5   Robots in the Microwave Filter Tuning

An present in production practice the filter tuning is performed by skilled and experienced technicians who are responsible for finding the positions of tuning screws that guarantee the shape of reflection characteristics matching requirements specified for the device. The manual tuning process is the only option when there is no algorithm that specifies which tuning element should be adjusted and how much. However, when the filter inverse model is available a fully automated tuning environment may be designed and implemented.

As already mentioned, there are two different attitudes to filter tuning: sequential and parallel. The parallel one has a major advantage over the sequential – the ability to change position of all tuning elements at the same time, considerably speeding up the tuning process. However, to change more than one element at the same time we need to have a dedicated robot head that interconnects multiple stepper motors to filter tuning elements. Practically this model is not flexible because different filter types require a different construction of the head.

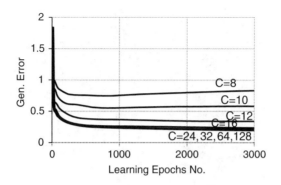

Figure 2.30 Generalization error of the ANN trained using the first C principal components. © 2011 IEEE. Reprinted, with permission, from [18], Fig. 3.

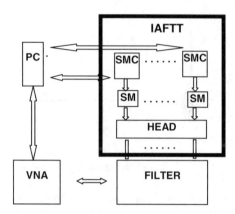

Figure 2.31   The tuning environment [23] – block diagram (PC – the computer used for ANN processing and reading characteristics from the VNA and stepper motor control, SMC – stepper motor controller, SM – stepper motor, HEAD – interconnects stepper motors and filter screws, VNA – vector network analyser, FILTER – microwave filter).

The other option is to use the single arm robot. That results in the much slower tuning process (because we may change the position of only one screw at a time) but is a much more flexible solution, because the same device may be configured for different filters.

The developed test environment is fully automated. It consists of the PC set (used for ANN processing), a vector network analyser (VNA) and a mechanical system responsible for tuning screw control. It must contain stepper motors with stepper motor controllers. The block diagram of the parallel system is presented in Figure 2.31.

In the parallel setup the IAFTT robot (Figure 2.32) was used to collect ANN learning elements and to tune the filter. It is possible for this robot to change all tuning elements simultaneously.

On the other hand, the one-arm SCARA robot (Figure 2.33) is used in the sequential tuning environment. Fortunately the one-arm robot configuration may also be used in the parallel tuning algorithm.

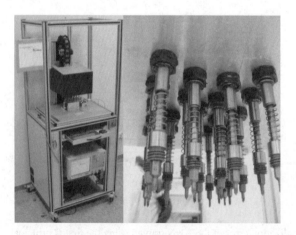

Figure 2.32   A photo of the IAFTT robot controlling all tuning elements simultaneously.

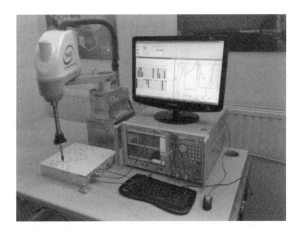

Figure 2.33   A photo of a one-arm SCARA robot controlling one tuning element at a time.

2.6   Conclusions

In this chapter the basic idea about the artificial neural network microwave filter tuning was outlined. Two attitudes towards the tuning algorithms with possible advantages and disadvantages of both of them were presented. The experiments were performed and proved that the suggested methods work in practice – even for relatively complex filters (of high order and complex topology). All tuning experiments, described in the present chapter, were performed with the use of the reflection characteristic of filters. The proposed tuning methods can work with the other filter characteristic representations like the transmission S_{21}. This may be necessary if we want to tune filters with transmission zeros (cross-coupled filters). In general, ANN input vectors can be defined as concatenation of vectors of more than one different filter characteristic.

There are still many open problems related to the optimization of the suggested solutions, especially to the learning process, which requires a lot of learning data samples that are not always possible to retrieve. Some novel, interesting ideas showing the possible ways to solve these problems were also given.

Acknowledgement

This work was supported by the Polish National Ministry for Science & Higher Education (Decision 736/N-COST/2010/0), under the project name 'New optimization methods and their investigation for the application in physical microwave devices which require tuning', performed within the COST Action RFCSET IC0803.

References

1. Dunsmore, J. (1999) Tuning band pass filters in the time domain. IEEE MTT-S Int. Microwave Symposium Digest, pp. 1351–1354.
2. Dunsmore, J. (1999) Simplify filter tuning in the time domain. *Microwaves RF*, **38** (4), 68–84.
3. Thal, H.L. (1978) Computer-aided filter alignment and diagnosis. *IEEE Transactions on Microwave Theory and Techniques*, **26** (12), 958–963.

4. Harscher, P., Vahldieck, R. and Amari, S. (2001) Automated filter tuning using generalized low-pass prototype networks and gradient-based parameter extraction. *IEEE Transactions on Microwave Theory and Techniques*, **49** (12), 2532–2538.

5. Ness, J.B. (1998) A unified approach to the design, measurement, and tuning of coupled-resonator filters. *IEEE Transactions on Microwave Theory and Techniques*, **MTT-46**, 343–351.

6. Meng, W. and Wu, K.-L. (2006) Analytical diagnosis and tuning of narrowband multicoupled resonator filters. *IEEE Transactions on Microwave Theory and Techniques*, **54** (10), 3765–3771.

7. Harscher, P. and Vahldieck, R. (2001) Automated computer-controlled tuning of waveguide filters using adaptive network models. *IEEE Transactions on Microwave Theory and Techniques*, **49** (11), 2125–2130.

8. Ibbetson, D. (2000) A synthesis based approach to automated filter tuning. Microwave Filters and Multiplexers (Ref. 2000/117), IEE Colloquium on 28 November 2000, pp. 11/1–11/3.

9. Miraftab, V. and Mansour, R.R. (2002) Computer-aided tuning of microwave filters using fuzzy logic. *IEEE Transactions on Microwave Theory and Techniques*, **50**, 2781–2788.

10. Cameron, R.J., Kudsia, C.M. and Mansour, R.R. (2007) *Microwave Filters for Communication Systems*, John Wiley & Sons. Ltd.

11. Haykin, S. (2009) *Neural Networks and Learning Machines*, 3rd edn, Prentice Hall, Upper Saddle River, NJ.

12. Michalski, J.J. (2011) Inverse modeling in application for sequential filter tuning. *Progress in Electromagnetics Research*, **115**, 113–129.

13. Michalski, J.J., Kacmajor, T., Gulgowski, J. and Mazur, M. (2011) Consideration on artificial neural network architecture in application for microwave filter tuning. Progress in Electromagnetics Research Symposium Proceedings, Marrakesh, Morocco, 20–23 March 2011, pp. 1313–1317.

14. Vapnik, V.N. and Chervonenkis, A. (1971) On the uniform convergence of relative frequencies of events to their probabilities. *Theory of Probability and Its Applications*, **6**, 264–280.

15. Cameron, R.J. (1999) General coupling matrix synthesis methods for Chebyshev filtering functions. *IEEE Transactions on Microwave Theory and Techniques*, **47** (4), 433–442.

16. Michalski, J.J. (2010) Artificial neural networks approach in microwave filter tuning. *Progress in Electromagnetics Research M*, **13**, 173–188.

17. Kacmajor, T. and Michalski, J.J. (2011) Neuro-fuzzy approach in microwave filter tuning. IEEE International Microwave Symposium, Baltimore, MD.

18. Kacmajor, T. and Michalski, J.J. (2011) Principal components analysis in application for filter tuning algorithm. Proceedings of IMWS Conference, Sitges, Barcelona, Spain, 15–16 September 2011, pp. 684–686.

19. Michalski, J.J. and Kacmajor, T. (2011) Filter tuning algorithm with compressed reflection characteristic by Daubechies D4 wavelet transform. Proceedings of EuMW 2011, Manchester, UK, 9–14 October 2011, pp. 778–781.

20. Daubechies, I. (1992) *Ten Lectures on Wavelets, CBMS-NSF Conference Series in Applied Mathematics*, SIAM Ed.

21. Mallat, S. (2009) *A Wavelet Tour of Signal Processing*, 3rd edn, Elsevier.

22. Pearson, K. (1901) On lines and planes of closest fit to systems of points in space. *Philosophical Magazine*, **2** (6), 559–572.

23. Michalski, J.J. (2007) Intelligent automatic filter tuning tool (IAFTT) is registered with 'Priority certificate # 2516' (Hannover Cebit 2007), and patent pending 'European Patent Application no. P382895 assigned by Polish National Patent Office'.

3

Wideband Directive Antennas with High Impedance Surfaces

Anne Claire Lepage, Julien Sarrazin and Xavier Begaud
Institut Mines-Télécom, Telecom ParisTech, Paris, France

3.1 Introduction

In recent years, there has been a growing interest in applying artificial materials, known as metamaterials, to antennas. This generic term covers a variety of definitions: left-handed material (LHM), high impedance surface (HIS), epsilon-near-zero (ENZ), and mu-near-zero (MNZ). A common thread to all these definitions is that these materials derive their unique properties not from their composition but from their structure. They are mostly composed of a periodic arrangement of materials, patterns. This spatial periodicity naturally induces a spectral selectivity. This narrow bandwidth is, in addition to losses, one of the main limitations for metamaterials applications.

The objective of this chapter is to demonstrate that it is possible to design wideband antennas with metamaterials. Among the above-mentioned variety of metamaterials, it focuses on high impedance surfaces (HISs), introduced by Sievenpiper in Reference [1]. These surfaces can be used in order to improve antennas by reducing their thickness and making them unidirectional rather than bidirectional. Thus, designing unidirectional antennas is required on many platforms (aircrafts, unmanned aerial vehicles, etc.) in order to obtain outward radiation and preserve the interior of any electromagnetic pollution. Furthermore, for integration and mechanical constraints, antennas have to be low profile. To achieve these properties, most common solutions consist in locating the antenna above a reflector or an absorbent cavity. The solution with an absorbing cavity is simple but half of the radiated power is lost. Absorbents are heavy and their features are difficult to reproduce. Moreover, the cavity is sized at a quarter of a wavelength at the lowest operating frequency and can become very bulky for low frequency applications. Another efficient technique is to use a reflector made

Microwave and Millimeter Wave Circuits and Systems: Emerging Design, Technologies, and Applications,
First Edition. Edited by Apostolos Georgiadis, Hendrik Rogier, Luca Roselli, and Paolo Arcioni.
© 2013 John Wiley & Sons, Ltd. Published 2013 by John Wiley & Sons, Ltd.

of a good electrical conductor to retrieve the radiated power lost in the first solution. This technique is optimal in the middle of the bandwidth where a constructive interference phenomenon is obtained by locating the reflector at a quarter wavelength (at center frequency) from the antenna. However, this solution is inherently limited bandwidth and can rarely exceed the octave and leads to quite thick structures.

It therefore appears difficult to design a unidirectional antenna while achieving a simultaneously wide bandwidth and compactness, but among metamaterials, the so-called HIS have remarkable characteristics and can exhibit two different properties. Indeed, within a limited frequency bandwidth, these periodic structures can exhibit on the one hand an electromagnetic bandgap (EBG) in which surface wave propagation is forbidden along the structure. On the other hand, they are able to reflect electromagnetic waves without any phase shift for the electric field, which makes them behave like an artificial magnetic conductor (AMC). Both the EBG and AMC behaviors may or may not occur at the same frequency. Furthermore, some geometries do not exhibit EBG characteristics whereas they act as an AMC. In this chapter, only the AMC property is considered since it is the one that allows low profile unidirectional antennas to be obtained by using the metamaterial as a reflector. In fact, while perfect electric conductors (PEC) impose a reflection phase of π, perfect magnetic conductors (PMC) do not introduce any phase shift. Artificial magnetic conductors reproduce the PMC behavior at a given frequency and, at about this specific frequency, constructive interferences between incident and reflected electric fields can occur. It is therefore possible to locate the antenna closer to the AMC reflector. Consequently, the antenna becomes unidirectional and thin. However, the main challenge consists in preserving the wideband properties of the antenna.

This chapter investigates the possibility of using AMC to achieve wideband antennas. An introduction to AMC is firstly presented in Section 3.2. Its characteristics are detailed in order to explain how these metamaterials are used with antennas. The main limitations related to wideband aspects are highlighted. Sections 3.3 and 3.4 propose two different approaches to deal with these limitations. Section 3.3 gives some insights on how to optimize the AMC design with a wideband bow-tie antenna. Furthermore, the addition of lumped elements on the AMC is considered in order to reduce side lobes of the radiation pattern. Section 3.4 illustrates the way that an AMC can be modified to increase antenna gain. This has been made possible with a technique based on surface current observation. Thanks to this technique, a low profile wideband antenna using a hybrid AMC is presented.

3.2 High Impedance Surfaces (HIS) Used as an Artificial Magnetic Conductor (AMC) for Antenna Applications

3.2.1 AMC Characterization

An artificial magnetic conductor surface is usually composed of a periodic arrangement of unit cells. The classical method to characterize such an AMC has been proposed by Sievenpiper in 1999 [1] and is called the reflection phase method. This method consists in illuminating the surface to be characterized by a TEM wave at normal incidence. Then, the phase difference between the incident electric field and the reflected one is compared. The reference plane at which this phase difference is determined is usually the AMC surface or the plane at which the radiating element will be located. Constructive interference occurs

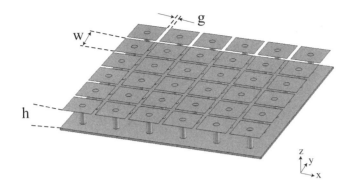

Figure 3.1   Mushroom structure ($W = 8$ mm, $g = 1$ mm, $h = 4.8$ mm, $\varepsilon_r = 1$).

when the phase difference in the AMC plane is between $-\pi/2$ and $+\pi/2$ and this defines the bandwidth of the artificial magnetic conductor. The graph of the evolution of the phase difference as a function of frequency is represented by the 'phase diagram'. It can be determined using analytical models, computational methods or experimental test-beds.

Most of analytical models are based on the transmission line theory. The HIS is described with an equivalent circuit model. This implies constraints on the shape of the unit cell, which has to be simple, and also on the periodicity of the HIS, which has to be small compared to the wavelength. Several models have been developed in References [2] to [5] for normal or oblique incidences. The advantage of using analytical models is that they enable one to provide the phase diagram in a very short time. The main drawback is the accuracy of the model.

Computational methods consist in simulating a single cell of the HIS and applying appropriate periodic boundary conditions to obtain the behavior of the infinite surface [6]. Figure 3.1 presents a mushroom pattern. Each cell is composed of a square metallic patch above a grounded substrate. The patch is connected to the ground with a metallic via hole. This HIS has been simulated with a CST Microwave Studio using methods described in Reference [7]. The phase diagram is presented in Figure 3.2. As described earlier, the incident wave is a TEM wave with normal incidence. It can be deduced from Figure 3.2 that the phase difference is null at 6.3 GHz. At this frequency, the surface behaves like a perfect magnetic conductor. The $\pm 90°$ bandwidth ranges from 4.6 to 8.5 GHz.

The bandwidth of this structure strongly depends on the height h. The greater the height, the wider the bandwidth, but the structure becomes bulky.

Regarding the mushroom structure, the simulation demonstrates that the presence of metallic vias has no influence on the phase diagram as long as the incident wave is normal to the surface. However, for this pattern, vias are mandatory when EBG properties are required. In coming sections, no via is used for simplicity reasons and because only the AMC behavior is considered in the design of low profile directive antennas.

By applying the Floquet theorem, it is even possible to characterize the surface for various incident angles [5]. Therefore numerical methods enable quite accurate characterization to be provided, but can be time consuming, especially when complex designs are involved.

Regarding the experimental characterization, several methods can be employed to determine the phase diagram. The first one presented in Reference [3] consists in using two horn

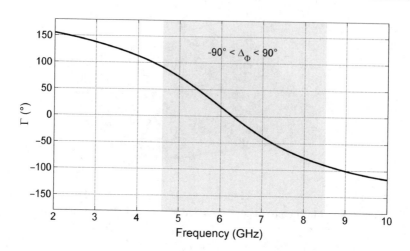

Figure 3.2 Phase diagram of the mushroom structure.

antennas located in an anechoic chamber as shown in Figure 3.3. The transmitting horn illu-
minates the surface to be characterized and the reflected wave is collected by the receiver
horn. The two antennas are located at an equal distance from the surface and an absorber
takes place between the two in order to avoid any coupling effect. A preliminary measurement is done with a metallic surface, which is taken as a reference. This method enables the
phase diagram to be determined at normal incidence but also for different values of the incident angle by rotating the surface under test and moving the receiving antenna [8].

   Another method uses a so-called 'waveguide simulator'. This measurement technique has
been first developed to characterize antenna arrays and frequency selective surfaces [9], but
can be applied to other periodic surfaces like the AMC. It consists in using an oversized
rectangular waveguide terminated with the AMC surface [7]. Thanks to the image principle,

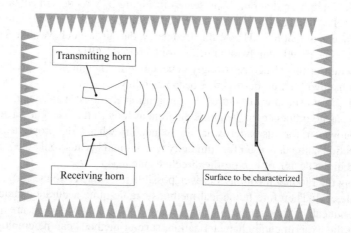

Figure 3.3 Phase diagram: measurement test bed using two horns.

this device simulates an infinite surface. The main advantage of this method is that it is easier to implement than the first one. Moreover, it does not require any anechoic chamber. However, it enables measurement only over the frequency band of the TE10 mode and incident angle depends of the frequency. Finally, the size of the AMC surface must be equal to the size of the waveguide and must be composed of an integer number of unit cells.

3.2.2   Antenna over AMC: Principle

As has been mentioned, the AMC can be used as a perfect magnetic conductor to reflect the electric field in-phase. So it can be located close to an antenna to behave as an efficient reflector. On the contrary, if a classical perfect electric conductor is used as a reflector with an antenna, theoretically it has to be located at a quarter wavelength distance away from the antenna to lead to constructive interferences between the field radiated by the antenna and the field reflected by the PEC surface. This is shown in Figure 3.4(a). While a PEC surface introduces a 180° phase shift in the reflected wave path, a PMC surface does not add any additional phase shift. Therefore it can be located very close to the antenna and still leads to constructive interferences, as presented in Figure 3.4(b).

To achieve an AMC-based antenna, the first step is to design the AMC to operate in the desired bandwidth. Then, the method consists in locating a planar antenna operating within the same bandwidth, parallel and close to the AMC. Theoretically, constructive interferences occur. However, it is important to keep in mind that the phase diagram has been calculated without the antenna and with a plane wave at normal incidence. These conditions are very different from practical ones. The next step is to take into account the coupling between the antenna and the AMC by simulating the whole structure. This methodology is followed in the next sections and limitations are discussed, especially for wideband applications.

3.2.3   AMC's Wideband Issues

The operational bandwidth of an antenna is commonly defined by its impedance matching bandwidth but not only that. The radiation pattern has to be taken into account as well. An efficient radiation in a given direction is often a limiting criterion in addition to a reflection coefficient below $-10\,\mathrm{dB}$. As has been mentioned earlier, an AMC leads to constructive interferences over a limited bandwidth defined in Reference [1] by a $\pm 90°$ reflected phase criteria. Thus, if the antenna bandwidth is equal or larger than the AMC bandwidth, one would expect that the operational bandwidth would be limited by the AMC one. Similarly, if

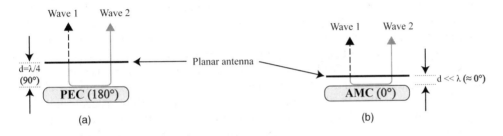

Figure 3.4   Planar antenna with (a) PEC reflector and (b) AMC reflector.

the AMC bandwidth is equal or larger than the antenna one, one would expect that the operational bandwidth would be limited by the antenna one. However, in practice, the behavior of an antenna over an AMC is a rather complex phenomenon. An AMC is a resonating surface and the antenna is often a resonator. Thus, some interactions take place between both and some authors study their influences.

In Reference [6], the behavior of a dipole antenna over a high impedance surface composed of square patches with metallic vias (mushrooms) is studied. This surface exhibits both AMC and EBG properties. It is found that the dipole antenna can be matched as long as its resonance frequency ranges within the AMC bandwidth defined by a new criterion: $90° \pm 45°$. This observation is different from what was expected from the theory mentioned in Reference [1]. This study shows that it is possible to design an efficient narrow bandwidth antenna when its band overlaps the AMC one. But what to do when a wide bandwidth is desired?

In Reference [10], a wide bandwidth is achieved with a simple dipole antenna over a mushroom-based AMC. By optimizing properly the interaction between the impedances of the dipole and its image through the AMC design, a bandwidth of 38% is obtained.

To increase the bandwidth further, different antennas should be used intuitively. Rather than a narrowband dipole, a wideband dipole could be considered. This has been done in Reference [11] for example. Two different structures are studied over an AMC, a diamond antenna and an open sleeve dipole. It is shown that if antenna and AMC designs are optimized together rather than separately, interesting wideband results can be obtained. A $-10\,dB$ impedance bandwidth of 67% is achieved. However, by taking into account the radiation pattern, the effective bandwidth is reduced to 36% around 5.6 GHz. This band has been defined by the authors by considering a gain greater than 6 dB. From this study, it appears that though it is possible to achieve an impedance matching over a wide bandwidth by optimizing the AMC and the antenna together, it is more difficult to achieve high gain toward one direction over the same bandwidth. It therefore appears that using AMC for wideband applications presents some issues regarding the radiation pattern.

These issues have been addressed in Reference [12]. Authors demonstrate that for wideband antennas, limited bandwidth AMCs do not necessarily avoid obtaining a wideband operation of the antenna over the AMC. With an appropriate design of the antenna element combined with the AMC, the return loss is independent of the reflection phase criteria. The operational bandwidth is then limited by the degradation of the radiation pattern that may occur at some frequencies. An example of a folded bow-tie dipole over a square-patch-based AMC is presented. Though the impedance bandwidth of the antenna can exceed 50%, its operational bandwidth is limited to 40% due to the drop of the gain. The main beam splits into two side beams thereby presenting a minimum of radiation in the broadside direction.

Mechanisms involved in the radiation of AMC-based antennas have been investigated in References [13] and [14]. By studying the behavior of a dipole antenna over an AMC composed of square patches, authors identified two different kinds of resonance. One is due to the resonance of the antenna modified by the presence of the AMC. The other one is due to resonances of the AMC itself. These resonances can be used for broadening the antenna bandwidth. However, the radiation pattern of these resonances strongly depends on the size of the AMC. When the size is greater than a wavelength in the effective medium, side lobes appear and the main beam may split into two beams. So the same beam splitting effect observed previously in Reference [12] is described here in Reference [13].

Wideband issues have been clearly identified in the literature. Limitations related to the radiation of AMC-based antennas have been explained as well. However, few techniques that could help antenna designers to improve antennas based on AMC surfaces have been reported [15]. That is why the following sections investigate some design optimization techniques to deal with these issues.

3.3 Wideband Directive Antenna Using AMC with a Lumped Element

The purpose of this section is to demonstrate how an AMC reflector enables the gain of a wideband linear polarized antenna to be increased. The classical bowtie antenna is proposed as the radiating element. Firstly, the features of the bowtie in free space without any AMC reflector are described. The second part deals with the design of the AMC reflector using the phase diagram. Then, the performances of the bowtie antenna above the AMC are presented. In the last part, an alternative solution is proposed with the use of lumped elements to improve the radiation pattern over a wider frequency band.

3.3.1 Bow-Tie Antenna in Free Space

The bow-tie antenna [16] is a planar version of the biconical antenna. Its bandwidth depends significantly on the flare angle α. As with any dipole, the antenna feed has to be balanced. To increase the mechanical rigidity and thus facilitate the fabrication of a reproducible prototype, the proposed bow-tie is etched on a substrate with a thickness of 1.6 mm. The relative permittivity is 3.7. There is no ground plane. The geometry of the antenna is presented in Figure 3.5. Dimensions are $d = 27$ mm and $\alpha = \pi/2$. The input impedance of the bow-tie is close to 175 Ω.

Simulations have been performed with CST Microwave Studio^{TM} software using the time domain transient solver. As expected, the impedance bandwidth is very large: it starts at 2.2 GHz until at least 6 GHz using the criteria $|S_{11}| \leq -10$ dB. Regarding the radiation pattern, it is bidirectional along the Oz axis at the beginning of the band as shown in Figure 3.6 at 3 GHz. For higher frequencies, several side lobes appear and the broadside realized gain decreases (see Figure 3.7). It should be remembered that the realized gain, by definition, takes into account the reflection coefficient $|S_{11}|$ of the antenna.

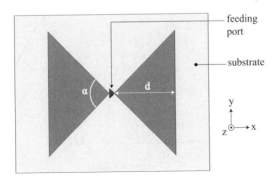

Figure 3.5 Geometry of the bow-tie antenna.

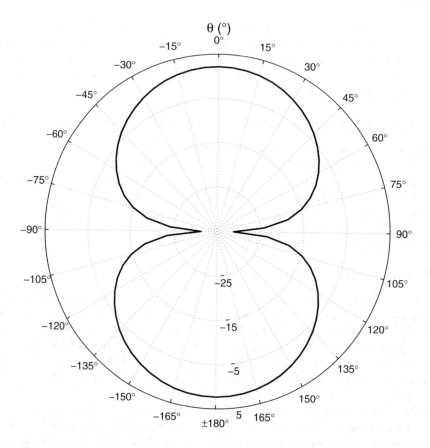

Figure 3.6   Realized gain (dB) of the bow-tie at 3 GHz in the E-plane (*XZ* plane).

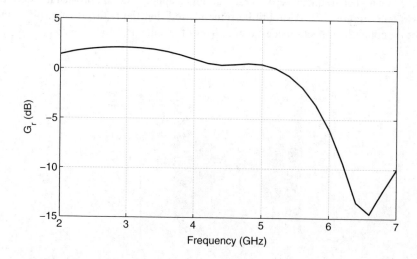

Figure 3.7   Evolution of the broadside gain (*Oz* direction) over the frequency band.

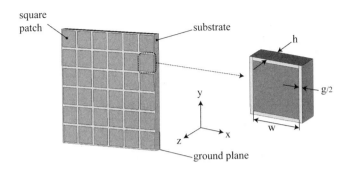

Figure 3.8 Unit cell geometry.

3.3.2 AMC Reflector Design

The AMC reflector is composed of metallic square patches etched on a grounded substrate. There is no via between the patch and the ground plane, which helps to reduce the cost and facilitate the realization. The unit cell geometry is presented in Figure 3.8.

Simulations have been performed with different values of the AMC parameters [17]. Results show that for a constant gap value, when the patch width decreases, the null reflection phase frequency increases and so does the fractional bandwidth of the AMC. Moreover, for a constant ratio w/g, when the gap width decreases, the null reflection phase frequency increases and also the fractional bandwidth of the AMC.

In order to design an AMC whose frequency band is included within the operational band of the bow-tie, the following dimensions have been chosen: $w = 8.2\,\text{mm}$, $g = 0.4\,\text{mm}$, $h_{AMC} = 3.2\,\text{mm}$. The substrate is the same as that of the bow-tie ($\varepsilon_r = 3.7$). The bandwidth of this AMC is deduced from the phase diagram depicted in Figure 3.9 and goes from 3.57 to 4.67 GHz.

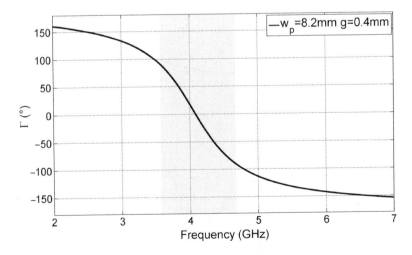

Figure 3.9 Phase diagram of the AMC.

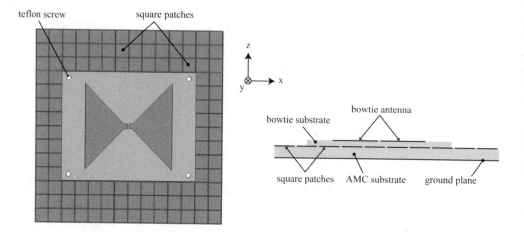

Figure 3.10   Geometry of the bow-tie above the AMC reflector: front view and side view.

3.3.3   Performances of the Bow-Tie Antenna over AMC

The bow-tie antenna is now located above the previous AMC made up with 14×14 unit cells. The port impedance is now 100 Ω. The complete structure described in Figure 3.10 has a total size of $120 \times 120\,mm^2$ with a thickness of $h = 4.8\,mm = \lambda/20$ at 3 GHz, where λ is the wavelength in free space. In Figure 3.10, one can see four teflon screws enabling the different layers to pile up. It has been verified that these elements do not disturb the return loss and the radiation pattern.

The reflection coefficient (see Figure 3.11) shows that the bow-tie's bandwidth has been reduced. The lowest frequency is now 2.9 GHz instead of 2.2 GHz for the bow-tie in free space. The coupling between the bow-tie and the AMC reflector leads to

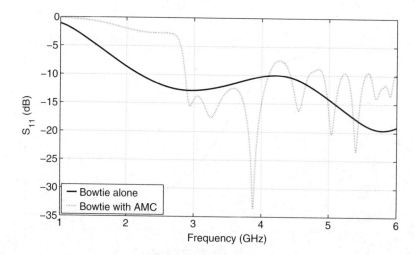

Figure 3.11   Simulated reflection coefficient of the bow-tie above the AMC reflector (normalized to $Z_0 = 100\,\Omega$).

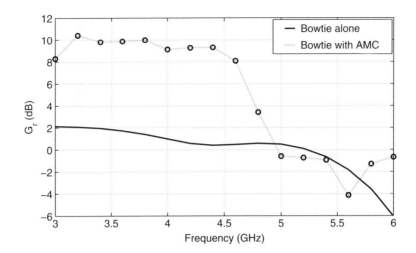

Figure 3.12 Simulated broadside realized gain of the bow-tie above the AMC reflector.

additional resonances, which enable an acceptable value of the magnitude of S_{11} to be maintained until 6 GHz.

From 3 to 4.6 GHz, the AMC reflector has significantly increased the broadside realized gain, which is greater than 8 dB (Figure 3.12). However, for higher frequencies, the gain strongly decreases due to the split of the radiation pattern.

This first result proves that designing the radiating element and the AMC reflector independently enables a structure to be obtained that provides a broadside gain higher than 6 dB over a bandwidth of 42% with a thickness of only λ/20 at the lowest frequency. In the next section, it is demonstrated that it is possible to limit the decrease of the gain at higher frequencies and thus to obtain a directive structure over a wider bandwidth.

3.3.4 AMC Optimization

In order to improve radiation patterns beyond 4.6 GHz, the AMC reflector is designed again to shift the null reflection phase at higher frequencies. Thereby, new dimensions are: $w_2 = 7.3$ mm, $g_2 = 1$ mm, $h_{AMC2} = 3.2$ mm. The operational frequency band of this optimized AMC deduced from its phase diagram is 4.3–6 GHz. The complete antenna (bow-tie above AMC) keeps approximately the same bandwidth and is matched from 3 to 6 GHz.

Compared to the previous antenna, the broadside gain with the optimized AMC has been significantly improved beyond 4.6 GHz, as shown later in Figure 3.16. However, the side lobe level remains high at higher frequencies. In fact, some important currents appear at the edges of the ground plane, as shown in Figure 3.13 at 5 GHz.

In order to minimize the propagation of surface currents at higher frequencies and thus to reduce the side lobe level, SMD resistors (100 Ω) are connected between patches at the edges of the AMC reflector. This technique has already been used in Reference [18] but SMD resistors were soldered to all patches, making the AMC surface behave like a wideband absorber. In this case, the position and the number of resistors have been optimized in order to both keep the radiation behavior in the broadside direction and also limit the side lobe level.

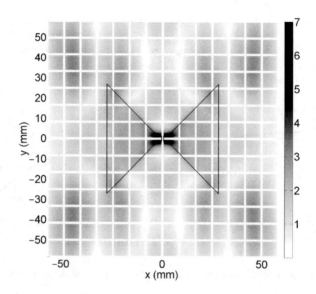

Figure 3.13   Current distribution on the AMC plane (*XY*) at 5 GHz: magnitude of $\mathbf{J}_x$ (in A/m).

A prototype of this structure has been realized and measured (Figure 3.14). The proto-type is fed by a taper balun designed to transform a 100 Ω balanced line to a 50 Ω unbalanced line [19].

Figure 3.15 presents the simulated and measured reflection coefficient of the bow-tie above the optimized AMC reflector with SMD resistors. It should be pointed out that adding SMD resistors does not change the return loss of the structure composed of the bow-tie above the optimized AMC. The difference between simulation and measurement can be explained by the fact that the balun has not been taken into account in the simulation in order not to increase the computation time.

Figure 3.16 presents the broadside gain from 3 to 6 GHz for three structures: the bow-tie above the AMC reflector previously designed and the bow-tie above the optimized AMC with and without the SMD resistors. We observe that the optimized AMC enables the gain to stay more stable over the band. The drop in the gain at 3 GHz was expected due to the fact that this frequency is out of the operational band of the new AMC, but it stays at an accept-able value (2 dB). The optimized AMC has significantly improved the gain beyond 4.6 GHz especially at 5.2 GHz, which corresponds to the null reflection phase of the AMC. It can also be noticed that SMD resistors contribute to increase the gain of 4 dB at the end of the band and thus to improve the stability of the gain over the band.

The effect of SMD resistors on surface currents is shown in Figure 3.17. Compared to Figure 3.13, one can see that currents at the edges of the AMC reflector have been reduced.

Figure 3.18 compares the realized gain in the E-plane (*ZX* plane) and in the H-plane (*ZY* plane) of the antenna with the optimized AMC with and without resistors at 5.8 GHz. It can be noticed that, in both planes, side lobe levels are largely reduced thanks to resistors while the broadside gain increases. Moreover, Figure 3.18 presents a comparison between simulation and measurement, which demonstrates the validity of the proposed method.

Figure 3.14   Prototype of the bow-tie above the optimized AMC with SMD resistors.

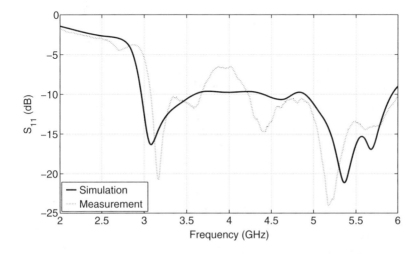

Figure 3.15   Reflection coefficient of the bow-tie above the optimized AMC reflector with SMD resistors.

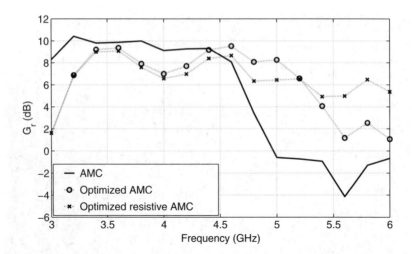

Figure 3.16 Broadside realized gain of the bow-tie above the optimized AMC reflector.

3.4 Wideband Directive Antenna Using a Hybrid AMC

The complex problem to achieve a wideband operation with an antenna over an artificial magnetic conductor has been addressed in Section 3.2. In particular, obtaining a high gain over a broadband is a challenge because of the beam splitting phenomenon occurring at certain frequencies. A method based on the phase diagram observation has been presented in the previous section in order to increase the bandwidth along which the antenna maintains a main beam in the broadside direction. In this section, an alternative procedure is given.

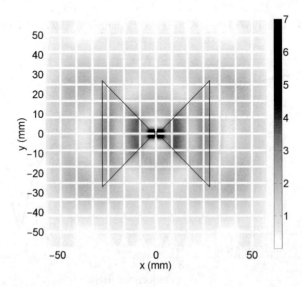

Figure 3.17 Current distribution in the AMC plane with resistors (*XY*) at 5 GHz: magnitude of $\mathbf{J}_x$ (in A/m).

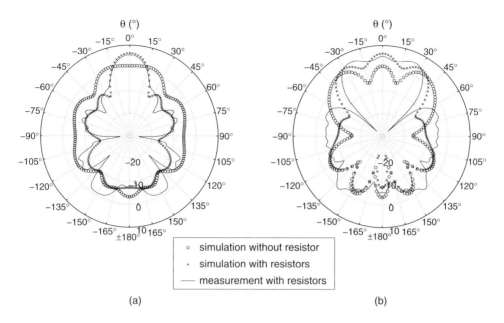

Figure 3.18 Realized gain (dB) of the bow-tie over the optimized AMC at 5.8 GHz: (a) E-plane and (b) H-plane.

A method based on the current's interpretation is applied to explain the radiation behavior of an AMC-based antenna and to identify the structure's areas contributing to the beam splitting. This approach is helpful in order to find a solution to cancel this unwanted effect. Consequently, a modified AMC is proposed that is able to maintain a formed radiation pattern over a wide frequency bandwidth.

3.4.1 Performances of a Diamond Dipole Antenna over the AMC

3.4.1.1 Diamond Dipole Antenna in Free Space

This section starts with analyzing a diamond dipole antenna alone, as presented in Figure 3.19. This antenna has already been used above for the periodical AMC in Reference [11] and difficulties to achieve a wideband operation have already been addressed. The diamond dipole structure is printed on a Teflon substrate of thickness $h = 0.8$ mm, relative permittivity $\varepsilon_r = 2.2$ and losses $\tan \delta = 0.0009$, and has the following dimensions according to Figure 3.19: $a = 1.5$ mm, $b = 3$ mm, $s = 0.5$ mm and $l = 8$ mm.

The structure is simulated with an input port impedance of $50\,\Omega$ by using the CST Microwave Studio. A relative bandwidth Δ_f of 28% (defined for $|S_{11}| < -10$ dB) is obtained at about 5 GHz (from 4.5 to 5.7 GHz). The realized gain in the broadside direction is shown in Figure 3.20, where the antenna bandwidth Δ_f is also indicated. It ranges between 1.1 and 1.6 dB over the whole bandwidth. The radiation pattern at 5 GHz in the E-plane is given in Figure 3.21. As expected, the antenna exhibits a typical dipole pattern which is quite stable over the bandwidth. The polarization is linear along the x axis (according to Figure 3.19). These results are taken as references for the next section where the classical AMC is introduced.

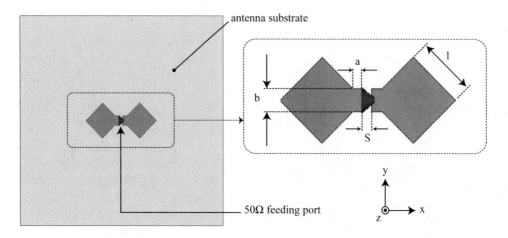

Figure 3.19 Diamond dipole antenna geometry.

3.4.1.2 Diamond Dipole Antenna with a Classical AMC

Once the reference antenna has been characterized, the AMC might be designed. The pattern chosen is the same as in the previous section, the square patch without via. The geometry of the AMC is shown in Figure 3.8. Its dimensions are found by simulating a unit cell using PEC/PMC boundary conditions in order to generate a TEM incident wave. The design is done in such a way that the resonance occurs at the antenna operating frequency, for example 5 GHz. So the dimensions are found to be: $w = 7.4\,\text{mm}$, $g = 1\,\text{mm}$, $h = 3.2\,\text{mm}$ with $\varepsilon_r = 4.1$, $\tan \delta = 0.0009$, where w is the width of the square patch, g is the gap width, h is the substrate height, ε_r is the relative permittivity of the substrate and $\tan \delta$ its losses. The simulated reflection phase is null at 5 GHz

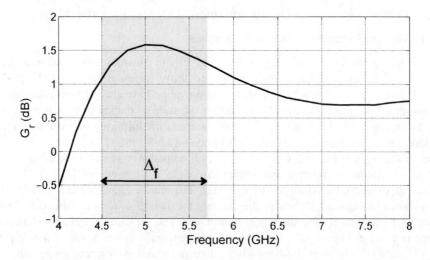

Figure 3.20 Realized gain of the simulated diamond dipole antenna (Δ_f is the antenna bandwidth defined for $|S_{11}| < -10\,\text{dB}$).

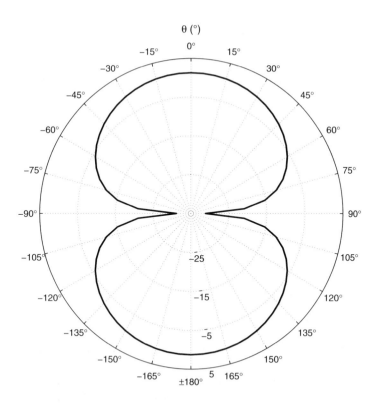

Figure 3.21 Radiation pattern of the simulated diamond dipole antenna at 5 GHz in the E-plane (*XZ* plane).

and the constructive bandwidth defined by the ±90° criteria ranges from 4.24 GHz up to 5.84 GHz (32%). Consequently, it completely overlaps the diamond dipole antenna bandwidth previously designed. Theoretically, and by neglecting the coupling between the square patches and the antenna, the AMC should reflect the electric field in phase with the electric field radiated by the diamond dipole. Thus, constructive interferences should occur over the whole antenna bandwidth.

The diamond dipole antenna is located above the AMC as presented in Figure 3.22. A distance $h_{air} = 2$ mm is observed between the antenna's substrate and the AMC. So the antenna's overall thickness is 6 mm ($\lambda_0/10$ at 5 GHz). This distance, h_{air}, has been obtained by optimization in order to improve the impedance matching.

The whole structure is simulated with AMC of four different sizes. All the surfaces analyzed are square. The number of patches constituting each of them ranges from 6 × 6 up to 12 × 12. The input impedance of the antenna is now 75 Ω, which provides better performances in terms of impedance matching when the AMC is located below the dipole. Results in terms of the reflection coefficient are compared in Figure 3.23. The case without AMC is also shown as a reference. The achieved −10 dB bandwidths of the different examples are summarized in Table 3.1. While the bandwidth of the antenna alone is 28%, the bandwidth of AMC-based antennas goes up to 51% for the surface with 12 × 12 square patches. As mentioned in Section 3.2, using AMC with a resonating antenna can lead to an enhancement

Table 3.1 Frequency bandwidths achieved with different sizes of the AMC

	Lower frequency	Higher frequency	Bandwidth	Bandwidth
Antenna alone	4.50 GHz	5.90 GHz	1.40 GHz	28%
AMC 6 × 6	4.35 GHz	6.65 GHz	2.30 GHz	42%
AMC 8 × 8	4.35 GHz	7.15 GHz	2.80 GHz	49%
AMC 10 × 10	4.45 GHz	7.15 GHz	2.70 GHz	47%
AMC 12 × 12	4.40 GHz	7.40 GHz	3 GHz	51%

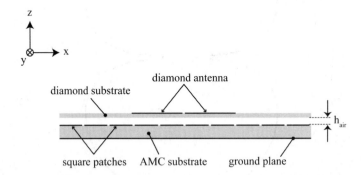

Figure 3.22 Geometry of the AMC-based diamond dipole antenna ($w_p = 7.4$ mm, g = 1 mm).

of the frequency bandwidth. The lower frequency of the bandwidth remains nearly constant at 4.4 GHz for any AMC size, whereas the upper frequency becomes higher as the AMC size increases (except between the 8×8 and 10×10 AMC). When observing reflection coefficients in Figure 3.23, we notice that when the number of square patches increases, more resonances appear in the upper frequency band. Thanks to those, using a greater number of patches can lead to an enhanced bandwidth (under proper impedance matching conditions).

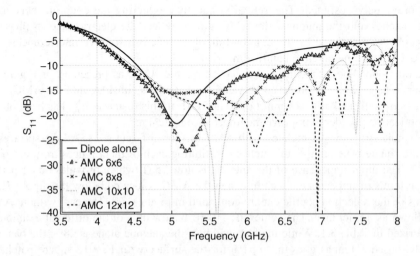

Figure 3.23 Simulated reflection coefficient of the antenna with and without the AMC.

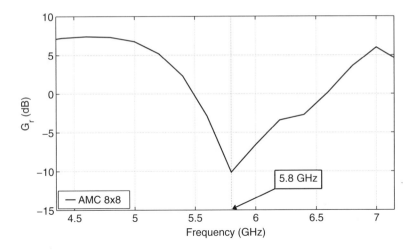

Figure 3.24 Broadside realized gain of the 8×8 AMC-based antenna.

Regarding the obtained frequency bandwidths with respect to AMC sizes, the antenna with the 8×8 AMC is now considered. The surface of the AMC is $67.2\,\text{mm}^2$ or $(1.1\lambda_0^2)$ at $5\,\text{GHz}$. The realized gain provided by this solution is shown in Figure 3.24 over the whole antenna bandwidth, that is, from 4.35 to 7.15 GHz. The gain goes up to $7.4\,\text{dB}$ at $4.6\,\text{GHz}$. However, it decreases drastically down to $-10\,\text{dB}$ at $5.8\,\text{GHz}$. To understand this behavior, the radiation pattern at $5.8\,\text{GHz}$ in the E-plane is drawn in Figure 3.25. The reason for the gain drop appears clearly: the radiation pattern splits into two main beams, thereby exhibiting a null of radiation in the broadside direction. Therefore avoiding this effect is compulsory in order to achieve a high gain over a wide band. Thus, in the next section, a method to analyze the origin of the beam splitting is presented.

3.4.2 Beam Splitting Identification and Cancellation Method

The behavior of the antenna is studied by analyzing its surface current. In fact, it is well known that a surface on which a uniform current is circulating exhibits a directive radiation pattern with a main beam in the direction normal to the surface (broadside direction). Furthermore, the larger the surface, the greater the directivity. However, when the surface becomes large compared to the wavelength, the current distribution may be not uniform and the current phase may range between $0°$ and $360°$, thereby inducing some opposite phase current. Consequently, destructive interference may occur with null directions in the radiation pattern. This effect is particularly undesirable when the null is in the broadside direction.

To understand if this effect happens, the surface current of the 8×8 AMC-based antenna previously designed is observed. The current is considered in the AMC plane, where it is mainly concentrated and spread over the larger surface (compare to the antenna where the current is highly localized). The surface current $\mathbf{J}_s$ lying in the XY plane is determined from the magnetic field $\mathbf{H}$ with the boundary condition on the patch's metallic surface:

$$\bar{\mathbf{J}}_s = -H_y \cdot \hat{x} + H_x \cdot \hat{y}$$

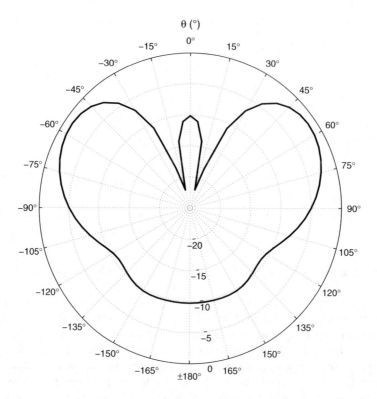

Figure 3.25   Radiation pattern of the 8×8 AMC-based antenna at 5.8 GHz in the E-plane (peak gain of 6.5 dBi).

Furthermore, as the antenna is linearly polarized along the x axis, only the current directly contributing to this polarization is considered. Thus the expression is reduced to

$$\bar{\mathbf{J}}_x = -H_y \cdot \hat{x}$$

The surface current $\mathbf{J}_x$ responsible for the main component of the radiation pattern is directly related to the y component of the magnetic field. The magnitude and the phase of $\mathbf{J}_x$ are plotted in Figure 3.26(a) and (b) respectively, at 5.8 GHz, the frequency at which a null appears in the broadside direction. From Figure 3.26(a), we can observe that the current is not uniformly spread. The maximum concentration of current is close to the antenna. Some other minor maxima can be distinguished as well. However, this nonuniform current distribution does not lead necessarily to a radiation pattern with null directions. It also depends on the phase of the current. This phase is depicted in Figure 3.26(b). We can see that the phase ranges between 0 and 360°. Thus, in some areas, the current is in opposite phase, thereby inducing some destructive interference. Because of this effect, at the frequency of 5.8 GHz, the main beam is split in the broadside direction. To identify clearly which parts of the surface are responsible for the beam splitting, a methodology is developed.

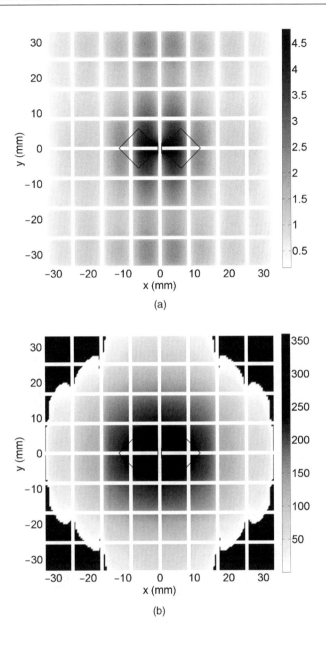

Figure 3.26 Current distribution of the 8 × 8 AMC-based antenna in the AMC plane (*XY*) at 5.8 GHz: (a) magnitude of $\mathbf{J}_x$ (in A/m) and (b) phase of $\mathbf{J}_x$ (in degrees).

1. **Strongest current identification.** The area where the current is the strongest is considered as the reference area. To define this reference, a threshold empirically chosen at half the current maximum value is taken. For this case, the maximum value of the current amplitude $|\mathbf{J}_x|$ is 4.8 A/m. Therefore, any area along which the circulating current is

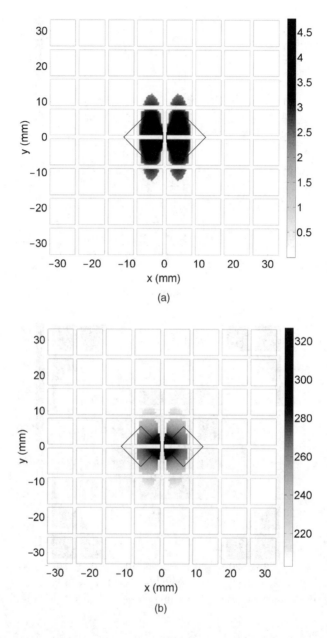

Figure 3.27 Distribution of the strongest current $\mathbf{J}_x$ of the 8×8 AMC-based antenna in the AMC plane (XY) at 5.8 GHz: (a) magnitude of $\mathbf{J}_x$ (in A/m) and (b) phase of $\mathbf{J}_x$ (in degrees).

greater 2.4 A/m is part of the reference area. This reference area is shown in Figure 3.27 (a), where the current amplitude ranges between 2.4 and 4.8 A/m. One can see that this area lies in the proximity of the dipole antenna.

2. **Destructive phase interval determination.** While considering the current $\mathbf{J}_x$ only in the reference area, the phase is plotted in Figure 3.27(b). As can be observed, the phase

ranges between 202° and 326°. By assuming that constructive interferences occur when currents are in phase ±90°, the phase range is extended from 112° to 416° (416° = 56° for phase wrapped between 0° and 360°). Consequently, a 'destructive interval' is defined between 56° and 112°. This means that wherever the current has a phase ranging between 56° and 112° on the AMC surface, this current contributes to destructive interference in the broadside direction and so to the beam splitting effect.

3. **Destructive current identification.** The phase of the current lying in the 'destructive interval' is plotted in Figure 3.28(a). Parts of the surface that are responsible for the beam splitting are now clearly identified. To further refine the localization of this area, it is weighted by the amplitude of the current. This is done by considering the current only wherever it is greater than a given threshold, here chosen empirically at a quarter of the maximum value. By doing so, it is assumed that the current whose phase lies in the destructive interval has an effect somehow proportional to its amplitude. In other words, the 'destructive current' having high amplitude contributes more to the beam splitting than the 'destructive current' having low amplitude. This final identification leads to the current phase shown in Figure 3.28(b). One can see that the current that mainly contributes to the beam splitting is localized in the middle of the AMCs edges along the x axis. By modifying the AMC structure in these areas, it may therefore be possible to change the current distribution and consequently avoid its phase ranging in the 'destructive interval'. This perspective is studied in the next section.

3.4.3   Performances with the Hybrid AMC

Since the problematic current responsible for the beam splitting is mainly localized on the last rows of patches along the x axis (one for positive y and one for negative y), it is natural to try to modify the geometry of these last rows. The first idea is to remove them completely in order to obtain a 6 × 8 AMC-based antenna. Thus, since the last rows are removed, no current can propagate along them. This could suppress the beam splitting effect at 5.8 GHz. However, while doing so, the size of the whole system is reduced and one can expect that the directivity decreases as well. Another idea is to short-circuit the patches of these last rows by filling the gap between them with metal in order to cancel their resonance. Thus the last rows are no longer composed of square patches but are a metallic surface as presented in Figure 3.29. With this modification, the phase of the circulating current along the x-oriented AMC edges is affected. Theoretically, since resonating patches act as a perfect magnetic conductor (PMC), replacing those with a perfect electric conductor (PEC) should introduce a 180° phase shift. Consequently, one can expect to suppress the beam splitting effect. Since both PEC and PMC surfaces are composing the surface, the solution is referred as a hybrid AMC-based antenna.

Such a solution has been simulated and results in terms of the reflection coefficient are shown in Figure 3.30 along with those of the antenna alone and the antenna with the classical 8 × 8 AMC. Though the shapes of the curves between antennas with the classical AMC and with the hybrid AMC are not similar, the −10 dB frequency bandwidths are identical. Therefore the proposed solution also has a wide bandwidth, $\Delta_f = 49\%$, ranging from 4.35 to 7.15 GHz.

Realized gains are compared in Figure 3.31 between the solutions based on the hybrid AMC, the classical AMC, and the diamond dipole alone. The realized gain obtained with the proposed hybrid solution is greater than the one obtained with the diamond dipole alone over

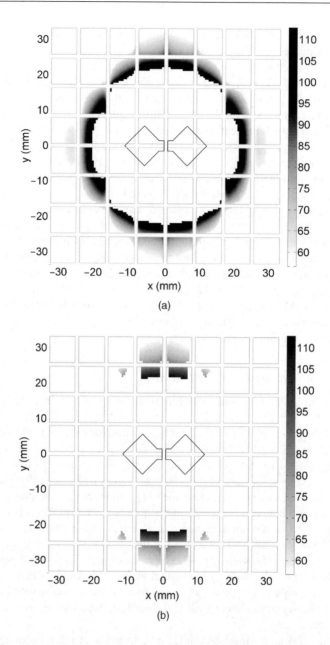

Figure 3.28 Phase of the problematic current $\mathbf{J}_x$ of the 8×8 AMC-based antenna in the AMC plane (XY) at 5.8 GHz: (a) full problematic current and (b) strongest problematic current (both in degrees).

the whole frequency bandwidth. It goes up to 9 dB at 5 GHz while with the diamond dipole alone it goes up to 1.6 dB at the same frequency. Furthermore, it does not decrease drastically like the classical AMC-based solution and remains greater than 5 dB along the frequency bandwidth except in the upper part, from 6.8 to 7.15 GHz, where it goes down to 0.6 dB. It

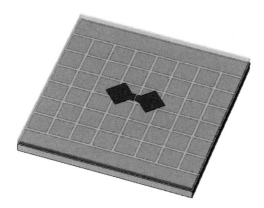

Figure 3.29 Hybrid AMC-based diamond dipole antenna (patches on the *y* edges are connected with each other in order to form a perfect metallic surface).

can also be noticed that except in this upper part of the band, the realized gain achieved with the hybrid AMC is always greater than the one achieved with the classical 8×8 AMC while both solutions have the same surface size.

To illustrate what happens at 5.8 GHz with the hybrid solution, the radiation pattern is shown in Figure 3.32. The main beam of the antenna no longer splits in the broadside direction and has a peak gain of 6.3 dBi. Thanks to the AMC modifications, the surface current might be more uniformly distributed than with the classical AMC. In order to prove this assumption, the magnitude and the phase of the surface current along the hybrid AMC is plotted in Figure 3.33(a) and (b) respectively. These figures are to be compared with Figure 3.26(a) and (b) from the classical AMC case. From Figures 3.26(a) and 3.33(a), it can be seen that the distribution of the current is more concentrated in the center of the AMC with the hybrid solution than with the classical one. However,

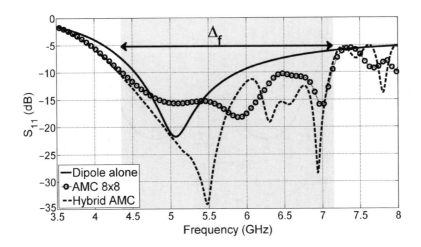

Figure 3.30 Reflection coefficient of the hybrid AMC-based antenna compared to the classical solutions.

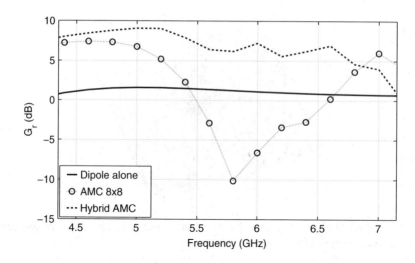

Figure 3.31 Realized gain of the hybrid AMC-based antenna compared to the classical solutions.

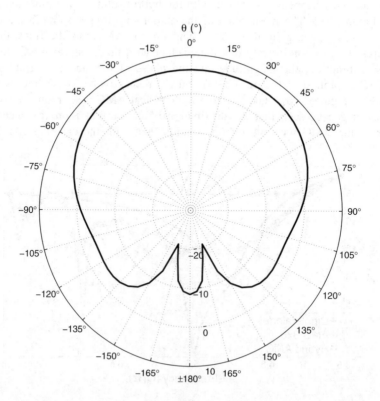

Figure 3.32 Radiation pattern of the co-polarization radiated by the hybrid AMC-based antenna at 5.8 GHz in the E-plane (peak realized gain of 6.3 dBi).

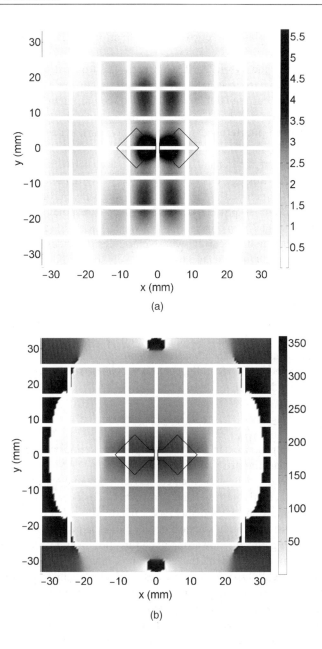

Figure 3.33 Current distribution of the hybrid AMC-based antenna in the AMC plane (*XY*) at 5.8 GHz: (a) magnitude of $\mathbf{J}_x$ (in A/m) and (b) phase of $\mathbf{J}_x$ (in degrees).

from the phase shown in Figure 3.33(b), it is difficult to determine qualitatively if this distribution is more suitable to obtain a main beam toward the broadside direction than the one of the classical AMC shown in Figure 3.26(b). In fact, the phase is also distributed over the entire 360° range. In order to give more insight into the hybrid solution's behavior, the same problematic current identification method used in the previous section is applied.

Firstly, regions where the hybrid AMC current is the strongest are identified by keeping the same threshold than earlier at half its maximum magnitude value. Secondly, in these regions, the phase of the current is plotted and can be seen in Figure 3.34(a) (this figure is to be compared with Figure 3.27b). The plotted phase ranges between 161° and 322°. By considering constructive interferences occurring when currents are in phase ±90° and by wrapping the phase between 0° and 360°, a 'destructive interval' is defined between 52° and 72°. Thirdly, the phase of the current lying in this 'destructive interval' is plotted and is shown in Figure 3.34(b) (this figure is to be compared with Figure 3.28a). One easily notices that not much current is included in the 'destructive interval'. Furthermore, if the current is considered only where its amplitude is above a given threshold, such as a quarter of its amplitude maximum, as has been done earlier between Figure 3.28(a) and (b), nothing would be displayed. This means that the current whose phase lies in the destructive interval does not contribute much to the radiation because of its relatively low amplitude.

Consequently, by applying the same identification method as previously with the same relative thresholds, with the classical AMC some problematic current areas have been identified as responsible for the beam splitting while with the hybrid AMC no problematic area has been found. Thus, this explains why no beam splitting is observed with the proposed hybrid solution. The observation of the current appears to be a good tool to use in order to understand the radiation behavior of an AMC-based antenna at particular frequencies within a given bandwidth. From an understanding of this behavior, some possible solutions can be found by the antenna's designers.

3.5 Conclusion

In this chapter, high impedance surfaces (HISs) are used to improve wideband antenna performances. In particular, the artificial magnetic conductor (AMC) behavior is utilized in order to achieve a reflector that can be located very close to antennas, thereby leading to low profiles. AMCs are inherently narrow band. However, it has been shown that the bandwidth of AMC-based antennas can exceed the AMC bandwidth alone in terms of impedance matching. Nevertheless, some issues regarding the radiation pattern are reported. In particular, maintaining a high gain value toward a given direction over a wide frequency range is a difficult task. Regarding this problem, this chapter presented two different approaches.

The first approach highlighted the fact that designing the antenna and the AMC separately does not lead to optimal performances and that adjustment in AMC designs is necessary. A methodology has been presented for that purpose and has been used to design a wideband AMC-based bow-tie antenna. Though a wide bandwidth has been obtained in terms of impedance matching, the frequency range over which the gain level is high is narrower. As a consequence, it reduces the operational antenna bandwidth. Therefore a method based on side lobe reduction using lumped resistors has been proposed. With this method, the gain has been successfully kept at a higher level over a wider bandwidth.

The second approach has been illustrated using a diamond dipole antenna. Unwanted radiation behaviors over the frequency range have been efficiently suppressed by modifying the AMC pattern. A method to identify the problematic current circulating over the AMC has

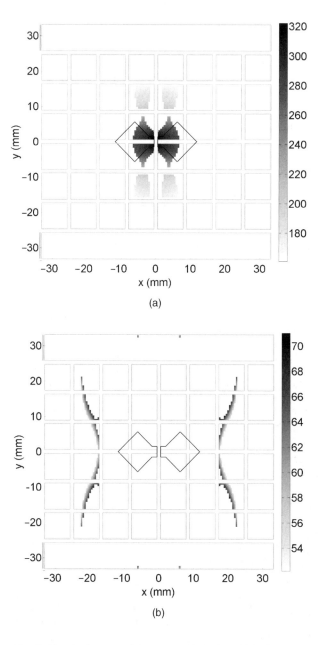

Figure 3.34 Phase distribution in degrees of the current $\mathbf{J}_x$ in the AMC plane (*XY*) at 5.8 GHz plotted (a) where the current $\mathbf{J}_x$ has the highest amplitude and (b) where the current is problematic (i.e., where the phase lies in the destructive interval, thereby leading to destructive interferences and so to beam splitting).

been developed. Then, from this identification, the AMC surface has been modified in order to cancel problematic currents. Thus, a hybrid AMC has been developed that makes the antenna radiating with a high gain over its entire impedance bandwidth.

Some solutions to overcome, to some extent, limitations regarding the achievability of a wide bandwidth with an AMC-based antenna have been detailed in this chapter. By increasing antenna impedance bandwidths, the AMC offers a great potential in designing wideband radiating structures, but only as long as the radiation is well controlled.

Acknowledgments

The authors would like to thank researchers involved in the presented topic: Christopher Djoma, Michaël Grelier, Fabrice Linot, Raquel Planas, Ludovic Schreider, and Aïta Thior, with the support of Michel Jousset, Michèle Labeyrie, Jean-Marc Lemener, Stéphane Mallégol, Bernard Perpère, Christian Renard, and Michel Soiron.

References

1. Sievenpiper, D.F. (1999) High-Impedance Electromagnetic Surfaces. PhD Thesis, University of California.
2. Yang, F. and Rahmat-Samii, Y. (2008) *Electromagnetic Band Gap Structures in Antenna Engineering*, Cambridge University Press.
3. Sievenpiper, D.F., Zhang, L., Broas, R. *et al.* (1999) High impedance electromagnetic surfaces with a forbidden frequency band. *IEEE Transactions on Microwave Theory and Techniques*, **47**, 2059–2074.
4. Tretyakov, S. and Simovski, C. (2003) Dynamic model of artificial reactive impedance surfaces. *Journal of Electromagnetic Waves and Applications*, **17** (1), 131–145.
5. Luukkonen, O., Simovski, C., Granet, G. *et al.* (2008) Simple and accurate analytical model of planar grids and high-impedance surfaces comprising metal strips or patches. *IEEE Transactions on Antennas and Propagation*, **56** (6), 1624–1632.
6. Yang, F. and Rahmat-Samii, Y. (2003) Reflection phase characteristics of the EBG ground plane for low profile wire antenna applications. *IEEE Transactions on Antennas and Propagation*, **51** (10), 2691–2703.
7. Linot, F., Cousin, R., Begaud, X. and Soiron, M. (2010) Design and measurement of high impedance surface. 4th European Conference on Antennas and Propagation.
8. Luukkonen, O., Alitalo, P., Simovski, C. and Tretyakov, S. (2009) Experimental verification of analytical model for high impedance surfaces. *Electronics Letters*, **45** (14), 720–721.
9. Pearson, R., Phillips, B., Mitchell, K. and Patel, M. (1996) Application of waveguide simulators to FSS and wideband radome design. IEE Colloquium on Advances in Electromagnetic Screens, Radomes and Materials, London, pp. 7/1–7/6.
10. Azad, M.Z. and Ali, M. (2008) Novel wideband directional dipole antenna on a mushroom like EBG structure. *IEEE Transactions on Antennas and Propagation*, **56** (5), 1242–1250.
11. Akhoondzadeh-Asl, L., Kern, D.J., Hall, P.S. and Werner, D.H. (2007) Wideband dipoles on electromagnetic bandgap ground planes. *IEEE Transactions on Antennas and Propagation*, **55** (9), 2426–2434.
12. Best, S.R. and Hanna, D.L. (2008) Design of a broadband dipole in close proximity to an EBG ground plane. *IEEE Antennas and Propagation Magazine*, **50** (6), 52–64.
13. Costa, F., Luukkonen, O., Simovski, C.R. *et al.* (2011) TE surface wave resonances on high-impedance surface based antennas: analysis and modeling. *IEEE Transactions on Antennas and Propagation*, **59** (10), 3588–3596.
14. Costa, F., Luukkonen, O., Simovski, C.R. *et al.* (2011) Accuracy of homogenization models for finite high-impedance surfaces located in the proximity of a horizontal dipole. Metamaterials, 2011, Barcelona, pp. 143–145.
15. Mateos, R.M., Craeye, C. and Toso, G. (2006) High-gain low profile antenna. *Microwave and Optical Technology Letters*, **48** (12), 2615–2619.

16. Bailey, M. (1984) Broad-band half-wave dipole. *IEEE Transactions on Antennas and Propagation*, **32** (4), 410–412.

17. Grelier, M., Lepage, A.C., Begaud, X. *et al.* (2012) Design methodology to enhance high impedance. 3rd International Conference on Metamaterials, Photonic Crystals and Plasmonics, META'12, Paris.

18. Schreider, L., Begaud, X., Soiron, M. *et al.* (2007) Broadband Archimedean spiral antenna above a loaded electromagnetic band gap substrate. *IET Microwaves, Antennas and Propagation*, **1**, 212–216.

19. Vahdani, M. and Begaud, X. (2006) A directive ultra wideband sinuous slot antenna. First European Conference on Antennas and Propagation, Nice.

4

Characterization of Software-Defined and Cognitive Radio Front-Ends for Multimode Operation

Pedro Miguel Cruz and Nuno Borges Carvalho
Instituto de Telecomunicações and Universidade de Aveiro, Aveiro, Portugal

4.1 Introduction

Software-defined radios (SDRs) [1] are now being accepted as the most probable solution for resolving the need for integration between actual and future wireless communication standards. SDRs take advantage of the processing power of modern digital processor technology to replicate the behavior of a radio circuit. Such a solution allows inexpensive, efficient interoperability between the available standards and frequency bands, because these devices can be improved, updated and their operation changed by a simple change in software algorithms.

The ultimate goal for an SDR architecture is to push the digitization closest to the antenna as much as possible, thus providing an increased adaptation and reconfigurability in the digital domain by the use of current digital signal processors (DSP, FPGA, etc.) capable of treating the incoming signals correctly. A common implementation for the SDR concept as proposed in Reference [1] is shown in Figure 4.1.

This SDR concept is also the basis for cognitive radio (CR) approaches [3] in which the underneath concept imposes strong changes in terms of both complexity and flexibility of operation due to its potential adaptation to the air interface. A promising application for this CR technology is to implement a clever management of spectrum occupancy by the use of opportunistic radios, where the radio will adapt and employ spectrum strategies in order to

Microwave and Millimeter Wave Circuits and Systems: Emerging Design, Technologies, and Applications,
First Edition. Edited by Apostolos Georgiadis, Hendrik Rogier, Luca Roselli, and Paolo Arcioni.
© 2013 John Wiley & Sons, Ltd. Published 2013 by John Wiley & Sons, Ltd.

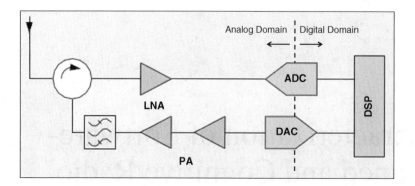

Figure 4.1 Typical implementation for an ideal software-defined radio [1]. © 2010 IEEE. Reprinted, with permission, from [2].

take profit from portions of spectra that are not being used by other radio systems at a given moment.

In that sense, these new developments will impose huge impairments in the design of radio receivers and characterization techniques that will allow the behavior of the radio receiver to be represented in order to minimize interferences and nonidealities and thus to increase the dynamic range, bandwidth, and so on. Moreover, this chapter provides a good foundation for supporting RF engineers in the overall receiver design to achieve better radio solutions.

This chapter will start first by presenting the most used receiver strategies for radio communications and concentrate on the usability for SDR/CR systems. Afterwards, the chapter will focus on techniques for modeling and characterization of SDR receivers, where the proposed behavioral model format allows the identification of different memory taps depending on the observed nonlinear mechanisms for a correct understanding of such wireless systems. Then, a specific practical application of the proposed behavioral model will be presented, when considering a band-pass sampling receiver with a quadrature-phase shift keying (QPSK) modulated wireless signal.

4.2 Multiband Multimode Receiver Architectures

For SDR/CR applications several receiver architectures may be used, ranging from common super-heterodyne, zero-IF, and low-IF designs to band-pass sampling approaches, but also recent proposals of six-port interferometers and direct RF sampling with analog decimation. All these are valid and practical receiving architectures, but some are gaining visibility over the others mainly because of the actual advancements in ADC/DAC technology and the enormous increase in the capabilities of digital signal processors.

The basic review that is done here is mostly based on References [2], [4], and [5]. Starting with the well-known super-heterodyne receiver (Figure 4.2), where the received signal at the antenna is translated to an intermediate frequency (IF) using a down conversion mixer, band-pass filtered and amplified. This is followed by a second stage for down conversion to baseband based on an I/Q demodulator and then converted to the digital domain to be treated. This architecture is now adopted mostly for higher-RF and millimeter-wave frequency designs [6, 7], for instance in point-to-point microwave links or short radar communications. However, super-heterodyne receivers have a considerable number of problems

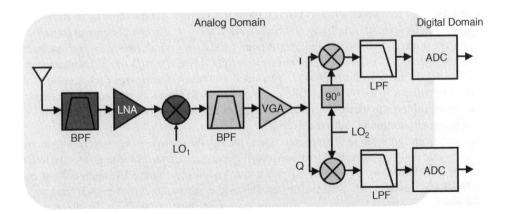

Figure 4.2   A super-heterodyne receiver architecture. © 2010 IEEE. Reprinted, with permission, from [2].

regarding their use in SDR/CR applications, the most relevant being the mandatory use of an image-reject filter in front of the mixer and the filter construction for a specific channel. Therefore, the super-heterodyne configuration is not attractive for use in SDR/CR receivers due to its complicated expansion for multicarrier multiband receivers.

The second approach is the zero-IF receiver [8, 9] (see Figure 4.3), which eliminates the IF stage of the previous super-heterodyne architecture, down converting the desired RF band directly to baseband. The received signal is selected at RF by a band-pass filter (BPF) and amplified by a low-noise amplifier (LNA), being afterwards directly down converted to DC by an I/Q demodulator and finally converted to the digital domain using an ADC. A substantial reduction in the number of analog components compared to a super-heterodyne architecture is evident. Moreover, in this architecture, a high level of integration is guaranteed thanks to its simplicity and relaxed specification requirements for the RF band-pass filter. These advantages make it a common architecture for multiband receivers such as the

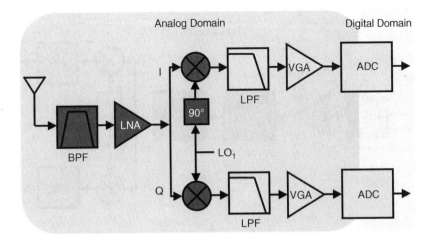

Figure 4.3   A zero-IF architecture. © 2010 IEEE. Reprinted, with permission, from [2].

one described in Reference [9] and for complete transceiver architectures as in References [10] and [11]. Despite its simplicity and higher integration, many components of the zero-IF receiver are more complex to design. Additionally, the direct translation to DC raises several problems such as DC offset [12], local oscillator (LO) leakage, I/Q impairments, second-order intermodulation products (that are generated around DC), and large flicker noise of the mixer [13], which can easily corrupt the received baseband signal. Some techniques to minimize these problems associated with its advantages and integration currently make this the most used configuration in radio receivers.

A similar configuration to the previous one is the low-IF receiver [14], wherein the incoming RF signal is down converted to a nonzero IF frequency instead of going directly to DC. This can be achieved by a complex RF down conversion (quadrature I/Q approach) or by a real RF down mixing. This architecture combines the advantages from zero-IF and super-heterodyne receivers. After a few filtering and amplification steps as in the previous approaches, the signal is converted to the digital domain by using relatively robust ADCs, which increase total power consumption because of the required higher conversion rate. Moreover, in this solution an image suppression block at the digital domain is commonly employed to cancel undesired effects from the image frequency problem now reintroduced. Again, a high level of integration is possible and the use of digital signal processing is allowed to execute the reception of several contiguous channels.

Another feasible alternative is the band-pass sampling receiver (BPSR) [15, 16] shown in Figure 4.4. The incoming signal is filtered by an RF band-pass filter that can be a tunable filter or a bank of filters, and subsequently amplified by a wideband LNA. Afterwards, the signal is sampled and converted to the digital domain by a high sampling rate ADC and digitally processed. The underlying idea is that energy from DC to the input analog bandwidth of the sampling circuit present in the ADC is folded back to the first Nyquist zone (NZ) $[0, f_S/2]$ because of the inherent sampling properties. It is essential to emphasize the importance of RF band-pass signal filtering because it must reduce all signal energy

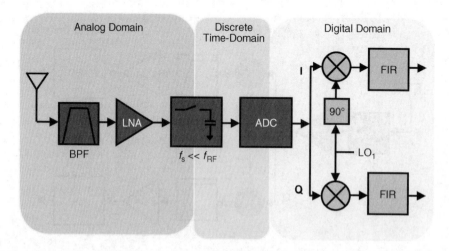

Figure 4.4 A band-pass sampling receiver. © 2010 IEEE. Reprinted, with permission, from [2].

(essentially noise) outside the desired NZ that would otherwise be aliased and thus degrade the attainable signal-to-noise ratio (SNR). Moreover, this filtering procedure has to separate the signals present in each NZ in order to avoid overlap of signals at the output. One advantage is that the needed sampling frequency and associated processing rate are proportional to the information bandwidth, rather than to the carrier frequency. However, a few critical requirements exist such as the fact that the analog input bandwidth of the sampling circuit must include the RF carrier band and the clock jitter dependence, which can be serious problems with modern ADCs.

A different possibility includes a quite old technique known as the six-port interferometer (SPI), which is being evaluated to become a practical architecture for SDR/CR receivers [17]. In the past this technique was mainly used for instrumentation and measurement applications [18–20]. Nevertheless, quite recent works demonstrate the use of an SPI radio receiver with some modifications to operate at millimeter-wave frequencies for QPSK and binary phase-shift keying (BPSK) modulated signals [21]. An SPI demodulator eliminates the use of down converting mixers and obtains directly the baseband information with a decoder (by means of a new phase spectrum demodulation scheme) from the four interferometer output signals. On the other hand, the support of quadrature amplitude-modulated (QAM) signals by the SPI radio are the object of ongoing research and development efforts. The possible operation of the SPI radio at very high transmission rates (large bandwidths) by using mostly passive devices and its low-cost implementation can be confirming factors for these SPI radio approaches.

Finally, other architectures proposed for SDR/CR receiver applications involve the utilization of direct RF sampling techniques with discrete-time analog signal processing and decimation to allow the signal to be received properly, such as the ones developed in References [22] and [23]. These designs are still in a very immature stage but should be further studied due to their potential efficiency in implementing reconfigurable receivers.

4.3   Wideband Nonlinear Behavioral Modeling

From the previously discussed architectures one of the most promising for SDR/CR applications is the BPSR design because of its approximation to the initial idea from Mitola [1]. In order to use such an approach in real SDR/CR configurations, a correct characterization of the front-end should be performed as a way to construct digital equalizers that could increase the receiving signal quality and, thus, maximize dynamic range, bandwidth, and so on. Moreover, the simulation of such a huge and complex architecture (an entire BPSR) is quite computer intensive, mainly when the objective is to simulate RF signals modulated with high bandwidth excitations.

In that way, the main motivation of this section is to give a brief overview of the BPSR architecture operation and then propose a suitable wideband behavioral model, accompanied by the respective parameter extraction procedure, to cover RF/IF and baseband frequency responses as presented in Reference [24], within the first and over several different NZs.

4.3.1   Details of the BPSR Architecture

This section aims to provide a more profound detail about the account of the BPSR operation. This architecture has its key element in the ADC component (commonly in a pipeline structure) that contains a sample-and-hold circuit, which in theory will down convert the incoming signals as a mixer module, followed by the quantizing scheme based on a pipeline approach [25]. As mentioned above, the BPSR design is an approach that permits all of the energy from DC to the input analog bandwidth of the ADC to be folded back to the first NZ. This process occurs without any mixing down conversion because a sampling circuit is somehow replacing the mixer module. Actually, this is one of the most interesting components of this architecture, because it allows an RF signal of higher frequency to be sampled by a much lower clock frequency.

The basic concept is depicted in Figure 4.5, in which it is observed that all the input signals present in the allowable bandwidth of the sampling circuit are folded back to the first NZ (Figure 4.5, bottom). Furthermore, the signals are down converted and fall over each other if no filtering has been used previously. This folding process occurs for all the available signals at the input of the circuit as well as for any nonlinearity that may have been generated previously (e.g., in the LNA or even in the specific sampling circuit). In addition, based on some relationships [15], it is possible to predict the resulting IF folded frequencies, f_{fold}, using the

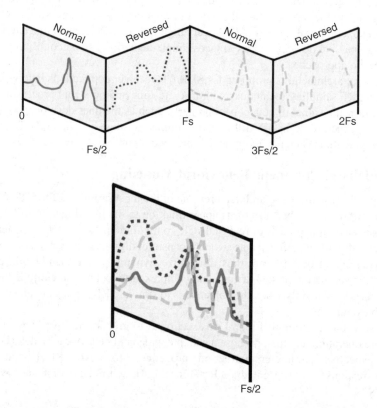

Figure 4.5 Process of folding that occurs in the sample-and-hold circuit: (top) entire input spectrum bandwidth and (bottom) output from the circuit with all the signals folded back and overlapped in the first NZ. © 2011 IEEE. Reprinted, with permission, from [24].

following formulation:

$$\text{if} \quad \text{fix}\left(\frac{f_C}{f_S/2}\right) \quad \text{is} \quad \begin{cases} \text{even,} & f_{\text{fold}} = \text{rem}(f_C, f_S) \\ \text{odd,} & f_{\text{fold}} = f_S - \text{rem}(f_C, f_S) \end{cases} \quad (4.1)$$

where f_C is the carrier frequency, f_S is the sampling frequency, fix(a) is the truncated portion of argument a and rem(a,b) is the remainder after division of a by b.

In order to better understand the operation of the explained BPSR in different NZs let us assume that the BPSR, as shown in Figure 4.4, is sampled by a clock of 100 MHz and excited first by a signal excitation present in the first NZ (e.g., 14 MHz) and then excited by a signal excitation situated in the second NZ (e.g., 78 MHz). For instance, if we consider a system that generates three harmonics, for the first excitation frequency and taking into account the frequency folding phenomena, the fundamental and respective harmonics will fall within the first NZ. However, the same will not happen for the second excitation frequency, where the baseband will fall on the first NZ, the fundamental and second harmonic will fall in the second and fourth NZs, respectively, and are folded back in the reversed way, obtained with Equation (4.1). Regarding the third harmonic, it will fall in the fifth NZ and is folded back in the normal mode.

In that sense, any model that may be used to describe the behavior of such architecture should have in mind that the operation over a huge bandwidth has to be covered and accompanied by different dynamic effects, which can be represented by different memory taps in the nonlinear model.

4.3.2   Proposed Wideband Behavioral Model

As was seen in the last section, to represent the BPSR nonlinear behavior effectively requires that the produced behavioral model should be wideband and take into account the NZ when the signal is sampled. In addition, several spectral components may appear in the first NZ case of the low-frequency baseband nonlinearities (defined by an even-order nonlinear product), with high-frequency components also possibly appearing at higher NZs where they are folded back to the resultant ADC bandwidth. This will impose conditions where the dynamic response of the BPSR will have time constants of highly different orders, with some at the RF time frame and others inside the baseband time frame.

In that sense, an appropriate behavioral model that produces the required mathematical description for describing the nonlinear behavior of the BPSR could be supported on the Volterra series theory [26], due to its good performance in this type of mildly nonlinear scenarios. The Volterra series conditions represent a combination of linear convolution and nonlinear power series providing a general structure to model nonlinear systems with memory. As such, it can be used to describe the relationship between the input and output of the considered BPSR, which may present a nonlinear behavior having memory effects. This relationship can be written as

$$y(t) = \sum_{n=0}^{\infty} \int_{-\infty}^{\infty} \cdots \int_{-\infty}^{\infty} h_n(\tau_1, \ldots, \tau_n) x_{in}(t - \tau_1) \cdots x_{in}(t - \tau_n) d\tau_1 \cdots d\tau_n \quad (4.2)$$

where $x_{in}(t)$ and $y(t)$ are the input and output signal waveforms, respectively, and $h_n(\tau_1, \ldots, \tau_n)$ is the nth-order Volterra kernel.

The applicability of such an RF time-domain Volterra series model to account with all these nonlinearities at once is complex because of the complicated model structure, which leads to an exponential increase in the number of coefficients for higher degrees of nonlinearities and memory lengths. Furthermore, the overall system description can behave very differently because, for instance, the even-order coefficients can generate signals at very high frequencies (such as in the case of the second harmonic) and at baseband frequencies near DC.

In that sense, the Volterra approach as presented in Equation (4.2) is not optimum for this situation since it uses the same descriptor for the second harmonic as for the baseband responses and thus does not provide the required flexibility. In fact, this problem was observed in the work presented in Reference [27], where a good approximation was achieved at higher frequencies but it had some problems at lower frequencies and vice versa.

To overcome this issue the Volterra series model can be applied in a low-pass equivalent format [28], in which a selection of each nonlinear cluster (baseband, fundamental, second harmonic, etc.) is firstly made and its particular complex envelope is then digitally obtained. As a result, the Volterra low-pass equivalent behavioral model is applied individually to each complex envelope cluster, taking into consideration the nonlinearity that has originated it. Actually, it can be seen as a model extraction based on the envelope harmonic balance method, where each cluster is addressed individually [29]. This low-pass equivalent conversion is exemplified in Figure 4.6, which considers a third-order degree nonlinear scenario.

As illustrated in Figure 4.7, the resulting model will be a collection of different sub-models obtained and extracted individually for each nonlinear cluster. Generally, this begins with the application of different Volterra operators in the extracted complex envelopes, followed by an up conversion of each cluster to the correct carrier frequency and finally summed together to create the resulting model output. In this way, the input of the proposed model will be the complex envelope of the desired excitation signal, which will then produce a real waveform representing the output of the nonlinear component/system.

Thus, as an example, the baseband and second harmonic arise from a second-order multiplication and are represented in this circumstance as

$$\tilde{y}_{BB}(k) = \tilde{h}_0 + \sum_{q_1=0}^{Q_{A1}} \sum_{q_2=q_1}^{Q_{A2}} \tilde{h}_{2,BB}(q_1, q_2)\tilde{x}(k - q_1)\tilde{x}^*(k - q_2) \tag{4.3}$$

$$\tilde{y}_{2Harm}(k) = \sum_{q_1=0}^{Q_{C1}} \sum_{q_2=q_1}^{Q_{C2}} \tilde{h}_{2,2Harm}(q_1, q_2)\tilde{x}(k - q_1)\tilde{x}(k - q_2) \tag{4.4}$$

where h_0 is the DC value of the output, $h_{2,BB}$ and $h_{2,2Harm}$ are the second-order Volterra kernels for the baseband and second harmonic responses, respectively. The character $\sim$ refers to a complex signal or value and the symbol $*$ means the complex conjugate.

For the proposed modeling strategy it can be noted that different memory lengths are used on the baseband and second harmonic components (represented in Equations (4.3) and (4.4)

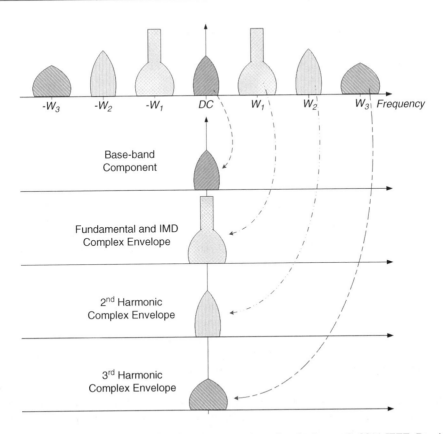

Figure 4.6 Diagram of the low-pass equivalent conversion of each cluster. © 2011 IEEE. Reprinted, with permission, from [24].

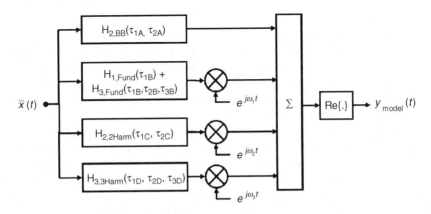

Figure 4.7 Proposed design for the wideband behavioral model. © 2011 IEEE. Reprinted, with permission, from [24].

by Q_{A1}/Q_{A2} and Q_{C1}/Q_{C2}), which provides an augmented flexibility to these models. As regards the fundamental signal and associated intermodulation distortion, these arise from a first-order function combined with a third-order nonlinear product:

$$
\begin{aligned}
\tilde{y}_{Fund}(k) &= \sum_{q_1=0}^{Q_{B1}} \tilde{h}_{1,Fund}(q_1)\tilde{x}(k-q_1) \\
&+ \sum_{q_1=0}^{Q_{B1}}\sum_{q_2=q_1}^{Q_{B2}}\sum_{q_3=q_2}^{Q_{B3}} \tilde{h}_{3,Fund}(q_1,q_2,q_3)\tilde{x}(k-q_1)\tilde{x}(k-q_2)\tilde{x}^*(k-q_3)
\end{aligned}
\tag{4.5}
$$

In the same line, the third harmonic arises uniquely from a third-order degree polynomial:

$$
\tilde{y}_{3Harm}(k) = \sum_{q_1=0}^{Q_{D1}}\sum_{q_2=q_1}^{Q_{D2}}\sum_{q_3=q_2}^{Q_{D3}} \tilde{h}_{3,3Harm}(q_1,q_2,q_3)\tilde{x}(k-q_1)\tilde{x}(k-q_2)\tilde{x}(k-q_3)
\tag{4.6}
$$

Moreover, when higher orders are requested more Volterra kernels should be determined. When extracting the kernels for each nonlinear cluster it is desirable to include all the possible contributions for each specific case, since it will deeply affect the extraction performance. For example, if expecting a component/system with fifth-order nonlinearities, then the third harmonic will not be exclusively characterized by a polynomial of third order but also including the contributions from a fifth-order coefficient.

In summary, it should be emphasized that in each cluster any nonlinear order and memory depth can be used and thus clearly different approaches can be employed. Also, the newly proposed behavioral model scheme has the feasibility to be extended and applied into multi-carrier nonlinear components/systems, as demonstrated in Reference [30].

4.3.3 Parameter Extraction Procedure

This section is devoted to describing the parameter extraction procedure, which has been employed in a BPSR design similar to that in Figure 4.4. The constructed laboratory proto-type of this BPSR architecture (device under test, DUT) considered several band-pass filters to select the desired NZ to be modeled connected to a wideband (2–1200 MHz) LNA, which has a 1-dB compression point close to +11 dBm, an approximated gain of 23 dB, and a noise figure near to 5 dB. This is then followed by a commercially 10-bit pipeline ADC that has a linear input range of around +10 dBm (2 Vpp for a 50-Ω source) and an analog input bandwidth (−3 dB bandwidth of the sampling circuit) of 160 MHz. This ADC component was then sampled using a sinusoidal clock of 90 MHz.

Evaluating the DUT at such a clock frequency will virtually create several NZs of 45 MHz ($f_s/2$) each at the output of the DUT. In this sense, the chosen excitation carrier frequencies are 11.5 MHz for the first NZ and 69 MHz for the second NZ. To measure the described DUT correctly, a laboratory setup based on the mixed-domain test bench proposed in Reference [31] was used, shown briefly in Figure 4.8. As illustrated in Reference [2] it is specifically dedicated to mixed-domain radio front-ends (SDR/CR) characterization.

As was widely discussed in Reference [24], it is quite difficult to have a setup for mixed-domain measurements with synchronized samplers between the different domains. The

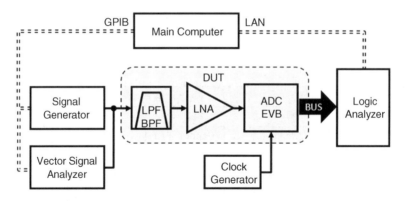

Figure 4.8 The experimental test bench proposed in Reference [31]. © 2011 IEEE. Reprinted, with permission, from [24].

solution for this situation was to embed a triggering pulse in the input signal followed by the waveform excitation of interest. In this way, all the measurements will be corrected accordingly to that trigger signal and become fully synchronous. Further details about this practice can be seen in Reference [24].

In addition to this, the treatment of the measured signals revealed in certain situations a huge corruption of these measurements by noise (instrumentation noise and noise generated in the DUT components), which is very close to the small distortion products desired to be modeled, making the parameter extraction impractical. Once again to minimize this issue a new approach was pursued consisting of the following steps:

1. Apply a fast Fourier transform (FFT) to the output RF time-domain signals.
2. Select only the desired frequency bins [27, 32] taking into account the nonlinearity order considered and construct a noise-free signal, only with the selected frequency components, for each cluster to be extracted.
3. Afterwards, apply an inverse fast Fourier transform (IFFT) in order to obtain a cleaner (without undesired frequency components and out-of-band noise) time-domain signal for each cluster.
4. Calculate the complex envelopes (e.g., using the Hilbert transform) for each cluster of the rearranged output signals.
5. Apply the low-pass equivalent Volterra series model, expressions (4.3) to (4.6), into these new output signals using also the measured input complex envelope and obtain the desired low-pass complex Volterra kernels.
6. Up convert each output complex signal to the corresponding cluster center frequency, depending on the resultant frequency from Equation (4.1), and finally assess the model performance.

Figure 4.9 shows a generalized flow diagram for the overall parameter extraction procedure. Such an approach allows, in step 5, the selection of nonlinear orders and memory taps that are more convenient in each specific cluster, reducing in some sense the required number of parameters. As well, it is important to notice that when the signal is within an even-order NZ, the output signal at the output of the DUT will appear rotated (reversed) (see Figure 4.5).

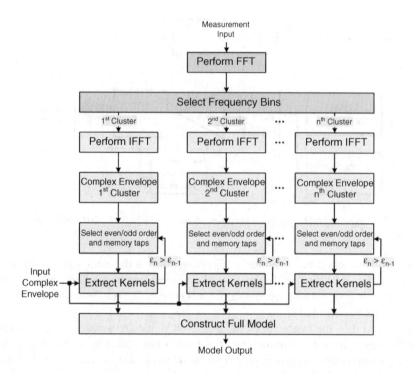

Figure 4.9 Flowchart diagram of the kernel extraction procedure. © 2011 IEEE. Reprinted, with permission, from [24].

Thus, in these circumstances an inversion of the signal is required, prior to the extraction of the particular cluster behavioral model.

Taking into consideration a few assumptions about the input signals, the extraction process of the low-pass complex Volterra kernels was based on a least-squares technique, expressed by

$$\mathbf{H} = \left(\mathbf{X}^{\mathrm{T}}\mathbf{X}\right)^{-1}\mathbf{X}^{\mathrm{T}}\mathbf{Y} \tag{4.7}$$

where $\mathbf{X}$ and $\mathbf{Y}$ are the input complex signal matrix and the output signal vector, respectively, and $\mathbf{H}$ is the vector of complex kernels being searched. This least-squares extraction is then executed for each one of the previously selected clusters.

As an example, if the complex parameters are being investigated for a baseband cluster with a memory length of Q taps, the input signal matrix ($\mathbf{X}$) will be designed in the following way:

$$\mathbf{X} = \begin{bmatrix} 1 & \tilde{x}(0)\tilde{x}^*(0) & \tilde{x}(0)\tilde{x}^*(-q) & \cdots & \tilde{x}(-Q)\tilde{x}^*(-Q) \\ \vdots & \vdots & \vdots & & \vdots \\ 1 & \tilde{x}(n)\tilde{x}^*(n) & \tilde{x}(n)\tilde{x}^*(-q) & \cdots & \tilde{x}(n-Q)\tilde{x}^*(n-Q) \\ \vdots & \vdots & \vdots & & \vdots \\ 1 & \tilde{x}(N)\tilde{x}^*(N) & \tilde{x}(N)\tilde{x}^*(N-q) & \cdots & \tilde{x}(N-Q)\tilde{x}^*(N-Q) \end{bmatrix} \tag{4.8}$$

and the complex output at baseband frequencies (**Y**) is defined as

$$\mathbf{Y} = \begin{bmatrix} \tilde{y}_{BB}(0) & \cdots & \tilde{y}_{BB}(n) & \cdots & \tilde{y}_{BB}(N) \end{bmatrix}^{\mathrm{T}} \tag{4.9}$$

where Q represents the memory length and N is the number of captured samples for both input and output complex envelope signals.

Afterwards, the seek vector of complex kernels (**H**) for the baseband cluster is calculated using Equation (4.7), which is actually composed of the following Volterra operators:

$$\mathbf{H} = \begin{bmatrix} \tilde{h}_0 & \tilde{h}_{2,BB}(0,0) & \tilde{h}_{2,BB}(0,q) & \cdots & \tilde{h}_{2,BB}(Q,Q) \end{bmatrix}^{\mathrm{T}} \tag{4.10}$$

This process is then executed for each individual cluster and then the final response of the behavioral model is achieved by employing the design depicted in Figure 4.7, wherein each cluster is up converted to the exact carrier frequency, based on Equation (4.1), as shown in the following expression:

$$y(k) = Re\{\tilde{y}_{BB}(k) + \tilde{y}_{Fund}(k)e^{j\omega_1 t} + \tilde{y}_{2Harm}(k)e^{j\omega_2 t} + \tilde{y}_{3Harm}(k)e^{j\omega_3 t}\} \tag{4.11}$$

As a final remark about the proposed behavioral model and respective parameter extraction strategy, it should be emphasized that great care should be taken when choosing carrier frequencies, signal bandwidths, and so on, due to the folding process that happens in the addressed DUT; the model extraction will become not valid if different clusters fall within overlapping frequency bins.

4.4 Model Validation with a QPSK Signal

In order to evaluate the performance of the proposed behavioral model for a BPSR, a QPSK modulated signal with a symbol rate of around 1 Msymbol/s filtered with a square-root raised cosine(RRC) filter with a roll-off factor of $\alpha = 0.25$, which determined a signal PAPR of approximately 5.4 dB, has been applied. These results were previously presented in Reference [24]. It used the laboratory setup shown in Figure 4.8 to perform the various measurements. The extraction of the seek parameters was executed in part of the captured input and each cluster output complex envelopes. After that, an equal number of remain samples were used to assess the accuracy of the complete behavioral model when compared with the obtained measurement results.

4.4.1 Frequency Domain Results

The obtained results are shown in Figures 4.10 and 4.11 for the two different NZs evaluated. Looking at the figures, it can be said that the proposed behavioral model and its associated parameter extraction procedure estimate the unknown parameters well and produce good results for the two NZ signals. Moreover, in Figure 4.10 the different memory depths (taps) for each nonlinear cluster can be checked for different NZs.

In order to be more precise in this evaluation, the integrated power within the frequency band of the fundamental signal, lower and upper adjacent channels, baseband component,

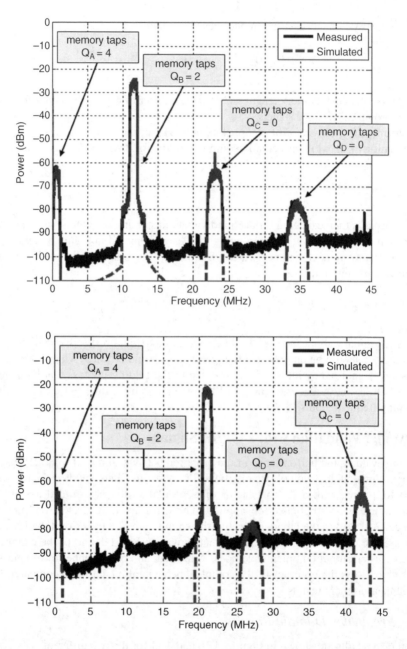

Figure 4.10 Entire bandwidth (smoothed) of measured and modeled outputs for a QPSK signal centered at 11.5 MHz (top) and 69 MHz (bottom). © 2011 IEEE. Reprinted, with permission, from [24].

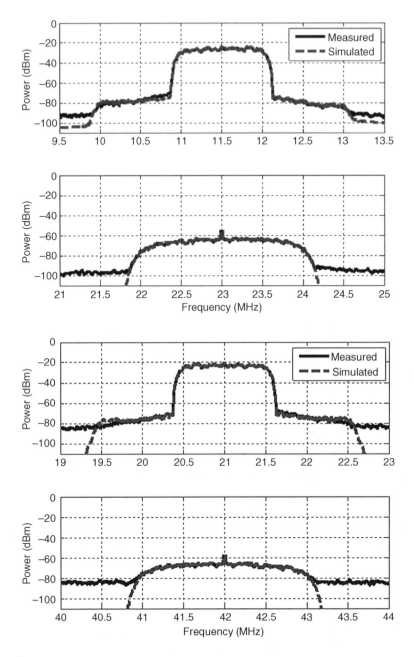

Figure 4.11 Spectrum of measured and modeled results, at the carrier band and second harmonic band for a QPSK signal centered at 11.5 MHz (top) and 69 MHz (bottom). © 2011 IEEE. Reprinted, with permission, from [24].

Table 4.1 Measured and modeled integrated powers for the QPSK excitation © 2011 IEEE. Reprinted, with permission, from [24]

	first NZ ($f_c = 11.5$ MHz)		second NZ ($f_c = 69$ MHz)	
	Measurement (dBm)	Model (dBm)	Measurement (dBm)	Model (dBm)
Baseband	−41.7	−42.9	−44.6	−45.4
Fundamental	−3.01	−3.02	0.40	0.38
Adjacent channel (lower)	−53.3	−54.6	−52.7	−52.5
Adjacent channel (upper)	−55.9	−56.0	−50.3	−51.2
Second harmonic	−38.5	−38.5	−40.8	−40.9
Third harmonic	−51.6	−52.2	−52.8	−53.0

and second and third harmonics were calculated. These results are revealed in Table 4.1 for the two NZs evaluated. It is clear that there is a good approximation to the DUT measurements presented by the proposed behavioral model.

Another figure-of-merit commonly used to express the error of a given model is the normalized mean square error (NMSE), [33]. The comparison between the complete measured output signals and the proposed behavioral model results reached NMSE values of −33.0 dB for the first NZ excitation and −32.9 dB for the second NZ excitation.

These demonstrated results validate in some sense this behavioral model proposed for BPSR application.

4.4.2 Symbol Evaluation Results

In order to validate the presented behavioral model even further, a digital version of a QPSK demodulator was implemented in order to obtain the symbol information (around 1000 symbols) from the previously measured and modeled QPSK signals, evaluated in the two different NZs.

Figure 4.12 illustrates the obtained normalized constellation diagrams for each NZ addressed. The figure verifies once again the good performance of the proposed behavioral model and the respective parameter extraction procedure. These assumptions are fully confirmed by the values presented in Table 4.2, where a good matching in terms of the root mean square (rms) EVM and also the peak EVM is observed.

Table 4.2 Measured and modeled EVM values for the QPSK excitation © 2011 IEEE. Reprinted, with permission, from [24]

	first NZ ($f_c = 11.5$ MHz)		second NZ ($f_c = 69$ MHz)	
	Measurement	Model	Measurement	Model
EVM rms	4.23%	4.97%	6.85%	5.2%
EVM peak at symbol	16.39% (703)	16.15% (703)	22.13% (898)	19.24% (898)

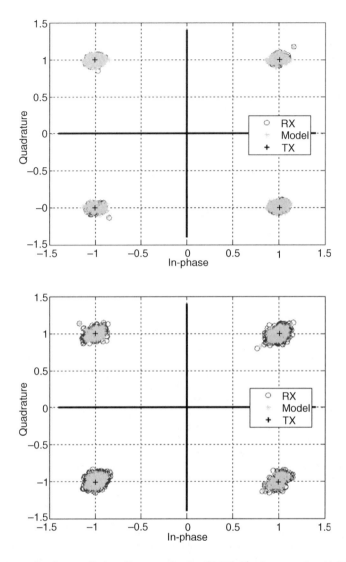

Figure 4.12 Normalized constellation diagrams for the QPSK signal centered at 11.5 MHz (top) and 69 MHz (bottom). © 2011 IEEE. Reprinted, with permission, from [24].

References

1. Mitola, J. (1995) The software radio architecture. *IEEE Communications Magazine*, **33** (5), 26–38.
2. Cruz, P.M., Carvalho, N.B. and Remley, K.A. (2010) Designing testing software-defined radios. *IEEE Microwave Magazine*, **11** (4), 83–94.
3. Mitola, J. and Maguire, G.Q. (1999) Cognitive radio: making software radios more personal. *IEEE Personal Communications*, **6** (4), 13–18.
4. Giannini, V., Craninckx, J. and Baschirotto, A. (2008) *Baseband Analog Circuits for Software Defined Radio*, Springer.

5. Puvaneswari, M. and Sidek, O. (2004) Wideband analog front-end for multistandard software defined radio receiver. IEEE International Symposium on Personal, Indoor and Mobile Radio Communications, September 2004, Barcelona, pp. 1937–1941.

6. Brandon, D., Crook, D. and Gentile, K. (2009) The Advantages of Using a Quadrature Digital Upconverter in Point-to-Point Microwave Transmit Systems. Application Note AN-0996, Analog Devices, Inc.

7. Lockie, D. and Peck, D. (2009) High-data-rate millimeter-wave radios. *IEEE Microwave Magazine*, **10** (5), 75–83.

8. Abidi, A.A. (2007) The path to the software-defined radio receiver. *IEEE Journal of Solid-State Circuits*, **42** (5), 954–966

9. Giannini, V., Nuzzo, P., Soens, C. *et al.* (2009) A 2 mm^2 0.1-to-5 GHz SDR receiver in 45 nm digital CMOS. IEEE International Solid-State Circuits Conference, February 2009, San Francisco, CA, pp. 408–409

10. Craninckx, J., Liu, M., Hauspie, D. *et al.* (2007) A fully reconfigurable software-defined radio transceiver in 0.13 μm CMOS. IEEE International Solid-State Circuits Conference, February 2007, San Francisco, CA, pp. 346–347

11. Ingels, M., Soens, C., Craninckx, J. *et al.* (2007) A CMOS 100 MHz to 6 GHz software defined radio analog front-end with integrated pre-power amplifier. European Solid State Circuits Conference, September 2007, Munich, pp. 436–439

12. Svitek, R. and Raman, S. (2005) DC offsets in direct-conversion receivers: characterization and implications. *IEEE Microwave Magazine*, **6** (3), 76–86.

13. Park, J., Lee, C-H., Kim, B-S. and Laskar, J. (2006) Design analysis of low flicker-noise CMOS mixers for direct-conversion receivers. *IEEE Transactions on Microwave Theory and Techniques*, **54** (12), 4372–4380

14. Adiseno, I.M. and Olsson, H. (2002) A wide-band RF front-end for multiband multistandard high-linearity low-IF wireless receivers. *IEEE Journal of Solid-State Circuits*, **37** (9), 1162–1168.

15. Akos, D.M., Stockmaster, M., Tsui, J.B.Y. and Caschera, J. (1999) Direct bandpass sampling of multiple distinct RF signals. *IEEE Transactions on Communications*, **47** (7), 983–988.

16. Vaughan, R.G., Scott, N.L. and White, D.R. (1991) The theory of bandpass sampling. *IEEE Transactions on Signal Processing*, **39** (9), 1973–1984.

17. Bosisio, R.G., Zhao, Y.Y., Xu, X.Y. *et al.* (2008) New-wave radio. *IEEE Microwave Magazine*, **9** (1), 89–100.

18. Engen, G.F. (1977) The six-port reflectometer: an alternative network analyzer. *IEEE Transactions on Microwave Theory and Techniques*, **25** (12), 1075–1080.

19. Ghannouchi, F.M. and Bosisio, R.G. (1992) An automated millimeter-wave active load-pull measurement system based on six-port techniques. *IEEE Transactions on Instrumentation and Measurement*, **41** (6), 957–962.

20. Hoer, C.A. (1972) The six-port coupler: a new approach to measuring voltage, current, power, impedance, and phase. *IEEE Transactions on Instrumentation and Measurement*, **21** (4), 466–470.

21. Li, J., Bosisio, R.G. and Wu, K. (1994) A six-port direct digital millimeter wave receiver. IEEE MTT-S International Microwave Symposium Digest, May 1994, San Diego, CA, pp. 1659–1662.

22. Muhammad, K., Ho, Y.C., Mayhugh, T. *et al.* (2005) A discrete time quad-band GSM/GPRS receiver in a 90 nm digital CMOS process. IEEE Custom Integrated Circuits Conference, September 2005, San Jose, CA, pp. 809–812.

23. Staszewski, R.B., Muhammad, K., Leipold, D. *et al.* (2004) All-digital TX frequency synthesizer and discrete-time receiver for Bluetooth radio in 130-nm CMOS. *IEEE Journal of Solid-State Circuits*, **39** (12), 2278–2291.

24. Cruz, P.M. and Carvalho, N.B. (2011) Wideband behavioral model for nonlinear operation of bandpass sampling receivers. *IEEE Transactions on Microwave Theory and Techniques*, **59** (4), 1006–1015.

25. Kester, W. (2008) ADC Architectures V: Pipelined Subranging ADCs. *Tutorial MT-024*, Analog Devices, Inc.

26. Schetzen, M. (1980) *The Volterra and Wiener Theories of Nonlinear Systems*, John Wiley & Sons, Ltd.

27. Cruz, P.M. and Carvalho, N.B. (2010) Modeling band-pass sampling receivers nonlinear behavior in different Nyquist zones. IEEE MTT-S International Microwave Symposium Digest, May 2010, Anaheim, CA, pp. 1684–1687.

28. Pedro, J.C. and Carvalho, N.B. (2003) *Intermodulation Distortion in Microwave and Wireless Circuits*, Artech House.

29. Ngoya, E. and Larcheveque, R. (1996) Envelop transient analysis: a new method for the transient and steady state analysis of microwave communication circuits and systems. IEEE MTT-S International Microwave Symposium Digest, June 1996, San Francisco, CA, pp. 1365–1368.

30. Cruz, P.M. and Carvalho, N.B. (2011) Multi-carrier wideband nonlinear behavioral modeling for cognitive radio receivers. European Microwave Integrated Circuits Conference, October 2011, Manchester, pp. 414–417.

31. Cruz, P.M., Carvalho, N.B., Remley, K.A. and Gard, K.G. (2008) Mixed analog–digital instrumentation for software-defined-radio characterization. IEEE MTT-S International Microwave Symposium Digest, June 2008, Atlanta, GA, pp. 253–256.

32. Pedro, J.C. and Carvalho, N.B. (1999) On the use of multitone techniques for assessing RF components' inter-modulation distortion. *IEEE Transactions on Microwave Theory and Techniques*, **47** (12), 2393–2402.

33. Muha, M.S., Clark, C.J., Moulthrop, A.A. and Silva, C.P. (1999) Validation of power amplifier nonlinear block models. IEEE MTT-S International Microwave Symposium Digest, June 1999, Anaheim, CA, pp. 759–762.

5

Impact and Digital Suppression of Oscillator Phase Noise in Radio Communications

Mikko Valkama[1], Ville Syrjälä[1], Risto Wichman[2] and Pramod Mathecken[2]

[1]*Tampere University of Technology, Tampere, Finland*
[2]*Aalto University, Espoo, Finland*

5.1 Introduction

Building compact-size and low-cost yet high-performance, flexible and reconfigurable radio transceivers for future wireless systems is generally a very challenging task. Using dedicated hardware particularly designed and optimized for only a single application or part of the radio spectrum yields only limited solutions, especially in terms of radio flexibility and re-configurability. Also, to keep the overall size and cost of the radio parts feasible, especially in multiantenna multiradio scenarios, the cost and size of individual radios are strongly limited. This implies that various circuit imperfections and impairments are expected to take place in the used radio transceivers, especially in the radio frequency (RF) analogue electronics [10]. This is also further catalysed by decreasing supply voltages and increasing electronics miniaturization. This is discussed at a general level, for example in References [1] to [4]. Good examples of such RF imperfections are, for example, mirror-frequency inter-ference due to I and Q branch amplitude and phase mismatches (I/Q imbalance), intermodulation and harmonic distortion due to mixer and amplifier nonlinearities, timing jitter and nonlinearities in sampling and analogue-to-digital (A/D) converter circuits, and oscillator phase noise. These RF impairments, if not properly understood and taken into account, can easily limit the performance of the radio transceivers, and thereby the perform-ance of the corresponding wireless radio link and system. The RF impairment effects are becoming increasingly important in the evolution of radio communications when more and

Microwave and Millimeter Wave Circuits and Systems: Emerging Design, Technologies, and Applications,
First Edition. Edited by Apostolos Georgiadis, Hendrik Rogier, Luca Roselli, and Paolo Arcioni.
© 2013 John Wiley & Sons, Ltd. Published 2013 by John Wiley & Sons, Ltd.

more complex, and thus more sensitive, high-order modulated wideband communications waveforms are being deployed in the emerging radio systems, compared to legacy narrowband binary modulation-based radio communications.

While there are various different types of RF impairments related to underlying radio electronics, this chapter concentrates on analysis and digital signal processing-based suppression of oscillator phase noise [5–8]. In general, phase noise has a rather complicated character and a large impact to the radio transceiver and link performance, especially in OFDM and other multicarrier type systems, and therefore analysis and compensation of phase noise in multicarrier systems continues to be a relevant active research topic. We first present the modelling of time-varying phase noise for free running and phase locked loop oscillators. Thereafter, we explain the effect of phase noise in OFDM systems and analyse the performance of OFDM in time-varying channels in the presence of phase noise. The last part of the chapter explores different algorithms for the compensation of phase noise in the digital baseband.

In general, as will be explained in detail in the forthcoming sections, phase noise appears in the form of multiplicative noise relative to the ideal phase noise-free signal. Thus the general impact of phase noise is that it broadens the spectral content of the signals (through spectral convolution). From an individual multicarrier waveform (like OFDM) point of view, this yields intercarrier interference (ICI) between the neighbouring subcarriers. This can be seen as an in-band problem, essentially increasing the in-band noise floor of a single radio waveform. In a more general context, however, phase noise also causes interference between the neighbouring RF channels or bands, which can be seen as an out-of-band problem dealing with multiple RF signals. Such an effect can be even more troublesome, compared to in-band problems, due to different RF power levels (dynamic range) of different RF carriers. In this chapter, the focus is on the phase noise-induced in-band problems. In terms of presentation, this chapter follows the presentations in References [9] and [10].

5.2   Phase Noise Modelling

This section gives a very short introduction to phase noise modelling and is based on the phase noise modelling given in References [11] and [12] and summarized in Reference [8].

There are many nonidealities related to oscillators, such as carrier frequency offset and phase offset. However, the most complex of the nonidealities is the time-varying phase noise denoted here by $\phi(t)$. A real oscillator-generated signal with phase noise can be written as

$$v(t) = A \cos(\omega_c t + \phi(t)) \qquad (5.1)$$

Here, A is the amplitude of the oscillating signal and ω_c is the nominal angular oscillation frequency. The phase noise modelling focuses on the modelling of the time-varying phase noise component $\phi(t)$. The modelling in this chapter is based on the simple mathematical free-running oscillator (FRO) model and on more complex phase-locked loop (PLL) oscillator model, since in practice PLL oscillators are used in communications devices.

5.2.1   Free-Running Oscillator

The FRO model is simple and easy to use in simulations and mathematical analysis. It is based on the assumption that the phase noise process is a so-called Brownian motion process (also known as the Wiener process or random-walk process). It can be written as

$$\phi(t) = \sqrt{c}\, B(t) \tag{5.2}$$

where $B(t)$ is the time-varying standard Brownian motion and c is the diffusion rate, which basically is the inverse of the relative oscillator quality. What makes FRO easy to use in simulations is the simple generation of the sampled version of Equation (5.2):

$$\phi_n = \sqrt{c}\, B(nT_s) \equiv \sqrt{c}\, B_n \tag{5.3}$$

where T_s is the sampling interval. From the definition of the standard Brownian motion, $B(nT_s) - B((n+1)T_s) \sim N(0, T_s)$, where $N(\mu, \sigma^2)$ denotes the normal distribution with expectation value μ and variance σ^2. This effectively means that the sampled FRO process can be generated as the cumulative sum of normal distributed random variables with zero mean and variance cT_s. Thus, it is possible to characterize the whole phase noise process with just a single parameter c.

To map the parameter c to a more easily measurable parameter, let us study the power spectral density (PSD) of the FRO. This is because the decay of the oscillator PSD is commonly used to characterize the oscillator phase noise properties. Specifically, a 3-dB bandwidth is used in this context, and it can be calculated as a point where the PSD has decayed to half of its maximum. If we assume that the oscillation frequency ω_c is relatively large and the diffusion rate c is relatively small (which they are in practice), we can approximate the one-sided PSD of the noisy oscillator signal as [11]

$$S_{v,ss}(\Delta\omega) \approx \frac{c/2}{(c/2)^2 + (\Delta\omega)^2} \tag{5.4}$$

Here, $\Delta\omega = \omega - \omega_c$ is the frequency difference from the nominal oscillation frequency. This corresponds to the well-known Lorentzian spectrum as discussed, for example, in Reference [11]. From this it is simple to calculate the 3-dB bandwidth β of the oscillator process of Equation (5.1) as

$$\beta = \frac{c}{4\pi} \tag{5.5}$$

Now, instead of characterizing the phase noise process with c, we can use the 3-dB bandwidth of the oscillator, which is a more tangible quantity.

5.2.2 Phase-Locked Loop Oscillator

There are various ways to model PLL oscillators. In this section, the model introduced in Reference [12] is used. It models a PLL oscillator that takes into account white and flicker $(1/f)$ noises [13] in its free-running (FR) voltage-controlled oscillator (VCO) and only white noise in its FR crystal oscillator (CO). The VCO model is based on the work done in Reference [14]. In the oscillator model, first a one-sided PSD of a baseband equivalent VCO is generated according to Reference [14] as

$$S_{a,ss}(\Delta\omega) = \frac{c_w + c_f S_f(\Delta\omega)}{\left(\dfrac{c_w + c_f S_f(\Delta\omega)}{2}\right)^2 + (\Delta\omega)^2} \tag{5.6}$$

Here, the PSD of the flicker noise is

$$S_f(\Delta\omega) = \frac{2\pi}{|\Delta\omega|} - \frac{4}{\Delta\omega}\tan^{-1}\left(\frac{\gamma_c}{\Delta\omega}\right) \tag{5.7}$$

$$c_w = \frac{2\pi}{\Delta\omega_f - \Delta\omega_w}\left[\Delta\omega_f^3 \times 10^{\frac{L(\Delta\omega_f)}{10}} - \Delta\omega_w^3 \times 10^{\frac{L(\Delta\omega_w)}{10}}\right] \tag{5.8}$$

and

$$c_f = \frac{2\pi(\Delta\omega_w - \Delta\omega_f)}{\Delta\omega_w \Delta\omega_f}\left[\Delta\omega_f^2 \times 10^{\frac{L(\Delta\omega_f)}{10}} - \Delta\omega_w^2 \times 10^{\frac{L(\Delta\omega_w)}{10}}\right] \tag{5.9}$$

On these, γ_c is a frequency corner point at which the flicker noise PSD essentially deviates from the nominal $1/f$ slope and $\Delta\omega_w$, $\Delta\omega_f$, $L(\Delta\omega_w)$ and $L(\Delta\omega_f)$ can be attained from the circuit simulator or one-sided PSD spot measurements of the VCO oscillator. $L(\Delta\omega_w)$ is a measurement at a white noise dominated region of the oscillator spectrum at offset $\Delta\omega_w$ from the nominal oscillation frequency and $L(\Delta\omega_f)$ is a measurement at a flicker noise dominated region of the oscillator spectrum at offset $\Delta\omega_f$ from the nominal oscillation frequency. The corresponding PSD of the CO is generated also with Equation (5.6), but without flicker noise. The equation for the PSD of the CO can thus be written as

$$S_{a,ss}(\Delta\omega) = \frac{c_{w,CO}}{\left(\dfrac{c_{w,CO}}{2}\right)^2 + (\Delta\omega)^2} \tag{5.10}$$

Here, $c_{w,CO}$ is given by Equation (5.8), but naturally from the measurements of the CO. Equation (5.10) closely resembles the PSD of the FRO model in Equation (5.4), because the used CO is a high-quality FRO with relatively low nominal oscillation frequency. However, Equation (5.10) is the PSD for the baseband equivalent oscillator and, furthermore, maps the oscillator measurements to the PSD through the measurement parameter $c_{w,CO}$. Now, to generate the actual PSD of the PLL oscillator $v(t) = Ae^{j(\omega_c t + \phi(t))}$ actually needed in the baseband simulations, we need to combine the PSD of the CO and VCO. In this work, combination is done according to the work in Reference [15].

We know that the PSD of the complex exponential of the phase noise approximately equals the PSD of the actual phase noise $\phi(t)$ at frequencies higher than the oscillator 3-dB bandwidth [16]. For our oscillator model, as justified in Reference [15], we use this approximation in general. Thus, we can generate the phase noise by shaping the spectrum of white Gaussian noise to correspond to the baseband equivalent version of $S_{a,ss}(\Delta\omega)$, namely

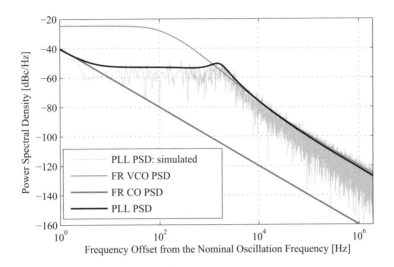

Figure 5.1 An example PSD of the CO and VCO used in the design of a PLL. Theoretical and simulated PSDs of the resulting PLL are also depicted. © 2009 IEEE. Printed, with permission, from [8].

$S_{a,ss}(\omega)$. An example of the CO, VCO and PLL PSDs with parameters $L(\Delta\omega_w) = -120\,\mathrm{dBc/Hz}$ at $\Delta\omega_w = 1\,\mathrm{MHz}$, $L(\Delta\omega_f) = -76$ dBc/Hz at $\Delta\omega_f = 10\,\mathrm{kHz}$, $L_{CO}(\Delta\omega_{w\_CO}) = -160$ at 1 MHz and $\gamma_c = 2.15\,\mathrm{kHz}$ is depicted in Figure 5.1. For more details and discussion refer to References [12] and [14].

5.2.3 Generalized Oscillator

To generalize the oscillator model even further, the nonconstrained form of the given PLL oscillator model is presented. We generate the phase noise in the same way with the help of a spectral masque, but now without tight constraints on the oscillator phase noise spectrum. Practically, this means that we generate white Gaussian noise, transfer it to the frequency domain with a discrete Fourier transform (DFT), then filter the signal with an arbitrary phase noise spectral masque and the filtered result is then finally transformed back to the time domain with an inverse discrete Fourier transform (IDFT). Naturally in this model, phase noise is generated in blocks since we cannot have an infinite length DFT/IDFT pair. The actual spectral masque can be obtained, for example, directly from the circuit simulator or from measurements of an oscillator.

For the purpose of later analysis, let us derive a connection between the phase noise spectral masque and average powers of the spectral components of the phase noise complex exponential. The spectral components can be calculated easily with the DFT. The N-point DFT of the phase noise complex exponential is

$$J_k(m) = \frac{1}{\sqrt{N}} \sum_{n=0}^{N-1} e^{j\phi_n(m)} e^{-j2\pi nk/N} \qquad (5.11)$$

Here, $k = 0, 1, \ldots, N - 1$ is the index of the spectral component and m is the index of the spectral DFT block. With a small phase approximation, namely $e^{j\phi_n(m)} \approx 1 + j\phi_n(m)$, and when keeping unit variance we can write

$$e^{j\phi_n(m)} \approx \frac{1 + j\phi_n(m)}{\sqrt{1 + \sigma_\phi^2}} \tag{5.12}$$

where σ_ϕ^2 is the average power of the phase noise $\phi_n(m)$, which can be derived from the spectral masque as [17]

$$\sigma_\phi^2 = \frac{\sigma_w^2}{N} \sum_{k=0}^{N-1} \lambda_k^2 \tag{5.13}$$

Here, σ_w^2 is the variance of the time-domain white Gaussian noise from which the phase noise is generated and λ_k is the spectral masque multiplier for the kth spectral component. Now by combining Equations (5.11) and (5.12), we can write

$$J_k(m) \approx \frac{1}{\sqrt{N\left(1 + \sigma_\phi^2\right)}} \sum_{n=0}^{N-1} [1 + j\phi_n(m)] e^{-j2\pi nk/N} \tag{5.14}$$

When $k \neq 0$, this can be written as

$$J_k(m) \approx \frac{j}{\sqrt{N\left(1 + \sigma_\phi^2\right)}} \sum_{n=0}^{N-1} \phi_n(m) e^{-j2\pi nk/N} = \frac{j\Phi_k(m)}{\sqrt{1 + \sigma_\phi^2}} \tag{5.15}$$

where $\Phi_k(m)$ is the kth frequency bin of the N-point Fourier transform of $\phi_n(m)$. So finally

$$k \neq 0: \quad E\left[|J_k(m)|^2\right] \approx \frac{\sigma_w^2 \lambda_k^2}{1 + \sigma_\phi^2} = \frac{\sigma_w^2 \lambda_0^2}{1 + \frac{\sigma_w^2}{N} \sum_{k=0}^{N-1} \lambda_k^2} = \frac{\psi_k^2}{1 + \frac{1}{N} \sum_{k'=0}^{N-1} \psi_{k'}^2} \tag{5.16}$$

For the DC bin, that is when $k = 0$, we can write

$$J_0(m) \approx \frac{N + j \sum_{n=0}^{N-1} \phi_n(m)}{\sqrt{N\left(1 + \sigma_\phi^2\right)}} = \frac{\sqrt{N} + j\Phi_0(m)}{\sqrt{1 + \sigma_\phi^2}} \tag{5.17}$$

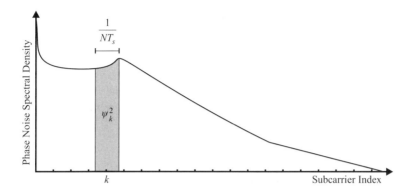

Figure 5.2 An example power spectral density function of the phase noise. © 2011 IEEE. Printed, with permission, from [17].

so we have

$$k = 0: \quad E\left[|J_0(m)|^2\right] \approx \frac{N + \sigma_w^2 \lambda_k^2}{1 + \sigma_\phi^2} = \frac{N + \sigma_w^2 \lambda_0^2}{1 + \frac{\sigma_w^2}{N}\sum_{k=0}^{N-1}\lambda_k^2} = \frac{N + \psi_0^2}{1 + \frac{1}{N}\sum_{k=0}^{N-1}\psi_k^2} \qquad (5.18)$$

Here, $\psi_k^2 = \sigma_w^2 \lambda_k^2$ is the energy of the phase noise around the kth subcarrier as depicted in Figure 5.2. Quantity ψ_k^2 can be simply connected to the practical oscillator PSD measurements by

$$\psi_k^2 = \sigma_w^2 \lambda_k^2 \approx \mathrm{PSD}_\phi\left(k\frac{1}{T_s}\right)\frac{1}{T_s} \qquad (5.19)$$

Here, PSD_ϕ is the PSD of the phase noise process at the frequency given by the argument. Equation (5.19) connects the powers of the spectral components of the phase noise complex exponential to the tangible PSD values.

5.3   OFDM Radio Link Modelling and Performance under Phase Noise

Direct conversion architecture is used in most commercial mobile devices of today, and therefore impairments that are not so tightly considered architectural weaknesses are interesting. One of the most interesting impairments in this context is phase noise. Its effect on DCR is especially interesting since phase noise is a very big problem in mobile OFDM receivers, in which DCR is usually used. Furthermore, emerging cellular systems like LTE and LTE-Advanced use OFDM.

First, this section describes the effect of the phase noise on a general I/Q signal in DCR. After that the section focuses on the phase noise effect on OFDM, followed by SINR and capacity analyses of the OFDM radio link impaired by receiver phase noise.

5.3.1   Effect of Phase Noise in Direct-Conversion Receivers

A general bandpass signal contains two low-frequency components, namely the inphase (I) and quadrature (Q) components. Such a bandpass signal is typically written as $s_I(t)\cos(\omega_c t) - s_Q(t)\sin(\omega_c t)$, where ω_c denotes the signal centre frequency and $s_I(t)$ and $s_Q(t)$ are the I and Q components. The so-called baseband equivalent signal, in turn, is a complex-valued signal $s(t)$, whose real and imaginary parts are the I and Q components, that is

$$s(t) = s_I(t) + js_Q(t) \tag{5.20}$$

I/Q modulation is thus a modulation technique in which the above formalism is utilized such that the I and Q components are low-frequency message signals, being then modulated into orthogonal cosine and sine carriers. In the case of digital transmission, these I and Q components, and thereon the corresponding bandpass signal, are carrying the transmitted bits, while the more detailed mapping from bits to message waveform(s) depends on the applied data modulation. On the receiver side, the I and Q components are recovered by I/Q demodulation.

In a direct-conversion receiver (DCR), the received signal from the target centre frequency is I/Q down-converted directly to baseband, and lowpass filtering implements most of the receiver selectivity. This is typically divided between both analogue and digital filters. Now, when the I/Q down-conversion stage suffers from phase noise, we essentially end up having a complex observation of the form

$$\hat{s}(t) = \hat{s}_I(t) + j\hat{s}_Q(t) = s_I(t)[\cos(\phi(t)) + j\sin(\phi(t))] + js_Q(t)[\cos(\phi(t)) + j\sin(\phi(t))]$$
$$= s_I(t)e^{j\phi(t)} + js_Q(t)e^{j\phi(t)} = s(t)e^{j\phi(t)} \tag{5.21}$$

Therefore, in the DCR the phase noise effect on the signal waveform (namely the baseband equivalent effect of phase noise) can be seen as a multiplication with a complex exponential that has phase noise as its argument. Thus phase noise appears as multiplicative noise. Of course, after the sampling, the signal with phase noise is

$$\hat{s}_n = s(nT_s)e^{j\phi_n} \tag{5.22}$$

where T_s is the sampling interval, n is the sampling index and $\phi_n = \phi(nT_s)$.

5.3.2   Effect of Phase Noise and the Signal Model on OFDM

In the frequency domain, the effect of phase noise can be seen as spread of the received signal spectral contents. From an individual waveform point of view, the corresponding effect on the constellation of a single carrier signal is just the corresponding time-varying phase rotation of the constellation, as can be seen in Figure 5.3(a), and, if small, the effect on the signal quality is only very minor. However, the phase noise can still be a serious problem in the case of single-carrier waveforms if the spectral content of the possibly much

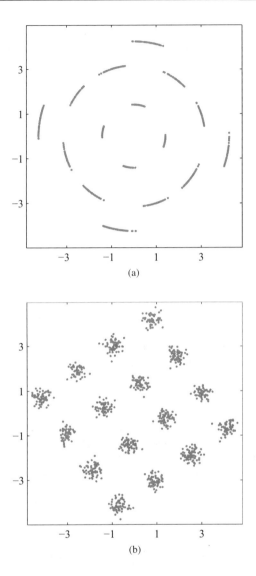

Figure 5.3 (a) Single-carrier 16QAM signal (1024 symbols) with phase noise and (b) OFDM with 16QAM subcarrier modulation (1024 subcarriers) with phase noise. In both cases the FRO oscillator with 100 Hz 3-dB bandwidth due to phase noise is assumed. I and Q components of the signals are presented in the horizontal and vertical axes respectively.

stronger neighbouring channel signal is spread on top of the weak desired signal. This depends on the assumed RF signal dynamic range and the amount of RF filtering in general.

For multicarrier signals, in turn, the effect of phase noise is much more complex and severe, even without any neighbouring channels, as depicted in Figure 5.3(b) for OFDM. This is the focus in the rest of this chapter. In addition to the rotation that every OFDM symbol experiences (as in single carrier symbols), OFDM symbols suffer from intercarrier interference (ICI) because phase noise spreads the energy of every subcarrier on top of the other subcarriers [18–20]. In a constellation it is seen as a spread around the ideal rotated constellation points. The effects of the phase noise on OFDM signals have been studied and analysed, for example, in References [9], [11], [19] and [21] to [24].

In general, for an OFDM configuration with N subcarrier symbols $k = 0, 1, \ldots, N - 1$, the mth OFDM symbol is generated from subcarrier modulated symbols $X_k(m)$ with the help of the N-point IDFT. The n th sample of such an OFDM symbol can be written as

$$x_n(m) = \frac{1}{\sqrt{N}} \sum_{k=0}^{N-1} X_k(m) \, e^{j2\pi kn/N} \tag{5.23}$$

where sampling indices are $n = 0, 1, \ldots, N - 1$. Now if the sampling interval is T_s, the symbol length in seconds is NT_s and the sampling frequency is $F_s = 1/T_s$. In practice, to exploit the long symbol duration in the OFDM signal, the cyclic prefix is also implemented to mitigate the impacts of multipath components of the radio channel. This is done by reconstructing the OFDM symbol so that the last G samples of Equation (5.23) are transmitted first and then the OFDM waveform without the prefix. This lengthens the OFDM symbol to $N + G$ samples, which is $(N + G)T_s$ seconds. The cyclic prefix effectively makes the OFDM signal immune to intersymbol interference, so at the receiver after ideal up-conversion at the transmitter, the time-invariant multipath channel, ideal down-conversion and removal of the cyclic prefix, the mth received OFDM symbol can be written as

$$\mathbf{r}_m = (\mathbf{h}_m * \mathbf{x}_m) + \mathbf{z}_m \tag{5.24}$$

where $\mathbf{h}_m$ is the $(D \times 1)$ multipath channel impulse response vector, $\mathbf{x}_m = [x_0(m), x_1(m), \ldots, x_{N-1}(m)]^{\mathrm{T}}$, operator $*$ denotes circular convolution between the elements of the operated vectors, and $\mathbf{z}_m$ is the $(N \times 1)$ vector of white Gaussian noise samples. Now the expression in (5.24) can be simplified with the help of the circular convolution matrix [25] as

$$\mathbf{r}_m = \mathbf{H}_m \mathbf{x}_m + \mathbf{z}_m \tag{5.25}$$

where $\mathbf{H}_m$ is the $(N \times N)$ circular convolution matrix corresponding to the channel impulse response vector $\mathbf{h}_m$.

The previous formulation assumed ideal oscillators in the transmitter and receiver. With phase noise included in up-converting and down-converting oscillators in the transmitter and receiver, respectively, we can write the received signal as

$$\mathbf{r}_m = \mathrm{diag}\left(e^{j\boldsymbol{\phi}_{m,R}}\right) \mathbf{H}_m \, \mathrm{diag}\left(e^{j\boldsymbol{\phi}_{m,T}}\right) \mathbf{x}_m + \mathbf{z}_m \tag{5.26}$$

Here, $\mathrm{diag}(\cdot)$ is a function that creates a diagonal matrix out from its input vector and $\boldsymbol{\phi}_{m,T}$ and $\boldsymbol{\phi}_{m,R}$ are vectors consisting of transmitter and receiver phase noise samples $\phi_{n,X}(m)$, $X \in \{T, R\}$, respectively, so that $\boldsymbol{\phi}_{m,T} = [\phi_{0,X}, \phi_{1,X}, \ldots, \phi_{N-1,X}]^{\mathrm{T}}$, $X \in \{T, R\}$. In this chapter, a reasonable channel delay spread is assumed. This means that the channel coherence bandwidth is relatively high. From this stems the fact that in approximations we are able to change the order of the diagonal phase noise matrices and the cyclic channel matrix in Equation (5.26). We are therefore able to approximately model all the phase noise either as transmitter or receiver phase and mark the combined phase noise term as $\phi_m = \phi_{m,T} + \phi_{m,R}$ [9, 6, 11].

5.3.3 OFDM Link SINR Analysis under Phase Noise

In the frequency domain (after the receiver FFT), the signal model for the received signal in Equation (5.26) without transmitter phase noise (or all phase noise referred to thereceiver side) can be written as

$$R_k(m) = X_k(m)H_k(m)J_0(m) + \sum_{l=0, l \neq k}^{N-1} X_l(m)H_l(m)J_{k-l}(m) + \sqrt{N}Z_k(m) \qquad (5.27)$$

where $H_k(m)$ is the channel transfer function, $J_k(m)$ is the frequency-domain phase noise complex exponential defined in Equation (5.11) and $Z_k(m)$ is the frequency-domain additive white Gaussian noise. $Z_k(m)$ is multiplied by $\sqrt{N}$ because with the DFT scaling used in Equations (5.11) and (5.23) the ideal oscillator response is amplitude multiplication by $\sqrt{N}$. The scaling used in Equation (5.11) is also used for the phase noise here. If we assume that the common phase error (CPE) is easily mitigated and the ICI is the only contribution of noise due to phase noise, we can derive the signal-to-interference-and-noise ratio (SINR) into the form [17]

$$\gamma_k = \frac{E\left[|X_k(m)H_k(m)J_0(m)|^2\right]}{E\left[\left|\sum_{l=0, l \neq k}^{N-1} X_l(m)H_l(m)J_{k-l}(m) + \sqrt{N}Z_k(m)\right|^2\right]} \qquad (5.28)$$

Here $E[\cdot]$ is the statistical expectation operator. Now, if we make natural assumptions that (1) $X_k(m)$, $H_k(m)$, $J_k(m)$ and $Z_k(m)$ are mutually statistically independent and stationary, (2) that for $\forall k$: $X_k(m)$ are independent of each other and (3) that $E[X_k(m)] = 0$, and with assumption that noise power, average channel power response and average transmitted signal power are subcarrier independent, namely

$$\forall k : \quad E\left[|Z_k(m)|^2\right] = \sigma_z^2 \qquad (5.29)$$

$$\forall k : \quad E\left[|H_k(m)|^2\right] = \sigma_h^2 \qquad (5.30)$$

and

$$\forall k : \quad E\left[|X_k(m)|^2\right] = \sigma_x^2 \qquad (5.31)$$

we can rewrite Equation (5.28) as

$$\gamma_k = \frac{\sigma_x^2 \sigma_h^2 E\left[|J_0(m)|^2\right]}{\sigma_x^2 \sigma_h^2 \sum_{l=1}^{N-1} E\left[|J_l(m)|^2\right] + N\sigma_z^2} = \frac{E\left[|J_0(m)|^2\right]}{\sum_{l=1}^{N-1} E\left[|J_l(m)|^2\right] + \frac{\sigma_z^2}{\sigma_x^2 \sigma_h^2}} = \frac{E\left[|J_0(m)|^2\right]}{\sum_{l=1}^{N-1} E\left[|J_l(m)|^2\right] + \frac{N}{\rho}}$$

(5.32)

where ρ is the received signal-to-noise ratio (SNR). Now from Parseval's theorem and linearity of the expectation value operator it is found that

$$\sum_{k=0}^{N-1} E\left[|J_k(m)|^2\right] = N$$

(5.33)

By using Equation (5.33), we can rewrite quation (5.32) in more simple form as

$$\gamma_k = \gamma = \frac{E\left[|J_0(m)|^2\right]}{N - E\left[|J_0(m)|^2\right] + \frac{N}{\rho}}$$

(5.34)

which is subcarrier independent and only depends on second-order statistics of the CPE. Now by using Equation (5.18), we can approximate the above as

$$\gamma \approx \frac{N + \psi_0^2}{\sum_{k=1}^{N-1} \psi_k^2 + \frac{1}{\rho}\sum_{k=0}^{N-1} \psi_k^2 + \frac{N}{\rho}}$$

(5.35)

This is relatively simple formula for the SINR in the general OFDM case with a general oscillator.

In Figure 5.4, the theoretical formula (5.35) is compared to results given by the OFDM link simulator. In the simulator, the OFDM signal with 1024 subcarriers is generated, and the signal is then passed through a channel. After that additive noise is added if applicable, and the down-converting oscillator with PLL phase noise in the receiver is modelled. Then the SINR is calculated.

5.3.4 OFDM Link Capacity Analysis under Phase Noise

In this section we shall derive closed-form expressions for the link capacity of an OFDM system impaired by phase noise and subject to Rayleigh fading. For a system model with the received signal comprising the desired signal part plus the noise part and assuming independence between the two, with each drawn from a Gaussian distribution, the ergodic capacity is typically employed to evaluate the throughput. To determine the ergodic capacity it is necessary to average the instantaneous capacity over phase noise and channel realizations, but since ICI is not a Gaussian random variable in general [23] the task is not straightforward.

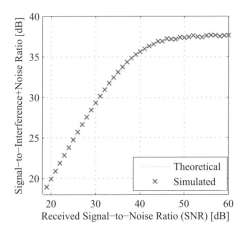

Figure 5.4 OFDM system (1024 subcarriers) performance under phase noise from a PLL type oscillator with $L(\Delta\omega_w) = -120$ dBc/Hz, $L(\Delta\omega_f) = -76$ dBc/Hz and $\gamma = 2.15$ kHz. Equation (5.35) is used and the SINR is given as a function of the received SNR.

To this end, we employ an alternative SINR expression instead of Equation (5.33), reflecting the instantaneous SINR (i.e. the SINR for a given channel and phase noise realizations) and determine its probability density function to determine the ergodic capacity [9, 26].

We start by determining the instantaneous capacity by fixing the phase noise process and the channel and assuming that the symbol alphabet $X_k, \forall k$ are Gaussian i.i.d. (independent and identically distributed) random variables that are also independent with the Gaussian receiver noise. Under such conditions, we see that for this fixed realization of the phase noise and channel, the ICI along with the receiver noise is Gaussian and, hence, instantaneous capacity is applicable and is given by

$$C_k = \log_2(1 + \gamma_k) \tag{5.36}$$

where

$$\gamma_k = \frac{|H_k|^2 |J_0|^2}{\left(\displaystyle\sum_{l=0, l \neq k}^{N-1} |H_l|^2 |J_l|^2 \right) + \sigma_Z^2/\sigma_X^2} \tag{5.37}$$

In the above equations, we have dropped the OFDM symbol index m without loss of generality. If the 3-dB bandwidth of the phase noise process is small compared with the subcarrier spacing and the channel coherence bandwidth is much larger than the subcarrier spacing, the following approximation holds [9]:

$$\sum_{l=1, l \neq k}^{N-1} |H_l|^2 |J_l|^2 \approx |H_k|^2 \sum_{l=1}^{N-1} |J_l|^2 \tag{5.38}$$

Thus, the SINR is simplified to

$$\gamma_k = \frac{|H_k|^2 |J_0|^2}{|H_k|^2 \left(\sum_{l=1}^{N-1} |J_l|^2 \right) + \sigma_Z^2 / \sigma_X^2} = \frac{1 - Y}{Y + \sigma_Z^2 / \left(|H_k|^2 \sigma_X^2 \right)} \tag{5.39}$$

where we have used the fact that

$$|J_0|^2 = 1 - \sum_{l=1}^{N-1} |J_l|^2 \tag{5.40}$$

and have defined

$$Y = \sum_{l=1}^{N-1} |J_l|^2 \tag{5.41}$$

We have dropped the subcarrier index k without loss of generality. From Equation (5.39), we see that the SINR and, hence, capacity in Equation (5.36) is a random variable whose distribution depends on the distribution of Y and of the channel $|H_k|^2$.

With the knowledge of the distributions of the random variables in Equation (5.39), statistical measures of capacity can be derived. For example, the average capacity can be derived by using Equation (5.39) in Equation (5.36) and sequentially averaging over the distributions of Y and the channel. We assume the channel to be Rayleigh faded. It is shown in Reference [9] that for small ratios of phase noise 3-dB bandwidth to subcarrier spacing, Y can be characterized as a sum of correlated gamma variables with a well-defined probability density function. In Figure 5.5, the probability density function plots of Y for a Wiener phase noise process are shown for two different phase noise levels. We clearly see from the figures that for a fixed subcarrier spacing, increasing the phase noise 3-dB bandwidth broadens the distribution of Y and thus we could expect the SINR in Equation (5.39) or (5.35) and, hence, capacity in Equation (5.36) to decrease.

Using the probability density function of Y derived in Reference [9] for a Wiener phase noise process, the capacity averaged over the distribution of Y while keeping the channel fixed is given as

$$\overline{C} = \log_2(1 + \Upsilon) - K \sum_{k=0}^{\infty} \varsigma_k \log_2 \left(\Upsilon^{\frac{1-c_k}{c_k}} + b_k \Upsilon^{\frac{1}{c_k}} \right) \tag{5.42}$$

where $\Upsilon = |H_k|^2 \sigma_Z^2 / \sigma_X^2$. The coefficients ς_k, c_k are obtained from the parameters that characterize the distribution of Y. Equation (5.42) represents the capacity for a static channel. The first term in Equation (5.42) represents the AWGN capacity for a static channel while the second term arises because of the presence of phase noise, which results in an overall reduction of the capacity. In the absence of phase noise the second term reduces to zero and the capacity reduces to the traditional AWGN capacity. Averaging Equation (5.42) over the distribution of $|H_k|^2$, the average capacity for an OFDM system impaired by Wiener phase

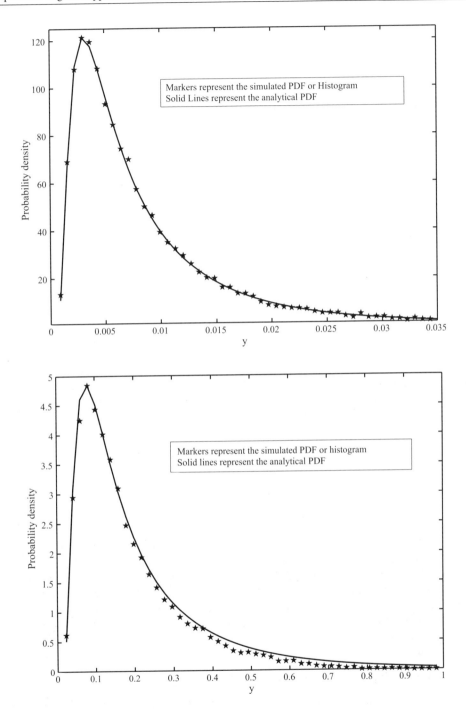

Figure 5.5 PDF plots of Y for two different Wiener phase noise 3-dB bandwidths. The first plot is for a phase noise 3-dB bandwidth equal to 80 Hz and the second plot is for a value of 2 kHz. Bandwidth is 625 kHz, $N = 32$ and subcarrier spacing is 19 kHz.

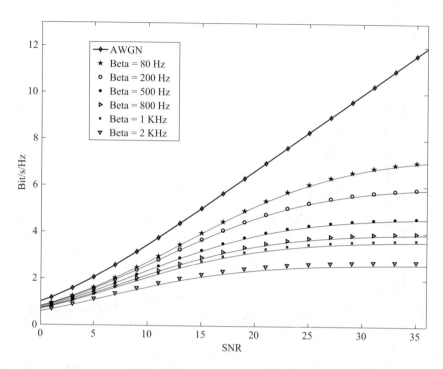

Figure 5.6 Average capacity $\overline{\overline{C}}$ plots. The solid lines represent the analytical results and markers represent the simulations. Bandwidth is 20 kHz, $N = 1024$ and subcarrier spacing is 19.5 kHz. Beta denotes the 3-dB bandwidth of Wiener phase noise.

noise is given by [21]

$$\overline{\overline{C}} = \log_2(e) \left[\dfrac{e^{\frac{1}{b_k \overline{\overline{Y}}}} -}{K \displaystyle\sum_{k=0}^{\infty} \varsigma_k \log_2\left(Y^{\frac{1-c_k}{c_k}} + b_k Y^{\frac{1}{c_k}} \right)} \right] \tag{5.43}$$

The first term in the above equation represents the capacity in a Rayleigh fading channel and the second term is due to the presence of phase noise, which causes the overall capacity to reduce from the Rayleigh fading channel capacity. Figure 5.6 shows the average capacity plots as a function of the SNR for different 3-dB bandwidths of the phase noise process. As can be seen from the figure, with the presence of phase noise the capacity reduces from the AWGN case where the phase noise is absent. Also, the capacity decreases with increasing levels of phase noise.

5.4 Digital Phase Noise Suppression

The task of phase noise estimation and mitigation in OFDM radios is one of the main topics of this chapter. The phase noise effect on OFDM signals is very severe and of a complicated nature, and there have not been good algorithms to mitigate the phase noise from OFDM signals until recently. In the literature, firstly, only CPE mitigation was considered, for example in Reference [19]. This was a natural approach as the CPE has potentially a very serious effect on the OFDM signal if a free-running oscillator is used. However, the CPE is hardly a problem in a wireless communications link, because in conventional channel estimation techniques the effect of the CPE is usually estimated as a part of the channel. Furthermore, if the phase noise is just jitter around the nominal phase, as is the case, for example in an PLL oscillator, the CPE alone is merely a minor problem, because constellation rotation caused by the CPE depends on the average phase error from the nominal phase during the duration of one OFDM symbol and is thus very small. However, if the channel is not fading, as, for example, with a plain additive white Gaussian noise channel assumption, the CPE estimation becomes interesting because it is simpler than doing the channel estimation. Furthermore, if the channel is assumed known, the CPE estimation is interesting for the analysis of ICI mitigation algorithms.

The mitigation of the ICI part of the phase noise has also received extensive attention in the recent literature. In Reference [27], a general phase noise mitigation technique mitigating CPE and ICI has been considered. A more advanced iterative algorithm for phase noise mitigation, where also the low-pass nature of the phase noise process is taken into account, was proposed in Reference [6]. The iterative part of the technique was further improved in Reference [28] by coding, and the actual phase noise compensation part was further improved in Reference [5].

This section reviews the latest phase noise mitigation algorithms in the literature. The first derivative estimates ICI using linear interpolation of CPE estimates first proposed in Reference [8]. The idea is simply to interpolate CPE estimates from the middle sample of one OFDM symbol to the middle samples of the adjacent OFDM symbols. Another idea proposed in Reference [8], and based on the same problem setting as the technique in Reference [5], was invented independently during the publication process of Reference [5]. It is rather simple as well. It was noticed that the algorithm in Reference [6] provides poor phase noise estimates at the beginning and at the end of each OFDM symbol. Due to the continuous nature of the phase noise, the phase noise estimates can however be interpolated over the interval in which the estimates are poor. Another iterative time-domain phase noise mitigation algorithm, which has first been published in Reference [7], is also reviewed. The idea is to detect the received signal after only the CPE of the phase noise is mitigated and then to reconstruct the phase noise-free signal. This signal is then used for a phase noise estimation by comparing it to the signal with phase noise still present.

5.4.1 State of the Art in Phase Noise Estimation and Mitigation

First, this section presents a simple CPE mitigation technique proposed in Reference [29], as it is usually needed in iterative ICI mitigation techniques to get the initial detection result. Then, some more advanced algorithms are reviewed and the ICI mitigation algorithm proposed in Reference [6] is presented in detail, followed by the detailed presentation of the improvement to the algorithm proposed in Reference [5].

5.4.1.1 CPE Estimation and Mitigation

CPE is a multiplication of all the subcarrier symbols by the same complex number within an OFDM symbol [11, 20]. CPE estimation techniques therefore merely estimate the common complex multiplier for all the subcarriers of an OFDM symbol. Such estimation techniques have, for example, been proposed in References [19], [29] and [30]. The techniques are similar and basically just solve the problem of a CPE estimation by averaging the calculated CPE values at pilot subcarriers using a least squares (LS) estimation. Therefore, directly from Reference [29], the LS solution for the CPE during the mth OFDM symbol can be written as

$$\hat{J}_0(m) = \frac{\sum\limits_{k \in S_p} R_k(m)X_k^*(m)H_k^*(m)}{\sum\limits_{k \in S_p} |X_k(m)H_k(m)|^2} \tag{5.44}$$

Here, S_p is a set of pilot subcarriers. In this chapter, the LS estimate of Equation (5.44) is always used for the CPE estimation, because of its computation simplicity. After computing the LS estimate for the individual OFDM symbols, the mitigation of the CPE is straightforward. Just the division of the received subcarrier symbol values with the corresponding CPE estimate is required.

5.4.1.2   ICI Estimation and Mitigation

The ICI is the more complex part of the phase noise effect, as is also its estimation. When the CPE is mitigated from the signal, the remaining time-domain phase noise contribution for individual OFDM symbols is just the same as before with only one exception: the mean of the remaining phase noise is approximately zero. The zero-mean phase noise causes the ICI. The problem of estimating and mitigating the ICI has been widely studied in the literature. Some examples are the studies in References [5] to [9], [12], [13], [23], [27] to [29] and [31] to [51]. Of these, in References [35], [37] and [38] the phase noise is compensated jointly with either channel and/or other transceiver impairments, such as an IQ imbalance. Examples of time-domain phase noise mitigation algorithms are given, for example, in References [34] and [38]. The algorithm in Reference [34] is based on estimating the most dominant discrete cosine transform terms of the phase noise. The algorithm is then enhanced in Reference [33] by iteratively using the nonpilot symbols in the phase noise estimation process. Another time-domain algorithm based on Kalman tracking is proposed in Reference [31]. The frequency-domain ICI mitigation algorithm proposed in Reference [6] is based on two simple but significant ideas, that is iterative estimation of the phase noise and exploiting the knowledge that most information of the phase noise can be recovered from the first few terms of its discrete Fourier transform. Its performance is improved in References [5] and [28] by means of improving the estimation algorithm and exploiting the information given by channel coding, respectively.

From all the available ICI mitigation algorithms, the algorithm of Reference [6] with its expansion in Reference [5] are presented as the state-of-the-art algorithms in more detail, because of their good performance. Another reason is the fact that the algorithm of Reference [6] is easily modified without practical performance loss to an LS-based algorithm, which does not require prior knowledge about the statistics of the phase noise. The idea is based on the frequency-domain signal model presented in Equation (5.27). With the assumption that most of the phase noise effect is indeed in its first u spectral components, we are able to rewrite Equation (5.27) as

$$R_k(m) = \sum_{l=-u}^{u} X_{k-l}(m)H_{k-l}(m)J_l(m) + \psi_l(m) = \sum_{l=-u}^{u} A_{k-l}(m)J_l(m) + \psi_l(m) \quad (5.45)$$

where $\psi_l(m)$ denotes the term with additive noise and remaining ICI outside of the first u spectral components, and for simplicity we have substituted $X_k(m)H_k(m)$ with $A_k(m)$. Now if this equation is written only for subcarriers $k \in \{l_1, l_2, \ldots, l_P\}$, where $P \geq 2(u+1)$, we can write a solvable matrix equation as

$$\begin{bmatrix} R_{l_1} \\ R_{l_2} \\ \vdots \\ R_{l_P} \end{bmatrix} = \begin{bmatrix} A_{l_1+u}(m) & A_{l_1+u-1}(m) & \cdots & A_{l_1-u}(m) \\ A_{l_2+u}(m) & \ddots & \ddots & \vdots \\ \vdots & \ddots & \ddots & \vdots \\ A_{l_P+u}(m) & \cdots & \cdots & A_{l_P-u}(m) \end{bmatrix} \begin{bmatrix} J_{-u}(m) \\ J_{-u+1}(m) \\ \vdots \\ J_u(m) \end{bmatrix} + \begin{bmatrix} \psi_{l_1}(m) \\ \psi_{l_2}(m) \\ \vdots \\ \psi_{l_P}(m) \end{bmatrix} \quad (5.46)$$

This can be written compactly as $\mathbf{R}_{m,P} = \mathbf{A}_{m,u}\mathbf{J}_{m,u} + \mathbf{\Psi}_{m,P}$, where $\mathbf{R}_{m,P}$ and $\mathbf{\Psi}_{m,P}$ are $(P \times 1)$ vectors, $\mathbf{J}_{m,u}$ is a $(2u+1 \times 1)$ vector and $\mathbf{A}_{m,u}$ is a $(P \times 2u+1)$ matrix. From this it is easy to use an LS estimation to estimate the most prominent ICI components $\mathbf{J}_{m,u}$ as

$$\hat{\mathbf{J}}_{m,u} = \left(\mathbf{A}_{m,u}^H \mathbf{A}_{m,u}\right)^{-1} \mathbf{A}_{m,u}^H \mathbf{R}_{m,P}. \quad (5.47)$$

In order to compute the above estimate, $\mathbf{A}_{m,u}$ needs to be known. In the algorithm, the detection results after CPE mitigation are used and then for the next iteration the detection result from the previous iterations are used. Channel estimates and prior channel information are also needed for the computation of the estimate. Instead of the previous LS solution, we could use also, for example, the minimum mean-square error (MMSE) estimation, but we would then need to know more about the statistics of the phase noise process in order to use it. Furthermore, the simulations by the author showed that the performance improvements are almost nonexistent. This is why in this chapter only the LS version of the estimation algorithm is used.

After estimation of the most prominent spectral components of the phase noise, the actual mitigation can simply be done by taking deconvolution between the received signal and the estimated phase noise in Equation (5.47). The resulting phase noise compensated signal can then be used as seen fit.

5.4.1.3 Modified ICI Estimation

Using a heavily truncated Fourier transform to estimate a nonperiodic signal such as the phase noise sequence is a problem, because the DFT assumes that the time-domain signal is periodic, whereas the phase noise process is not. However, this is not a problem with the nontruncated version of the DFT, but truncating the DFT makes the periodicity assumption very prominent in the corresponding time-domain signal [5, 8].

In Reference [5], a solution is proposed to the truncated DFT in the ICI estimation algorithm of Reference [6]. The proposed solution is based on mapping the received time-domain signal vector so that the edge parts of an OFDM symbol are mapped to the centre part of the vector. The mapping can be done with time-domain multiplication with permutation matrices of the form

$$\mathbf{p}_r = \left[\mathbf{e}_{N/2+1}, \mathbf{e}_{N/2+2}, \ldots, \mathbf{e}_N, \mathbf{e}_{N-1}, \ldots, \mathbf{e}_{N/2} \right]^{\mathrm{T}}$$

$$\mathbf{p}_l = \left[\mathbf{e}_{N/2}, \mathbf{e}_{N/2-1}, \ldots, \mathbf{e}_1, \mathbf{e}_2, \ldots, \mathbf{e}_{N/2+1} \right]^{\mathrm{T}} \tag{5.48}$$

Here, $\mathbf{p}_r$ and $\mathbf{p}_l$ are the size $(N \times N)$ permutation matrices for the right and left edges, respectively; $\mathbf{e}_n$ is a unit vector of length N, which has unity as its nth element and other elements are zeros. This time-domain mapping corresponds to mapping the truncated frequency-domain phase noise estimate with Fourier transforms $\mathbf{P}_r$ and $\mathbf{P}_l$ of the permutation matrices $\mathbf{p}_r$ and $\mathbf{p}_l$. Therefore what we actually need to do is to multiply the truncated phase noise estimate $\hat{\mathbf{J}}_{m,u}$ (filled with zeros to be size $N \times 1$) with $\mathbf{P}_r$ and $\mathbf{P}_l$. When carrying out the actual calculations in the receiver, naturally only nonzero elements of the $\hat{\mathbf{J}}_{m,u}$ are interesting, so $\mathbf{P}_r$ and $\mathbf{P}_l$ can be modified accordingly to be smaller. The resulting sequence is then transformed to the time domain, and the samples corresponding to the estimated phase noise are picked and used as phase noise estimates at the edges, instead of using the estimates provided by $\hat{\mathbf{J}}_{m,u}$. In Reference [5], the edge substitution window on the both sides was proposed to be around 6% of the OFDM symbol length, which also gave the best results in the simulations [8].

Compared to the algorithm of [6], the complexity is increased because we need some extra computations. Most burdensome computation results from the fact that signal processing is done in the time domain, so the signal needs to be transformed to the frequency domain again. This results in one extra DFT.

5.4.2   Recent Contributions to Phase Noise Estimation and Mitigation

In this section, first, a very simple technique to improve the CPE estimates with linear interpolation is presented. This is followed by the presentation of the technique to improve the performance of the state-of-the-art technique of Reference [6]. Both of these techniques were first published in Reference [8]. Then, the time-domain phase noise mitigation technique is presented, first published in Reference [7].

5.4.2.1   ICI Estimation Technique Using CPE Interpolation (LI-CPE)

The idea of the first ICI estimation technique introduced in this chapter, called LI-CPE, is based on the fact that the CPE estimate for an OFDM symbol approximates the mean of the phase noise sequence during that symbol. Because of this, the CPE corresponds to the exact value of the phase noise most likely in the middle of the OFDM symbol. Then, if we interpolate between the CPE values of adjacent OFDM symbols, from the middle of one OFDM symbol to the middle of adjacent OFDM symbols, we should have a crude estimate of the phase noise with the CPE and ICI taken into account. To improve the estimate a little, we can scale the interpolation result by replacing the DC bin of the estimated phase noise with the original CPE estimate. This is logical since we know that CPE estimates should be

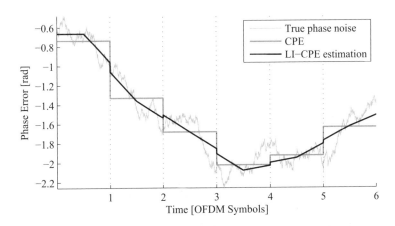

Figure 5.7 Illustration of the phase noise sequence and corresponding CPE estimate and the estimation result given by the LI-CPE technique. © 2009 IEEE. Printed, with permission, from [8].

relatively reliable. Interpolation, on the other hand, has changed the CPE estimate, so replacing it with more reliable one gives an improved estimate. Illustration of the technique is shown in Figure 5.7. In the figure, the small hops at the OFDM symbol boundaries result from the CPE replacement.

The interpolation in this technique can be done in various ways. In this chapter, however, only linear interpolation is used. The reason for this choice is not only the fact that linear interpolation is very simple but also the fact that if we interpolate between two points of a random walk (Wiener) process, the linear interpolation is actually approximately the optimum way to do the interpolation [38].

It should be noted that using this technique imposes a delay of one OFDM symbol, since we cannot get the full interpolation result before the estimation of the CPE of the next symbol has been done.

5.4.2.2 Iterative ICI Estimation Technique Using Tail Interpolation (LI-TE)

The second ICI estimation technique is based on the ICI estimation algorithm of Reference [6]. Before having access to Reference [5], the authors of this chapter also noticed the poor performance of the algorithm of Reference [6] at the OFDM symbol boundaries. This is clear when plotting the time-domain phase noise estimate versus the actual phase noise sequence shown in Figure 5.8. The peaks at the OFDM symbol boundaries are very clear and relatively wide, and the values of the estimates during these peaks differ from the underlying phase noise sequence very clearly. Mathematically the effect results from the truncation of the DFT done in the algorithm of Reference [6].

To combat the peaking effect of the algorithm of Reference [6] effectively, this chapter introduces a new technique, called LI-TE, for improving the estimates at the OFDM symbol boundaries. Simulations have shown that already continuing the last reliable estimate of the phase noise to the edge of the OFDM symbol gives impressive performance improvement. To further improve the estimate, once again, interpolation is used. In the LI-TE technique,

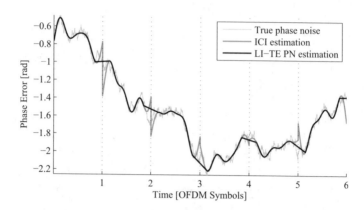

Figure 5.8 Illustration of the phase noise sequence and the corresponding ICI mitigation result given by the algorithm of Reference [6] and by the introduced LI-TE technique. © 2009 IEEE. Printed, with permission, from [8].

interpolation is done from the last reliable phase noise estimate of the previous OFDM symbol to the first reliable estimate of the phase noise of the current OFDM symbol. So once again the idea is very simple and, like in LI-CPE, only linear interpolation is used. According to empirical analysis, the optimal interpolation window is around 15% of the total OFDM symbol length at both edges of the OFDM symbol. In this empirical analysis, the length of cyclic prefix was assumed to be 6.15% of the total OFDM symbol length.

It should be noted that like LI-CPE also LI-TE imposes a delay to the system. LI-TE needs the ICI estimation results given by the algorithm of Reference [6] for the previous and next OFDM symbol, so delay of one OFDM symbol results in the first iteration. The delay can increase when the number of iterations increases if the LI-TE is also used to the next OFDM symbol prior to interpolation. This is not necessary, because the interpolation can be done based on the nonperfect phase noise estimate of the next OFDM symbol as well. However, the best performance is obtained if LI-TE is done for the next OFDM symbol as well as in the iteration loops. This results in one extra delay length of one OFDM symbol per iteration. Already two or three iterations give performances very near to the maximum performance the technique is capable of [7, 8, 39], so the actual number of iterations does not need to be high. Also, being relatively computationally complex, using many iterations is not very feasible anyway. In terms of complexity, the technique is similar to that in Reference [5].

In Reference [36], the authors present a phase noise compensation scheme using the iterative method proposed in Reference [6]. A Bayesian approach is used in arriving at the estimates of the spectral components of the phase noise. The authors show that the real and imaginary parts of the spectral components can be characterized as the sum of two random variables. The first random variable follows a Gaussian distribution while the second has a distribution that is a weighted sum of gamma distributions. Using this a priori knowledge, estimates of the spectral components are derived. The obtained estimates require knowledge of the transmitted symbols. Thus, the compensation method operates in an iterative fashion where symbol estimates are used for arriving at estimates of the spectral components of the phase noise process.

5.4.2.3   Channel Estimation Aspects in the Introduced Techniques

Both the introduced techniques, LI-CPE and LI-TE, rely on interpolation of phase noise estimates between adjacent OFDM symbols. However, if a channel estimation is done in the
conventional way, OFDM symbol by OFDM symbol, the CPE is also estimated at the same
time, and when the channel equalization is done, the CPE information is also lost. CPE information is vital for the LI-CPE technique and LI-TE also needs the CPE information to work
optimally, but it still manages to cut the bad peaks in the ICI estimates at the symbol boundaries even when the CPE information is lost. Here, a channel estimation technique that retains
the CPE information is reviewed. It works with a quasistatic channel and was first proposed
in Reference [39].

In the reviewed channel estimation algorithm, channel estimates for the pilot subcarriers
are first needed. These can be attained, for example, by using the zero forcing principle as

$$\hat{\mathbf{H}}_{m,pilots} = \mathbf{R}_{m,pilots} \circ /\mathbf{P}_m \tag{5.49}$$

Here, $\circ/$ is a point-by-point division operator, $\hat{\mathbf{H}}_{m,pilots}$ is a $(P \times 1)$ vector (P is the number of
pilots) consisting of the estimate of the channel frequency responses for the pilot subcarriers,
which are in $(P \times 1)$ vector $\mathbf{P}_m$ and $\mathbf{R}_{m,pilots}$ is a $(P \times 1)$ vector of the received subcarrier
symbols corresponding to the pilot subcarriers. Now if we assume that the channel is quasi-
static for the duration of K OFDM symbols, we know that, without CPE, the partial channel
estimates $\hat{\mathbf{H}}_{m,pilots}$ should be the same for the current and adjacent $K - 1$ OFDM symbols.
Using this, it is trivial to estimate the relative CPE from the partial channel estimates. The
relative CPE in the lth OFDM symbol with respect to the CPE in the mth OFDM symbol
(we assume that the relative CPE is nonexistent in the m th OFDM symbol) can then be
written as

$$\hat{\mathbf{J}}_{l,0,rel} = \hat{\mathbf{H}}_{l,pilots} \circ /\hat{\mathbf{H}}_{m,pilots} \tag{5.50}$$

As stated before, this is the relative CPE. We are actually not interested in the absolute CPE
value before channel equalization, but we are indeed interested in the relative CPE value.
This is because the ICI estimation methods LI-CPE and LI-TE rely on the relative CPE information. We now have P estimates of the same CPE for each OFDM symbol, so to get the
final CPE estimate for the m th OFDM symbol we can take the mean of the all CPE estimates
within one OFDM symbol as

$$\hat{J}_{0,rel}(m) = \overline{\hat{\mathbf{J}}}_{m,0,rel} \tag{5.51}$$

where $\bar{\mathbf{x}}$ denotes taking the mean of the elements of vector $\hat{\mathbf{J}}$. The CPE can then be taken
out from the channel estimates before the channel equalization to retain the CPE infor-
mation in the OFDM symbols. This also allows averaging of the channel estimated
within the quasistatic windows. Naturally if the channel can be assumed to be static during the K OFDM symbols, it is beneficial to average the channel estimates. However, this
would not be possible if the CPE is not removed first from the channel estimates. After
averaging, the partial channel estimates can be used in any way seen fit for channel

estimation. Here, the total channel estimate is constructed by using linear interpolation between the adjacent pilot subcarriers.

It should be noted that for LI-CPE and LI-TE algorithms, the minimum quasistatic case ($K = 3$) is already sufficient to make them work very well.

5.4.2.4 Iterative Time-Domain Phase Noise Mitigation Algorithm

The time-domain algorithm (called here the Syrjälä algorithm), presented next, was first proposed in Reference [7] and is depicted in Figure 5.9. The idea is to use the time-domain signal, which is reconstructed from the detection results without phase noise, and compare that reconstructed signal to the time-domain signal, which still has phase noise present. Since the phase noise effect in the time domain is a simple multiplication by the time-varying complex exponential, estimating it is very simple.

Let us start from the received time-domain signal of Equation (5.26), which is corrupted by the noise oscillator at the transmitter, channel and the noisy oscillators at the receiver. Now, as already discussed, Equation (5.26) can be approximated as

$$\mathbf{r}_m \approx \mathbf{H}_m \operatorname{diag}\left(e^{j\Phi_m}\right)\mathbf{x}_m + \mathbf{z}_m \tag{5.52}$$

After DFT, channel equalization and CPE mitigation and IDFT, Equation (5.52) can be written as

$$
\begin{aligned}
\mathbf{y}_m &\approx \left(\hat{\mathbf{H}}_{m,CPE}\right)^{-1}\left[\mathbf{H}_m \operatorname{diag}\left(e^{j\Phi_m}\right)\mathbf{x}_m + \mathbf{z}_m\right] \\
&\approx \operatorname{diag}\left(e^{j\left(\Phi_m - \hat{\phi}_{m,CPE}\right)}\right)\mathbf{x}_m + \left(\hat{\mathbf{H}}_{m,CPE}\right)^{-1}\mathbf{z}_m
\end{aligned}
\tag{5.53}
$$

Here, $\hat{\mathbf{H}}_{m,CPE}$ is the ($N \times N$) convolution matrix of channel estimates, which also includes the CPE estimate $\hat{\phi}_{m,CPE}$. Already before the IDFT we are able to do symbol detection in the

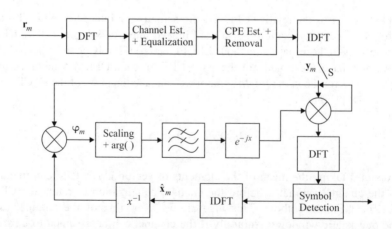

Figure 5.9 The first introduced time-domain phase noise mitigation algorithm (Syrjälä). The switch S passes through the OFDM symbol and then opens. It is open until all the iterations have been completed and the next OFDM symbol is taken in. © 2011 IEEE. Printed, with permission, from [7].

first iteration, so Figure 5.9 is a little misleading on this part. After the symbols are detected, an estimate of the time-domain OFDM waveform $\hat{\mathbf{x}}_m$ is reconstructed from the detected symbols with IDFT. This symbol is approximately the input signal of the iterative part of the algorithm $\mathbf{y}_m$ without the phase noise and additive channel noise effect. The signal $\mathbf{y}_m$ is then divided by the signal $\hat{\mathbf{x}}_m$, and the result is a very crude estimate of the phase noise term, given as

$$
\begin{aligned}
\boldsymbol{\varphi}_m &\approx \operatorname{diag}^{-1}(\hat{\mathbf{x}}_m)\left[\operatorname{diag}\left(e^{j\left(\boldsymbol{\Phi}_m-\hat{\phi}_{m,CPE}\right)}\right)\mathbf{x}_m + \left(\hat{\mathbf{H}}_{m,CPE}\right)^{-1}\mathbf{z}_m\right] \\
&\approx \operatorname{diag}\left(e^{j\left(\boldsymbol{\Phi}_m-\hat{\phi}_{m,CPE}\right)}\right) + \operatorname{diag}^{-1}(\hat{\mathbf{x}}_m)\left(\hat{\mathbf{H}}_{m,CPE}\right)^{-1}\mathbf{z}_m
\end{aligned}
\tag{5.54}
$$

Here we need reciprocals of the elements in vector $\hat{\mathbf{x}}_m$, which can however have zero elements. In the case of a zero element, the resulting value from the inverse operation is set to zero. This selection is done because it forces the estimation algorithm to ignore zero values in $\hat{\mathbf{x}}_m$. If $\mathbf{x}_m$ had zero values, they would anyway be affected very heavily by the additive noise, so it is better to ignore them in the estimation process. At this point, we use the knowledge that the phase noise complex exponential and the actual phase noise sequences are steep low-pass processes, so we filter signal $\boldsymbol{\varphi}_m$ to improve the estimate. Prior to filtering, however, we scale the signal and take its argument. The scaling is done to give more weight to the samples that probably have more amplitude. This is beneficial because the additive noise affects the high-amplitude samples very heavily and has only a mild effect on low-amplitude samples. The dynamics of the signal is high because the OFDM signal is well known to have a high peak-to-average power ratio. The scaling is done according to the amplitudes of the reconstructed signal as

$$
\mathbf{q}_m = \frac{\sqrt{2}|\hat{\mathbf{x}}_m|^2}{N_a/N}
\tag{5.55}
$$

This is an $(N \times 1)$ vector of scaling factors, where $|\hat{\mathbf{x}}_m|^2$ is a vector consisting of squared absolute values of the elements of the vector $\hat{\mathbf{x}}_m$ and N_a is the number of active subcarriers. With this scaling, it is assumed that each active subcarrier has unit average power and that the channel power response is also unity. The scaling indeed weights the high-amplitude signal samples exponentially. After scaling, taking an argument and low-pass filtering, the estimate of the phase noise (without the estimated part of the CPE) can be written as

$$
\operatorname{LPF}\{\operatorname{diag}(\mathbf{q})\arg(\boldsymbol{\varphi}_m)\} \approx \boldsymbol{\Phi}_m - \hat{\phi}_{m,CPE}
\tag{5.56}
$$

Here, the $\arg(\cdot)$ function gives the argument of a complex exponential. This is then used as an argument of the inverse complex exponential function, and the time-domain input signal of the algorithm $\mathbf{y}_m$ is sample by sample multiplied by this. The result is then discrete Fourier transformed and the symbols are detected. The algorithm can be used iteratively by using these now much improved symbol detection results as a basis for reconstructing a new time-domain reference signal $\hat{\mathbf{x}}_m$.

5.4.3 Performance of the Algorithms

5.4.3.1 Parameters of the Simulation and the Simulation Setup

To compare the performance of the presented phase noise mitigation algorithms, symbol error rate (SER) simulations are run with the following simulation setup. First, a 3GPP-LTE downlink type OFDM signal with 1024 subcarriers is created with 15 kHz subcarrier spacing, 300 active 16 QAM modulated subcarriers on the both sides of the centre subcarrier are created and null carriers are added so that the total of 1024 subcarriers are in place. After the signal creation the cyclic prefix of length 63 samples is added to the OFDM symbols. The transmitter phase noise is then modelled to the signal. After this, the signal is put through a channel, either the additive white Gaussian noise (AWGN) channel or the extended ITU-R vehicular A multipath channel [40]. The multipath channel is assumed to be quasistatic for the duration of 12 OFDM symbols. After the channel, receiver phase noise is modelled and the OFDM signal is inverse discrete Fourier transformed. Then, the channel and CPE are estimated and equalized. In the channel estimation, three cases are simulated: (1) perfect channel information, (2) a traditional pilot subcarrier-based LS algorithm with linear interpolation to estimate the missing channel frequency response and (3) the advanced channel estimation scheme proposed in Reference [39]. For the perfect channel information case, only 16 subcarriers are considered as pilots (for CPE estimation), and for practical channel estimation cases 66 pilots are used. The channel and CPE equalization is then followed by the ICI mitigation algorithms presented and proposed.

All the iterative ICI mitigation algorithms are iterated three times. For the Petrovic algorithm [6], and thus for Bittner [5] and LI-TE algorithms [8, 10, 39], the parameter $u = 3$. For the Bittner algorithm 70 samples and for LI-TE algorithms 155 samples at both edges of the OFDM symbol are used for tail estimation. For the Syrjälä algorithm [7, 10], the used low-pass filters to separate the phase noise estimate from the noise are designed with the Remez algorithm and are of the order of 200 and 350 for the AWGN channel and extended ITU-R vehicular A channel, respectively.

5.4.3.2 Analysis of Results of Simulation

In Figures 5.10 and 5.11, the simulation results are presented for all the presented phase noise mitigation algorithms in the additive white Gaussian noise channel case. Clearly the performance of the Syrjälä algorithm is superior to that of the other algorithms over the simulated received SNR and phase noise 3-dB bandwidth regions. Also the LI-TE technique does a good job in phase noise mitigation. In Figures 5.12 and 5.13, the performances of two of the best algorithms are compared in the extended ITU-R vehicular A multipath channel case with different levels of prior channel knowledge. Only two of the best algorithms were selected because otherwise the figures would have been too busy. Furthermore, the relative performance difference is the same compared to the AWGN case, with one exception: the LI-TE and LI-CPE methods have very good gain when using the proposed advanced channel estimation scheme. As depicted in the figures, LI-TE even outperforms the Syrjälä algorithm in the low-phase noise region when advanced channel estimation is used. However, the Syrjälä algorithm then regains its place as the best performing algorithm again when the phase noise error starts to dominate at higher phase noise regions.

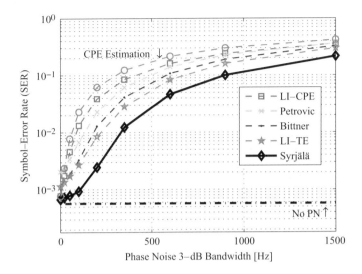

Figure 5.10 Simulated SER versus phase noise 3-dB bandwidth. The AWGN channel with a fixed received SNR of 18 dB.

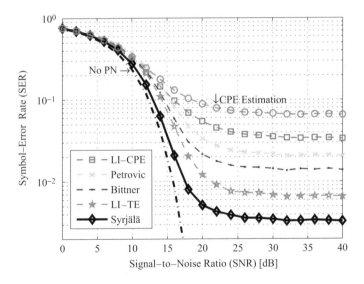

Figure 5.11 Simulated SER versus received SNR bandwidth. The AWGN channel and a fixed phase noise 3-dB bandwidth of 300 Hz.

5.5 Conclusions

Phase noise has a serious effect on performance of OFDM transceivers. In this chapter, the effect was analysed in terms of the SINR and capacity. The SINR analysis was done for arbitrary oscillators, and the result was a very simple formula that can be used in the design of an oscillator used in OFDM receivers. The performance of OFDM systems impaired by

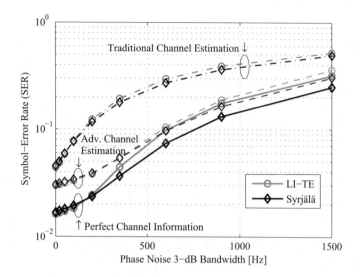

Figure 5.12 Simulated SER versus phase noise 3-dB bandwidth. The extended ITU-R vehicular A channel with a fixed received SNR of 26 dB. Three different channel estimation approaches are used.

phase noise was also characterized in terms of the link capacity. Analytical expressions of average capacity of an OFDM radio link impaired by phase noise were presented. For a given signal-to-noise ratio, the capacity decreases, in a nonlinear fashion, as the phase noise 3-dB bandwidth increases. Also, phase noise mitigation algorithms for the OFDM signal were presented and their performances were compared to each other. The phase noise mitigation

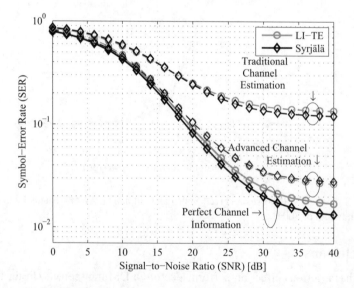

Figure 5.13 Simulated SER versus received SNR bandwidth. The extended ITU-R vehicular A channel and a fixed phase-noise 3-dB bandwidth of 300 Hz. Three different channel estimation approaches are used.

algorithms, which also consider the intercarrier-interference effect provide a significant increase in performance when compared to algorithms considering only the common phase error effect.

Acknowledgements

The authors of this chapter would like to express their warmest appreciation to the following entities, which in means of funding have made part of the work presented in this chapter possible: Tampere University of Technology Graduate School, Jenny and Antti Wihuri Foundation, HPY Research Foundation, Finnish Foundation for Technology Promotion, Yrjö and Tuula Neuvo Foundation, Ulla Tuominen Foundation, the Centre of Excellence in Smart Radios and Wireless Research (SMARAD), the Graduate School in Electronics, Telecommunication and Automation (GETA), the Academy of Finland, the Finnish Funding Agency for Technology and Innovation and EUREKA CELTIC E!3187 B21C-Broadcasting for the twenty-first century. The authors would like to thank DSc (Tech.) Stefan Werner and MSc (Tech.) Taneli Riihonen of Aalto University, Department of Signal Processing and Acoustics, and Lic. (Tech.) Jukka Rinne and MSc (Tech.) Nikolay N. Tchamov, Department of Communications Engineering, Tampere University of Technology, for their valuable suggestions.

References

1. Bagheri, R. *et al.* (2008) Software-defined radio receiver: dream to reality. *IEEE Communications Magazine*, **44** (8), 111–118.
2. Fettweis, G., Löhning, M., Petrovic, D. *et al.* (2005) Dirty-RF: a new paradigm. Proceedings of the International Symposium on Personal, Indoor and Mobile Radio Communications 2005 (PIMRC'05), Berlin, Germany, September 2005, vol. 4, pp. 2347–2355.
3. Valkama, M. (December 2010) RF impairment compensation for future radio systems, Chapter 15, in *Multi-Mode/Multi-Band RF Transceivers for Wireless Communications: Advanced Techniques, Architectures, and Trends* (eds G. Hueber and R.B. Staszewski), John Wiley & Sons, Ltd, UK.
4. Valkama, M., Springer, A. and Hueber, G. (2010) Digital signal processing for reducing the effects of RF imperfections in radio devices – an overview. Proceedings of the IEEE International Symposium on Circuits and Systems (ISCAS'10), Paris, France, May–June 2010, pp. 813–816.
5. Bittner, S., Zimmermann, E. and Fettweis, G. (2008) Exploiting phase noise properties in the design of MIMO-OFDM receivers. Proceedings of the IEEE Wireless Communications and Networking Conference 2008 (WCNC'08), Las Vegas, NV, March 2008, pp. 940–945.
6. Petrovic, D., Rave, W. and Fettweis, G. (2007) Effect of phase noise on OFDM systems with and without PLL: characterization and compensation. *IEEE Transactions Communications*, **55** (8), 1607–1616.
7. Syrjälä, V. and Valkama, M. (2011) Receiver DSP for OFDM systems impaired by transmitter and receiver phase noise. Proceedings of the IEEE International Conference on Communications 2011 (IEEE ICC'11), Kyoto, Japan, June 2011.
8. Syrjälä, V., Valkama, M., Tchamov, N.N. and Rinne, J. (2009) Phase noise modelling and mitigation techniques in OFDM communications systems. Proceedings of the Wireless Telecommunications Symposium 2009 (WTS'09), IEEE, Prague, Czech Republic, April 2009.
9. Mathecken, P., Riihonen, T., Werner, S. and Wichman, R. (2011) Performance analysis of OFDM with Wiener phase noise and frequency selective fading channel. *IEEE Transactions on Communications*, **59** (5), 1321–1331.
10. Syrjälä, V. (June 2012) Analysis and Mitigation of Oscillator Impairments in Modern Receiver Architectures. DSc (Tech.) Thesis, Department of Communications Engineering, Tampere University of Technology.
11. Schenk, T. (2006) RF Impairments in Multiple Antenna OFDM: Influence and Mitigation. PhD Dissertation, Technische Universiteit Eindhoven, 291 pp. ISBN: 90-386-1913-8.

12. Tchamov, N.N., Rinne, J., Syrjälä, V. *et al.* (2009) VCO phase noise trade-offs in PLL design for DVB-T/H receivers. Proceedings of the IEEE International Conference on Electronics, Circuits and Systems 2009 (ICE-CS'09), Yasmine Hammamet, Tunisia, December 2009, pp. 527–530.
13. Brownlee, M., Hanumolu, P.K., Mayaram, K. and Moon, U. (2006) A 0.5-GHz to 2.5-GHz PLL with fully differential supply regulated tuning. *IEEE Journal of Solid-State Circuits*, **41** (12), 2720–2728.
14. Demir, A. (2002) Phase noise and timing jitter in oscillators with colored-noise sources. *IEEE Transactions on Circuits and Systems I: Fundamental Theory and Applications*, **49** (12), 1782–1791.
15. Tchamov, N.N., Syrjälä, V., Rinne, J. *et al.* (2012) System- and circuit-level optimization of PLL designs for DVB-T/H receivers. *Analog Integrated Circuits and Signal Processing Journal*. DOI: 10.1007/s10470-011-9823-2.
16. Demir, A. (2006) Computing timing jitter from phase noise spectra for oscillators and phase-locked loops with white and $1/f$ noise. *IEEE Transactions on Circuits and Systems – I: Regular Papers*, **53** (9), 1869–1884.
17. Syrjälä, V., Valkama, M., Zou, Y. *et al.* (2011) On OFDM link performance under receiver phase noise with arbitrary spectral shape. Proceedings of the IEEE Wireless Communications and Networking Conference 2011 (IEEE WCNC'11), Cancun, Quintana-Roo, Mexico, March 2011.
18. Armada, A. and Calvo, M. (1998) Phase noise and sub-carrier spacing effects on the performance of an OFDM communication system. *IEEE Communications Letters*, **2** (1), 11–13.
19. Robertson, P. and Kaiser, S. (1995) Analysis of the effects of phase noise in orthogonal frequency division multiplex (OFDM) systems. Proceedings of the IEEE International Conference on Communications 1995 (ICC'95), Seattle, WA, June 1995, vol. 3, pp. 1652–1657.
20. Tomba, L. (1998) On the effect of Wiener phase noise in OFDM systems. *IEEE Transactions on Communications*, **46** (5), 580–583.
21. Krondorf, M., Bittner, S. and Fettweis, G. (2008) Numerical performance evaluation for OFDM systems affected by phase noise and channel estimation errors. Proceedings of the Vehicular Technology Conference Fall 2008 (VTC'08-Fall), Calgary, Canada, September 2008.
22. Rutten, R., Breems, L.J. and van Veldhoven, R.H.M. (2008) Digital jitter-cancellation for narrowband signals. Proceedings of the IEEE International Symposium on Circuits and Systems 2008 (ISCAS 2008), Seattle, WA, May 2008, pp. 1444–1447.
23. Schenk, T., van der Hofstad, R. and Fledderus, E. (2007) Distribution of the ICI term in phase noise impaired OFDM systems. *IEEE Transactions on Wireless Communications*, **6** (4), 1488–1500.
24. Yih, C.-H. (2008) BER analysis of OFDM systems impaired by phase noise in frequency-selective Rayleigh fading channels. Proceedings of the Global Telecommunications Conference 2008 (GLOBECOM'08), New Orleans, LA, December 2008.
25. Goldsmith, A. (2005) *Wireless Communication*, Cambridge University Press, p. 672, ISBN: 978-0521837163.
26. Mathecken, P., Riihonen, T., Tchamov, N.N. *et al.* (2012) Characterization of OFDM radio link under PLL-based oscillator phase noise and multipath fading channel. *IEEE Transactions on Communications*, **99**, 1–8.
27. Wu, S. and Bar-Ness, Y. (2004) OFDM systems in the presence of phase noise: consequences and solutions. *IEEE Transactions on Communications*, **52** (11), 1988–1997.
28. Bittner, S., Rave, W. and Fettweis, G. (2007) Joint iterative transmitter and receiver phase noise correction using soft information. Proceedings of the IEEE International Conference on Communications 2007 (ICC'07), Glasgow, Scotland, June 2007, pp. 2847–2852.
29. Wu, S. and Bar-Ness, Y. (2002) A phase noise suppression algorithm for OFDM-based WLANs. *IEEE Communications Letters*, **6** (12), 535–537.
30. Schenk, T., Tao, X.-J., Smulders, P. and Fledderus, E. (2004) Influence and suppression of phase noise in multi-antenna OFDM. Proceedings of the IEEE Vehicular Technology Conference 2004 Fall (VTC'04-Fall), Los Angeles, CA, September 2004, pp. 1443–1447.
31. Bittner, S., Frotzscher, A., Fettweis, G. and Deng, E. (2009) Oscillator phase noise compensation using Kalman tracking. Proceedings of the International Conference on Acoustics, Speech and Signal Processing 2009 (ICASSP'09), Taipei, Taiwan, April 2009, pp. 2529–2532.
32. Bittner, S., Zimmermann, E. and Fettweis, G. (2007) Iterative phase noise mitigation in MIMO-OFDM systems with pilot aided channel estimation. Proceedings of the Vehicular Technology Conference 2007 Fall (VTC'07-Fall), Baltimore, MD, September 2007, pp. 1087–1091.
33. Bhatti, J., Noels, N. and Moeneclaey, M. (2012) Phase noise estimation and compensation for OFDM systems: a DCT-based approach. Proceedings of the International Symposium on Spread Spectrum Techniques and Applications 2010, Taichung, Taiwan, October 2012, pp. 93–97.

34. Casas, R., Biracree, S. and Youtz, A. (2002) Time domain phase noise correction for OFDM signals. *IEEE Transactions on Broadcasting*, **48** (3), 230–236.

35. Corvaja, R. and Armada, A. (2009) Join channel and phase noise compensation for OFDM in fast-fading multipath applications. *IEEE Transactions on Vehicular Technology*, **58** (2), 636–643.

36. Mathecken, P., Riihonen, T., Werner, S. and Wichman, R. (2011) Accurate characterization and compensation of phase noise in OFDM receivers. 45th Annual Asilomar Conference on Signals, Systems, and Computers (ACSSC), Pacific Grove, California, November 2011.

37. Rinne, J. and Renfors, M. (1996) An equalizations method for orthogonal frequency division multiplexing systems in channels with multipath propagation, frequency offset and phase noise. Proceedings of the Global Telecommunications Conference 1996 (GLOBECOM'96), London, UK, November 1996, pp. 1442–1446.

38. Zou, Q., Tarighat, A. and Sayed, A.H. (2007) Compensation of phase noise in OFDM wireless systems. *IEEE Transactions on Signal Processing*, **55** (11), 88.

39. Syrjälä, V. and Valkama, M. (2010) Analysis and mitigation of phase noise and sampling jitter in OFDM radio receivers. *International Journal of Microwave and Wireless Technologies*, **2** (2), 193–202.

40. Sorensen, T.B., Mogersen, P.E. and Frederiksen, F. (2005) Extension of the ITU channel models for wideband (OFDM) systems. Proceedings of the IEEE Vehicular Technology Conference Fall 2005 (VTC'05-Fall), Dallas, TX, September 2005, pp. 392–396.

41. Abidi, A. (2006) Evolution of a software-defined radio receiver's RF front-end. Proceedings of the Radio Frequency Integrated Circuits (RFIC) Symposium 2006, San Francisco, CA, June 2006.

42. Arkensteijn, V.J., Klumperink, E.A.M. and Nauta, B. (2006) Jitter requirements of the sampling clock in software radio receivers. *IEEE Transactions on Circuits and Systems – II: Express Briefs*, **53** (2), 90–94.

43. Carley, L.K. and Mukherjee, T. (1995) High-speed low-power integrating CMOS sample-and-hold amplifier architecture. Proceedings of the Custom Integrated Circuits Conference, Santa Clara, CA, May 1995, pp. 543–546.

44. Ho, Y.-C., Staszewski, R.B., Muhammad, K. *et al.* (2006) Charge-domain signal processing of direct RF sampling mixer with discrete-time filter in Bluetooth and GSM receivers. *EURASIP Journal on Wireless Communications and Networking*, **2006**, 1–14.

45. Karvonen, S., Riley, T. and Kostamovaara, J. (2003) On the effect of timing jitter in charge sampling. Proceedings of the International Symposium on Circuits and Systems 2003 (ISCAS'03), Bangkok, Thailand, May 2003, vol. 1, pp. 737–740.

46. Mathecken, P. (March 2011) Performance Analysis of OFDM with Wiener Phase Noise and Frequency Selective Fading Channel. MSc Thesis, School of Science and Technology, Faculty of Electronics, Communications and Automations, Department of Signal Processing and Acoustics, Aalto University.

47. Muhammad, K. *et al.* (2004) A discrete-time Bluetooth receiver in a 0.13 μm digital CMOS process. Proceedings of the International Solid-State Circuits Conference, San Francisco, CA, February 2004.

48. Muhammad, K., Staszewski, R.B. and Leipold, D. (2005) Digital RF processing: towards low-cost reconfigurable radios. *IEEE Communications Magazine*, **43** (8), 105–113.

49. Onunkwo, U., Li, Y. and Swami, A. (2006) Effect of timing jitter on OFDM-based UWB systems. *IEEE Journal on Selected Areas in Communications*, **24** (4), 787–793.

50. Razavi, B. (1997) Design considerations for direct-conversion receivers. *IEEE Transactions on Circuits and Systems – II: Analog and Digital Signal Processing*, **44** (6), 428–435.

51. Shinagawa, M., Akazawa, Y. and Wakimoto, T. (1990) Jitter analysis of high-speed sampling systems. *IEEE Journal of Solid-State Circuits*, **25** (1), 220–224.

6

A Pragmatic Approach to Cooperative Positioning in Wireless Sensor Networks

Albert Bel Pereira, José López Vicario and Gonzalo Seco-Granados
Department of Telecommunications and Systems, Universitat Autònoma de Barcelona, Barcelona, Spain

6.1 Introduction

Location estimation in wireless sensor networks has become an important field of interest from researchers [1–3]. This is due to the demand for knowledge of the node position by the majority of applications. In environmental monitoring, such as fire or agriculture control, a basic premise to give sense to all the data measured is to know their location; if not known, data could be considered as meaningless information.

The main purpose of a localization algorithm is to estimate those positions of nodes with unknown coordinates from the following information: a priori knowledge of some node positions and intersensor measurements. Hence, the majority of existing localization methods applied in a wireless sensor network (WSN) tries to achieve the best accuracy considering the restrictions that this kind of network imposes.

Although nowadays there are many methods of localization in wireless networks, such as the global positioning system (GPS), a localization method suitable to be used in a WSN must take into account the resource constraints imposed by the nodes, such as energy consumption and the costs of transmission and computing hardware. The increase in terms of size and cost of energy required by the GPS hardware makes this method unsuitable to be applied in WSN. It could only be used to obtain a priori knowledge of the positions of reference nodes that are present in the network.

Microwave and Millimeter Wave Circuits and Systems: Emerging Design, Technologies, and Applications,
First Edition. Edited by Apostolos Georgiadis, Hendrik Rogier, Luca Roselli, and Paolo Arcioni.
© 2013 John Wiley & Sons, Ltd. Published 2013 by John Wiley & Sons, Ltd.

The objective of this chapter is to provide a review on cooperative schemes for WSN. Among all of them, special emphasis is given to received signal strength (RSS)-based techniques as these provide suitable solutions for practical implementation and take care of the restrictions in terms of cost and complexity. Since the accuracy of RSS methods depends on the suitability of the propagation models, cooperative localization algorithms that dynamically estimate the path loss exponent are also described. The chapter is also interested in the cost, in terms of energy consumption, that a localization algorithm has. In that sense, in order to achieve a trade-off between energy consumption and accuracy, a node selection criterion is proposed. In addition, practical examples based on real WSN deployments are presented.

The rest of the chapter is organized as follows. In Section 6.2 the two main categories of internode measurements, range-free and range-based, are presented. Furthermore, a first classification of the localization algorithms between cooperative and noncooperative is also presented. In Section 6.3 several localizations algorithms, divided between centralized and distributed, are presented. In Section 6.4 the two parts that compose an RSS-based algorithm, measurement and location update phases, are presented. Later, in Section 6.5 the different node selection mechanisms existing in the literature are presented. Furthermore, two different methods are proposed and an energy model is presented. In addition, the localization algorithm with joint node selection and path loss exponent estimation is presented. Finally, Sections 6.6 and 6.7 present, respectively, the simulation and experimental results obtained from the algorithm proposed.

6.2 Localization in Wireless Sensor Networks

Nowadays, the existing algorithms are able to locate nodes inside a WSN. Choosing between the different approaches depends on requirements that the final application demands. In order to differentiate between the great number of existing methods different classifications have been created.

The first classification divides existing methods into two main categories: range-based and range-free approaches. These approaches differ in the way of obtaining internode distances.

In the context of a WSN, a second classification is cooperative versus noncooperative algorithms. These classifications will be presented in the following sections.

6.2.1 Range-Free Methods

Range-free methods are based on connectivity information. These are the simplest measurements that an algorithm can do. The basic idea is to decide if a node is connected to an adjacent node or not. The connectivity information is usually obtained from RSS measurements. Considering perfect circle radio coverage, any node that receives an RSS above a threshold is supposed to be connected to the receiving node. Also a node can be considered inside the coverage range of another node by measuring the number of received packets. The commonest algorithms based on connectivity measurements are centroid, DV-Hop or APIT [4–6].

6.2.1.1 Centroid

One of the simplest methods used is the centroid. The mean of the coordinates of all anchors (nodes with known location) becomes the position estimation of each listening node. A

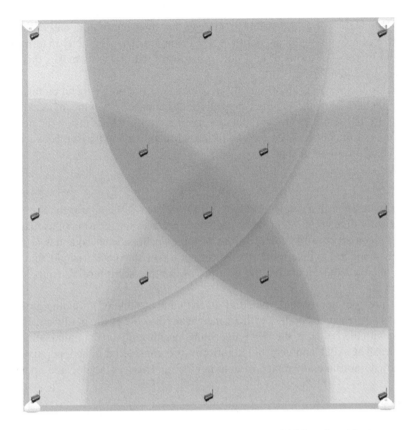

Figure 6.1 Centroid method with the different possibilities of positioning.

possible improvement could be the introduction of weights in order to achieve higher accuracy. An application of this method in a WSN is presented in Reference [4]. The authors considered an unconstrained outdoor environment. The results presented are for a network formed by four anchor nodes with an ideal spherical transmission range (see Figure 6.1). In that case an average localization error of 1.83 m is obtained. The major problem of this method, applied to a WSN, is that the accuracy is directly related to the number of anchor nodes. It could be increased if the number of overlapped reference nodes is increased. Nevertheless, the inclusion of more anchor nodes increases the cost of the network.

6.2.1.2 DV-Hop

The distance vector (DV)-Hop algorithm [6] is another method that is based on connectivity measurements. The DV-Hop is more complex than the centroid but more accurate estimations are achieved. The algorithm starts with a broadcast message sent by all the anchors. Nodes in the network count how many hops exist between them and all the anchor nodes inside the network. Each node selects the shortest path to every anchor node. Once having these hop distances, anchors broadcast the average distance per hop to neighbours (achieved when a message from an anchor is received by another anchor). With both measurements a trilateration or multilateration method is used to obtain the final location.

The DV-Hop algorithm is a distributed strategy that does not require any extra hardware and is capable of providing global coordinates to the nodes. The simplicity in terms of calculation and no increase in terms of cost due to hardware requirements makes this algorithm suitable to be used in a WSN.

Compared to the previous algorithm, the accuracy does not only depend on the connectivity with the anchors. Now the accuracy depends on the connectivity with any node (anchor or not anchor). In that sense, reducing the necessity of having anchors inside the network will reduce the cost of the network. At the end of the subsection a comparison between the three range-free algorithms will be made.

6.2.1.3 APIT

The approximate point in triangle (APIT) test [5] is an area-based approach. The APIT algorithm is basically divided into four steps: each node receives the location of as many anchors as possible; given all possible combinations of three anchors, each node has to form triangles; then the node has to determine whether or not it is within each triangle; and finally the position is obtained by calculating the centroid of the intersection of all triangles selected (see Figure 6.2).

As in the previous approaches, the APIT algorithm is capable of providing global coordinates (thanks to the anchor nodes) without increasing the cost of the hardware. However, in comparison with the DV-Hop, the APIT algorithm depends only on communications with the anchor nodes. The APIT and centroid algorithms are considered noncooperative methods, because they do not take advantage of communication between nonlocated nodes. Hence,

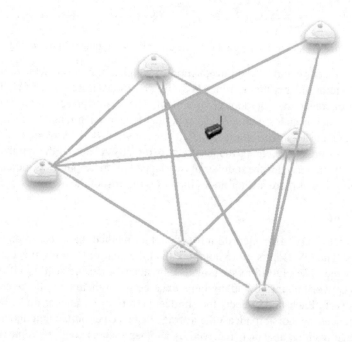

Figure 6.2   APIT estimation.

the accuracy of the solution will depend, in part, on the density of anchor nodes or having long-range transmission range anchor nodes able to be sensed at further distances.

6.2.1.4 Comparison of the Three Methods

A comparison of these three methods is presented in Reference [5]. Authors analyse the dependence of the accuracy of the three methods with different parameters such as anchor density, radio range ratio or nonlocated node density. The results reflect that no single algorithm can be used in all scenarios. The centroid algorithm has the largest localization error, but it is not dependent on features such as node density. Furthermore, this is the algorithm with less exchange of information and is simpler to implement. The DV-Hop algorithm requires a higher exchange of messages through the network and it is necessary to have a higher density of nodes in order to obtain good values of accuracy. The DV-Hop achieves a mean error equal to the radio range R. The APIT algorithm obtains similar results in terms of accuracy but nonlocated nodes have to have more anchor nodes inside their radio range. The parameter that badly affects all the algorithms is the anchor node range. As the transmission range of the anchors increases, the error also increases. The APIT algorithm achieves a mean error 0.75 times the radio range R.

6.2.2 Range-Based Methods

The range-based classification groups all the methods that estimate the internode distances or angles with the use of range information. Range information helps the algorithm to achieve a better accuracy in distance estimates than range-free approaches.

In this subsection the different signal metrics used to obtain distance estimates are presented. These distances are then used to determine node position.

6.2.2.1 Time of Arrival (TOA)

The first range-based measurement is the time of arrival (TOA) [7]. The TOA is based on measuring the difference between the sending time of a signal at the transmitter and the receiving time at the receiver. The major problem of this method is the possible lack of synchronization between nodes. Errors of about 2% are achieved over a communication range of 3–6 m [8]. Moreover, the node clocks resolution should be of the order of nanoseconds (in radio frequency (RF) 1 ns translates to 0.3 m [9]). Although it is a range measure usually used in wireless or satellite networks [10] (in which base stations and mobile nodes are synchronized), a recent trend uses time measurement approaches with ultra wideband signals (UWB) [11–13]. The UWB signal achieves a high accuracy because the transmitted pulses has a wide bandwidth and hence a very short pulse waveform. With the recovery of this transmitted pulse it is possible to estimate the distance between the receiver and transmitter.

6.2.2.2 Time Difference of Arrival (TDOA)

Another existing time measurement approach avoids the necessity of having an entire synchronized network. This method is known as the time difference of arrival (TDOA) and two different ideas are presented [14].

The first TDOA method [11] is based on the measurement of the difference between the arrival times of a signal sent by a transmitter at two receivers. This method assumes that the receiver locations are known and the two receivers are perfectly synchronized. For that reason, it is mostly used in cellular networks where the complexity of base stations is considerably relaxed.

Another method also based on a difference of arrival times is presented in References [15] and [16]. The basic idea is to eliminate the necessity of having a synchronized network, neither the senders nor the transmitters. In order to achieve its purpose the method uses a combination of two kinds of signal, for example RF and ultrasonic signals, which have different velocities of transmission. One transmitter sends two kinds of signal to a unique receiver. The time difference between the first and the second signal is used as an estimate of the one-way acoustic propagation time. In this case nodes require extra [16] hardware in order to be able to transmit and receive different signals.

The first case is more appropriate in cellular networks due to the fact that base stations have fewer requirements in terms of complexity or cost. It is possible to synchronize them; hence they could act as the receivers at the time of estimating the TDOA. On the other hand, the second TDOA approach could be more suitable to be used in a WSN. The major con is the necessity of including extra hardware in the entire network.

Both approaches obtain a good accuracy. In Reference [16] the results show an average error of the distance estimates between 29 and 8 cm. However, both methods increase the complexity of the network and the cost of nodes.

6.2.2.3 Round-Trip Time of Arrival (RTOA)

The round-trip time (RTT) method [17] avoids the synchronization constraint that the TOA or first TDOA methods impose, nor the hardware requirements of the second TDOA method. The measurement starts when a node A sends a packet to a node B. When node B receives the packet, it retransmits it to node A. At the end, node A receives the packet; hence, it can calculate the propagation time because the difference between the sending time and the receiving time at node A is twice the propagation time plus the processing time at node B (obtained from specifications or estimated at the calibration time).

Numerical results based on different experimental setups are presented in Reference [17]. The results show that RTT measurements give a root mean square (RMS) error between a minimum of 75 cm and a maximum of 2.51 m. The difference in accuracy compared to that achieved by TOA or TDOA measurements is remarkable. Furthermore, this technique needs a higher exchange of packets in order to estimate the internode distance.

6.2.2.4 Received Signal Strength Indicator (RSSI)

In this case, the distance between two nodes is obtained by using the measured power of the received signal [18]. The RSS is measured during normal transmission. Hence, this technique does not impose any extra requirement in terms of complexity or hardware. Compared to the AOA- or TOA-based approach, this technique has become the most inexpensive. However, distance estimates obtained through RSS measurements suffer from errors induced by shadowing and multipath effects. Usually, RSS-based distance estimations are based on the well-known radio propagation path loss model. This model assumes that the power decays proportionally to the distance $1/_{d^\alpha}$, where α is the path loss exponent. In order to include the shadowing effects the power received is modelled as a lognormal variable (Gaussian if it is expressed in dB), resulting in

$$P_{R_x}(dBm) = P_0(dBm) - 10\alpha\log_{10}(d) - \upsilon_i \tag{6.1}$$

where P_0 is the power received at a reference distance (usually 1 metre) and υ_i represents the shadowing effects modelled as a Gaussian with zero mean and variance $\sigma_{\upsilon_i}^2$ expressed in dB.

The adoption of lognormal modelling is usually motivated by experimental results such as those provided by References [19] and [20]. Some results in terms of accuracy are presented in Reference [21]. The authors carried out a measurement campaign by means of using TelosB motes. The average error achieved in the measured distances is 2.25 m, for a distance between 1 and 8 meters. The RSS-based estimates achieve worse accuracy compared to that achieved with time measurements. However, the major advantage is that RSS provides a lower complex solution. Moreover, distance estimates could achieve better accuracy if an accurate model is used.

6.2.2.5 Angle of Arrival (AOA)

Angle or direction of arrival, AOA and DOA [9], are those methods that rely on the measurements of the angle between senders and receivers. In Reference [9] two different ways of estimating distances are presented. The first method [22], which is the most common, estimates the angle of arrival by means of using a sensor array. Each sensor requires two or more sensors placed at a known location with respect to the centre of the node. The angle is estimated following the same approach as in a time-delay estimation. The measurement consists of two phases. At the first phase, anchors transmit their location and a short omnidirectional pulse. Then they transmit a beacon with a rotating radiation pattern. Taking advantage of beamforming techniques, the anchors are able to transmit directional pulses every T seconds and to change the direction of the signal by a constant angular step $\Delta\beta$. Sensors have to register the arrival time between the first omnidirectional signal and the time of arrival of the pulse with maximal beacon power. This difference in time (Δt) allows the sensor to estimate the angle of arrival as

$$\beta = \Delta\beta\frac{\Delta t}{T} \tag{6.2}$$

The accuracy achieved in Reference [22] is an average error of 2 m in a scenario with 6 anchors and 100 nonlocated nodes uniformly distributed in a 50 m $\times$ 50 m area. Increasing the anchors produces mean errors in the localization below 1.5 m.

In References [23] and [1], RSS measurements from directional antenna arrays on each node were also used to estimate arrival angles. Accuracy errors below 1 metre are achieved compared to those achieved with distance-based algorithms. However, the increase in cost and size of the nodes makes the AOA a more complex solution although it achieves good results in terms of localization accuracy.

6.2.2.6 Radio Interferometry

This technique [24] is based on exploiting interfering radio waves. Although it is also based in RSS measurements, the procedure of extracting the distance is more complex. The basis of radio interferometry is to utilize two transmitters (two reference nodes with known location) to create an interference signal and compare the phase offset at two receivers. By measuring this relative phase offset at different carrier frequencies, it is possible to obtain a linear combination of the distances between both transmitters and receivers. The accuracy obtained is considerably increased. Results achieved in Reference [24] show that more than 50% of the range measurements achieve accuracy lower than a quarter of the wavelength. In Reference [25] a tracking algorithm based on radio interferometry measurements is presented. The mean absolute error achieved with the mobile experiment is between 0.94 and 1.96 m. The error achieved with a stationary experiment is between 0.54 and 0.83 m. A high accuracy is achieved with the measurement of RSS signals (no extra hardware is required). On the other hand, the method presents some requirements, such as synchronization of some nodes and signal processing units able to estimate the carrier offset.

6.2.3   Cooperative versus Noncooperative

Once distance estimates are obtained the next step is to estimate node locations inside the network. This second phase, usually called the location-update phase, could also be classified as cooperative or noncooperative.

Noncooperative approaches estimate the nonlocated node positions by only considering the anchor node positions. In this case, the accuracy of these approaches mainly depends on the density of the anchor nodes or the usage of long-range transmission anchors. Using this kind of technique in large-scale networks could be critical. On the other hand, in small-scale networks, where nodes have a high probability of having direct communication with anchor nodes, the use of a noncooperative technique could become a great option.

In cooperative algorithms, there are no restrictions in the communication between any nodes inside the network. Nodes are able to obtain information from more nodes than only anchors. With this strategy the number of anchor nodes in the network can be reduced. Hence, cooperative localization can offer increased accuracy and coverage.

6.3   Cooperative Positioning

The cooperative algorithms have become an accurate approach for localization algorithms in low-cost and low-complex sensor networks. Cooperating with the entire, and not only with the anchor node, network can increase the accuracy, in terms of error, of the final position estimate. An important purpose is to achieve a good trade-off in terms of accuracy versus network complexity and cost, and the cooperative approaches are a kind of algorithm that could achieve a good trade-off. Furthermore, a cooperative algorithm is a scalable solution because it does not only depend on the number of anchors nodes; it also takes advantage of all the nodes inside the network (anchor or nonlocated).

Cooperative approaches could be classified into two categories: centralized versus distributed. Although this classification could also be applied to noncooperative algorithms, the discussion is here presented as differences among such alternatives, emphasized in a cooperative system.

Centralized approaches involve those algorithms in which one node becomes a central unit. This central unit is the one that has to recollect all data and it is the only one that computes all the position estimates. On the other hand, in a distribution algorithm each node is responsible for estimating its own position, relative to its neighbours.

In the following subsection different alternatives of both approaches are presented, discussing which are their strong points and when they are the appropriate algorithms to be used.

6.3.1 Centralized Algorithms

In some applications it is appropriate to implement a centralized WSN due to the nature of the application, for example environmental monitoring, where the information should be controlled in a central point. In this kind of application, a unique central processor controls all the data. In this situation, the use of centralized algorithms in order to locate the nodes inside the network should be useful.

On the one hand, centralized algorithms provide good accuracy, because the central node has information of the entire network. Moreover, these nodes are less restricted in terms of complexity and hence a more complex algorithm could be implemented. On the other hand, all the information collected at the network must be transmitted to the central node; hence the traffic is increased and the scalability is reduced.

6.3.1.1 Multidimensional Scaling

Multidimensional scaling (MDS) [1] was originally developed for use in mathematical psychology and has many variations. Although MDS is normally developed in a centralized fashion, distributed-based algorithms exist in the literature. The most usual approach is the MDS-MAP (maximum a posteriori) [26], which is a direct application of the simplest kind of multidimensional scaling: classic metric MDS. The basic idea is to arrange objects in a space of a certain number of dimensions trying to reproduce the dissimilarities observed in the objects. Adapting to a localization algorithm, the objects are the nodes and the dissimilarities are the distance estimates. By means of using the law of cosines and linear algebra the MDS is able to reconstruct the relative positions of the points based on the internode distances. The last step of an MDS algorithm is transforming the relative map obtained to an absolute map based on the knowledge of the absolute position of some anchors.

Clearly, MDS has potential in the sensor localization domain. It is possible to construct a relative map without knowing any absolute position. Results that depend on different features are shown in Reference [26]. They show that the accuracy of the MDS-MAP is highly dependent on the connectivity. In order to reduce the error, the algorithm needs a high density of nodes (a minimum number of 12 cooperating nodes). Another important point is the high accuracy achieved when range information is used instead of connectivity information. In conclusion, it is possible to achieve an error less than half the range radio of the nodes with a number of 12 cooperating nodes and 4 anchors in order to change from a relative to an absolute map. A major drawback is the necessity of having all the information in a central node. This problem can be reduced by means of using map-stitching techniques [27].

6.3.1.2 Semidefinite Programming

Semidefinite programming (SDP) is a subfield of the convex optimization. SDP basically consists in minimizing a linear function subject to the constraint that an affine combination of symmetric matrices is positive semidefinite. Such a constraint is nonlinear and non-smooth, but convex, so semidefinite programs are convex optimization problems. The major problem of applying these techniques is that the localization problem is a nonconvex problem. Hence, the basic idea of an SDP algorithm is to convert the nonconvex quadratic distance constraints into linear constraints by introducing a relaxation to remove the quadratic term in the formulation. Three different approaches applying SDP algorithms are presented in Reference [28]. The best result achieved is an error of 5% of the radio range. This accuracy is highly affected by the noise factor, although it is maintained below a value of 20% of the radius range having a value equal to or higher than 0.3. Moreover, when the size of the network increases the solution of a large SDP becomes more complex. This problem can be solved by means of dividing the network into several clusters, reducing the complexity of the entire network.

6.3.1.3 Maximum-Likelihood Estimation

The maximum-likelihood estimation (MLE) is a centralized localization algorithm [19]. As it occurs with MDS it can be solved in a distributed fashion. MLE is a popular statistical method used for fitting a statistical model to data and providing estimates for the model's parameters. One of its advantages is its asymptotic efficiency. This method applied to a WSN is developed in Reference [19]. Simulations carried out in a scenario with 40 nonlocated nodes achieve a root mean square of 2.1 m. Although it is possible to achieve a good accuracy two major problems appear when this method is used. The first is that if the method is not initialized with a good starting point it is possible not to achieve the global maxima. The second one is that if data measurements deviate from the statistical model assumed, the results obtained may not be optimal.

6.3.2   Distributed Algorithms

In the distributed algorithm each node processes all the data that it collects from the network. They themselves are responsible for estimating their own coordinates. This is possible because nodes share their position information. A distributed algorithm is usually considered more efficient, in terms of computational cost, and scalable. On the other hand, distributed algorithms have lower accuracy compared to that achieved with a centralized algorithm due to the fact that the calculus is done at the nodes, which have less computational capacities.

6.3.2.1   Lateration

Instead of using angles to estimate the position, lateration methods use distances. These methods compute the nonlocated node positions by using the estimated distances to reference nodes nearby. The node position is obtained by means of calculating the intersection point of the circles with radius equal to the estimated distances and centred at reference node positions (see Figure 6.4). This technique is significantly affected by the errors on the distance measurement (ranging). Once a nonlocated node estimates its position, it becomes a new reference node that will help the rest of the nodes, giving to nonlocated nodes the possibility of exchanging information with other nonlocated nodes. Hence a lateration method can be considered as a cooperative localization algorithm.

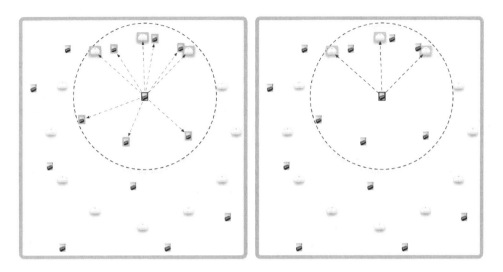

Figure 6.3   Cooperative and non-cooperative approaches

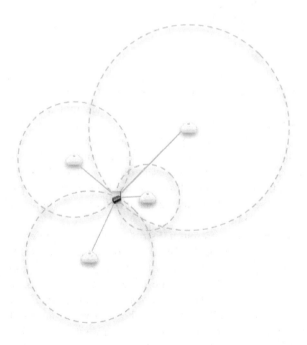

Figure 6.4   Lateration example.

6.3.2.2   Non-Bayesian Estimators

The non-Bayesian estimators [29] are one of the categories inside the distributed algorithms. The basic idea is the minimization of a cost function, such as LS (least squares) or ML (maximum likelihood), in a distributed way. Each node estimates its own position following a three-step algorithm. The procedure starts with the distribution of their coordinates to their neighbours. Then each node estimates the internode distance, by means of using range metrics, with those nodes from which the node has received their coordinates. Finally, nodes recalculate their position estimates by means of using both data. These steps are repeated until a convergence. The problem with these methods is the possibility of not converging to a global minimum. This fact is related to the use of a good starting point. If an initial point near to the final solution is not used the algorithm could not converge to a global solution.

The cost functions of both methods are

$$C_{LS}(x) = \sum_{i=1}^{N_2} \sum_{j \in S_i} \left\| z_{j \to i} - f\left(\mathbf{x}_i - \mathbf{x}_j\right) \right\|^2 C_{ML}(x) = \sum_{i=1}^{N_2} \sum_{j \in S_i} \left(\ln\left(\frac{z_{j \to i}}{f\left(\mathbf{x}_i - \mathbf{x}_j\right)} \right) \right)^2 \quad (6.3)$$

where $z_{j \to i}$ is the distance estimate between node i and node j and $f(\mathbf{x}_i - \mathbf{x}_j)$ is a function that estimates the distance between node i and j using the coordinates of them.

The major difference between both methods is that the ML approach takes advantage of the statistics of noise sources and the LS approach does not. A comparison between both methods is shown in Reference [30]; more concretely, the LS approach is a weighted one. All results presented show that the ML approach achieves better accuracy in terms of positioning errors. The mean error values obtained with the ML algorithm are between 1.2 and 1.5 m in comparison to that obtained with a WLS algorithm (values between 1.4 and 1.75 m).

6.3.2.3   Bayesian Estimators

Other methods that are used in the localization algorithms are those that are based on the Bayesian estimators. These methods have been firstly used in localization algorithms for robots [31]; nowadays many works have used them in sensor localization and tracking algorithms.

The basic idea is that, given sensors measurements z, what is the probability of being at position x ($p(x|z)$)? This posterior density over the random x conditioned to all received measurements z is usually called belief [25]. By means of computing the belief it is possible to obtain the position estimate. The authors present different algorithms of obtaining the position estimates using a Bayesian approach such as the Kalman filter or particle filter. Results show that both methods achieve good accuracy but in terms of complexity the particle filter implementation is the best option.

In Reference [29], the authors present a distributed approach called factor graphs (FGs), which is a successive refinement method used to estimate the probability density of sensor network parameters. Each sensor initializes their belief. Then, all nodes broadcast their beliefs to neighbours. Each node updates their belief with their own belief and the information extracted from the neighbour beliefs. Hence, nodes iteratively refine their position. Results obtained in Reference [29] show that the distributed approach achieves similar results obtained with a centralized approach. Results presented show that 90% of the nodes achieve an error below 1 m, as compared to the results achieved by the non-Bayesian LS algorithm in which only the 40% achieve error values below 1 m.

6.4 RSS-Based Cooperative Positioning

Several methods have been presented in previous sections. The first classification is between range-free and range-based methods. Both methods are suitable to be used in a localization algorithm for a WSN. However, the choice between one or another is based on the requirements of the final application. In order to achieve a good trade-off between accuracy and cost one of the most appropriate options is an RSS-based range measurement. RSS-based methods present less accurate measurement results but they are the simplest methods that could be applied in a WSN.

A cooperative approach is most suitable for large-scale networks, compared to a noncooperative approach. The accuracy of the noncooperative methods is highly dependent on the density of the anchor nodes. Also a high density of neighbour nodes, obtained with a cooperative approach, gives more robustness to the localization algorithm.

Finally, the last choice is between a centralized and a distributed algorithm. As discussed before, centralized approaches give a higher accuracy due to the possibility of developing a more complex algorithm. Distributed approaches need to be able to be computed in each node so they have to be as simple as possible. On the other hand, a centralized approach needs a higher traffic exchange, because all data has to be sent to the central node, which limits the capability of scaling the network.

In order to start the discussion of an RSS-based cooperative algorithm let us consider a wireless sensor network with N nodes. There are N_1 nodes, whose exact locations are known (anchor nodes). The rest of the nodes, $N_2 = N - N_1$, do not know their position (nonlocated nodes). Those algorithms are normally divided into two steps. The first one is the measurement phase in which the algorithm uses some range measurement in order to obtain distance estimates. The second one is the location update phase, in which by means of using the estimates obtained at the first phase and the node state information, the algorithm computes the position estimates.

6.4.1 Measurement Phase

As commented above, this chapter focuses on the RSS-based cooperative approach. In this kind of algorithm the first phase of the algorithm consists in obtaining internode distances, in this case by means of RSS measurements.

RSS-based measurements are a very attractive ranging method for practical implementation because they do not need extra hardware to be measured. The most common sources of error that affect RSS-based distance estimations are shadowing and multipath signals, which complicate the modelling of the channel that nodes need to know a priori. Usually, RSS measurements are modelled through the well-known radio-propagation path loss and shadowing model [19]. Received power is modelled as a lognormal distributed random variable with a distance-dependent mean. Hence, power received in node j from a signal transmitted by node i, P_{ij}, is expressed as

$$RSS_{ij} = P_{ij} = P_0 = 10\alpha_{ij} \log_{10} d_{ij} - v_{ij} \tag{6.4}$$

where P_0 is the power received in dBm at 1 m distance, d_{ij} is the distance between nodes i and j in metres, parameter α_{ij} is the path loss exponent, that is the rate at which the power decreases with distance, and $v_{ij} \sim N(0, \sigma_v^2)$ represents lognormal shadow-fading effects, where the value of the standard deviation σ_v depends on the characteristics of the environment. The small-scale fading effects are diminished [19] by time averaging; hence they do not affect the distribution of v_{ij}. Since static scenarios are considered, the major sources of error are shadowing and path loss.

In Reference [20], the authors discuss that the lognormal distribution is often used to explain the large-scale variations of the signal amplitudes in multipath fading environments. References inside Reference [20] present the validity of this model for modelling an indoor radio channel. Some results show that the lognormal fits better than the Rayleigh model. Furthermore, large-scale variations of data collected at 900 MHz, 1800 MHz and 2.3 GHz for transmission into and within buildings were found to be lognormal.

Given the received power RSS_{ij} in Equation (6.4), the density of P_{ij} is [32]

$$f_{P|\gamma}(P_{ij}|\gamma) = \frac{\frac{10}{\log 10}}{\sqrt{2\pi\sigma_{dB}^2}} \frac{1}{P_{ij}} \exp\left(-\frac{1}{8}\left(\frac{10\alpha}{\sigma_{dB}\log 10}\right)^2 \left(\log\left(\frac{d_{ij}^2}{d_0\left(\frac{P_0}{P_{ij}}\right)^{\frac{2}{\alpha}}}\right)\right)^2\right) \tag{6.5}$$

It is worth noting that P_0 and P_{ij} are not expressed in dBm. Hence, an ML estimate of the distance d_{ij} could be derived as [32]

$$\delta_{ij} = 10^{\frac{P_0 - RSS_{ij}}{10\alpha_{ij}}} \tag{6.6}$$

An important result of the lognormal model is that RSS-based distance estimates have variance proportional to their actual range [9]. The standard deviation in decibels is considered constant with range. This consideration means that the multiplicative factors are constant with range; hence, this explains the multiplicative error present in RSS-based distance estimates.

6.4.2 Location Update Phase

Once the relative distances between nodes are obtained, the main goal is to estimate the location of the nonlocated nodes with the help of anchor nodes and the rest of the nodes in the network.

The position estimates for each nonlocated node are obtained by means of the least squares (LS) criterion. The localization algorithm has to obtain the set of nonlocated node positions that minimize the difference between the estimated distances at the first phase and the distances computed using such position estimates. In particular, the problem consists in minimizing the following cost function:

$$C_{LS}(x) = \sum_{i=1}^{N_2} \sum_{j \in S_i} \left(\delta_{ij} - d_{ij}(\mathbf{x}_i - \mathbf{x}_j)\right)^2 \tag{6.7}$$

where $d_{ij}(\mathbf{x}_i, \mathbf{x}_j) = \|\mathbf{x}_i - \mathbf{x}_j\|$ is the distance between nodes i and j, calculated with the estimated position (or real coordinates if node j is an anchor) of nodes i and j, S_i is the group of nodes (anchor and nonlocated) that cooperates in the position estimation of nonlocated node i and $\mathbf{x}$ are the coordinates of the nodes.

The cost function is minimized with the optimization of the node coordinates. The minimization will be obtained by means of calculating the derivative of Equation (6.7) with respect to $\mathbf{x}_i$:

$$\frac{\partial C_{LS}}{\partial x_i} = \sum_{j \in S_i} \frac{(\delta_{ij} - d_{ij})^2}{\partial x_i} + \sum_{k \in S_i} \frac{(\delta_{ki} - d_{ki})^2}{\partial x_i} \tag{6.8}$$

Wireless channels are usually not considered reciprocal. For that reason measurements of the second summation should be omitted. As a result, the cost function adopted by each node can be rewritten as

$$C_{DLS}(x_i) = \sum_{j \in S_i} \left(\delta_{ij} - d(\mathbf{x}_i - \mathbf{x}_j) \right)^2 \tag{6.9}$$

A distributed cost function is found, so each node is responsible for obtaining the minimization of this cost function. Many methods could be applied in order to obtain this minimization. A gradient descent is one of the simplest approaches that have a lower computational complexity. Hence, the distributed cost function in Equation (6.9) is iteratively minimized. These algorithms have the possibility of not reaching a global minimum when a good starting point is not used. Even so, it is a simple method with a low computational complexity.

The gradient of the cost function is

$$\nabla_{x_i} C_{DLS}(x_i) = \nabla_{x_i} \left(\sum_{j \in S_i} \left(\delta_{ij} - \|x_i - x_j\| \right)^2 \right) = \sum_{j \in S_i} \left(\delta_{ij} - \|x_i - x_j\| \right) e_{ij} \tag{6.10}$$

where $e_{ij} = (x_i - x_j/\|x_i - x_j\|)$ is the unit vector that takes the orientation between the node i and node j. So, the estimate of $\mathbf{x}_i$ can be iteratively computed by using the gradient descent algorithm as follows:

$$\hat{x}_i(t+1) = \hat{x}_i(t) + \gamma \sum_{j \in S_i} \left(\delta_{ij} - d_{ij} \right) e_{ij} \tag{6.11}$$

where γ is the step length factor.

This algorithm becomes a simple, low computational approach that obtains position estimates in a cooperative and distributed way by means of using RSS measurements.

In Figure 6.5, a WSN scenario of 50 m × 50 m with 24 anchor nodes and 30 nonlocated nodes is considered. The mean absolute error obtained in the position estimates as a function of the number of cooperating nodes is presented. As commented previously, the higher the number of cooperating nodes, the lower the error obtained. However, it is remarkable that the error is not monotonically decreasing and the error is saturated for high values of cooperating nodes. This is basically due to the effect commented previously: having a further node causes a higher error in a distance estimate. On the other hand, there also exists a relation with the presence of more anchor nodes cooperating in the estimation of coordinates. For that reason, in the following section a node selection mechanism is presented to limit the number of

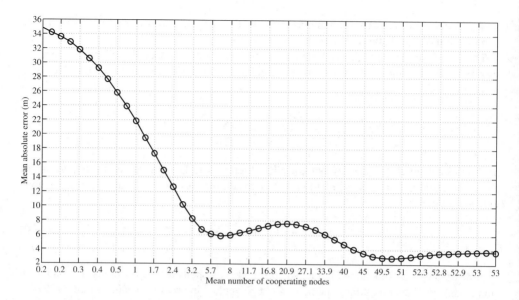

Figure 6.5   Mean absolute error versus mean number of cooperating nodes (24 anchor nodes).

cooperating nodes by selecting those nodes that provide an appropriate error value in terms of position accuracy. By doing so, the number of cooperating nodes, those within group S_i, is reduced and energy efficiency can be improved. For example, the difference in error obtained with a number of cooperating nodes equal to 6 and 50 is minimal but the number of cooperating nodes is considerably lower; that is energy consumption can thus be significantly reduced.

6.5   Node Selection

A cooperative approach allows the algorithm to achieve better accuracy results. On the other hand, the computational cost is increased because nodes receive more information from more cooperating nodes. Furthermore, a higher packet exchange in the network will be necessary; hence there exists a higher probability of losing a packet due to a collision.

Considering also that this chapter focuses on RSS-based distance estimates, one needs to take into account that the error that suffers these estimates is multiplicative to distance. Having a cooperative approach allows each node to cooperate with more nodes that could be further away. Hence, the distance estimates that a node would use should have a higher error. It is also shown in Figure 6.5 that reducing the number of cooperating nodes does not seriously affect the accuracy of the localization algorithm.

In order to reduce the use of a great number of cooperating nodes, that is having a lower number of nodes inside each group S_i, the node selection mechanism is presented. The major purpose behind the node selection is to reduce the packet exchange inside the network, and thus the reduction of computational effort done by each node and the energy consumption of them. Some works have presented different approaches in order to select those nodes that will cooperate with each nonlocated node.

In Reference [33], the authors present a noncooperative microgenetic algorithm (MGA) in order to select nodes and improve the localization in a WSN. The adaptation of the MGA presented is based on three steps: firstly, the construction of a small population of chromosomes based on the best values of position estimation; secondly, a genetic operator called a descend-based mutation is applied; and, finally, a second genetic operator called a crossover operator is used. The basic idea is to first select best nodes (done at the first step) and then apply both genetic operators to the chromosomes (aka nodes) in order to converge to a final solution. The results obtained are, in mean, 0.2 times the node range R.

In Reference [34] the authors propose to select nodes by means of using the Crámer–Rao bound (CRB) instead of using the closest nodes. The algorithm calculates the CRB of all the reference nodes that a node receives and selects those nodes with a lower CRB. Results obtained show that with a number of eight reference nodes the location error is, approximately, 7 m if RSS-based distance measurements are used. On the other hand, 0.8 m of error is achieved if TOA-based distance estimates are used.

In Reference [35], the authors present a censoring method based, also, on the CRB. The algorithm censors those nodes with an unreliable estimation. Based on the CRB calculation, the authors propose a criterion that will reflect the quality of the information that a node transmits as well as the geometry of the positions of the anchors and nonlocated nodes. Three methods of censoring are presented in the work: the first one is the one in which a node can censor itself, that is each node can decide not to broadcast its own information; the second one blocks the reception of information from the neighbours considered not reliable; and, finally, the last one is created to avoid an unnecessary transmission when a node is censored by all its neighbours, that is a node receives the order not to transmit because all its neighbours have censored it.

All these censoring methods are executed through the calculation of the CRB, and by comparing it with a threshold imposed by the algorithm. With the inclusion of all these censoring methods, the authors obtain a reduction in complexity and in network traffic, while the position accuracy is maintained.

All methods achieve a high reduction of messages exchanged in their results. A comparison between distance-based selection and CRB-based selection is done in Reference [34]. On the one hand, better results are achieved when the CRB is used to select cooperating nodes; for example differences between 2.5 m with three cooperating nodes and 0.3 m with 10 cooperating nodes are presented in the location error. On the other hand, the distance-based selection does not require any extra calculation at the time of deciding which nodes are the best to cooperate.

Finally, a node selection least squares (NS-LS) location algorithm is presented in Reference [36]. The idea of obtaining a good trade-off in terms of position accuracy versus energy consumption is maintained. As discussed in Reference [36], the derivation of the optimal selection criterion is not possible. For that reason, the authors presented a suboptimal scheme based on the received power threshold (RSS_{th}). In other words, only nodes with an RSS higher than RSS_{th} are allowed for cooperation. In a simple scheme this criterion becomes suitable for a hardware restricted WSN. In particular, the choice of the RSS_{th} value was designed to ensure a minimum number of anchor nodes inside the cooperating node group S_i. In accordance with this value N_m, different trade-off points of energy consumption versus accuracy can be achieved. Results showed that $N_m = 3$ allows the algorithm to achieve an excellent trade-off. Concerning the relation between RSS_{th}

and N_m, the authors in Reference [36] derived an analytical procedure to obtain the required threshold that assures the desired value of N_m. Considering a uniform distribution of the nonlocated nodes, the mean number of anchor nodes inside a circumference of radius r_{th} is obtained as

$$N_m \approx \sum_{j=1}^{N_1} \pi r_{th}^2 \frac{1}{A} = \frac{N_1 \pi r_{th}^2}{A} \tag{6.12}$$

where A is the total area of the considered scenario and N_1 is the total number of anchor nodes in the scenario. An appropriate received power threshold (RSS_{th}) must be selected in order to ensure that a number N_m of anchor nodes are inside the coverage range radius. Taking r_{th} as the coverage range radius, a relation between the RSS received at a distance r_{th} is obtained by means of using the path loss propagation model. Hence, a relation between RSS_{th} and N_m could be established by means of

$$RSS_{th} = \left(\frac{N_1 \pi P_0^{2/\alpha}}{N_m A} \right)^{\alpha/2} \tag{6.13}$$

With this relation it is possible to fix a threshold that assures a mean number of anchor nodes. With this threshold nodes are able to discard those nodes that have an RSS below this value; hence the algorithm is able to reduce the number of cooperating nodes.

The algorithm achieves worse results in terms of accuracy (a degradation between 0.23 and 0.44 m). On the other hand, the reduction achieved in terms of energy consumption is above 78% compared to a nonselection algorithm.

6.5.1   Energy Consumption Model

Wireless sensor network nodes rely on low data rates, a very long battery life (several months or even years) and very low computational complexity associated with the processing and communication of the collected information across the WSN. In order to maintain the battery life, the reduction of the energy consumption is an important point in a WSN.

The energy consumption of a node is directly related to the number of messages transmitted or received. The purpose of using node selection mechanisms is to reduce the packet exchange between nodes inside the network. The reduction of packet exchanges becomes an important function in the reduction of energy consumption. Nodes only interact with their group of cooperating nodes. Hence, the energy consumed will depend on the required number of transmissions. An energy consumption model is presented in order to reflect the effects produced in the network.

Each node i needs to create its own group of cooperating nodes S_i (see Figure 6.6). At the beginning, each node i sends a broadcast message with its coordinates $\mathbf{x}_i$. Only those nodes that receive this message answer with their node id and their location coordinates. With the received messages, each nonlocated node can create its own S_i group. Once these groups are created, the exchange of messages is only done between cooperating nodes. At this time, the total amount of energy consumed by the network follows the

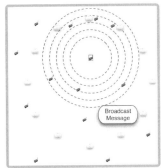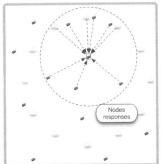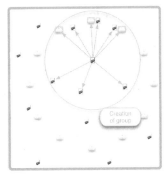

Figure 6.6   Creation of group S_i.

model presented in Reference [37]:

$$\varepsilon = \left(\mu_{R_x} - \mu_{T_x}\right) \left(\sum_{i=1}^{N_2} N_{S_i} - N_2 \right) \kappa \tag{6.14}$$

where κ is the number of iterations of the algorithm, N_{S_i} is the number of nodes inside S_i, and μ_{T_x} and μ_{R_x} (e.g. a value equal to 400 nJ/s is used in Reference [37]) are the energy consumption values dedicated for peer to peer transmission and reception procedures, respectively. It is noticeable that this model presents some differences compared with the model presented in Reference [37]. It is supposed that the energy per transmission is always the same instead of having an energy consumption depending on time. The purpose behind a node selection mechanism is the reduction in the traffic. Hence, the energy consumption model presented only reflects the impact that selection mechanisms could produce. Furthermore, it has only taken care of the energy consumption at the transmission and reception times.

Energy consumption is an increasing function of the number of cooperating nodes (N_{S_i}). The introduction of a node selection mechanism reduces the number of cooperating nodes; hence the energy consumption will decrease.

In the following subsection two low complex mechanisms suitable to be used in a WSN localization algorithm are proposed. Later, in Section 6.5.3, the joint path loss and node selection localization algorithm is proposed. The different blocks that form the proposed on-line path loss and node selection localization algorithm are presented in detail.

6.5.2   Node Selection Mechanisms

Which is the best criterion to use in order to select the nodes that will cooperate in the location-update phase? With this question in mind, and taking a look at existing methods presented at the beginning of the section, two node selection criteria are presented.

The continuous idea that is maintained through the entire chapter has an important purpose: to present a low complex localization algorithm. For that reason, two node selection criteria are presented. The introduction of a node selection should not suppose an increase in the computational complexity of the algorithm. Both node selection mechanisms are related by the fact that a node can do without any extra measurement requirements.

In order to take the advantage of these measurements, both node selection mechanisms will depend on the path loss estimates obtained by the location algorithm. With those selection mechanisms a reduction in the number of nodes that cooperate in the location algorithm, as well as a reduction in energy consumption of the network, is achieved.

6.5.2.1 Low Path Loss Selection

The first selection criterion is the selection of those nodes that have the lowest values of the path loss exponent estimated by the localization algorithm. The idea behind this node selection mechanism is to select those nodes that have the best channel conditions. When the path loss exponent takes a high value the conditions of propagation are worse.

The method works as follows. Given all the estimates of $\hat{\alpha}_{ij}$, the first selection mechanism selects those nodes that have the lowest values for the path loss exponents of the nodes inside the coverage of node i:

$$\hat{\alpha}_{i1} \leq \hat{\alpha}_{i2} \leq \cdots \leq \hat{\alpha}_{in}, \qquad 1, \ldots, n \in S_i$$

where $\hat{\alpha}_{i1}$ and $\hat{\alpha}_{in}$ are the lowest and the highest exponents, respectively. Those nodes with the lowest values are selected:

$$S_i^{NS} = \{i1, i2, \ldots, in_\alpha\}$$

with n_α standing for the number of selected nodes.

6.5.2.2 Low Distance Selection

The second selection criterion selects those nodes with lower values of distance estimates, that is those nodes closer to the nonlocated nodes. RSS-based distance estimates have an error multiplicative to the distance; hence the selection of closer nodes will choose those nodes with a lower distance estimate. The higher the distance, the higher the error of the distance estimates. For that reason the mechanism tries to reduce this effect by selecting the closest nodes to the nonlocated node i.

Given now all the distance estimates of the nodes inside the coverage of node i:

$$\delta_{i1} \leq \delta_{i2} \leq \cdots \leq \delta_{in}, \qquad 1, \ldots, n \in S_i,$$

where in this case δ_{i1} and δ_{in} are the lowest and highest distance estimate, respectively. The new group of cooerating nodes becomes

$$S_i^{NS} = \{i1, i2, \ldots, in_\delta\}$$

with n_δ standing for the number of selected nodes.

6.5.2.3 Selection Mechanisms Performance

In Figure 6.7, an example of both node selection mechanisms is shown. Both nodes could not use nearer nodes; hence they have the possibility of having a higher error on distance

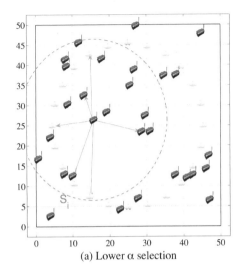
(a) Lower α selection

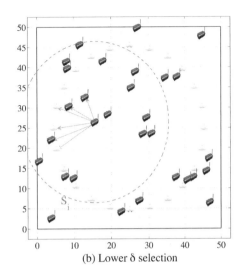
(b) Lower δ selection

Figure 6.7   Example of both node selection methods.

estimates. The only mechanism that could have a higher probability of selecting closest nodes is the lower distance criterion. Only with perfect channel knowledge is it possible to achieve a perfect selection of the closest nodes.

Two important points now have to be discussed: which of the suitable methods is better to use and how many nodes has to be used in order to obtain the best results?

Selection of the Criterion

The first results of the performance are presented in Figure 6.8. Both methods are used in the same scenario and for different numbers of nodes for n_α and n_δ. Low path loss selection selects those nodes with the best-estimated channel conditions. The major problem is that these nodes could be further away from the node i. As RSS-based range measurements are

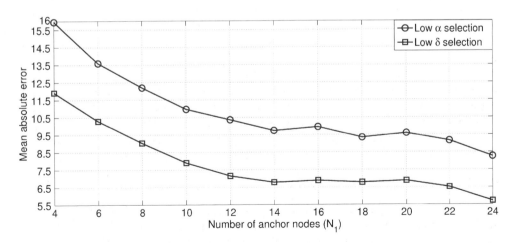

Figure 6.8   Mean absolute error versus $N_1 (n_\alpha = n_\delta = 6)$.

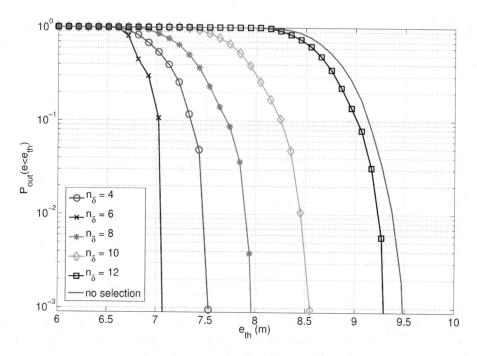

Figure 6.9   Outage probability ($N_1 = 18$).

used, the error in the distance estimates becomes higher. The probability of achieving a worse accuracy is increased, as reflected in the results shown. The low distance selection achieves better results in terms of accuracy.

Selection of the Number of Cooperating Nodes

This is the second important point when a node selection mechanism is included in the localization algorithm. In the previous presented works both methods has defined how a node collaborates or not in the location-update phase. In the first work they only allow cooperation to those nodes that are above a CRB threshold. In the second work, authors have analytically extracted a way in which they assure, in mean, that at least three cooperating nodes are anchors.

Simulation results are presented in order to decide which is the suitable number of cooperating nodes, because it is not straightforward to obtain the value of cooperating nodes that optimize the system behaviour. In principle, the optimum number of n_δ is scenario dependent. In the scenario considered here, results presented in Figure 6.9 show that a value equal to 6 is generally the best choice. This means that actually the optimum n_δ does not depend on the fine-grained distribution of the nodes, but rather on general parameters of the scenario.

6.5.3   Joint Node Selection and Path Loss Exponent Estimation

During the entire chapter the different blocks of a localization algorithm have been presented. A presentation of an RSS-based distributed algorithm has been done at the beginning, followed with the introduction of a node selection mechanism in order to reduce the cooperating nodes. Moreover, obtaining RSS-based distance estimations depends on the knowledge of a transmission model. This model is usually obtained with a previous measurement campaign. However, the idea of repeating a measurement campaign in every possible scenario in which the algorithm has to work is not a good decision. For that reason, an on-line path loss estimation is introduced in order to dynamically estimate the transmission model and allow the algorithm to estimate the model that best fits the current scenario. In this subsection is presented all the steps that the localization has to do in order to obtain position estimates. Algorithm 6.1 presents the different stages of the node localization algorithm with an on-line path loss estimation. A more in-depth explanation of the different blocks of the algorithm present in Algorithm 6.1 is presented in the following subsections.

Algorithm 6.1   LS Localization Algorithm with On-Line Path Loss Estimation

---

Discovering Cooperating Group S_i:
node i transmits a broadcast message
nodes able to receive respond with their id and their coordinates
Previous Coordinate Estimation:

for $i = 1$ to N_2 **do**

$$\hat{\mathbf{x}}(t = 1) = \sum_{a=1}^{n_{anch}} \mathbf{x}_a \left| \frac{RSS_{ia}}{\sum_{a=1}^{n_{anch}} RSS_{ia}} \right|$$

end for
On-Line Path Loss-Node Selection-Least Squares algorithm
for $i = 1$ to t_{iter1} **do**
    Coordinate Estimation:
    for $t(= 1$ to t_{iiCT2} **do**
      for $i = 1$ to N_2 **do**

$$\hat{\mathbf{x}}_i(t) = \hat{\mathbf{x}}_i(t-1) + \gamma_x \sum_{j \in S_i} \left(\delta_{ij} - \hat{d}_{ij} \right) \mathbf{e}_{ij}$$

      end for
    end for
    Path Loss Estimation:
    for $t' = 1$ to t_{iter2} **do**
      for $i = 1$ to N_2 **do**
        for $j = 1$ to N_{Si} **do**

$$\hat{\alpha}_{ij}(t) = \hat{\alpha}_{ij}(t-1) - \gamma_\alpha \log(10) \frac{P_0 - RSS_{ij}}{5} \frac{1}{\alpha_{ij}^2} \delta_{ij} \left(\delta_{ij} - \hat{d}_{ij} \right)$$

$$\delta_{ij} = 10^{\frac{P_0 - P_{ij}}{10\hat{\alpha}_{ij}}}$$

        end for
      end for
    Low Distance Node Selection
    if $t == 1$ **then**
      $S_i = \{i1, i2, \ldots, in_\delta\}$ with $\delta_{i1} < \delta_{i2} < \cdots < \delta_{in\delta} < \cdots < \delta_{i_{N_{S_i}}}$
      $N_{S_i} = n_\delta$
    end if
  end for
end for

6.5.3.1   Discovering Cooperating group S_i

The first necessary step is to discover which nodes are inside the radio range of each non-located node i. In order to do that, each nonlocated node broadcasts a message with their node id. Nodes that are able to receive this message send an answer to the nonlocated sender, with their node id and their position. As a result, each node can form its own group and can know the number of anchor and nonlocated nodes that are inside its radio range.

6.5.3.2   Initial Coordinates Estimation

The next step is giving an initial value to the position estimates of the nonlocated nodes. As discussed before, a gradient descent approach could not converge to the global solution if a biased initial value is used. Hence, it is important to give a good starting point in order to converge to a global solution. In that sense many works present different options. The authors of Reference [38] present the idea of combining two approaches: obtaining a previous estimate by means of using an MDS algorithm and then implementing a refinement algorithm, such as an ML non-Bayesian approach. On the one hand, this solution obtains good results in terms of accuracy. On the other hand, this method is more complex compared to other solutions. Moreover, it is possible to give a random initialization to all nonlocated nodes. It is a low complex but inaccurate method. Following with the general purposes presented previously, the starting point procedure has to attain a good trade-off between accuracy and complexity.

A simple method to initialize each $\hat{x}_i(0)$ is the use of a centroid method. Once each node forms its own cooperating group S_i, all the nonlocated nodes are able to compute its initial position with a weighted centroid algorithm based on the use of anchors inside S_i. The computation becomes

$$\hat{x}_i(t=0) = \sum_{a=1}^{n_{anch}} x_a \left| \frac{RSS_{ia}}{\sum_{a=1}^{n_{anch}} RSS_{ia}} \right| \tag{6.15}$$

where x_a are the coordinates of the anchor $a \in S_i$.

It is possible to achieve a closer initial point with a weighted centroid algorithm; hence a better performance of the gradient descent approach could be achieved. Although the centroid algorithm presents low accuracy it is a simple method and is only used to obtain the initial coordinates that will be later refined with a non-Bayesian LS method.

6.5.3.3   On-line Path Loss Estimation and Node Selection Least Squares Algorithm (OLPL-NS-LS)

The introduction of this estimation is motivated by the necessity of having a good propagation model in order to have accurate distance estimates. Usually, the propagation model is achieved by doing a previous measurement campaign. With the introduction of an on-line path loss estimation, the localization algorithm is responsible for obtaining the node coordinates and the path loss estimation; hence the necessity of previous modelling is avoided. The next step presented in Algorithm 6.1 is the minimization of the LS cost function:

$$C_{DLS}(x, \alpha_{i1}, \alpha_{i2}, \ldots, \alpha_{in_\alpha}) = \sum_{j \in S_i} \left(\delta_{ij}(\alpha_{ij}) - d_{ij}(x_i - x_j) \right)^2 \tag{6.16}$$

The objective is to minimize the difference between both distances and optimize the node coordinates and also the set of path loss exponents. The node coordinates $(\mathbf{x}_i)$ and the set of path loss exponents $\left(\alpha_{ij} \forall j \in S_i\right)$ affect the computation of both distances, d_{ij} and δ_{ij}, respectively. In order to solve the cost function of Equation (6.16), a Gauss–Seidel algorithm is adopted in Reference [39]. This nonlinear algorithm is based on a circular iterative optimization with respect to one set of variables while maintaining the rest of the variables fixed. Hence, the minimizations are carried out successively for each component.

Considering a generic cost function F that depends on a set of variables β, the desired minimization of F is formally defined as [39]

$$\beta(t + 1) = \arg \min_{\beta_i} F\left(\beta_1(t + 1), \ldots, \beta_{i-1}(t + 1), \beta_i, \beta_{i+1}, \ldots, \beta_m(t)\right) \qquad (6.17)$$

At the time instant $t + 1$, the F function is minimized by means of optimizing the β_i component. Components between β_1 and β_{i-1} have already been optimized while components from β_{i+1} to β_m (m being the total number of components) have not yet been optimized. These last components must remain constant while the other components are being optimized. By using the Gauss–Seidel approach, it is possible to divide the optimization into two steps: firstly, a minimization of the cost function by means of optimizing the node coordinates (fixing the path loss exponents) could be carried out; secondly, another minimization is done by means of the optimization of the path loss exponents (fixing the nodes coordinates). As the convergence of the nonlinear Gauss–Seidel algorithm can be achieved using a descent approach (see Reference [9]), both minimizations could be carried out through a gradient descent mechanism. The basic idea is summarized in Figure 6.10.

Coordinates Estimation
The estimation is done following the algorithm described in the previous section. The algorithm minimizes, by means of a least squares criterion, the difference between the estimated distance and the distance calculated with node coordinates. With the use of a gradient descent approach the algorithm is able to converge to a minimum of the cost function presented before.

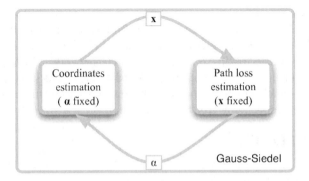

Figure 6.10 Optimization of the node coordinates and path loss exponents using a Gauss–Siedel algorithm.

Path Loss Estimation

This step is necessarily done after a previous position estimate. The algorithm has obtained the previous coordinates by means of using an arbitrary and equal value of the path loss exponent. At the second step, each nonlocated node estimates a path loss exponent for each link. Following the Gauss–Seidel approach, the cost function of Equation (6.16) is now minimized by means of optimizing the path loss exponents. Following the same methodology applied in Section 6.2.2, it is necessary to calculate the gradient of the cost function. The estimated distance

$$\delta_{ij} = f(\alpha_{ij}) = 10^{\frac{P_0 - RSS_{ij}}{10\alpha_{ij}}}$$

depends on the path loss exponent α_{ij}; hence, the cost function is minimized by means of calculating the derivate with respect to the path loss exponent of each individual link. In that case, the fixed variables are the coordinate estimates and the rest of the path loss exponents $(\alpha_{ik}\forall k \neq j)$. The gradient of the cost function is

$$\nabla_{\alpha_{ij}} C_{DLS}(x_i, \alpha_{S_i}) = \nabla_{\alpha_{ij}} \left(\left(10^{\frac{P_0 - RSS_{ij}}{10\alpha_{ij}}} - d_{ij}(x_i, x_j)\right)^2 \right)$$

$$= \log(10) \frac{P_0 - RSS_{ij}}{5} \frac{1}{\hat{\alpha}_{ij}^2} \delta_{ij}\left(\delta_{ij} - \hat{d}_{ij}\right) \tag{6.18}$$

Each node estimates their own path loss exponents for all the links in an iteratively fashion as

$$\hat{\alpha}_{ij}(t+1) = \hat{a}_{ij}(t) - \gamma_\alpha \log_{10}(10) \frac{P_0 - RSS_{ij}}{5} \frac{1}{\hat{\alpha}_{ij}^2(t)} \delta_{ij}\left(\delta_{ij} - \hat{d}_{ij}\right) \tag{6.19}$$

It is a distributed method that minimizes the cost function through an iterative gradient descent strategy. The algorithm is able to estimate the path loss exponents by means of using RSS measurements. The presented algorithm maintains the philosophy of having a low complex and low cost localization algorithm.

Low Distance Node Selection

Finally the node selection mechanism is applied. With this node selection the localization algorithm modifies the cooperating group formed at the beginning of the algorithm. Hence, the algorithm reduces the traffic exchange among the network, allowing the network to decrease the total amount of energy consumed. Based on the results shown in Figure 6.8, the algorithm selects n_δ number of nodes that have the lowest distance estimates. With this selection mechanism each cooperating group S_i is reduced.

6.6 Numerical Results

This section presents the performance of the presented location algorithm with an on-line path loss estimation and node selection. In order to evaluate the accuracy of the algorithm different simulation results are presented. The simulated scenario and the assumed simulation parameters are presented in Table 6.1.

Table 6.1 Simulation parameters

Simulation parameters	Parameter value
Size of sensor field	50 m × 50 m
Number of nonlocated nodes (N_2)	30
Path loss exponent α_{ij}	2–5
Standard deviation σ_υ	1 dB
First-metre RSS P_0	−50 dBm
Anchor radius	20.4 m
Energy consumption to transmit or receive μ_{Tx} or μ_{Rz}	400 nJ

Path loss exponents take values between a maximum value of 5 and a minimum value of 2 (the uniform distribution of the path loss exponents between 2 and 5 are based on experimental results obtained in Reference [40]). Hence, path loss values are simulated with a uniform distribution ($\alpha \in u(2,5)$). An initial value of the path loss equal to 3.5, which is the middle value of the random values used in the uniform distribution, is assumed. The experimental parameters are shown in Table 6.1.

Concerning the anchor node placement, the approach presented in Reference [41] is initially adopted, where the authors contend that the best anchor placement is a centred circumference with the radius equal to the root-mean-square (rms) of the nonlocated node distances to the centre.

As previously discussed, the use of a gradient descent algorithm requires a good starting point in order to achieve better results. The algorithm calculates the starting coordinates using a centroid method. The weighted centroid method depends on the number of anchor nodes inside the network.

In order to reflect the effect of using a great number of anchor nodes n_{anch}, the mean absolute error achieved by the localization algorithm is shown in Figure 6.11 for different values of anchor nodes used in the weighted centroid.

Better results are obtained when a number of anchor nodes equal to 1 is used. At first sight, this result could seem strange because, normally, it is better to use as many nodes as possible. However, at the initial time instant of our algorithm the path loss exponent values have still not been estimated. Hence, the higher RSS_{ij}, the lower the distance estimate δ_{ij}. If the number of anchor nodes n_{anch} has a greater value, the probability of having a further anchor node is increased. Hence, an initial position far away from the real position is estimated. Selecting only the anchor node with the highest RSS is the best option.

Although the distribution of the anchors in a centred circumference was demonstrated to be the best one in Reference [41], it could not be the optimal one when a centroid algorithm is used as the initial position estimate. The centroid method depends on the distribution of the anchors inside the network. For that reason, a grid-based positioning is proposed as an alternative.

The results achieved are shown in Figure 6.12(b). On this occasion it is necessary to use more anchor nodes in order to obtain better results. The centroid method achieves better results when the anchors are placed in a grid-based distribution. Hence, the starting point used in the LS algorithm will be closer to the final solution. In this case, when the centroid

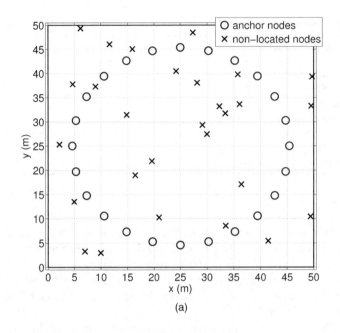

(a)

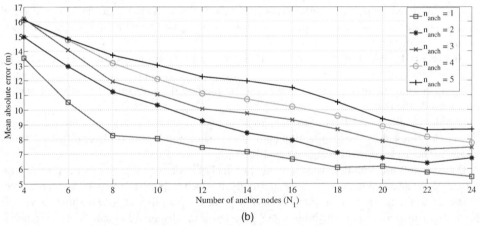

(b)

Figure 6.11 Circular-based anchor distribution: (a) simulation scenario, (b) mean absolute error versus N_1.

uses three anchor nodes, the localization algorithm achieves better results in terms of accuracy.

In the sequel, both scenarios will be simulated in order to compare the efficiency of the solution, comparing the accuracy when the algorithm estimates the path loss exponent and when it does not. Furthermore, a comparison between the presented algorithm and two other localization algorithms (a distributed and a centralized) will be presented.

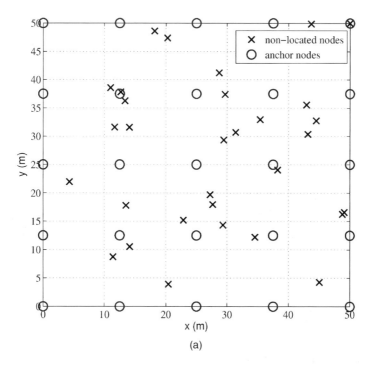

(a)

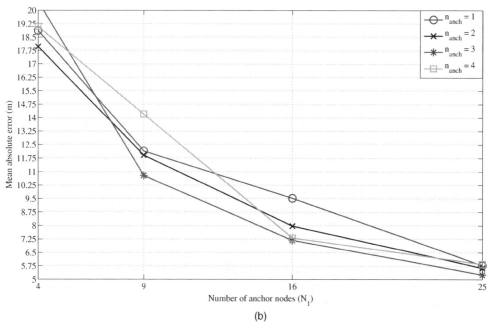

(b)

Figure 6.12 Grid-based anchor distribution: (a) simulation scenario, (b) mean absolute error versus N_1.

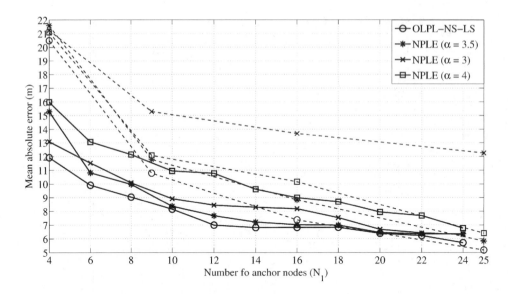

Figure 6.13 Mean absolute error versus N_1 (solid line: circular-based anchor distribution, dashed line: grid-based anchor distribution).

6.6.1 OLPL-NS-LS Performance

In order to simulate a more realistic scenario different path loss exponents are assumed. Each link has different values for the path loss exponent. With this assumption the importance of doing an on-line estimation of the path loss exponent is reflected in Figure 6.13.

With the proposed solution (OLPL-NS-LS) a gain in terms of position accuracy is achieved. With a centred anchor distribution, the gain oscillates between 2 and 0.5 m, compared to that achieved with the non path loss estimation (NPLE) algorithm. On the other hand, with results achieved using a grid-based anchor distribution the error gain oscillates between 7 and 0.7 m (not taking into account the results achieved with a fixed path loss equal to 3).

With a grid-based distribution, the centroid algorithm uses three anchors, but these nodes, when the density of of anchors is low, are, in mean, at a further distance from the non-located nodes. Hence, the starting point has low accuracy and this fact affects the final accuracy of the localization algorithm. When N_1 is increased the grid-based anchor distribution achieves better results in terms of position accuracy.

On the other hand, the best result is always achieved with the on-line path loss estimation and node selection least squares algorithm(OLPL-NS-LS). Different values of path loss exponents have been simulated. Hence having an on-line estimation of the path loss has a good influence on the localization performance accuracy. This estimation also makes possible the adaptation of the algorithm to possible changes in the scenario.

6.6.2 Comparison with Existing Methods

In this subsection, the OLPL-NS-LS algorithm is compared with two different existing solutions: a distributed method based on a maximum likelihood estimation (MLE) algorithm and a centralized algorithm based on multidimensional scaling (MDS). The on-line path loss estimation is applied to all the methods in order to achieve a fair comparison between them. Only the OLPL-NS-LS method presents the node selection method. The comparison between the three methods is carried out in terms of both energy consumption and positioning accuracy.

The MLE localization algorithm [30] used is based on the minimization of the following cost function carried out with a distributed iterative method:

$$C_{ML}(x) = \sum_{i=1}^{N_2} \sum_{j \in S_i} \left(\log_{10}\left(\delta_{ij}\left(\alpha_{ij}\right)\right) - \log_{10}\left(d_{ij}\left(x_i - x_j\right)\right) \right)^2$$

As presented before, the MDS algorithm is a simple centralized approach that builds a global map using classical MDS [26]. MDS works well on networks with relatively uniform node density, but less well on more irregular networks.

The simulations are carried out with both anchor distributions. The results achieved are shown in Figure 6.14. The OLPL-NS-LS algorithm outperforms the other methods in both cases. When the number of anchor nodes increases, the grid-based anchor distribution achieves a high accuracy compared to that achieved by the circular distribution.

Given a circular anchor distribution, the MLE localization algorithm achieves similar values to those achieved by the OLPL-NS-LS. It is important to remark that a gain between 0.5 and 1.5 m is obtained with the OLPL-NS-LS. This gain is achieved thanks to the selection algorithm. It is more remarkable than the gain achieved with respect to the MDS localization algorithm. On the one hand, the MDS method includes more distant nodes. Then nodes with a high error on their distance estimates are used. On the other hand, all possible nodes inside

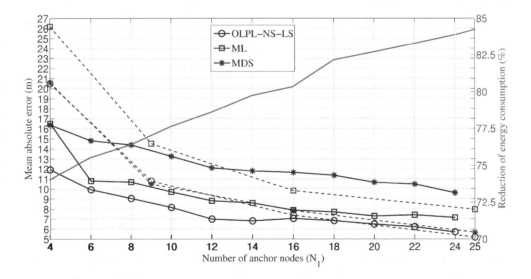

Figure 6.14 Mean absolute error versus N_1 (solid line: circular-based anchor distribution, dashed line: grid-based anchor distribution).

each group S_i are also used in the path loss estimation process. Probably nodes that are not near to node i would not have similar propagation conditions compared to those nodes that are closer.

The results achieved by the grid-based anchor distribution reflect the same tendency. The results of the OLPL-NS-LS present the best accuracy. However, in this case, the centralized MDS achieves better results than the MLE algorithm. The distribution of the anchor nodes in a grid along the scenario increases the probability of having closer anchor nodes in each S_i and hence better results are achieved when the number of anchor nodes inside S_i increases. The MDS achieves a better performance compared to that achieved with the centred distribution of the anchors. This reflects the importance of having more anchor nodes inside the cooperating group and also having nodes as close as possible. With a grid-based distribution the probability of having closer anchor nodes is higher. This higher probability is reflected in the better results obtained. The selection of the anchor node positions will depend on the possibility of having a higher density of nodes inside the network.

A node selection scheme allows the mean absolute error results to be reduced in an RSS-based localization algorithm. The reduction of the energy consumption is also important. According to the previous energy consumption model presented, the use of a reduced cooperating group S_i results in a lower consumption of energy. With the use of a number of cooperating nodes $n_\delta = 6$, the OLPL-NS-LS algorithm achieves a percentage of reduction between 74 and 83% compared to the energy consumed by a method without the node selection mechanism (see Figure 6.14).

6.7   Experimental Results

In this section experimental results are presented. The measurements taken in different indoor scenarios have been carried out with the Mica2 motes at 915 MHz of Crossbow [42]. Two different indoor scenarios are presented in Figure 6.16. In the first scenario the total number of nodes with an unknown position, N_2, is nine (see Figure 6.15a) and the number of anchor nodes, N_1, is four. These nodes are located in a 4.8 m $\times$ 4.8 m scenario. The second scenario (see Figure 6.15b) is composed of $N_2 = 20$, $N_1 = 6$ and $N_1 = 4$ in a scenario of 8 m $\times$ 12 m. Experimental results are only carried out with the centred-based anchor distribution. As the anchor nodes used in the network are of a low value, simulation results reflect that it is better to select the circular-based distribution instead of the grid-based one.

6.7.1   Scenario 1

The localization error achieved in the first scenario is shown in Figure 6.16. The outage probability presented in Figure 6.16(a) validates the result shown in Figure 6.8. The outage probability of having an error below an error threshold shows that the best result is achieved with six cooperating nodes. Figure 6.16(b) shows the results of including the path loss estimation inside the algorithm. Considering a fixed path loss exponent in order to model the propagation channel produces worse results in terms of accuracy. The experimental values achieved are also compared with simulation results that have the same conditions. On the one hand, the lowest localization error achieved has a value of n_δ equal to 6, the same as in

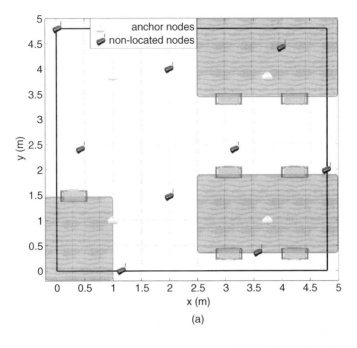

(a)

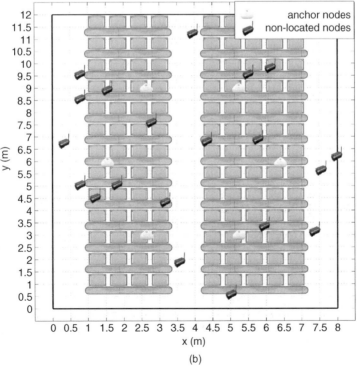

(b)

Figure 6.15   Experimental scenarios: (a) scenario 1, (b) scenario 2.

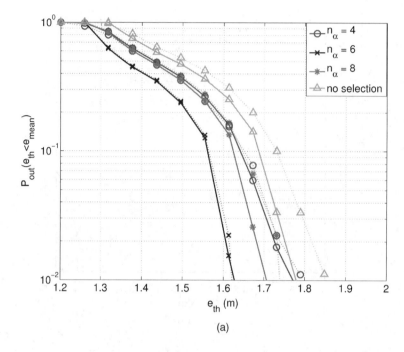

(a)

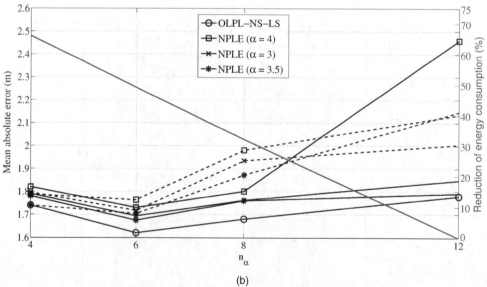

(b)

Figure 6.16 Results at scenario 1: (a) outage probability (solid line: experimental results, dashed line: simulation results), (b) mean absolute error versus n_δ (solid line: experimental results, dashed line: simulation results).

the large-scale case. On the other hand, simulation results are very similar to those achieved with the experimental scenario.

The best accuracy achieved is approximately 1.6 m. Furthermore, Figure 6.16(b) shows the percentage of reduction of the energy consumption compared to that consumed by an algorithm without node selection. With a value of n_δ equal to 6, the percentage of reduction in terms of energy consumption is about 50%. The comparison of behaviour between the OLPL-NS-LS and the NPLE with different values of the path loss exponent (α) gives the following results: the OLPL-NS-LS always achieves the best error values compared to those achieved with all the NPLE presented (see Figure 6.16b). The best behaviour is achieved with a fixed value of $\alpha = 3.5$. The differences in terms of mean absolute error oscillates between 0.2 cm when $\alpha = 3.5$ and 10 cm when $\alpha = 4$. The algorithm achieves a minimum gain of 5% and a maximum of 15%. Results of this experimental scenario show the gains obtained by considering the proposed on-line path loss estimation with respect to the case adopting an equal path loss exponent for all the links.

6.7.2   Scenario 2

As commented before, two different results are presented: one for a number of four anchor nodes and the other for six anchor nodes. Figure 6.17 shows both results. A similar behaviour is achieved when the simulation results are compared with the experimental results. As in the previous results, the best performance is achieved when the algorithm selects only six nodes in order to cooperate, validating the results achieved in Figure 6.9.

Although the number of nodes and the scenario dimensions have been increased, the performance of the algorithm (see Figure 6.17a) is similar to that achieved in the scenario shown in Figure 6.15(a). A higher scenario is considered but the probability of having closer nodes is increased. The accuracy achieved is equal to 1.7 m, on average.

Observing the results in Figure 6.17(b), one can see that the accuracy obtained is now 1.2 m. The benefit obtained with the increase in the number of anchor nodes is 0.5 m in the mean absolute error. Another important point is the difference of 0.5 m between the results achieved with the OLPL-NS-LS compared to an algorithm without path loss estimation results. Finally, the reduction in terms of energy consumption is, approximately, 75% in both scenarios.

In both figures, one can observe the benefits of the proposed OLPL-NL-LS approach when compared with the case of assuming a constant path loss exponent. A minimum gain of 18% (in the six anchor scenario) and a gain of 5% (in the four anchor scenario) is achieved. A better result in terms of position accuracy is always achieved with the OLPL-NS-LS.

6.8   Conclusions

A review of cooperative schemes for wireless sensor networks (WSNs) has been provided. Among all of them, a special emphasis has been given to the RSS-based measurement techniques, because they provide suitable solutions for practical implementations. Although RSS-based measurements are the simplest method, their accuracy depends on the suitability of the propagation model. Hence, a cooperative localization algorithm that dynamically estimates the path loss exponent has also been described. Furthermore, the reduction of the complexity and the message exchange through the node selection mechanism has been

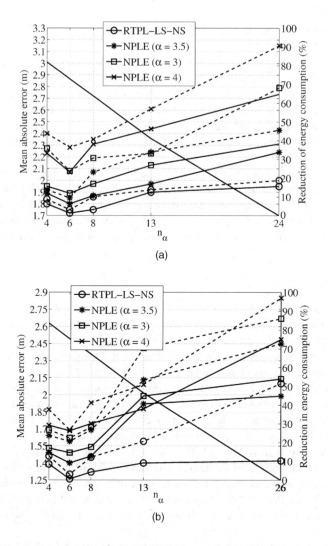

Figure 6.17 Mean absolute error versus n_δ (solid line: experimental results, dashed line: simulation results): (a) four anchor nodes (b) six anchor nodes.

presented. A good trade-off between localization accuracy versus energy consumption based on a node selection mechanism has been presented. In addition, practical examples based on a real WSN have been presented.

References

1. Bachrach, J. and Taylor, C. (2005) Localization in sensor networks, in *Handbook of Sensors Networks* (ed. J. Stojmenovic), John Wiley & Sons.
2. Boukerche F.A., Oliveira, H., Nakamura, E. and Loureiro, A. (2007) Localization systems for wireless sensor networks. *Wireless Communications*, **14**, 6–12.

3. Mao, G. and Fidan, B. (2009) *Localization Algorithm Strategies for Wireless Sensor Networks*, Information Science Reference.

4. Bulusu, N., Heidemann, J. and Estrin, D. (2000) GPS-less low-cost outdoor localization for very small devices. *IEEE Personal Communications*, **7** (5), 28–34.

5. He, T., Huang, C., Blum, B.M. *et al.* (2003) Range-free localization schemes for large scale sensor networks. MobiCom.

6. Niculescu, D. and Nath, B. (2003) DV based positioning in ad hoc networks. *Telecommunication Systems*, **22** (1), 267–280.

7. Li, R. and Fang, Z. (2010) LLA: a new high precision mobile node localization algorithm based on TOA. *Journal of Communications*, **5** (8), 604–611.

8. Yu, K., Guo, Y.J. and Hedley, M. (2009) TOA-based distributed localisation with unknown internal delays and clock frequency offsets in wireless sensor networks. *IET Signal Processing*, **3** (2), 106–118.

9. Patwari, N., Ash, J., Kyperountas, S. *et al.* (2005) Locating the nodes: Cooperative localization in wireless sensor networks. *IEEE Signal Processing*, **22**, 54–69.

10. Guvenc, I. and Chong, C.C. (2009) A survey on TOA based wireless localization and NLOS mitigation techniques. *IEEE Communications Surveys and Tutorials*, **11** (3), 107–124.

11. Mao, G., Fidan, B. and Anderson, B.D.O. (2007) Wireless sensor networks localization techniques. *Computer Networks*, **51** (10), 2529–2553.

12. Shimizu, Y. and Sanada, Y. (2003) Accuracy of relative distance measurement with ultra wideband system. IEEE Conference on Ultra Wideband Systems and Technologies, pp. 374–378.

13. Gezici, S., Tian, Z., Giannakis, G.B. *et al.* (2005) Localization via ultra-wideband radios: a look at positioning aspects for future sensor networks. *IEEE Signal Processing Magazine*, **22** (4), 70–84.

14. Boukerche, A., Oliveira, H.A.B., Nakamura, E.F. and Loureiro, A.A.F. (2007) Localization systems for wireless sensor networks. *IEEE Wireless Communications*, **14** (6), 6–12.

15. Priyantha, N.B., Chakraboty, A. and Balakrishnan, H. (2000) The cricket location-support system. 6th ACM International Conference on Mobile Computing and Networking, pp. 32–43.

16. Sallai, J., Balogh, G., Maróti, M. and Lédezci, Á. (2004) Acoustic ranging in resource-constrained sensor networks. ICWN, Las Vegas, NV.

17. Mazomenos, E., De Jager, D., Reeve, J. and White, N. (2011) A two-way time of flight ranging scheme for wireless sensor networks. 8th European Conference on Wireless Sensor Networks, Bonn, pp. 163–178.

18. Yang, J. and Chen, Y. (2009) Indoor localization using improved RSS-based lateration methods. IEEE Global Telecommunications Conference, pp. 1–6.

19. Patwari, N., O'Dea, R.J. and Wang, Y. (2001) Relative location in wireless networks. IEEE Vehicular Technology Conference Spring, pp. 1149–1153.

20. Hashemi, H. (1993) The indoor radio propagation channel. *Proceedings of the IEEE*, **81** (7), 943–968.

21. Kumar, P., Reddy, L. and Varma, S. (2009) Distance measurement and error estimation scheme for RSSI based localization in wireless sensor networks. 5th Conference on Wireless Communication and Sensor Networks, pp. 1–4.

22. Kulakowski, P., Vales-Alonso, J., Egea-López, E. *et al.* (2010) Angle-of-arrival localization based on antenna arrays for wireless networks. *Computer and Information Science*, **36** (6), 1181–1186.

23. Ash, J.N. and Potter, L.C. (2004) Sensor network localization via received signal strength measurements antennas. Allerton Conference on Communication, Control and Computing, pp. 1861–1870.

24. Maróti, M., Völgyesi, P., Dóra, S. *et al.* (2005) Radio interferometric geolocation. Proceedings of the 3rd International Conference on Embedded Networked Sensor Systems.

25. Fox, V., Hightower, J., Liao, L. *et al.* (2003) Bayesian filtering for location estimation. *IEEE Pervasive Computing*, **2** (3), 24–33.

26. Shang, Y., Ruml, W., Zhang, Y. and Fromherz, M.P.J. (2003) Localization from mere connectivity. Proceedings of the 4th ACM International Symposium on Mobile Ad Hoc Networking and Computing.

27. Kown, O.H. and Song, H.J. (2008) Localization through map stitching in wireless sensor networks. *IEEE Transactions on Parallel and Distributed Systems*, **19** (1), 93–105.

28. Biswas, P., Liang, T.C., Toh, K.C. *et al.* (2006) Semidefinite programming approaches for sensor network localization with noisy distance measurements. *IEEE Transactions on Automation Science and Engineering*, **3** (4), 360–371.

29. Wymeersch, H., Lien, J. and Win, M.Z. (2009) Cooperative localization in wireless networks. *Proceedings of the IEEE*, **97** (2), 427–450.

30. Denis, B., Pierrot, J.B. and Abou-Rjeily, C. (2006) Joint distributed synchronization and positioning in UWB ad hoc networks using TOA. *IEEE Transactions in Microwave Theory and Technology*, **54**, 1896–1911.
31. Thrun, S., Fox, D., Burgard, W. and Dellaert, F. (2001) Robust Monte Carlo localization for mobile robots. *Artificial Intelligence*, **128** (1), 99–141.
32. Patwari, N., Hero, A.O.I., Perkins, M. *et al.* (2003) Relative location estimation in wireless sensor networks. *IEEE Transactions on Signal Processing*, **51** (8), 2137–2148.
33. Tam, V., Cheng, K. and Lui, K. (2006) Using micro-genetic algorithms to improve localization in wireless sensor networks. *Journal of Communications*, **1** (4), 1–10.
34. Lieckfeldt, D., You, J. and Timmermann, D. (2008) Distributed selection of references for localization in wireless sensor networks. Proceedings of the 5th Workshop on Positioning, Navigation and Communication, pp. 31–36.
35. Das, K. and Wymeersch, H. (2012) Censoring for Bayesian cooperative positioning in dense wireless networks. *IEEE Journal on Selected Areas in Communications* (in press).
36. Bel, A., Vicario, J.L. and Seco-Granados, G. (2010) Node selection for cooperative localization: efficient energy vs. accuracy trade-off. Proceedings of the 5th IEEE International Symposium on Wireless Pervasive Computing, pp. 307–312.
37. Zou, Y. and Chakrabarty, K. (2003) Energy-aware target localization in wireless sensor networks. Proceedings of the 1st IEEE International Conference on Pervasive Computing and Communications, pp. 60–67.
38. Li, X. (2007) Collaborative localization with received-signal strength in wireless sensor networks. *IEEE Transactions on Vehicular Technology*, **56** (6), 3807–3817.
39. Bertsekas, D.P. and Tsitsiklis, J.N. (1997) *Parallel Distributed Computation: Numerical Methods (Optimization and Neural Computation)*, Athena Scientific.
40. Mazuelas, S., Bahillo, A., Lorenzo, R. *et al.* (2009) Robust indoor positioning provided by real-time RSSI values in unmodified WLAN networks. *IEEE Journal of Selected Topics in Signal Processing*, **3**, 821–831.
41. Ash, J. and Moses, R. (2008) On optimal anchor node placement in sensor localization by optimization of subspace principal angles. IEEE International Conference on Acoustics, Speech and Signal Processing, pp. 2289–2292.
42. Crossbow Technology [online], available from: HYPERLINK "http://www.xbow.com", http://www.xbow.com.

7

Modelling of Substrate Noise and Mitigation Schemes for UWB Systems[*]

Ming Shen, Jan Hvolgaard Mikkelsen and Torben Larsen
Aalborg University, Aalborg, Denmark

7.1 Introduction

The concept of Internet of Things (IoT) has, since initially being coined back in 1999 by Kevin Ashton [1], been one of the driving factors behind much of the research and development within the area of wireless communication. From initially being thought of as formation of a link between the then-new idea of RFID and the then-explosively growing Internet, IoT has spawned into a very diverse range of wireless communication system scenarios. Examples of systems descending from IoT are near field communication (NFC), personal area networks (PANs), body area networks (BANs), to mention but a few. On the basis of shared characteristics many of these systems can be contained within the concept of wireless sensor networks (WSNs), as illustrated in Figure 7.1.

The WSN concept plays host to an immensely broad range of applications and the revenue forecasts estimates that the WSN market is to reach 2.0 billion USD in 2021 [2]. With all its diverse application areas WSNs are going to have a significant impact on the ICT infrastructure and how we as humans interact with ICT systems.

One of the areas where the application of WSN-based technologies is expected to have the most significant societal impact is within the health care system. It is a generally accepted fact that one of the key challenges faced by modern welfare societies relates to human health care. The increasing life expectancy of people combined with a significantly greater

[*]This book chapter is mainly based on the work from Ming Shen's PhD study and the PhD dissertation 'Measurement, Modeling, and Suppression of Substrate Noise in Wide Band Mixed-Signal ICs'.

Microwave and Millimeter Wave Circuits and Systems: Emerging Design, Technologies, and Applications, First Edition. Edited by Apostolos Georgiadis, Hendrik Rogier, Luca Roselli, and Paolo Arcioni.
© 2013 John Wiley & Sons, Ltd. Published 2013 by John Wiley & Sons, Ltd.

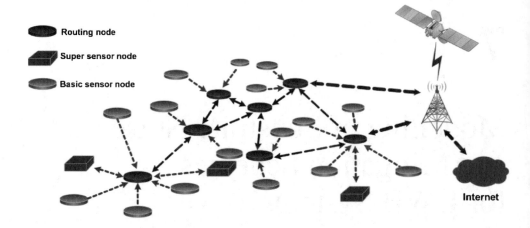

Figure 7.1 Overall concept of a wireless sensor network (WSN).

prevalence of chronic illnesses causes health expenses to increase with age, which places tremendous demands on the health system. One way to reduce cost is to use home health care scenarios for some of the more tedious tasks, such as screening and monitoring of people. In many cases specific patient groups, especially those disposed to certain diseases, such as cancer, are being monitored. To meet these challenges it is necessary to develop new technologies that can reduce the cost burden of the health care system. One example of a costly monitoring procedure is the screening of women for breast cancer. Second only to lung cancer, breast cancer is the most frequent form of cancer among women. It has been estimated that in the year 2000 there was a total of 350 000 new cases of breast cancer in Europe [3]. This number amounted to more than 25% of all new cancer cases in Europe in 2000. In addition it is estimated that breast cancer is responsible for more than 17% of all cancer related deaths across Europe [3]. It would clearly be of great societal value with a cost effective monitoring scheme that could provide an early detection of tumours that potentially could develop into breast cancer. Existing scanning systems, such as X-ray and MR, both require huge setups and involve powerful magnetic fields and/or dangerous radiation. In addition, running such systems require trained staff, which adds to the operational expenses.

Scanning systems based on ultra wideband (UWB) technologies have recently been applied in experimental setups aiming at detection of breast tumours, knee tissue tears and build-up of water/fluids in the body [4]. Owing to its wide frequency range UWB systems provide for very fine temporal resolution, which offers unique detection capabilities. UWB systems are ideal for miniaturization, making a whole new range of noninvasive medical applications possible. In addition UWB makes use of nonionized electromagnetic waves, which at the envisioned power levels are harmless to the human body [4]. From a medical point of view UWB-based detection technologies offer features that are ideal for home health care implementation. Consequently UWB-based systems potentially could become one of the absolute key technologies in leveraging the burden of the health care systems.

7.1.1 Ultra Wideband Systems – Developments and Challenges

During past years the semiconductor industry has continued its impressive progress. This development, which is largely driven by market demands for more powerful functionality at lower cost, has resulted in a device density that still obeys Moore's law, which states that the transistor density of integrated circuits doubles every 2 years [5]. This continued technology scaling has enabled the integration of more and more functionalities on a single chip. However, as the complexity of the systems increases so does the challenge of designing the required mixed-signal system-on-chips (SoCs). One of the more severe challenges relates to disturbances caused by substrate noise. Since SoCs integrate sensitive analogue/RF (radio frequency) circuits on the same die as high-speed digital processing circuits, the switching noise produced by the digital circuits can easily propagate through substrate and power supply rails to the analogue/RF circuits, degrading their performance.

The design issues related to substrate noise are expected to become even more challenging in the future. The reason behind this is twofold: (1) the trend is towards more and more digital functionality in the systems that inevitably will generate more noise and (2) to meet the ever increasing need for higher data rates wireless systems have to turn to more bandwidth efficient transmission schemes, which inevitably increases sensitivity towards noise.

Considering UWB-based systems, due to the ultra wideband nature of such systems combined with a very low power spectral density of the signal, the impact of substrate noise on UWB circuits is expected to be more detrimental than narrowband circuits. To overcome these problems, the substrate noise has to be characterized and subsequently efficient strategies for cancellation/suppression have to be developed.

7.1.2 Switching Noise – Origin and Coupling Mechanisms

In trying to understand the substrate noise generation and injection mechanisms of integrated circuits it is quite informative to first consider a single CMOS (complementary metal oxide semiconductor) inverter. When implemented using a lightly doped CMOS process the coupling and injection mechanisms of the basic inverter are as illustrated in Figure 7.2. These mechanisms include: (1) power/ground contacts to bulk coupling, (2) source/drain to bulk capacitive coupling and (3) impact ionization. A lumped element circuit model describing the electrical characteristics of the substrate is also given in the figure.

Unlike heavily doped substrates, where the substrate can be treated as a single node, lightly doped bulk substrates have to be modelled using a resistive network representation [6–8]. The N-well is, however, still modelled as a single node owing to its comparatively higher conductivity than the bulk. Among the coupling mechanisms shown in Figure 7.2, the

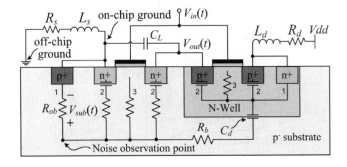

Figure 7.2 Cross-section view of a CMOS inverter stage and the equivalent circuit of the coupling mechanisms of the substrate noise.

couplings through the power supply and ground contacts – the n^+ node of the PMOS and the p^+ node of the NMOS, respectively – dominate the injection effects [6, 9].

Apart from just revealing the basic underlying noise generation and coupling mechanisms the inverter is also quite instrumental in describing the noise behaviour of large-scale digital designs. Most digital blocks include large-sized inverter stages to act as output buffers. During switching the comparatively larger buffer circuits are likely to dominate the switching noise contribution. Moreover, with many digital blocks integrated on the same die, there is a higher probability for these inverters to switch simultaneously. All of these switching noise contributions are injected either into the substrate or propagate via interconnects. Figure 7.3 illustrates how the switching noise may couple from the digital blocks to the sensitive analogue/RF circuits.

As analogue/RF circuits share the same die and in some cases even the same on-chip ground, the substrate noise can couple from areas with digital circuits either via the substrate or via the power rails to reach and disturb analogue/RF circuits. The coupled substrate noise could therefore lead to a significant performance deterioration of the analogue/RF circuit, such as degraded noise figures of low noise amplifiers.

To mitigate the effects of switching noise in mixed-signal integrated ultra wideband systems it is therefore highly important to study substrate noise related issues in circuits and to come up with suitable mitigation solutions for those issues. To reach this goal a number of tasks need to be accomplished: (1) impact evaluation, (2) modelling and (3) suppression of substrate noise.

7.2   Impact Evaluation of Substrate Noise

For the previously reported work on the effects of substrate noise on analogue/RF integrated circuits (ICs), almost all of them are focusing on narrowband applications [8–11]. It has not been fully investigated if substrate noise is a severe issue for wideband systems. It is also unclear how the substrate noise affects the wideband sensitive circuits. This section experimentally investigates the features of substrate noise in a wide frequency band from DC to 10 GHz, and attempts to evaluate the impact of substrate noise on the performance of wideband RF circuits. The vehicle used for the investigation is a 1–5 GHz low-noise amplifier (LNA) for UWB systems [12].

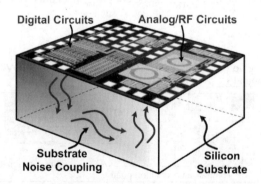

Figure 7.3 Illustration of the coupling mechanisms of substrate noise from the digital block to sensitive analogue/RF circuits.

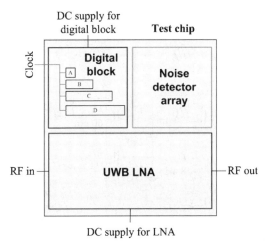

Figure 7.4 The schematic of the integrated circuit for the experiment.

7.2.1 Experimental Impact Evaluation on a UWB LNA

The schematic used for the experiment is shown in Figure 7.4. It consists of a digital block, a substrate noise detector array and a UWB LNA. The digital block plays the role of a substrate noise generator. The substrate noise detector array is used in the experiment for substrate noise measurements. The UWB LNA is used as a victim circuit for the experiment.

The detailed schematic of the digital block consisting of four buffer chains is shown in Figure 7.5. The buffer chain A contains six one-stage inverters.

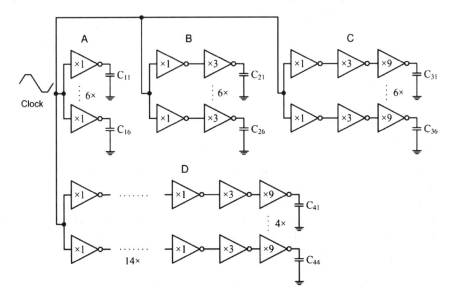

Figure 7.5 The schematic of the digital block consisting of four buffer chains for the generation of switching noise.

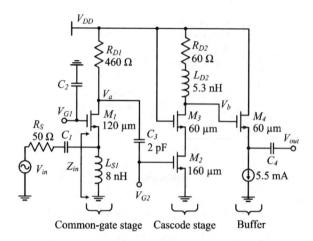

Figure 7.6   The schematic of the UWB LNA for noise impact evaluation. Reproduced from: Ming Shen (2010) Design and implementation of a 1–5 GHz UWB low noise amplifier in 0.18 μm CMOS. Analog Integrated Circuits and Signal Processing with kind permission from Springer Science+Business Media.

The width/length ratios of the NMOS and PMOS transistors of each inverter are 25 μm/0.18 μm and 50 μm/0.18 μm, respectively. The buffer chain B, C and D are multi-stage buffer chains with different stage numbers and sizing factors shown in Figure 7.5. C_{ij} represents the parasitic capacitance at the output of each buffer chain.

Figure 7.6 shows the schematic of the UWB LNA. It is a 1–5 GHz two-stage single ended amplifier with a source follower buffer for measurement purposes. In previously reported work, the study of substrate noise above 1 GHz is rare. For that reason the lower bound of the operation band in this design is set to 1 GHz aiming to obtain more results. This is done while maintaining the desired performance in the UWB band of 3–5 GHz.

7.2.1.1   Test Chip

The fabricated test chip is shown in Figure 7.7 and the PCB with the bonded chip for the measurements is shown in Figure 7.8.

7.2.1.2   UWB LNA

The UWB LNA is measured to make sure that it has a proper overall performance. The common-gate stage and the cascade stage consume 5 mA in total at a supply voltage of 1.8 V. The simulated and measured S-parameters are shown in Figure 7.9.

It can be seen that the measured gain is 11–13.7 dB, the measured $|S_{11}|$ is less than −12 dB and the measured $|S_{22}|$ is less than −10 dB in the frequency band of 1–5 GHz. The simulated and measured noise figures are shown in Figure 7.10 and the measured NF is 5.0–6.5 dB in 1–5 GHz. It can be seen that this UWB LNA has a fairly good overall performance.

7.2.2   Results and Discussion

In the experiment, square waves from a function generator Agilent® 33250a are used as the clock to drive the digital block. The substrate noise due to the switching noise generated by the digital block is measured at the noise detector array as shown in Figure 7.11(a). It can be

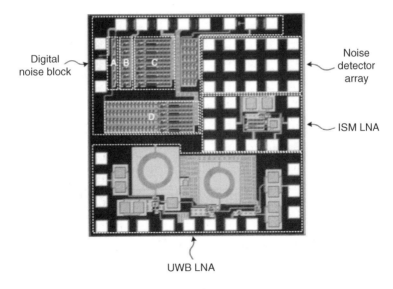

Figure 7.7 The microphotograph of the test chip.

seen that the major part of the substrate noise is located in the frequency band from DC to 2 GHz. However, there are also noise tones in the frequency band up to 10 GHz with magnitudes about 10 dB higher than the noise floor. The spectrum of the RF output of the LNA is also measured with the LNA turned on. In this case the LNA is fed with a sinusoidal input of −50 dBm at 3.88 GHz and the resulting spectrum is illustrated in Figure 7.11(b).

It is clear that the magnitudes of the substrate noise in 2–5 GHz are much higher than those measured at the noise detector. This indicates that the in-band substrate noise coupled at the input of the LNA has been amplified by the LNA. In practical applications the magnitude of the received UWB signal could be significantly lower than the magnitude of the substrate noise as the allocated EIRP of a UWB signal is lower than −41.3 dBm/MHz [13].

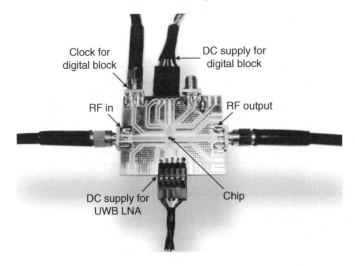

Figure 7.8 The PCB with the bonded chip for the measurements.

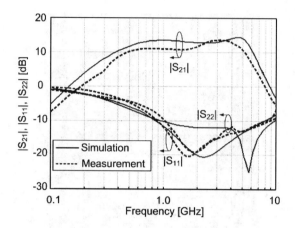

Figure 7.9 Measured and simulated S parameters versus frequency of the UWB LNA. Reproduced from: Ming Shen (2010) Design and implementation of a 1–5 GHz UWB low noise amplifier in 0.18 μm CMOS. Analog Integrated Circuits and Signal Processing with kind permission from Springer Science+Business Media.

The effects of the clock frequency on the substrate noise is investigated by measuring the spectrum at the RF out of the LNA, while varying the frequency of the digital clock. The measured results are shown in Figure 7.12, zoomed into the frequency band of 3.5–4.5 GHz for more detailed information. It can be seen that the magnitude of the substrate noise is increased when f_{clock} is increased from 10 to 50 MHz. This is due to the fact that the digital circuit switches faster (generates and injects more switching noise) with a higher clock frequency. This effect will be discussed with more detail in the subsequent subsection.

Apart from the original harmonics of the substrate noise, it can be seen in Figure 7.12 that some extra frequency components are also present. These components are caused by the intermodulation between the RF input signal and the substrate noise at the fundamental and higher order harmonics. To confirm this, the intermodulation components with two different RF input frequencies are measured and marked by circle symbols shown in Figure 7.13.

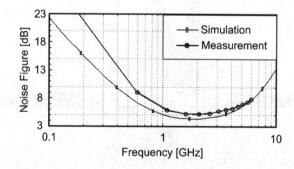

Figure 7.10 Measured and simulated noise figure versus frequency of the UWB LNA. Reproduced from: Ming Shen (2010) Design and implementation of a 1–5 GHz UWB low noise amplifier in 0.18 μm CMOS. Analog Integrated Circuits and Signal Processing with kind permission from Springer Science+Business Media.

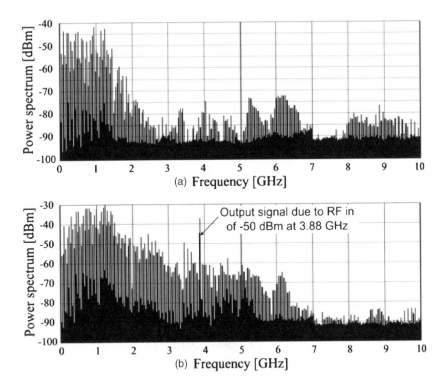

Figure 7.11 Measured power spectrum (a) from the substrate noise detector and (b) from RF out of the UWB LNA; f_{clock} is 50 MHz and the rising and falling time of the clock is 5 ns.

It can be seen that the noises at the harmonics of f_{clock} remain with insignificant changes, while the noise components due to intermodulation shift along with the RF input signal.

The impact of the substrate noise on the UWB LNA is studied by comparing the measured noise figure of the LNA when the digital block is turned off and turned on with three different values of f_{clock}. The measured results are shown in Figure 7.14.

It can be seen that the LNA has a smooth and flat noise figure around 6 dB in the measured frequency band when the digital block is turned off. When the digital block is turned on, significant deterioration of the noise figure is seen. It can also be seen that the deterioration is stronger for higher f_{clock}, which is consistent with the results shown in Figure 7.12.

7.2.3 Conclusion

In this section, the impact of substrate noise on UWB LNA is experimentally evaluated. Significant substrate noise is observed in the frequency band of DC to 10 GHz. It is shown by the results that the substrate noise can drastically deteriorate the performance of the UWB LNA in terms of a noise figure. The results clearly indicate the need for a good understanding of the generation and propagation of the substrate noise, as well as the need of effective approaches for reducing the substrate noise in the design of wideband mixed-signal integrated circuits.

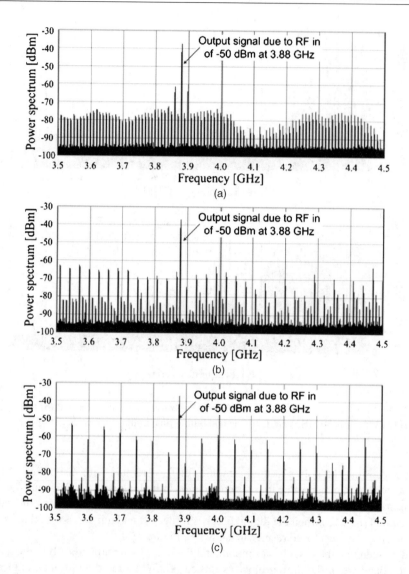

Figure 7.12 Measured power spectrum from RF out of the UWB LNA, while the digital block is driven with three different clock frequencies. The parameters for the clock are: (a) $f_{clock} = 10\,\text{MHz}$, (b) $f_{clock} = 30\,\text{MHz}$, (c) $f_{clock} = 50\,\text{MHz}$. The rising and falling times of the clocks are 5 ns for all of the three cases.

7.3 Analytical Modelling of Switching Noise in Lightly Doped Substrate

7.3.1 Introduction

Switching noise produced in digital circuits has gradually become one of the most important concerns in IC design due to increasingly higher density of digital design. The root cause here is the need for more and more digital functionalities combined with tight requirements for area in mixed-signal integrated circuits. As illustrated in Figure 7.15, the switching noise

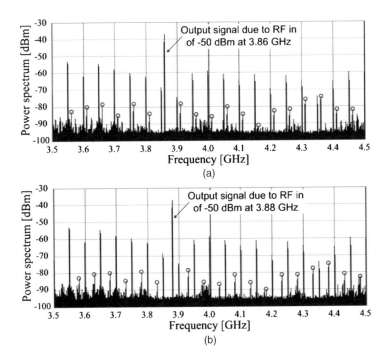

Figure 7.13 Measured power spectrum from RF out of the UWB LNA with two different RF input frequencies. The parameters for the RF input signal are: (a) $f_{RF} = 3.86\,\text{GHz}$, (b) $f_{RF} = 3.88\,\text{GHz}$. The magnitude of the RF input signal is $-50\,\text{dBm}$, f_{clock} is 50 MHz and the rising and falling time of the clock is 5 ns. The circle symbols mark the intermodulation components.

generated by digital circuits can propagate through the substrate to reach sensitive analogue/RF circuits and deteriorate their performance.

A number of methods have been proposed for achieving an efficient model that is able to capture the main features of the switching noise [6–9, 14–20]. For heavily doped substrates

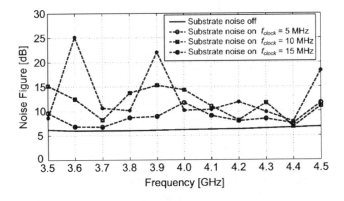

Figure 7.14 Measured noise figure versus frequency of the UWB LNA when the digital block is turned off and turned on with three different values of f_{clock}.

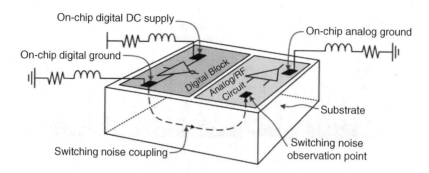

Figure 7.15 Switching noise coupling mechanism in mixed-signal circuits.

the substrate is simplified as one electrical node due to their low resistivity, and switching noise modelling methods using this simplification have been proposed [15, 16]. Featuring higher resistivity and better noise isolation properties, lightly doped substrates are widely used in mixed-signal designs [9, 14, 17]. Since the resistivity of lightly doped substrates is much higher than that of heavily doped substrates, the substrate can no longer be treated as a single electrical node. Hence, modelling methods applicable for heavily doped substrates fall short when lightly doped substrates are considered.

Conventional approaches to characterize switching noise in lightly doped substrate are typically highly dependent on SPICE simulations (SPICE is a simulation program with IC emphasis). One example is the computationally extensive approach where digital circuit blocks are simulated directly together with an extracted substrate network [6, 7]. While this approach may lead to accurate results it involves lengthy and time-consuming simulations. Therefore, in an attempt to reduce simulation complexity the use of macro models has been proposed [8, 9, 14, 17, 18]. Here the switching current from digital blocks is typically represented using an asymmetrical triangular waveform which, when combined with a lumped element equivalent network representing the substrate, allows for fast generation of SPICE simulation results. Both of the above approaches rely heavily on simulations but despite providing for accurate results neither offer sufficient insight into switching noise generation and propagation.

A slightly different approach is to make use of analytical methods where, based on mathematical analysis of waveform functions characterizing the switching current and a transfer function model of the switching noise propagation, closed-form expressions for the switching noise can be derived [20, 21]. While this method provides more insight into the propagation of switching noise the switching current source still needs to be characterized through SPICE simulations. Further, layout extractions are still needed to identify the resistances needed to derive the noise propagation transfer function.

To avoid the need for layout extractions the use of compact models to characterize the spreading resistance between arbitrary sized diffusion contacts on lightly doped substrates has been proposed [21]. By doing so the transfer function describing the propagation of the switching noise can be determined without the need for cumbersome layout extractions. The combination of such a compact model of the propagation transfer function and an analytic closed-form expression for the switching noise would provide the mixed-signal designer with a very powerful noise estimation tool and allow for proper measures to be taken early in the design flow. Such a tool would clearly be advantageous, however, as currently a feasible analytical model characterizing the switching current source is still unavailable.

This section presents two modelling methods aiming to provide feasible models for the switching noise of simple digital circuits and large-scale digital blocks, respectively. The switching noise generated by individual inverters is analytically investigated. An analytical model, named the GAP (Gaussian pulse) model, is proposed to characterize the switching noise of individual inverters. The model is validated by both SPICE simulations and measured results obtained from a test chip fabricated in a lightly doped CMOS process. The simulated and measured results are in good agreement with the proposed model. The GAP model is suitable only for simple circuits, so to extend its usability to include large-scale circuits also, the GAP model is complimented by a statistical analysis method. The extended method is verified by SPICE simulation results.

7.3.2   The GAP Model

The reason for focusing on the analysis of individual inverters is based on two facts:

1. The inverter is a basic and often used building block in digital circuits.
2. A typical large-scale digital circuit usually contains big size inverters that act as buffers. Due to the large sizes, these inverters generate significant switching noise, and therefore usually dominate the total generation of the switching noise in the digital circuit [19].

Thus, analysing the switching noise in an individual inverter is considered a good starting point for obtaining detailed understanding of the generation and injection of the switching noise. The proposed model approximates the switching current using a Gaussian pulse (GAP) function. The key parameters in the model are derived by solving the differential equation describing the output voltage of a capacitively loaded inverter. Combining the switching current model and the transfer function, which models the propagation of the switching current, the spectrum of the switching noise can be predicted.

The simplified switching noise coupling mechanisms involved in Figure 7.15 are illustrated by four blocks in Figure 7.16. The digital supply rail block and the digital block in Figure 7.16 are similar in structure to the coupling network used in Reference [8]. In the digital block, $I_s(t)$ represents the switching current generated by the digital block. C_{cir} is the circuit capacitance between the on-chip digital DC supply and the on-chip digital ground.

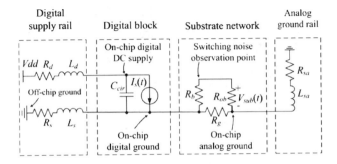

Figure 7.16   The switching noise coupling network.

The substrate network block and the analogue ground rail block are used to model the switching noise coupling to analogue circuits. R_d and L_d are the resistance and inductance of the bonding wire and interconnect connecting the on/off-chip digital DC supply, respectively. R_s, R_{sa}, L_s and L_{sa} are the resistances and inductances of the interconnects and bonding wires connecting the off-chip ground to the on-chip digital and analogue ground, respectively. R_g is the resistance between the on-chip digital and analogue grounds, R_{ob} is the substrate resistor between the noise observation point and the on-chip analogue ground and R_b is used to model the spreading resistance between the on-chip digital ground and the observation point on the substrate. Based on Figure 7.16 the resulting noise voltage at the observation point can be derived as

$$V_{sub}(j\omega) = H(j\omega)I_s(j\omega) = \frac{R_g R_{ob} Z_L I_s(j\omega)}{\left(j\omega C_{cir}\left(\dfrac{Z_L Z_{ea}}{Z_L + Z_{ea}} + Z_D\right) + 1\right)(Z_{ea} + Z_L)(R_b + R_{ob} + R_g)}$$

(7.1)

where $Z_D = R_d + j\omega L_d$, $Z_L = R_s + j\omega L_s$ and $Z_{ea} = R_{sa} + R_g(R_d + R_{ob})/(R_g + R_d + R_{ob}) + j\omega L_{sa}$. How to find the parameters in the transfer function $H(j\omega)$ has been provided in previous studies [8, 20]. The remaining challenge is to find a scalable analytical model for I_s.

7.3.2.1 The Switching Current of Individual Inverters

A typical schematic of a capacitively loaded inverter is shown in Figure 7.17(a). Here $I_p(t)$ is the drain current of the PMOS device, $I_n(t)$ is the drain current of the NMOS device and $I_{ch}(t)$ is the charging/discharging current. The input voltage (dash line) and output voltage (solid line) in a falling input transient of the inverter are shown in the top figure of Figure 7.17(b). As V_{in} falls, the PMOS is turned on at time t_{pon} and the capacitor load is charged towards a higher voltage. When $V_{out}(t)$ reaches V_{opsl} at t_{psl} and $V_{out}(t_{psl}) - V_{in}(t_{psl}) = |V_{tp}|$, where V_{tp} is the threshold voltage, the PMOS leaves the saturation region and enters the linear region. V_{of} is defined as the output voltage at the falling time t_f. The time when the output reaches 90% of the final value is defined as t_{tw}.

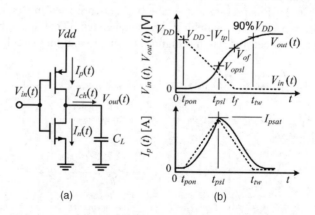

(a)   (b)

Figure 7.17 (a) The schematic of a capacitively loaded inverter and (b) the input/output voltage (dash/solid line in the top figure) and switching current (bottom figure) of a falling input transient.

The switching current I_p during the transient is shown by the solid line at the bottom of Figure 7.17(b). A number of modelling approaches of the output response and switching currents in CMOS inverters have been reported [22, 23]. However, to derive the expression of the switching current is no trivial task, and no comprehensive expressions for the switching current have been presented. To simplify the analysis task, a triangular waveform is often used to model the switching current, as shown by the dash line in the bottom figure in Figure 7.17(b) [20]. The triangular model has a peak value of I_{psat} at t_{psl}, and its magnitude is zero at $t \leq t_{pon}$ and $t \geq t_{tw}$. This model is simple and accurate for estimating the power spectral density (PSD) of the switching noise at low frequencies. However, this approach leads to significant estimation errors at high frequencies due to the abrupt simplifications of the switching current [8, 20]. Moreover, the expressions of the key parameters such as t_{psl} and t_{tw} are not provided. In this study, Gaussian pulse equations are used to model the switching currents. The resulting GAP model is thus described by the following equation:

$$I_p(t) = \begin{cases} I_{psat} \times \exp\left[\dfrac{-2\pi(t - t_{psl})^2}{(1.45 \times (t_{psl} - t_{pon}))^2}\right] & \text{for } 0 \leq t \leq t_{psl} \\[4mm] I_{psat} \times \exp\left[\dfrac{-2\pi(t - t_{psl})^2}{(2.0 \times (t_{tw} - t_{psl}))^2}\right] & \text{for } t > t_{psl} \end{cases} \tag{7.2}$$

The GAP model has the same peak current value as the triangular model at t_{psl} and its magnitude is 10 and 20% of I_{psat} at t_{pon} and t_{tw}, respectively. Compared with the triangular model, the GAP model has smoother transitions at the peak and bottom of the waveform. As a result, the GAP model produces less predicting errors at high frequencies. The expressions of the parameters in the GAP model are derived in the following part of this section. The expressions can also be used in the conventional triangular model.

In this analysis, the short circuit current during switching transients is neglected. This approximation is based on the fact that the charging/discharging currents contribute the major part of the switching current for capacitively loaded inverters in most of the present CMOS processes [22–24]. Under this assumption, the NMOS and PMOS are assumed to be off during a falling input transient and a rising input transient, respectively, and $I_s(t)$ only consists of $I_p(t)$ generated at the falling input transient. The discharging current generated at the rising input transient is flowing in the closed loop formed by the NMOS and C_L. Thus it has no contribution to $I_s(t)$ and is consequently neglected. Moreover, the analysis in this study is based on the well-known square-law MOSFET (metal oxide semiconductor field effect transistor) model. This simplification is necessary since higher order models are intractable for analytical manipulation. However, as shall be seen in the measurement section, the results derived based on the simple model are sufficiently accurate. In the square-law model the transistor drain current is expressed as

$$I_D = K_p\left[(V_{gs} - V_t)V_{DS} - \frac{V_{DS}^2}{2}\right] V_{DS} < V_{Dsat} \tag{7.3}$$

$$I_D = \frac{1}{2}K_p(V_{gs} - V_t)^2 V_{DS} \geq V_{Dsat} \tag{7.4}$$

where $V_{Dsat} = V_{GS} - V_t$, $K_p = \mu C_{ox}(W/L)$ and V_{DS}, V_{GS} are the drain-source, gate-source voltage respectively. For reasons of concise and general expressions, corresponding lower-case letters are used to denote voltages normalized by V_{DD} in the following analysis. For example, $v_{out}(t) = V_{out}(t)/V_{DD}$. The input voltage waveform is approximated to be a falling ramp with a falling time t_f and a slope $s_f = -1/t_f$

$$v_{in}(t) = \begin{cases} s_f \times (t - t_f) & \text{for } 0 \leq t \leq t_f \\ 0 & \text{for } t > t_f \end{cases} \tag{7.5}$$

This approximation is widely used due to its simplicity and effectiveness [22, 23]. With this t_{psl} and t_{tw} can be derived based on the expression of the output voltage, which can be found by solving the following differential equation:

$$C_L \frac{dV_{out}(t)}{dt} = I_p(t) \tag{7.6}$$

where $I_p(t)$ can be replaced by Equation (7.3) or (7.4) with corresponding terminal voltages. Categorizing the falling input transient as two cases where $v_{opsl} < v_{of}$ (slow input ramp) or $v_{opsl} \geq v_{of}$ (fast input ramp), the expressions for those parameters for each case are given as

Case A: $v_{opsl} < v_{of}$

$$t_{psl} = (v_{opsl} + p)/s_f \tag{7.7}$$

where $p = V_{tp}/V_{DD}$ and v_{opsl} can be found by solving the equation

$$v_{opsl} = \frac{K_p V_{DD}}{6s_f C_L}(v_{opsl} - 1)^3 \tag{7.8}$$

Furthermore, t_{tw} can be formed by

$$\begin{aligned} t_{tw} = t_f + [\ln((2(1 + p) - 0.1)/0.1) \\ - \ln\left(\frac{2p + 1 + v_{of}}{1 - v_{of}}\right)] \quad C_L/(K_p V_{DD}(1 + p)) \end{aligned} \tag{7.9}$$

where v_{of} can be determined from

$$\begin{aligned} v_{of}(t) = 1 - \exp\left(\frac{K_p V_{DD}(-1 - p)^2}{2s_f C_L}\right) \Big/ \Bigg[\frac{\exp\left(\frac{K_p V_{DD} v_{opsl}^2}{2s_f C_L}\right)}{v_{opsl}} \\ + \sqrt{\frac{K_p V_{DD}\pi}{8s_f C_L}} \text{erf}\left(\sqrt{\frac{K_p V_{DD} v_{opsl}^2}{2s_f C_L}}, \sqrt{\frac{K_p V_{DD}(-1 - p)^2}{2s_f C_L}}\right)\Bigg] \end{aligned} \tag{7.10}$$

Case B: $v_{opsl} \geq v_{of}$

$$t_{psl} = t_f - \frac{2pC_L}{K_p V_{DD}(1+p)^2} + \frac{1+p}{3s_f} \tag{7.11}$$

$$t_{tw} = t_{psl} + \frac{\ln[(2(1+p)-0.1)/0.1]C_L}{K_p V_{DD}(1+p)} \tag{7.12}$$

In addition, it is easy to derive $I_{psat} = K_p(s_f t_{psl} + v_t)^2/2$ and $t_{pon} = v_{tp}/s_f$. Since the remaining parameters in Equation (7.2) are given by Equations (7.7) to (7.12), the expressions for all parameters needed in the models are available.

7.3.2.2   Validation of the Model

The proposed GAP model is verified by comparing its predictions with HSPICE simulations (HSPICE is a device-level circuit simulator) on the schematic in Figure 7.17(a). A $0.18\,\mu\text{m}$ CMOS process with a $0.34\,\mu\text{m}$ option is used for the verification. Successful verifications with different capacitive loads, transistor sizes and falling times have been conducted. For the verification shown here, the widths of the PMOS and the NMOS are $0.68\,\mu\text{m}$ and $0.34\,\mu\text{m}$, respectively, and both of them have the same length of $0.34\,\mu\text{m}$. The HSPICE model used in the verification is level 49. The related parameters are: $K_p = 4.75 \times 10^{-5}$, $V_{tp} = -0.66\,\text{V}$, $V_{DD} = 3.3\,\text{V}$ and $t_f = 5\,\text{ns}$. To evaluate the model in both the case A and case B scenarios, C_L is set as 0.3 and 1 pF respectively. It should be noted that Equations (7.7) to (7.12) are functions of the term $K_p V_{DD}/C_L$, which means that the inverters having the same V_t, input voltages and $K_p V_{DD}/C_L$ generate switching currents with the same t_{pon}, t_{psl} and t_{tw}. Thus the example shown here represents a group of general scenarios. The simulated and modelled results for case A and case B are shown in Figures 7.18 and 7.19, respectively.

The normalized Fourier transforms of the switching currents are also shown. It can be seen that both the triangular model and the GAP model match the simulation results well at low frequencies and it is obvious that the estimation errors at high frequencies are clearly reduced in the GAP model, as expected.

7.3.2.3   Test Chip

To validate the proposed model experimentally, the model has been practically applied to predict the switching noise generated by a capacitively loaded inverter implemented using a lightly doped CMOS process ($20\,\Omega\,\text{cm}$). The microphotograph of the test chip is shown in Figure 7.20.

The PMOS and NMOS devices in the inverter are $600\,\mu\text{m}$ and $300\,\mu\text{m}$ wide, respectively, and both transistors have the same length of $0.34\,\mu\text{m}$. The reason for using such fairly large transistors is to generate enough switching noise to enable measurements. In this case, $K_p = 0.04$ and $V_{tp} = -0.7\,\text{V}$. A 20 pF capacitor is connected to the output of the inverter as a load. In the measurement, the inverter is driven using a periodic square wave signal as $V_{in}(t)$. It has a high and a low voltage level of 3.3 and 0 V, and it is fed using a

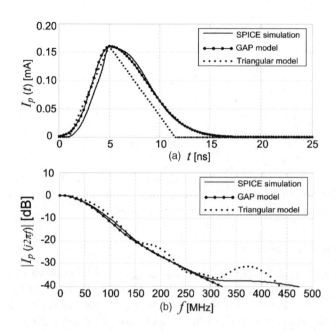

Figure 7.18 Modelled and simulated results for the capacitively loaded ($C_L = 0.3\,\text{pF}$) inverter in Case A: (a) time domain, (b) frequency domain.

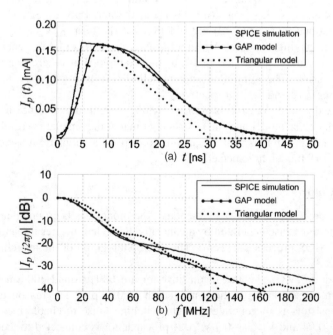

Figure 7.19 Case B: (a) the modelled and simulated switching current of the individual inverter in the time domain and (b) the frequency domain. $C_L = 1\,\text{pF}$ in this case.

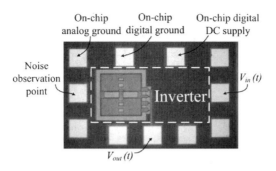

On-chip　　　　On-chip　　　On-chip digital
analog ground　digital ground　　DC supply

Noise
observation
point

$V_{in}(t)$

Inverter

$V_{out}(t)$

Figure 7.20　Microphotograph of the test chip.

G-S-G probe connected to a signal generator. The DC supply voltage is 3.3 V and it is fed to the on-chip digital supply and ground using DC probes. When the inverter is switching due to the square wave, the switching noise in the test chip is measured at the noise observation point using a G-S-G probe connected to a spectrum analyser. The observation point is an ohmic contact connected to the substrate [25].

During the measurements, the spectrum analyser is set to have a low displayed noise floor (around −90 dBm in this case) to measure as much as possible of the low-powered switching noise. Since the input signal is periodic, the switching noise is periodic as well. Thus the switching noise could be represented as a summation of harmonics in the frequency domain

$$V_{sub}(j\omega) = \sum_{k=-\infty}^{+\infty} \frac{2\pi H(j\omega) I_p(j\omega)}{T} \delta(\omega - k2\pi/T) \tag{7.13}$$

where $I_p(j\omega)$ is the Fourier transform of $I_p(t)$ in Equation (7.2). $H(j\omega)$ is expressed in Equation (7.1) and can be derived based on parameters extracted from the layout and bonding wires [8, 20]. T is the period of the input signal and k is an integer. Based on Parseval's theorem, the term

$$\left| H(jk2\pi/T) I_p(jk2\pi/T)/T \right|^2 \tag{7.14}$$

is the average power of the switching noise at its kth harmonic. In this case, $H(j\omega)$ is simplified as a purely resistive network since the measurement is on-wafer and no bonding wires and on-chip decoupling capacitors are used. Hence $H(j\omega)/T$ only affects the magnitude of the switching noise but not the spectral envelope, which is given by the term $I_p(j\omega)$. Thus the measured PSDs of the switching noise can be compared with the calculated magnitudes of the harmonics to verify the proposed model. Measurements with different signal frequencies (from 20 to 50 MHz) and different falling times have been conducted. By comparing v_{opsl} and v_{of} obtained from Equations (7.8) and (7.10), respectively, the GAP model of case A is applicable in this test. Thus corresponding equations of case A are used for obtaining $I_p(j\omega)$. A comparison of measurement results, the triangular model and the GAP model is shown in Figure 7.21.

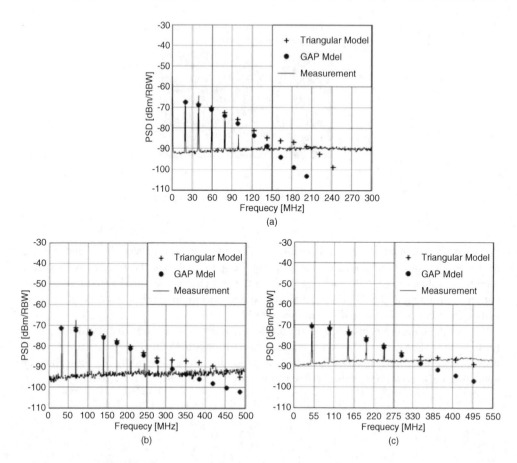

Figure 7.21 Measured PSDs and modelled harmonics of the switching noise: (a) $f_{clk} = 20\,\text{MHz}$, $t_f = 22\,\text{ns}$, RBW $= 500\,\text{kHz}$; (b) $f_{clk} = 35\,\text{MHz}$, $t_f = 9.3\,\text{ns}$, RBW $= 100\,\text{kHz}$; (c) $f_{clk} = 50\,\text{MHz}$, $t_f = 7.8\,\text{ns}$, RBW $= 1\,\text{MHz}$.

The harmonics with the highest magnitude of the models are normalized to the magnitude of the first harmonic of the measured switching noise. It can be seen that the GAP model matches the measured results quite well as it is within 4 dB for all measured switching noise components in the three cases. Moreover, the GAP model is more accurate than the triangular model at high frequencies. This is shown in Figure 7.21(a) at the frequency of 100 MHz and Figure 7.21(b) at the frequency of 285 MHz. For higher frequencies, the switching noise drops below the noise floor of the spectrum analyser. However, the harmonics of the triangular model are higher than the noise floor at these frequencies, which also indicates larger estimation errors in the triangular model.

7.3.3 The Statistic Model

In large-scale digital circuits, there are numerous gates switching with varying delays on the chip. For a synchronized digital circuit, the switching currents are aligned at the edge of the clock with random amplitudes and random time delays. For such a scenario the substrate

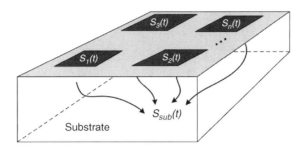

Figure 7.22   Coupling mechanism of the switching noise in large-scale digital blocks.

noise is a sum of these signals and cannot be simply modelled by a periodic current pulse as shown in Figure 7.22.

A statistic model is proposed to solve the problem. This model is based on the assumption that all switching current sources can be represented by a common pulse function yet with random magnitudes and random time delays. This assumption is based on the fact that identically sized inverters are widely used in large-scale digital blocks and the driving signal and load condition for these inverters are usually similar as well. Hence the switching currents generated by these inverters are expected to be similar to each other. Based on this assumption, the noise coupling mechanism shown in Figure 7.23(a) in large-scale digital blocks can be approximated as a multipath coupling model with a single noise source as shown in Figure 7.23(b). In the figure, $S_u(t)$ is the mean of the switching noises and $H_i(t)$ is the subtransfer function describing the random attenuation and delay of the switching noises.

Using the equivalent multipath model, the substrate noise can be expressed as $V_{sub}(f) = S_u(f)H(f)$, where $H(f)$ is the total transfer function of the coupling. $H(f)$ is the Fourier transform of the impulse response $h(t)$ of the noise coupling model shown in Figure 7.23(b); $h(t)$ can be expressed as

$$h(t) = \sum_{i=0}^{N} A_i \delta(t - T_i) \tag{7.15}$$

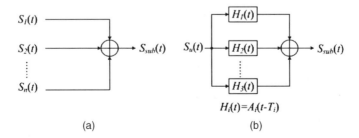

Figure 7.23   (a) The original switching noise and (b) the equivalent multipath coupling model.

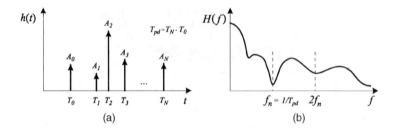

(a)   (b)

Figure 7.24 (a) Impulse response function of the equivalent multipath noise coupling model and (b) the coupling transfer function.

where A_i and T_i represent the amplitude and the time delay of the ith impulse, respectively. N is the number of the noise sources. Thus

$$H(f) = \sum_{i=0}^{N} A_i \exp(-j2\pi f T_i) \tag{7.16}$$

The illustrations of $h(t)$ and $H(f)$ are shown in Figure 7.24(a) and (b), respectively. Due to the time delay, $H(f)$ typically features high magnitudes at lower frequencies and remarkably lower magnitudes at frequency f_n and its multiples. It is easy to derive $f_n = 1/T_{pd}$, where T_{pd} is the time delay between the last and the first impulse. This means that the frequency band ($<f_n$) with the most server substrate noise can be denoted by only one parameter, T_{pd}. Comparing f_n with the frequency band where the sensitive analogue/RF circuits work, it can be estimated if the sensitive circuits are working in the most noisy frequency band, which is useful information for IC designers. In practical digital designs T_{pd} can be approximated by the maximal propagation delay of the circuits, and usually it can be easily obtained by compact equations or SPICE simulations (no extracted substrate network is needed, which is different from conventional simulation methods).

The equivalent multipath model is verified by SPICE simulation. An inverter chain block shown in Figure 7.25 is used for the simulation. A 0.18 μm CMOS process is used. Minimum width and length are used for all the NMOS in the inverters. PMOS in all the inverters is twice as large as the corresponding NMOS. C_{li} represents the parasitical capacitance at the

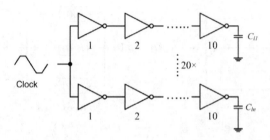

Figure 7.25 The schematic of the inverter chain block for simulation of the switching noise.

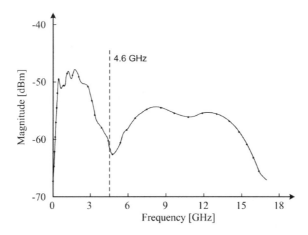

Figure 7.26   Simulated PSD of the switching noise generated by the circuit in Figure 7.25.

output of the inverters. Figure 7.26 shows the PSD of the switching noise in the inverter chain block. The simulated maximal propagation delay T_{pd} is 216 ps and the calculated notch location is 4.6 GHz. It can be seen that the calculated notch location is consistent with the simulation results.

7.3.4   Conclusion

This section presents two modelling methods for switching noise in both small- and large-scale digital circuits. The novel analytical model for a simple digital circuit provides a detailed understanding of the generation and propagation of switching noise in lightly doped CMOS technologies. The spectral envelope of the switching noise can be easily predicted using the proposed model. Closed-form expressions for calculating the parameters in the conventional triangular model are also provided. The model is validated by both SPICE simulations and measurement results from a test chip fabricated in a lightly doped CMOS process. Good agreement is found between the model and simulations as well as measurement results. With this model, the simulations needed when using traditional approaches can be avoided. This can help to achieve a scalable analytical switching noise model for digital blocks. This model especially holds the advantage for estimating the switching noise when big size buffers are dominating the noise generation. An equivalent multipath coupling model is also proposed to characterize the switching noise in large-scale synchronized digital circuits.

7.4   Substrate Noise Suppression and Isolation for UWB Systems

7.4.1   Introduction

Various methods have been proposed over time for the isolation and suppression of the substrate noise in mixed-signal ICs [26–29]. These methods can generally be divided into two categories. One is passive approaches, including physical separation, guard rings or deep N wells. The passive methods are practically feasible. However, they usually suffer the drawbacks of consuming large areas, uncertainty of the isolation level and high cost. The other

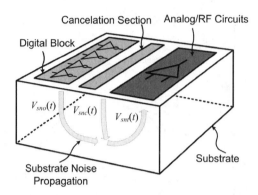

Figure 7.27 The propagation mechanism of the substrate noise from the digital block to the analogue/RF block and the illustration of the proposed active suppression technique.

method is active cancellation. Active cancellation is proposed to achieve low cost yet effective substrate noise suppression. The basic idea in this method is to cancel the substrate noise at the RF area by injecting an anti-phase signal produced by an active cancellation circuit. As shown in Figure 7.27, $V_{sno}(t)$ is the original substrate noise generated by the digital circuits. $V_{snc}(t)$ is the signal generated by the active cancellation section and $V_{snt}(t)$ is the total substrate noise beneath the RF circuit. If $V_{snc}(t) = -V_{sno}(t)$ at a location, $V_{sno}(t)$ can be totally cancelled and $V_{snt}(t)$ is reduced to zero at such a location. However, the result is not as expected. Due to the fact that the signal propagation in the substrate noise is layout dependent, the location where $V_{snt}(t)$ can be reduced is significantly based on the specific layout. If the layout is not properly designed, the substrate noise could be increased rather than reduced. This degrades the feasibility of the method. In addition, owing to the limited frequency band of the active cancellation circuit, this method is only feasible for low frequency substrate noise suppressions.

This section presents a novel active suppression technique for substrate noise in mixed-signal ICs. In the proposed active suppression technique, an active spectrum shaping section is used to generate extra switching currents to modify the shape of the original switching noise in both the time domain and frequency domain.

The relative time delays of these switching currents to the original switching current are realized by controllable delay lines. The pulse widths and magnitudes of the switching currents are also controllable. By manipulating the extra switching currents, the spectrum of the total switching noise can be shaped to have a suppressed magnitude at desired frequencies.

7.4.2 Active Suppression of Switching Noise in Mixed-Signal Integrated Circuits

In the presented technique, an active cancellation section is used to generate an extra switching current. This additional current is introduced with a specific time delay relative to the original switching current. The pulse width and magnitude of the switching current are also adjustable by biasing voltages. Due to the relative time delay, the extra switching current is

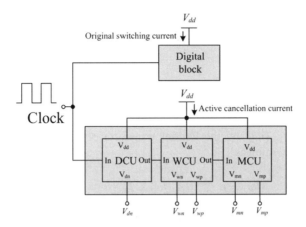

Figure 7.28   The circuit diagram of an active suppression circuit for the IC in Figure 7.27.

out-of-phase with respect to the original switching current at specific frequencies. Thus, by adjusting the time delay, pulse width and magnitude of the extra switching current, the original switching current can be suppressed at desired frequencies.

Figure 7.28 shows the diagram of the active cancellation section and its connection with the clock and DC supply. It comprises three subsections, the delay control unit (DCU), the width control unit (WCU) and the magnitude control unit (MCU). The MCU is used to generate the major part of the extra switching current, and the magnitude of the switching current can be controlled by the controlling voltage V_{mp} and V_{mn}. The DCU and the WCU are used to adjust the time delay and pulse width of the switching current, respectively. V_{dn} is for controlling the time delay, while V_{wn} and V_{wp} are for controlling the pulse width. The active cancellation section is driven by the same clock driving the digital circuit, so that the extra switching current can be aligned to the original switching current with the desired time delay.

7.4.2.1   Theory

The switching current generated by a digital block could be represented by a triangular waveform. One such current in a single cycle is shown as $I_{sno}(t)$ in Figure 7.29. The basic idea in this invention is to inject an extra current signal $I_{snc}(t)$ to cancel $I_{sno}(t)$ in the desired frequency band. $I_{snc}(t)$ has a relative delay of $T_1 - T_0$. Thus the total switching current is given as

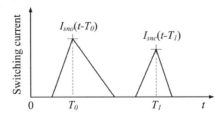

Figure 7.29   Illustration of switching currents involved in the proposed suppression method. The original switching current, I_{sno}, is combined with a cancellation current, I_{snc}, to achieve suppression.

$$I_t(t) = I_{sno}(t - T_0) + I_{snc}(t - T_1) \qquad (7.17)$$

and its Fourier transform is

$$I_t(j\omega) = I_{sno}(j\omega)e^{-j\omega T_0} + I_{snc}(j\omega)e^{-j\omega T_1} \qquad (7.18)$$

where $I_{sno}(j\omega)$ and $I_{snc}(j\omega)$ are the Fourier transform of $I_{sno}(t)$ and $I_{snc}(t)$, respectively. $I_t(j\omega)$ can be further written as

$$I_t(j\omega) = e^{-j\omega T_0}[I_{sno}(j\omega) + I_{snc}(j\omega)e^{-j\omega(T_1 - T_0)}] \qquad (7.19)$$

At the desired substrate noise suppression frequency

$$\omega_0 = (2k + 1)\pi/(T_1 - T_0), \quad \text{for } k = 0, 1, 2 \ldots \qquad (7.20)$$

it has

$$|I_t(j\omega_0)| = |I_{sno}(j\omega_0) - I_{snc}(j\omega_0)| \qquad (7.21)$$

Thus if $I_{snc}(j\omega_0)$ has a similar magnitude and pulse width as $I_{sno}(j\omega_0)$, I_{sno} could be significantly suppressed at ω_0.

7.4.2.2   Validation of the Method in Matlab

The proposed method has been validated by Matlab simulations. One of the simulations is shown in Figure 7.30.

The original switching current $I_{sno}(t)$ is shown by the solid line in Figure 7.30(a). It has a rising time of 100 ps and a falling time of 300 ps. Its magnitude is normalized to 1 A.

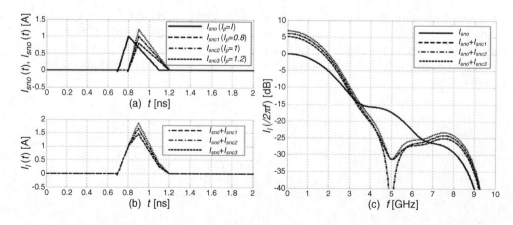

Figure 7.30   Simulated results of the suppression of substrate noise using cancellation currents with three different magnitudes.

The active cancellation current $I_{snc}(t)$ with three different magnitudes are also shown in Figure 7.30(a). The active cancellation currents have the same rising and falling time as the original current. In this simulation, the desired suppression frequency is at 5 GHz. Thus the relative delay between $I_{snc}(t)$ and $I_{sno}(t)$ is set as 100 ps. Figure 7.30(b) and (c) show the total switching current with three different cancellation currents in the time domain and frequency domain, respectively. It is clear that the original switching current is significantly suppressed at the desired frequency. The -10 dB suppression band is wider than 500 MHz for all three cases. The scenarios with different relative delays are also studied. The simulation results are shown in Figure 7.31.

The original switching current $I_{sno}(t)$ is shown by the solid line in Figure 7.31(a). It is the same as that in Figure 7.30(a). The active cancellation current $I_{snc}(t)$ with three different relative delays are also shown in Figure 7.31(a). The active cancellation currents have the same magnitudes as the original current. In this simulation, the desired suppression frequency is also at 5 GHz. Thus the desired relative delay between $I_{snc}(t)$ and $I_{sno}(t)$ is 100 ps. Figure 7.31(b) and (c) show the total switching current with three different cancellation currents in the time domain and frequency domain, respectively. The suppression of the original switching current is clear even though there are undesired offsets in the relative delay in I_{snc2} and I_{snc3}. A time-delay sensitivity demonstration of the conventional anti-phase cancellation approach with an undesired time delay is also shown in Figure 7.31. I_{anti} is the delayed anti-phase current to the original switching current. As the conventional anti-phase cancellation method does not control the time delay, the delay can vary drastically due to different circuit topologies, layouts and interconnects. In Figure 7.31(a), the time delay of I_{anti} is set the same as the delay of I_{snc2}. It can be seen in Figure 7.31(c) that the magnitude of the total switching current is significantly but undesirably increased in the frequency band of 2–8 GHz, which indicates the poor application potential of the anti-phase method for UWB applications unless the timing can be well controlled.

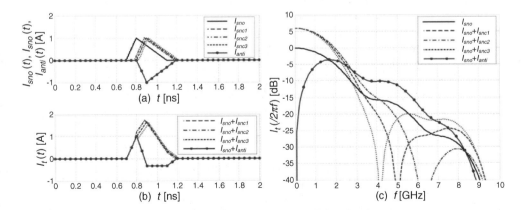

Figure 7.31 Simulated results of the suppression of substrate noise using cancellation currents with three different relative delays. The time delay between I_{sno} and I_{snc1} to I_{snc3} are 80 ps, 100 ps and 120 ps, respectively. The time delay between I_{sno} and I_{anti} is 100 ps.

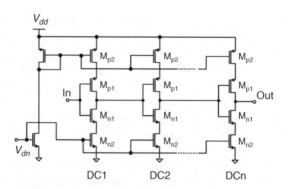

Figure 7.32   Simplified circuit of the delay control unit (DCU).

7.4.2.3   Sub-blocks in the Active Cancellation Section

There can be different circuit implementations of the three control units. The key issue is to ensure that the units must be independent of each other for feasible design and reliable performance. In the following part, some implementation examples are presented and the issue of unit dependency is discussed further.

One possible implementation of a delay control unit consists of a number of delay cells (DSs), as shown in Figure 7.32. By controlling the voltage V_{dn}, the propagation delay of the DCU can be adjusted to a desired value. The number of the delay cells is chosen based on the needed total delay time and the delay that each delay cell can introduce. Small transistors should be used in the delay cells to avoid undesired generation of significant switching currents in this subsection.

Another promising DCU implementation consisting of a number of inverters with tunable capacitive loads is shown in Figure 7.33. By controlling the voltage V_{dn}, the loads of the inverters can be changed and a desired time delay can be obtained. The number of the inverters is chosen based on the needed total time delay. Small transistors should be used in the delay cells to avoid undesired generation of significant switching currents in this subsection.

A simple WCU implementation consisting of one or a number of delay cells is shown in Figure 7.34. By controlling the voltage V_{wn} and V_{wp} the capacity of the WCU to drive the MCU can be adjusted so that the slope of the input signal to the MCU at the falling or rising edge can be adjusted. As a result, the pulse width of the switching current generated by the MCU can be adjusted to a desired value. The sizes of the transistors in the WCU are chosen based on the size of the MCU and the desired pulse width of the switching current.

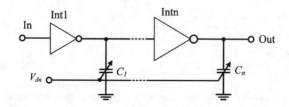

Figure 7.33   Simplified circuit of the delay control unit.

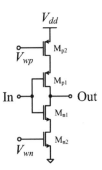

Figure 7.34 Simplified circuit of the width control unit (WCU).

A simple MCU implementation is shown in Figure 7.35. By controlling the voltage V_{mn} and V_{mp} the magnitude of the generated switching current can be adjusted.

7.4.2.4 Validation of the Method in SPICE

The proposed method is also validated by SPICE simulations. The schematic shown in Figure 7.28 is used for the simulation and the schematics shown in Figures 7.32, 7.34 and 7.35 are used for the DCU, WCU and MCU blocks, respectively. The digital block is composed of a number of inverter chains, which act as a switching noise source. The DCU, WCU and MCU are designed based on the information of the original switching current. The controlling voltages of the units are set such that the suppressed switching noise can be achieved at desired frequencies. The simulated results with and without the active suppression are shown in Figures 7.36 to 7.38. The switching current without active cancellation is probed at the DC supply of the digital block, while the switching current with active cancellation is probed from the main DC supply for both the digital block and the active cancellation circuits. From the simulations, it can be seen that the switching noise at the desired frequencies can be significantly suppressed. The minimal suppression is 10 dB, observed in Figure 7.36, and the best is 27 dB, observed in Figure 7.38. It is also clear that the proposed technique is valid for suppression of the switching noise at least from 100 MHz to 3.3 GHz, which is attractive for different applications.

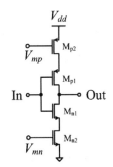

Figure 7.35 Simplified circuit of the magnitude control unit (MCU).

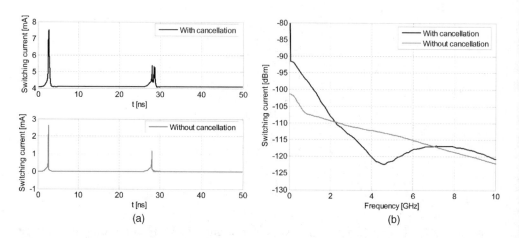

Figure 7.36 The simulated one-cycle switching current with and without active suppression in (a) the time domain and (b) the frequency domain. The designed relative delay in this case is 110 ps and the desired notch frequency is 4.5 GHz.

7.4.2.5 The Mutual Effects between the Design of DCU, WCU and MCU

When designing the active suppression circuits it is desired and needed to have an insignificant mutual effect between the design of controlling voltages for DCU, WCU and MCU. This subsection investigates these effects using SPICE simulations. The schematic used for the SPICE simulations is the same as that in the last subsection.

The performances of DCU, WCU and MCU used for the investigation are defined in Figure 7.39. The relative time delay is defined as the peak-to-peak time delay between the original switching current and the cancellation current at the falling-input edge. The peak value of the cancellation current is defined as I_p and the pulse width is the time span of the

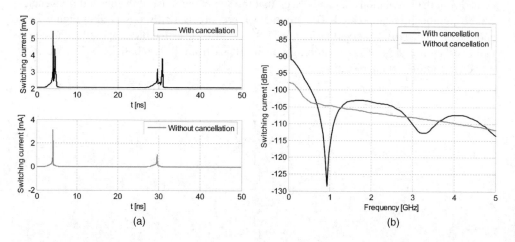

Figure 7.37 Simulated one-cycle switching current with and without active suppression in (a) the time domain and (b) the frequency domain. The designed relative delay in this case is 500 ps and the desired notch frequency is 1 GHz.

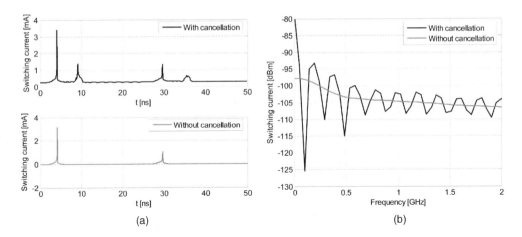

Figure 7.38 Simulated one-cycle switching current with and without active suppression in (a) the time domain and (b) the frequency domain. The designed relative delay in this case is 5 ns and the desired notch frequency is 0.1 GHz.

cancellation current at the value of $0.1 \cdot I_p$. It should be noted that the switching current at the falling-input edge of the magnitude controlling unit is much greater than the switching current at the rising-input edge since it consists of both short circuit current and charging current. Thus the greatest concern of this study is at the falling-input edge.

Figure 7.40 shows the simulated performance of the active cancellation circuits with different control voltages for DCU, WCU and MCU. It can be seen that V_{wn} has significant effects on the time delay when its value is small (Figure 7.40b). The reason for this is that decreasing V_{wn} to a small value close to the threshold voltage greatly reduces the driving capability of the width control unit, which increases the time delay. V_{wn} also has an effect on I_p. The explanation for this can be found in Section 7.3. It has been shown that changing the value of V_{wn} changes the slope of the signal at the output of WCU (input for MCU), which affects the magnitude of the active cancellation current. Apart from V_{wn}, other control voltages are shown to have insignificant mutual effects. Based on the SPICE simulation results, the sequence of the design for DCU, WCU and MCU can be proposed, which is to design V_{wn} first and then V_{mp} and V_{dn}.

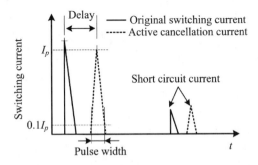

Figure 7.39 The parameter definitions in the SPICE simulation for investigating the mutual effects between the designs of DCU, WCU and MCU.

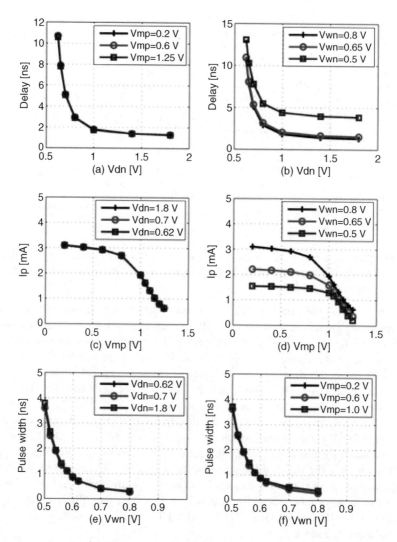

Figure 7.40 Simulated performance of the active cancellation circuit with different values of the control voltages for DCU, WCU and MCU. V_{dd} is 1.8 V, the clock frequency is 20 MHz and the rising/falling time of the clock is 5 ns. Other parameters in the simulations are: (a) $V_{mn} = 0.65$ V, $V_{wn} = 0.8$ V, $V_{wp} = 0.9$ V; (b) $V_{mn} = 0.65$ V, $V_{mp} = 0.2$ V, $V_{wp} = 0.9$ V; (c) $V_{mn} = 0.65$ V, $V_{wn} = 0.8$ V, $V_{wp} = 0.9$ V; (d) $V_{mn} = 0.65$ V, $V_{dn} = 0.7$ V, $V_{wp} = 0.9$ V; (e) $V_{mn} = 0.65$ V, $V_{mp} = 0.2$ V, $V_{wp} = 0.9$ V; (f) $V_{mn} = 0.65$ V, $V_{dn} = 0.7$ V, $V_{wp} = 0.9$ V.

7.5 Summary

This chapter focuses on three aspects of the substrate noise issues, the impact evaluation, modelling and suppression of substrate noise, with special focus on UWB circuits and systems. Efforts have been invested to derive an analytical switching noise model for simple digital circuits and a statistical model for large-scale digital blocks. In addition, an active suppression technique for reducing switching noise in digital circuits is also proposed.

The switching noise model presented in this chapter has originally been devised for CMOS technologies with lightly doped substrates. Satisfactory results have been achieved for standard CMOS technologies. The methodology can be applied to heavily doped substrates with a lightly doped epitaxial layer, silicon on insulator (SOI) and other technologies. The proposed active suppression technique could potentially be integrated in the design of IP cells. In this way, the substrate noise can be reduced at the very beginning of the entire design flow by setting specific design parameters in the IP cell generation. By doing so, the need for area consuming passive noise mitigation schemes is limited, whereby the overall cost of the implementation is reduced.

References

1. Ashton, K. (2009) That 'Internet of Things' Thing [Internet]. *RFID Journal*, July 2009, available from http://www.rfidjournal.com/article/view/4986.
2. IDTechEx Ltd. (2011) Wireless Sensor Networks 2011–2021 [Internet]. Report W0E16F1BB0CEN, July 2011, available from http://pdf.marketpublishers.com/idtechex/wireless_sensor_networks_2010_2020.pdf.
3. Tyczynski, J.E., Bray, F. and Maxwell Parkin, D. (2002) Breast Cancer in Europe [Internet]. ENCR Cancer Fact Sheets, Vol. 2, December 2002, available from http://www.encr.com.fr/DownloadFiles/breast-factsheets.pdf.
4. Paulson, C.N., Chang, J.T., Romero, C.E., Watson, J., Pearce, F.J. and Levin, N. (2005) Ultra-wideband radar methods and techniques of medical sensing and imaging [Internet]. SPIE International Symposium on Optics East, Boston, MA, US, October 2005, available from https://e-reports-ext.llnl.gov/pdf/325992.pdf.
5. Moore, G. (1965) Cramming more components onto integrated circuits. *Electronics Magazine*, **38** (8), April, 114–117.
6. Van Heijningen, M., Compiet, J., Wambacq, P. *et al.* (2000) Analysis and experimental verification of digital substrate noise generation for EPI-type substrates. *IEEE Journal of Solid-State Circuits*, **35** (7), 1002–1008.
7. Su, D., Loinaz, M., Masui, S. and Wooley, B. (1993) Experimental results and modeling techniques for substrate noise in mixed-signal integrated circuits. *IEEE Journal of Solid-State Circuits*, **28** (4), 420–430.
8. Soens, C., Van Der Plas, G., Badaroglu, M. *et al.* (2006) Modeling of substrate noise generation, isolation, and impact for an LC-VCO and a digital modem on a lightly-doped substrate. *IEEE Journal of Solid-State Circuits*, **41** (9), 2040–2051.
9. Mitra, S., Rutenbar, R., Carley, L. and Allstot, D. (1995) A methodology for rapid estimation of substrate-coupled switching noise. Proceedings of the IEEE Custom Integrated Circuits Conference, Santa Clara, CA, pp. 129–132.
10. Min, Xu., Su, D.K., Shaeffer, D.K. *et al.* (2001) Measuring and modeling the effects of substrate noise on the LNA for a CMOS GPS receiver. *IEEE Journal of Solid-State Circuits*, **36** (3), 473–485.
11. Checka, N., Wentzloff, D.D., Chandrakasan, A. and Reif, R. (2005) The effect of substrate noise on VCO performance. Proceedings of Digest Papers of the IEEE RFIC Symposium, pp. 523–526.
12. Shen, M., Tian, T., Mikkelsen, J.H. *et al.* (2011) Design and implementation of a 1–5 GHz UWB low noise amplifier in 0.18 μm CMOS. *Analog Integrated Circuits and Signal Processing*, **67** (1), 41–48.
13. FCC (2002) Revision of Part 15 of the Commission's Rules Regarding Ultra-Wideband Transmission System. Technical Report, ET-Docket, pp. 98–153.
14. Owens, B., Adluri, S., Birrer, P. *et al.* (2005) Simulation and measurement of supply and substrate noise in mixed-signal ICs. *IEEE Journal of Solid-State Circuits*, **40** (2), 382–391.
15. Samavedam, A., Sadate, A., Mayaram, K. and Fiez, T. (2000) A scalable substrate noise coupling model for design of mixed-signal ICs. *IEEE Journal of Solid-State Circuits*, **35** (6), 895–904.
16. Hazenboom, S., Fiez, T. and Mayaram, K. (2006) A comparison of substrate noise coupling in lightly and heavily doped CMOS processes for 2.4-GHz LNAs. *IEEE Journal of Solid-State Circuits*, **41** (3), 574–587.
17. Valorge, O., Andrei, C., Calmon, F. *et al.* (2006) A simple way for substrate noise modeling in mixed-signal ICs. *IEEE Transactions on Circuits and Systems I: Regular Papers*, **53** (10), 2167–2177.
18. Elvira, L., Martorell, F., Aragonés, X. and Luis González, J. (2004) A physical-based noise macromodel for fast simulation of switching noise generation. *Microelectronics Journal*, **35** (8), 677–684.
19. Badaroglu, M., Donnay, S., De Man, H.J. *et al.* (2003) Modeling and experimental verification of substrate noise generation in a 220-K gates WLAN system-on-chip with multiple supplies. *IEEE Journal of Solid-State Circuits*, **38** (7), 1250–1260.

20. Mendez, M., Mateo, D., Rubio, A. and Gonzalez, J. (2006) Analytical and experimental verification of substrate noise spectrum for mixed-signal ICs. *IEEE Transactions on Circuits and Systems I: Regular Papers*, **53** (8), 1803–1815.

21. Kristiansson, S., Ingvarson, F. and Jeppson, K.O. (2007) Compact spreading resistance model for rectangular contacts on uniform and epitaxial substrates. *IEEE Transactions on Electron Devices*, **54** (9), 2531–2536.

22. Hedenstierna, N. and Jeppson, K.O. (1987) CMOS circuit speed and buffer optimization. *IEEE Transactions on Computer-Aided Design*, **6** (2), 270–281.

23. Hamoui, A.A. and Rumin, N.C. (2000) An analytical model for current, delay, and power analysis of submicron CMOS logic circuits. *IEEE Transactions on circuits and systems – II: Analog and Digital Signal Processing*, **47** (10), 999–1007.

24. Kabbani, A. (2008) Modeling and optimization of switching power dissipation in static CMOS circuits. IEEE Computer Society Annual Symposium on VLSI, Montpellier, France, April 2008, pp. 281–285.

25. Shen, M., Tian, T., Mikkelsen, J.H. and Larsen, T. (2009) A measurement fixture suitable for measuring substrate noise in the UWB frequency band. *Analog Integrated Circuits and Signal Processing*, **58** (1), 11–17.

26. Yeh, W., Chen, S. and Fang, Y. (2004) Substrate noise-coupling characterization and efficient suppression in CMOS technology. *IEEE Transactions on Electron Devices*, **51** (5), 817–819.

27. Hsu, T.-L., Chen, Y.-C., Tseng, H.-C. *et al.* (2005) Psub guard ring design and modeling for the purpose of substrate noise isolation in the SOC era. *IEEE Electron Device Letters*, **26** (9), 693–695.

28. Dai, H. and Knepper, R.W. (2009) Modeling and experimental measurement of active substrate-noise suppression in mixed-signal 0.18-μm BiCMOS technology. *IEEE Transactions on Computer-Aided Design of Integrated Circuits and Systems*, **28** (6), 826–836.

29. Chao, H., Wuen, W. and Wen F K. (2008) An active guarding circuit design for wideband substrate noise suppression. *IEEE Transactions on Microwave Theory and Techniques*, **56** (11), 2609–2619.

Part Two

Applications

8

Short-Range Tracking of Moving Targets by a Handheld UWB Radar System

Dušan Kocur and Jana Rovňáková

Department of Electronics and Multimedia Communications, Faculty of Electrical Engineering and Informatics, Technical University of Košice, Košice, Slovak Republic

8.1 Introduction

Ultra-wideband (UWB) radar systems have a variety of potential applications including through wall and through fire detection and tracking of moving targets during security operations, protection of reservoirs, power plants and other critical infrastructures against a terrorist attack, detection of trapped people after an avalanche or earthquake, and so on. UWB radar emissions applied for these scenarios are at a relatively low frequency (typically between 100 MHz and 5 GHz). Additionally, the fractional bandwidth of the signal emitted by UWB radar is very large. The result is a UWB radar system with a fine resolution that also has the ability to penetrate many common materials. Then, such systems are able to detect a moving person by measuring changes in the impulse response of the environments [1]. In the above outlined applications, a short-range UWB radar (range up to 20–25 m) is usually applied. Therefore these scenarios can be referred to as short-range tracking.

The problems of short-range detection and tracking of moving persons have been studied, for example, in References [2] to [7]. However, the problem of multiple human tracking in a real complex environment has been less well addressed. Therefore, on the basis of rich experiences received at several measurement campaigns with the short-range UWB radars the complete signal processing procedure for tracking of multiple targets has been proposed. As the result of this processing, the positions of moving people are not seen as radar blobs localized in the scanned area, which is a typical result of the radar imaging methods [4–7], but the

Microwave and Millimeter Wave Circuits and Systems: Emerging Design, Technologies, and Applications,
First Edition. Edited by Apostolos Georgiadis, Hendrik Rogier, Luca Roselli, and Paolo Arcioni.
© 2013 John Wiley & Sons, Ltd. Published 2013 by John Wiley & Sons, Ltd.

person coordinates are analytically computed by the localization and tracking algorithms. In such a way the proposed procedure employs one-dimensional signal processing and because of this represents a computational less complex alternative to the imaging methods based on two-dimensional signal processing.

The key intention of this chapter is to provide a clear and fundamental description of the proposed signal processing procedure dedicated to short-range tracking of multiple moving targets by the handheld UWB radar system. In order to fulfil this intention, the chapter will have the following structure. Firstly, some requirements for practical handheld UWB radar systems will be outlined in the next section. Then the particular phases of the procedure, that is raw radar data preprocessing, background subtraction, weak signal enhancement, target detection, time-of-arrival (TOA) estimation, target localization and tracking will be described in Section 8.3. There the significance of the particular phases together with a review of signal processing methods, which can be applied for the phase task solution, will be presented. In Section 8.4, the procedure performance will be illustrated based on processing of real radar signals obtained by the short-range UWB radar within a measurement scenario with two persons moving behind a wall. Finally, conclusions and remarks concerning the next research in the field of detection and tracking of moving targets will be drawn in Section 8.5.

8.2   Handheld UWB Radar System

In the case of detection and short-range tracking of moving persons, portable (handheld) radars are applied by operators (e.g. security forces) immediately in the operation place. These devices have to operate in a stand-alone mode, so that the results of the object monitoring are provided to the operator immediately and he or she can change the radar location in a flexible way. A light weight of the devices is preferable. On the other hand, if they are applied for protection of critical infrastructures, the radars have to be small in size and located unobtrusively in monitored region/objects. These requirements ensure that these radar systems usually use only a small antenna array (at least one transmitting and two receiving antennas) necessary for motion detection and basic spatial positioning of the targets by trilateration methods. Other practical demands on handheld radars are for low power consumption and the possibility of operating independently or as part of a sensor network. In the case of the sensor network application, the wireless communication between sensor nodes is usually required.

The handheld prototype of the UWB pseudo-noise radar using a maximum-length binary sequence (referred to as an M-sequence) as the stimulus signal [8], which was created within the project RADIOTECT [9], is an example of the handheld UWB radar system fulfilling the requirements listed here. An experimental version of the M-sequence UWB radar system equipped with one transmitting (Tx) and two receiving horn antennas (Rx1, Rx2) is available at the authors' department (Figure 8.1). The signals obtained by using such a radar have become the basis for designing and testing the UWB radar signal processing procedure described in the next section.

8.3   UWB Radar Signal Processing

The presented signal processing procedure for short-range tracking of multiple moving targets consists of seven phases, namely raw radar data preprocessing, background subtraction,

Figure 8.1 The experimental version of the M-sequence UWB radar.

weak signal enhancement, detection, TOA estimation, target localization and tracking. In the next subsections, the phase significance followed by a number of source references devoted to particular algorithms are given. Additionally, within every processing phase, the main idea of a specific method providing a stable, good and robust performance for the considered application is outlined. Because of the limited range of the contribution, detail descriptions of these methods are not presented. Instead, a mathematical base of the method or its processing flowchart is shown.

8.3.1 Raw Radar Data Preprocessing

Raw radar signals obtained by the M-sequence UWB radar can be interpreted as a set of impulse responses $h(t, \tau)$ of the surroundings, through which the electromagnetic waves emitted by the radar were propagated. They are aligned to each other creating a two-dimensional picture called a radargram, where the vertical axis is related to the propagation time t of the impulse response and the horizontal axis is related to the observation time τ.

The intention of the raw radar data preprocessing phase is to remove or at least to decrease the influence of the radar system by itself to raw radar signals. In our contribution, we will focus on a time-zero setting. In the case of the M-sequence UWB radar, its transmitting antenna transmits M-sequences around periodically. The exact time instant at which the transmitting antenna starts emitting the first elementary impulse of the M-sequence (the so-called chip) is referred to as time-zero t_0. It depends, for example, on the cable lengths between transmitting/receiving antennas and transmitting/receiving amplifiers of radar, total group delays of radar device electronic systems, and so on, but especially on the chip position at which the M-sequence generator started to generate the first M-sequence. This position is randomly changed after every power supply reconnection. To set time-zero means to rotate all received impulse responses in such a way that their first chips correspond to the spatial position of the transmitting antenna. There are several techniques for finding the number of chips needed for such rotation of impulse responses [10].

The most often used method is that of utilizing signal cross-talk, that is the direct wave between transmitting and receiving antennas [10]. The time-zero t_0 is estimated as follows:

$$t_0 = t_{\text{meas}} - t_{\text{calc}} = t_{\text{meas}} - \frac{d}{c} \qquad (8.1)$$

where t_{meas} represents the cross-talk position found in the measured data and t_{calc} is the cross-talk position calculated on the basis of the known distance d between antennas Tx and Rx and the known velocity of propagation in the air c. The preprocessed radar signals $h_p(t, \tau)$ then have the following form:

$$h_p(t, \tau) = h(t - t_0, \tau) \qquad (8.2)$$

The significance of the time-zero setting follows from the fact that targets could not be localized correctly in the measured area without the correct time-zero setting.

8.3.2 Background Subtraction

As the components of the impulse responses due to the target are much smaller than that of the cross-talk between transmitting and receiving antennas or due to reflections from other large or metal static objects, it is impossible to identify any target in the raw or preprocessed radar signals. In order to be able to detect, localize and track the targets, the ratio of signal scattered by the target to noise has to be increased. For that purpose, background subtraction methods must be used. They help to reject especially the stationary and correlated clutter, such as antenna coupling, impedance mismatch response and ambient static clutter, and allow the response of a moving object to be detected.

Let us denote the signal scattered by the target as $s(t, \tau)$ and all other waves and noises jointly as background $b(t, \tau)$. Let us assume also that there is no jamming at the radar performance and the radar system can be described as a linear one. Then, the preprocessed radar signals can be simply modelled by the following expression:

$$h_p(t, \tau) = s(t, \tau) + b(t, \tau) \qquad (8.3)$$

As indicated by the name, the background subtraction methods are based on the idea of subtracting a background (clutter) estimation from preprocessed radar signals. Then, the result of the background subtraction phase can be expressed as

$$h_b(t, \tau) = h_p(t, \tau) - \hat{b}(t, \tau) = s(t, \tau) + [b(t, \tau) - \hat{b}(t, \tau)] \qquad (8.4)$$

where $h_b(t, \tau)$ represents a set of signals with a subtracted background and

$$\hat{b}(t, \tau) = [h_p(t, \tau)]_{\tau_1}^{\tau_2} \qquad (8.5)$$

is the background estimation obtained by $h_p(t, \tau)$ processing over the interval $\tau \in \langle \tau_1, \tau_2 \rangle$.

In the case of a scenario with moving targets, it can be seen very easily that $s(t, \tau)$ for $t =$ const. represents a nonstationary component of $h_p(t, \tau)$. On the other hand, $b(t, \tau)$ for $t =$ const. represents a stationary and correlated component of $h_p(t, \tau)$. Therefore, the methods based on an estimation of stationary and correlated components of $h_p(t, \tau)$ can be applied for the background estimation.

Following this idea, the methods such as basic averaging (mean, median) [11], exponential averaging [12], adaptive exponential averaging [12], adaptive estimation of Gaussian background [13], Gaussian mixture method [14], moving target detection by finite impulse response (FIR) filtering [15], moving target detection by infinite impulse response (IIR) filtering [16], prediction [17], principal component analysis [18], and so on, can be used for background subtraction. These methods differ in relation to assumptions concerning clutter properties as well as to their computational complexity and suitability for online signal processing.

For considered application, a noticeable result can be achieved by using the simple exponential averaging method, where the background estimation is given by

$$\hat{b}(t, \tau) = \alpha \hat{b}(t, \tau - \Delta\tau) + (1 - \alpha)h_p(t, \tau) \tag{8.6}$$

where $\Delta\tau$ represents a delay between successive impulse responses and $\alpha \in (0, 1)$ is a constant exponential weighing factor controlling the effective length of window over which the mean value and background of $h_p(t, \tau)$ is estimated.

8.3.3   Weak Signal Enhancement

It follows from the basic radar equation that the level of signal components scattered by a target and received by the radar depends among other things on the distance between transmitting antenna and target-receiving antenna. Then, the target located close to the radar antenna system is able to produce strong reflections whereas another target located far from the antenna system will reflect only weak signals. This effect strengthened by other phenomena (e.g. multipath reflections or mutual shadowing) can result in mimic target signatures or a target disappearing. These theoretical considerations have been confirmed experimentally by real radar signal processing. Whereas in single moving target scenarios, the target has been detectable almost in all observation time instants, in the case of multiple target scenarios the reflections only from the target situated most closely to a receiving antenna can usually be seen. In order to solve this problem, the enhancement of weak signals scattered by the target has been included in our signal processing procedure.

The weak signal enhancement can be achieved by several approaches. Filters can be ranked between more complicated ones, for example a nonlinear enhancement of weak signals using optimization theory (stochastic resonance), [19] or neural-network-based filters [20]. On the other hand, the method based on weighting of a processed impulse response with regard to distance from the radar antenna system (referred to as time gain method) and the automatic gain control (AGC) method are computationally less complex, but still can produce improved results for the considered application [21]. To such simple methods, the method referred to as the advance normalization method can be ranked as well [21].

The advance normalization method is based on serial searching of maxima υ_{max} in the interval (t_{Lmax}, t_{end}) and consequential normalization of a current signal in the interval

$(t_{L\max}, t_{N\max})$ as follows:

$$h_e(t, \tau) = \frac{h_b(t, \tau)}{v_{\max}} \tag{8.7}$$

where $h_e(t, \tau)$ is the enhanced signal, $t_{L\max}$ is the propagation time instant of the last found maximum, $t_{N\max}$ represents the propagation time instant of the new found maximum and t_{end} is the last propagation time instant of the whole signal. It has been shown in Reference [21] that the advance normalization can be used to advantage for the purpose of weak signal enhancement.

8.3.4   Target Detection

The detection phase is represented by a class of methods that determine whether a target is absent or present in examined radar signals. The solution of target detection task is based on statistical decision theory [22]. Detection methods analyse the enhanced signals $h_e(t, \tau)$ along a certain interval of propagation time $t \in \Xi_t = \{t_1, t_1 + \Delta t, \ldots, t_2\}$ and reach the decision as to whether a signal scattered from target $s(t, \tau)$ is absent (hypothesis H_0) or is present (hypothesis H_1) in $h_e(t, \tau)$. The hypotheses can be mathematically described as follows:

$$H_0: \quad h_e(t, \tau) = n_{BS}(t, \tau) \tag{8.8}$$

$$H_1: \quad h_e(t, \tau) = s(t, \tau) + n_{BS}(t, \tau) \tag{8.9}$$

where $n_{BS}(t, \tau)$ represents residual noise obtained by $h_p(t, \tau)$ processing by a proper background subtraction method. Following expressions (8.8) and (8.9), $n_{BS}(t, \tau)$ can be expressed as follows:

$$n_{BS}(t, \tau) = b(t, \tau) - \hat{b}(t, \tau) \tag{8.10}$$

A detector discriminates between hypotheses H_0 and H_1 based on a comparison of testing (decision) statistics $X(t, \tau)$ and threshold $\gamma(t, \tau)$. Then, the output of detector $h_d(t, \tau)$ is given by

$$h_d(t, \tau) = \begin{cases} 0 & \text{if } X(t, \tau) \le \gamma(t, \tau) \\ 1 & \text{if } X(t, \tau) > \gamma(t, \tau) \end{cases} \tag{8.11}$$

The detailed structure of a detector depends on the selected detection strategy and optimization criteria of detection [22]. The selection of detection strategies and optimization criteria results in a testing statistic specification and a threshold estimation method.

The most important groups of detectors applied for radar signal processing are represented by sets of optimum or suboptimum detectors. Optimum detectors can be obtained as a result of solution of an optimization task formulated usually by means of probabilities or likelihood functions describing the detection process. Here the Bayes criterion, maximum likelihood criterion or Neymann–Pearson criterion are often used as the basis for detector design.

However, the structure of an optimum detector could be extremely complex. Therefore, suboptimum detectors are also applied very often [23].

For the purpose of target detection using UWB radars, detectors with a fixed threshold, (N, k) detectors, detectors based on interperiod correlation processing (IPCP) detectors and constant false alarm rate (CFAR) detectors have been proposed [23, 24]. As well as detectors capable of providing a good and robust performance for detection of multiple moving targets by UWB radar, CFAR detectors can be especially useful. They are based on the Neymann–Person optimum criterion providing the maximum probability of detection for a given false alarm rate.

There are many different CFAR techniques [24–27]. The selection of a particular CFAR detector depends largely on the background noise and clutter models assumed to hold in the detection environment, although selection may depend as well on system constraints, such as the system's signal processing capability or its ability to store data. In addition to conventional and widespread kinds of CFAR the cell-averaging CFAR detector [26] can be classified. Its modification for the case of a Gaussian clutter model is the Gaussian CFAR detector presented in Reference [27] and is also convenient for our application. The main difference between it and the cell-averaging CFAR detector consists in utilization of an exponentially weighted moving filter for updating the decision statistic $X(t, \tau)$ and the threshold $\gamma(t, \tau)$.

8.3.5   Time-of-Arrival Estimation

For target positioning, quantities such as the received signal strength intensity (RSSI), the angle of arrival (AOA) and the time of arrival (TOA) are traditionally used. From them, the RSSI is the least adequate for the UWB case, since it does not profit from the fine space–time resolution of UWB signals and requires a site-specific path loss model. Estimation of the AOA, on the other hand, requires multiple antennas or at least an antenna capable of beamforming at the receiver. This requirement implies size and complexity needs that are often not compatible with the low-cost, small-size constraints associated with typical scenarios for UWB technology. Given the reasons above, the TOA stands out as the most suitable signal parameter to be used for positioning with UWB devices [28]. Therefore, the phase of TOA estimation is also included in the radar signal processing procedure, namely between the detection and localization phases.

If a target is represented by only one nonzero sample of the detector output for the observation time instant $\tau = \tau_k$, then the target is referred to as a simple target. However, in the case of the application analysed in this contribution, the radar range resolution is considerably finer than the physical dimensions of the target to be detected. As a result, the detector output for $\tau = \tau_k$ is not expressed by only one nonzero impulse at $t = TOA(\tau_k)$ expressing the target position by the TOA for the observation time τ_k, but the detector output is given by a complex binary sequence $h_d(t, \tau_k)$. The set of nonzero samples of $h_d(t, \tau_k)$ represent multiple reflections of electromagnetic waves from the target or false alarms. The multiple reflections due to the target are concentrated around the true target position at the detector outputs. In this case, the target is referred to as the distributed target. In the part of $h_d(t, \tau_k)$ where the target should be detected not only nonzero but also zero samples of $h_d(t, \tau_k)$ can be observed. This effect can be explained by a complex target radar cross-section due to the fact that the radar resolution is much finer than that of the target size and taking into account different shape and properties of the target surface. The set of false alarms is due to especially weak signal processing under a very strong clutter presence.

In order to simplify the target localization, such distributed targets are replaced by simple targets; that is the target position in every observation time instant is given by only one TOA. This phase of radar signal processing is referred to as the TOA estimation. In the literature, the most commonly reported methods for the TOA estimation are the correlation approach [29] and the energy collection based approach [30]. However, most of these methods cannot be directly used for the considered application. Therefore we proposed a novel algorithm entitled TOA association in Reference [31]. It enables one to estimate the TOA not only of the distributed targets but also of the multiple targets. Furthermore, it combines the TOA estimated from both receiving antennas of the handheld UWB radar to such couples from which only the positions of the potential true targets can be computed during the localization phase. This part of the algorithm represents a data-association phase (TOA-to-TOA association) and is responsible for the deghosting task solution. Its main idea consists in the utilization of known and short distance d between adjacent antennas (in a handheld radar it is usually less than 50 cm) and results in computation of a small maximal difference TOA_{Dmax} between the TOA estimated from both receivers and belonging to the same single target:

$$TOA_{Dmax} = \frac{2d}{c} + t_{corr} \qquad (8.12)$$

where c is the light propagation velocity and t_{corr} represents a small correcting factor added with the intention of taking into account errors due to measurement and previous phases of radar signal processing. The difference TOA_{Dmax} is calculated on the basis of the triangle inequality arising from the antenna layout and an arbitrary target position. A simplified flowchart of the whole TOA association method is shown in Figure 8.2, with a detailed description of the algorithm given in Reference [31]. The method output is formed by the

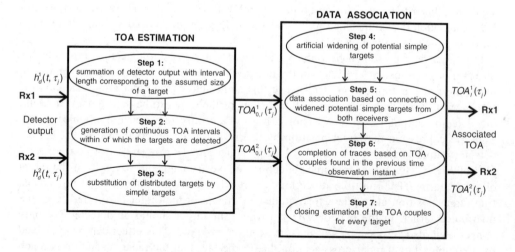

Figure 8.2 A simplified flowchart of the TOA association method.

associated TOA couples $\{TOA_l^1(\tau), TOA_l^2(\tau)\}$ for the lth found target ($l \in L$) in the observation time instant τ, where L represents the total number of targets.

8.3.6 Target Localization

The aim of the localization phase is to determine target coordinates in defined coordinate systems whereby the target locations estimated in consecutive time instants create the target trajectory. As the input of this radar signal processing phase, estimated TOA couples are used. Because the considered radar system consists of one transmitting and two receiving antennas, only the noniterative direct method of localization can be employed. In that case, the target coordinates are simply calculated by the trilateration methods as intersections of two ellipses formed on the basis of the estimated TOA couples and known coordinates of the transmitting and receiving antennas [32].

The main idea of the direct calculation method can be mathematically described in the following way. Let us assume that $TOA_l^k(\tau)$ is the TOA estimated at the observation time instant τ for the lth target ($l \in L$) and the receiving antenna Rx_k, $k = 1,2$ and c is the light propagation velocity. Then the measured distance d_l^k among the transmitting antenna Tx, target T_l and receiving antenna Rx_k is given by

$$d_l^k = cTOA_l^k(\tau) \tag{8.13}$$

For arbitrarily placed transmitting antenna $Tx = [x_T, y_T]$ and receiving antennas $Rx_k = [x_k, y_k]$, the most straightforward way of estimating the target position, that is determining its coordinates $T_l(\tau) = [x, y]$, is to solve a set of equations created by using Equation (8.13) and taking into account the known coordinates of the transmitting and receiving antennas. Then, the following set of nonlinear equations can be built up based on $TOA_l^k(\tau)$ measurements for the scenario with one transmitting and two receiving antennas [33]:

$$d_l^k = \sqrt{(x - x_T)^2 + (y - y_T)^2} + \sqrt{(x - x_k)^2 + (y - y_k)^2}, \quad k = 1,2 \tag{8.14}$$

Each range d_l^k and the pairs $Tx = [x_T, y_T]$ and $Rx_k = [x_k, y_k]$, $k = 1,2$, form two ellipses with the foci Tx and Rx and with the length of the main half-axis $a_l^k = d_l^k/2$. The coordinates of target $T_l(\tau) = [x, y]$ lie on the intersection of these ellipses. Because the ellipses are expressed by the polynomials of the second order, there are two solutions for their intersections. However, there is only one solution determining the desirable true coordinates of the target. Therefore, one of the obtained solutions has to be excluded for every moving target. Usually this can be done based on knowledge of a half-plane where the targets are located. One of the solutions can be eliminated also if the solution (target coordinates) is beyond the monitored area or it has no physical interpretation (e.g. complex roots of Equation (8.14)). The detailed description of the direct calculation method can be found, for example, in Reference [33].

8.3.7 Target Tracking

The particular locations of the targets $T_l(\tau) = [x, y]$ are estimated with a certain random error usually described by its probability distribution function. Taking into account this model of the target position estimation, the target trajectory can be further processed by tracking algorithms. They provide a new estimation of the target location based on foregoing positions of the target. Usually, the target tracking results in a decrease in the localization error and smoothing of the trajectory.

There is a fundamental distinction between single target tracking (STT) systems and multiple target tracking (MTT) systems. Because the STT systems are dedicated to a single target, there is no need to perform a complex data association function, such as that discussed later for an MTT system. However, consistency tests must be performed to ensure that the radar is still pointing at the target. The tracking filter may be an analogue device, but modern systems typically use Kalman filtering [34].

The extension of STT to MTT requires a complex data association logic in order to sort out the returning sensor data into the general categories of targets of interest, recurrent sources that are not of interest (such as background clutter) and false signals with little or no correlation over time. The gating, observation-to-track association and track maintenance functions are usually part of the overall data association function (Figure 8.3). Firstly, gating is used as a screening mechanism to determine which observations are valid candidates to update existing tracks. Gating is performed primarily to reduce unnecessary computations by the association and maintenance functions that follow. The association function takes the observation-to-track pairings that satisfied gating and determines which observation-to-track assignments will actually be made. Finally, track maintenance refers to the functions of track initiation, confirmation and deletion. Modern MTT systems typically combine data association and multiple Kalman filter models that are running in parallel [35]. One of the prospective implementations of the complex MTT system, which is also used in our signal processing procedure, is shown in Figure 8.3 and described in Reference [36].

8.4 Short-Range Tracking Illustration

The performance of the described signal processing procedure is illustrated by processing the radar signals acquired by the experimental M-sequence UWB radar (Figure 8.1) according to the scenario outlined in Figure 8.4. The basic parameters of the radar system used are listed

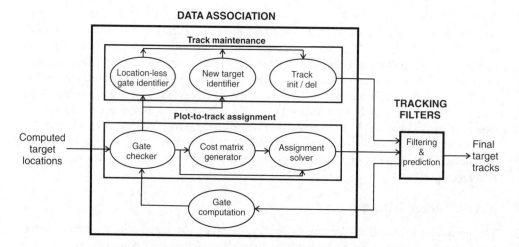

Figure 8.3   The implementation scheme of a modern MTT system.

(a)

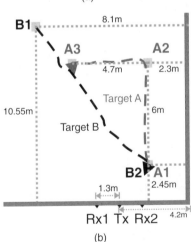

(b)

Figure 8.4 Measurement scenario: (a) interior photo, (b) scheme with the true target trajectories.

here. The system clock frequency for the radar device is about 4.5 GHz, which results in the operational bandwidth of about DC 2.25 GHz. The M-sequence order emitted by the radar is 9; that is the impulse response covers 511 samples regularly spread over 114 ns. This corresponds to an observation window of 114 ns leading to an unambiguous range of about 17 m. The 256 hardware averages of environment impulse responses are always computed within

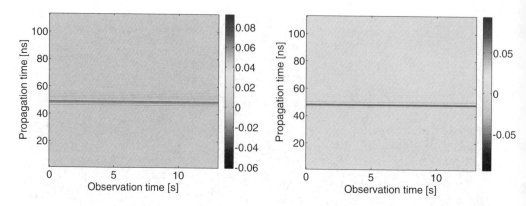

Figure 8.5 Radargram with raw radar signals. Receiving channel: (a) Rx1, (b) Rx2.

the radar head field programmable gate array FPGA to provide a reasonable data throughput and to improve the signal-to-noise ratio (SNR) by 24 dB. The additional software averaging can be provided by basic software of the radar device. In our measurement, the radar system was set in such a way as to provide approximately 10 impulse responses per second. The total power transmitted by the radar was about 1 mW.

In the measurement scenario, two persons were moving in a gymnasium (Figure 8.4a) behind a 24 cm thick wooden wall covered by tiles. The first person (labelled target A) was walking from the position A1 to the position A2 and consequently to A3 and at the same time the second person (target B) was walking from the position B1 to the position B2 (Figure 8.4b). The antenna radar system was located outdoors according to the layout outlined in the building scheme in Figure 8.4(b); that is all antennas were placed along a line with Tx in the middle of Rx1 and Rx2. There was no separation between the antennas and the wall.

The raw radar signals corresponding to the described scenario and obtained by both receiving channels are depicted in Figure 8.5. In these radargrams, only the cross-talk signal and the reflections of the emitted electromagnetic wave from the wall can be viewed, inasmuch as they are very strong in comparison with weak signals scattered by the moving

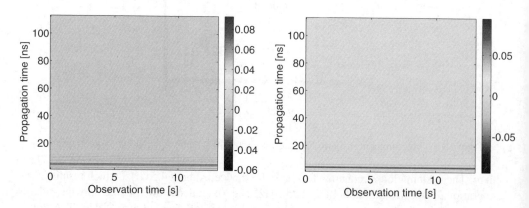

Figure 8.6 Preprocessed radargram. Receiving channel: (a) Rx1, (b) Rx2.

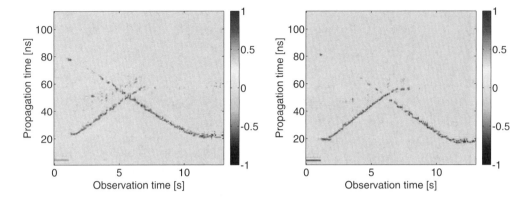

Figure 8.7   Radargram with subtracted background. Receiving channel: (a) Rx1, (b) Rx2.

targets. After utilization of the cross-talk signal for preprocessing (Figure 8.6), the background was estimated using exponential averaging.

After subtraction from the raw radar data, the primary traces, formed by the multiple reflections from moving human bodies, were produced in the radargrams (Figure 8.7). There the trace of target A started approximately at the propagation time of 20 ns and finished at around 60 ns, while the trace of target B gradually decreased from 80 ns to 20 ns (Figure 8.7a). A similarity of radargrams obtained from both receiving channels results from the small distances and symmetric positioning with respect to the transmitting antenna.

For enhancement of reflections from the target moving further from the radar system, advance normalization has been applied (Figure 8.8). In spite of the fact that the obtained signals are noticeably noisy, they enabled moving targets also to be detected in critical parts of the radargrams – in the examined scenario it is after a propagation time of 50 ns.

Detector outputs from the Gaussian CFAR detector are depicted in Figure 8.9. Here, not only the target traces but also some harmful artefacts have been highlighted. The shadowing effect, which appears as a time-shifted copy of the trace belonging to target A, is the most massive. With the high probability, the fitness centre full of metal equipment located behind

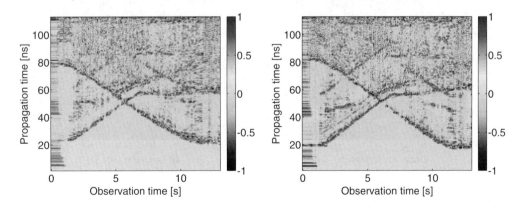

Figure 8.8   Radargram with enhanced signals. Receiving channel: (a) Rx1, (b) Rx2.

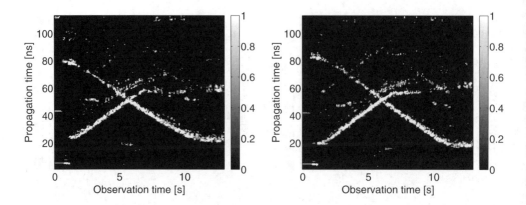

Figure 8.9   Radargram with detector output. Receiving channel: (a) Rx1, (b) Rx2.

the right wall of the gymnasium was the reason for the described phenomenon. Other visible artefacts are so-called cable reflections. They also copy the original target traces but are much weaker. In larger distances from the radar system they disappear completely. The last negative effect is a mutual shadowing of person A by person B approximately from observation time of 7 s. This results in the partial disappearance of the target trace belonging to person A in that time.

The huge amount of the false alarms has resulted in a huge amount of potential simple targets (Figure 8.10a), but most of them have been successfully suppressed within the association phase (Figure 8.10b). The previously mentioned similarity of the radargrams can be seen more accurately in Figure 8.10(a), where the radargrams from both channels are combined into one joint radargram. There the TOA belonging to the same target is depicted by white colour and the primary estimated and artificially widened TOA from Rx1 and Rx2 are outlined by dark grey and light grey colour, respectively (see the colour bar at the right side of the radargram in Figure 8.10a). Their conjunction implies that both receiving antennas

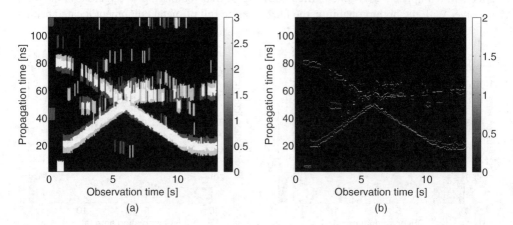

Figure 8.10   Combined radargram with TOA estimations: (a) primary estimations, (b) associated TOA couples.

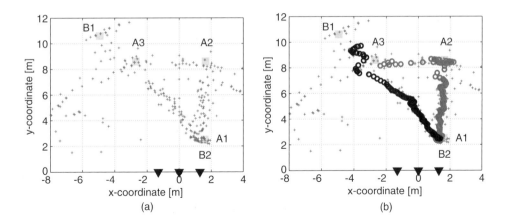

Figure 8.11 (a) Target locations with true reference positions, (b) target tracks with true reference positions.

captured the relevant reflections. Not associated TOAs are considered to be the false alarms. In this way the ghost generation is avoided.

Since the wall is thin and with small relative permittivity, its effect can be considered negligible in this case. The target locations computed on the basis of estimated TOA couples are depicted in Figure 8.11(a). Following the artefacts confirmed by both channels, the redundant amount of estimated positions appeared in the scanned area. However, the applied MTT system has correctly identified and preserved tracks of two targets. As can be seen from Figure 8.11(b), the estimated tracks correspond very well with the true target trajectories.

8.5 Conclusions

In this contribution, the complete radar signal processing procedure for short-range tracking of multiple targets by the handheld UWB radar has been described. The presented procedure consists of the application of the sequence of conveniently selected signal processing phases enabling the target coordinates to be computed analytically using the localization and tracking methods. This approach also allows a number of problems to be solved following from multiple target tracking (e.g. the deghosting problem, target positioning improvement by MTT, enhancement of weak signals from further targets, etc.) that have not been yet solved satisfactorily by the imaging methods. The experimental results obtained by the processing of measured radar signals confirm good performance properties of the proposed procedure.

The effect of mutual shadowing between people arising when a person moves near the radar antennas was briefly discussed in the last section. This effect results in the decreasing probability of target detection for multiple moving target scenarios if a single radar is used for target tracking. In the case of simpler measurement scenarios, the mutual shadowing can be suppressed by the methods of weak signal enhancement. This is the case for so-called partial shadowing, an example of which is the scenario examined in this chapter. In more complex situations (more targets, smaller areas, full furnished rooms, etc.), it is necessary to apply the UWB radar network instead of the single radar device. Then the proposed signal processing procedure can be easily extended to the low-complex method of data fusion with the centralized architecture. The centralized architecture of data fusion includes N radar systems and one fusion centre. The signals acquired by each radar are independently processed by the same procedure until the localization phase. Then the obtained target coordinates are time synchronized and transformed into the common coordinate system. A set of such data represents the input data for the described MTT system. The preliminary results obtained by such approach are described in Reference [37]. In the follow-up research, it is necessary to deal with the extension of the data fusion method for through wall tracking of multiple moving targets. The core of this problem will consist in the wall parameter estimation by the UWB radar and their application for wall-effect compensation.

Acknowledgement

This work was supported by the COST Action IC0803 'RF/Microwave Communication Subsystems for Emerging Wireless Technologies (RFCSET)' and by the Slovak Research and Development Agency under Contract LPP-0080-09.

References

1. Withington, P., Fluhler, H. and Nag, S. (2003) Enhancing homeland security with advanced UWB sensors. *Microwave Magazine, IEEE*, **4** (3), 51–58.
2. Chang, S., Sharan, R., Wolf, M. *et al.* (2010) People tracking with UWB radar using a multiple-hypothesis tracking of clusters (MHTC) method. *International Journal of Social Robotics*, **2** (1), 3–18.
3. Gurbuz, S.Z., Melvin, W.L. and Williams, D.B. (2007) Comparison of radar based human detection techniques. Proceedings of the 41st Asilomar Conference on Signals, Systems and Computers, pp. 2199–2203.
4. Gauthier, S.S. and Chamma, W. (2004) Surveillance through concrete walls. Proceedings of SPIE – C3I Technologies for Homeland Security and Homeland Defense III, pp. 597–608.
5. Nag, S., Barnes, M.A., Payment, T. and Holladay, G. (2002) Ultrawideband through-wall radar for detecting the motion of people in real time. Proceedings of SPIE – Radar Sensor Technology and Data Visualization, vol. 4744, pp. 48–57.
6. Ram, S.S. and Ling, H. (2008) Through-wall tracking of human movers using joint Doppler and array processing. *IEEE Geoscience and Remote Sensing Letters*, **5** (3), 537–541.
7. Narayanan, R. (2008) Through wall radar imaging using UWB noise waveforms. Proceedings of the IEEE International Conference on Acoustics, Speech and Signal Processing (ICASSP 2008), April 2008, pp. 5185–5188.
8. Sachs, J. (2004) M-sequence radar, in *Ground Penetrating Radar*, 2nd edn (ed. D.J. Daniels), The Institution of Electrical Engineers, London, UK.
9. Ultra wideband radio application for localisation of hidden people and detection of unauthorised objects. Project co-funded by the European Commission within the sixth framework programme, acronym: RADIOTECT, available from: http://www.radiotect.vub.ac.be/.
10. Yelf, R. (2004) Where is true time zero?Proceedings of the 10th International Conference on Ground Penetrating Radar, (1), pp. 279–282.
11. Piccardi, M. (2004) Background subtraction techniques: a review. Proceeedings of the IEEE – SMC International Conference on Systems, Man and Cybernetics, The Hague, The Netherlands.
12. Zetik, R., Crabbe, S., Krajnak, J. *et al.* (2006) Detection and localization of persons behind obstacles using M-sequence through-the-wall radar. Proceedings of the SPIE – Sensors, and Command, Control, Communications, and Intelligence (C3I) Technologies for Homeland Security and Homeland Defense.
13. Wren, C., Azarbayejani, A., Darrell, T. and Pentland, A. (1997) Pfinder: real-time tracking of the human body. *IEEE Transactions on Pattern Analyses and Machine Intelligence*, **19** (7), 780–785.
14. Stauffer, C. and Grimson, W. (2000) Learning patterns of activity using real-time tracking. *IEEE Transactions on Pattern Analyses and Machine Intelligence*, **22** (8), 747–757.
15. Nag, S. and Barnes, M. (2003) A moving target detection filter for an ultra-wideband radar. Proceedings of the IEEE Radar Conference, pp. 147–153.

16. Nag, S., Fluhler, H. and Barnes, H. (2001) Preliminary interferometric images of moving targets obtained using a time-modulated ultra-wide band through-wall penetration radar. Proceedings of the IEEE – Radar Conference, pp. 64–69.
17. Toyama, K., Krumm, J., Brumitt, B. and Meyers, B. (1999) Wallflower: principles and practice of background maintenance. Proceedings of the International Conference on Computer Vision, pp. 255–261.
18. Tipping, M.E. and Bishop, C.M. (1999) Mixtures of probabilistic principal component analyzers. *Neural Computation*, **11** (2), 443–482.
19. Xingxing, W., Zhong-Ping, J., Repperger, D. and Guo, Y. (2006) Nonlinear enhancement of weak signals using optimization theory. Proceedings of the IEEE International Conference on Mechatronics and Automation, pp. 66–71.
20. Hanseok, K. and Arozullah, M. (2000) Background noise suppression for signal enhancement by novelty filtering. *IEEE Transactions on Aerospace and Electronic Systems*, **36** (1), 102–113.
21. Rovňáková, J. and Kocur, D. (2010) Weak signal enhancement in radar signal processing. Proceedings of the 20th International Conference Radioelektronika, Brno, Czech Republic, April 2010, pp. 147–150.
22. Poor, H.V. (1994) *An Introduction to Signal Detection and Estimation*, Springer, Heidelberg.
23. Taylor, J.D. (ed.) (2001) *Ultra-Wideband Radar Technology*, CRC Press, Boca Raton, FL.
24. Minkler, G. and Minkler, J. (1990) *CFAR*, Magellan Book Company.
25. Rohling, H. (1983) Radar CFAR thresholding in clutter and multiple target situations. *IEEE Transactions on Aerospace and Electronics Sysems*, **19** (4), 608–621.
26. Urdzík, D. (2010) Performance comparison of CFAR detectors for UWB radars. Proceedings of the 8th International Conference, Králíky, Czech Republic, August–1 September 2010, pp. 158–161.
27. Dutta, P.K., Arora, A.K. and Bibyk, S.B. (2006) Towards radar-enabled sensor networks. Proceedings of the Fifth International Conference on Information Processing in Sensor Networks, Special Track on Platform Tools and Design Methods for Network Embedded Sensors, pp. 467–474.
28. Gezici, S., Tian, Z., Biannakis, G.B. *et al.* (2005) Localization via ultra-wideband radios. *IEEE Signal Processing Magazine*, **22** (4), 70–84.
29. Fleming, R., Kushner, C., Roberts, G. and Nandiwada, U. (2002) Rapid acquisition for ultra-wideband localizers. Proceedings of the IEEE Conference on Ultra Wideband Systems and Technologies (UWBST), Baltimore, MD, May 2002, pp. 245–249.
30. Rabbachin, A., Montillet, J.P., Cheong, P. *et al.* (2005) Non-coherent energy collection approach for TOA estimation in UWB systems. IST Mobile and Wireless Communications Summit, Dresden, Germany, June 2005.
31. Rovňáková, J. and Kocur, D. (2010) TOA estimation and data association for through wall tracking of moving targets. *EURASIP Journal on Wireless Communications and Networking, The Special Issue: Radar and Sonar Sensor Networks*, 420767, 1–11, DOI: 10.1155/2010/420767.
32. Paolini, E., Giorgetti, A., Chiani, M. *et al.* (2008) Localization capability of cooperative anti-intruder radar systems. *EURASIP Journal on Advances in Signal Processing*, **2008**, 94–122.
33. Švecová, M. (2007). Node Localization in UWB Wireless Sensor Networks. Thesis to the Dissertation Examination, Technical University of Košice, Košice, Slovakia.
34. Grewal, M.S. and Andrews, A.P. (2008) *Kalman Filtering: Theory and Practice Using MATLAB*, 3rd edn, Wiley-IEEE Press.
35. Blackman, S.S. and Popoli, R. (1993) *Design and Analysis of Modern Tracking Systems*, Artech House Publishers.
36. Khan, J., Niar, S., Rivenq-Menhaj, A. and Hillali, Y.E. (2008) Multiple target tracking system design for driver assistance application. Design and Architectures for Signal and Image Processing, November 2008.
37. Rovňáková, J. and Kocur, D. (2011) Short range tracking of moving persons by UWB sensor network. Proceedings of the 8th European Radar Conference (EuRAD), Manchester, UK, October 2011, pp. 321–324.

9

Advances in the Theory and Implementation of GNSS Antenna Array Receivers

Javier Arribas[1], Pau Closas[1], Carles Fernández-Prades[1], Manuel Cuntz[2], Michael Meurer[2] and Andriy Konovaltsev[2]

[1]*Center Tecnològic de Telecomunicacions de Catalunya (CTTC), Castelldefels, Spain*
[2]*German Aerospace Center (DLR), Wessling, Germany*

9.1 Introduction

This chapter focuses on recent developments in the field of antenna arrays applied to the design of robust, high-performance global navigation satellite system (GNSS) receivers. This topic deserved some attention in the past, with many theoretical contributions stressing its superiority and versatility in challenging scenarios with respect to single antenna receivers and mechanically steerable dishes. Nowadays it is of the utmost interest in applications with life-security content, such as precise airplane landing, which requires adequate countermeasures against jamming and sophisticated intentional interferences. This work mainly focuses on the architecture of a GNSS receiver with an adaptive antenna array. The specifics of the design of such receivers are discussed in detail covering both analog and digital signal processing blocks. Special attention is given to the design of such critical components as the antenna array, RF front-ends and A/D converters. The array signal processing techniques are also addressed describing two main strategies based on (1) the spatial filtering with the help of digital beamforming and (2) signal parameter estimation by statistical array processing. In the conclusion of this work, two working prototypes of GNSS receivers with adaptive antennas are described including the latest test results achieved with their help.

Microwave and Millimeter Wave Circuits and Systems: Emerging Design, Technologies, and Applications,
First Edition. Edited by Apostolos Georgiadis, Hendrik Rogier, Luca Roselli, and Paolo Arcioni.
© 2013 John Wiley & Sons, Ltd. Published 2013 by John Wiley & Sons, Ltd.

9.2   GNSS: Satellite-Based Navigation Systems

The application of radio waves to determine a position starts in the twentieth century with the patent of the first direction finding system in 1902 [1]. World War II spurred hyperbolic, ground-based low-frequency radio navigation systems such as Decca and LORAN. Satellite-based navigation, which now is usually referred to within the general framework of global navigation satellite systems (GNSSs), started in the early 1960s with the TRANSIT system, based on the fact that if the satellite's position were known and predictable, the Doppler shift could be used to locate a receiver on Earth. The Cold War arms race and associated military needs, mostly related to ballistic missiles guidance and the nuclear deterrence posture, required more accurate and reliable navigation systems. This made technology evolve rapidly: the Navstar-GPS program was set in 1973 and in 1974 the first atomic clock was put into orbit. Moving forward, the first experimental Block-I global positioning system (GPS) satellite was launched in 1978.

In 1983, U.S. President Ronald Reagan issued a directive guaranteeing that GPS signals would be available at no charge for civilian uses when the system became operational, opening the commercial market. Second-generation GPS satellites were launched beginning in 1989, and the system's full operational capability (FOC) was declared in April 1995. Since initially the highest quality signal was reserved for military use, the signal available for civilian use was intentionally degraded with a system function referred to as selective availability (SA), which deliberately introduced random errors for civilian GPS receivers. This changed when the U.S. President Bill Clinton ordered SA to be turned off at midnight May 1, 2000. This improved the potential precision of civilian GPS receivers from 300 to about 20 m.

In parallel, the former Soviet Union had begun the development of the GLONASS system in 1976, also with military endeavors, completing its satellite constellation in 1995. Following completion, the system rapidly fell into disrepair with the collapse of the Russian economy. Beginning in 2001, Russia committed itself to restoring the system, and by the time of this writing (October, 2011) it has been practically restored, with 24 satellites operational. GLONASS provides two types of navigation signals: a standard precision (SP) navigation signal and a high-precision (HP) navigation signal. SP positioning and timing services are available to all GLONASS civil users on a continuous, worldwide basis. Actually, on May 18, 2007, Russian President Vladimir Putin signed a decree reiterating the offer to provide free access to GLONASS civil signals.

European's Galileo program is the first guaranteed global positioning service under civilian control. The first stage of the Galileo program was agreed upon officially on May 26, 2003 by the European Union and the European Space Agency. By the end of 2013, it will have an initial constellation of 16 satellites: 4 in-orbit validation (IOV) and 12 FOC satellites. The European Commission announced that three of the five services offered by the system will be provided in early 2014: the open service (basic signal provided free of charge), the public regulated service (two encrypted signals with controlled access for specific users like governmental bodies, agencies, and organizations involved with defense, internal security, law enforcement, and critical transport, providing position and timing to specific users requiring a high continuity of service), and the search and rescue service (contributing to the international COSPAS-SARSAT cooperative system for humanitarian search and rescue activities).

China is also deploying a GNSS named COMPASS. Plans call for completion of a 14-satellite constellation by 2012. The constellation would consist of five geostationary or GEO, five inclined geosynchronous orbit (IGSO), and four medium-Earth orbit (MEO) satellites, to provide a regional service over eastern Asia. Long-range plans envision a 35-satellite constellation providing global service by 2020: 27 MEO, 3 IGSO, and 5 GEO.

These new deployments and system modernizations depict an unforeseen and close forthcoming scenario with a plurality of systems, satellite constellations, frequency bands, and new signal structures ready to be exploited. Multiconstellation/multifrequency GNSS receivers, complemented with regional information, promise dramatically improved positioning solutions and enhanced integrity, at the expense of posing a number of challenges to the navigation engineering community. Topics such as accuracy, precision, robustness, reliability, interoperability with other systems, shorter time to first fix, satellite selection, and coverage will be tacked from novel points of view, enabling unexplored business models and applications that only imagination can bound.

GNSS space vehicles broadcast a low-rate navigation message that modulates continuous repetitions of pseudorandom spreading codes, which in turn are modulating a carrier signal allocated in the L-band. The navigation message, after proper demodulation, contains among other information the so-called ephemeris, a set of parameters that allow the computation of the satellite position at any time. These positions, along with the corresponding distance estimations, allow the receiver to compute its own position and time, as we will see hereafter. The distance between the receiver and a given satellite can be computed by

$$\rho_i = c\left(t_i^{\text{Rx}} - t_i^{\text{Tx}}\right) \tag{9.1}$$

where $c = 299\ 792\ 458$ m/s is the speed of light, t_i^{Rx} is the receiving time in the receiver's clock, and t_i^{Tx} is the time of transmission for a given satellite i. Receiver clocks are inexpensive and not perfectly in sync with the satellite clock, and thus this time deviation is another variable to be estimated. The clocks on all the satellites belonging to the same system s, where $s = \{\text{GPS, Galileo, GLONASS, COMPAS, } \dots \}$, are in sync with each other, so the receiver's clock will be out of sync with all satellites belonging to the same constellation by the same amount $\Delta t^{(s)}$. In GNSS, the term *pseudorange* is used to identify a range affected by a bias, directly related to the bias between the receiver and satellite clocks. There are other factors of error: since propagation at speed c is only possible in the vacuum, atmospheric status affects the propagation speed of electromagnetic waves modifying the propagation time and thus the distance estimation. This can also be mitigated by differential systems, where a network of well-positioned, ground-fixed receivers measure the distortion provoked by the atmosphere and broadcast the corrections to the surrounding rover receivers.

For each in-view satellite i of system s, it is possible to write

$$\rho_i = \sqrt{\left(x_i^{\text{Tx}} - x\right)^2 + \left(y_i^{\text{Tx}} - y\right)^2 + \left(z_i^{\text{Tx}} - z\right)^2} + c\Delta t^{(s)} + \sigma_e \tag{9.2}$$

where $\left(x_i^{\text{Tx}}, y_i^{\text{Tx}}, z_i^{\text{Tx}}\right)$ is the satellite's position (known from the navigation message), (x, y, z) the receiver's position, and σ_e gathers the cumulative effect of all sources of error. Since the receiver needs to estimate its own three-dimensional position (three spatial unknowns) and its clock deviation with respect to the satellites' time basis, at least $3 + N_s$ satellites must be seen by the receiver at the same time, where N_s is the number of different navigation systems available (in view) at a given time.

Each received satellite signal, once synchronized and demodulated at the receiver, defines one equation such as the one defined in Reference [2], forming a set of nonlinear equations that can be solved algebraically by means of the Bancroft algorithm [2] or numerically, resorting to multidimensional Newton–Raphson and weighted least squares methods [3]. When a priori information is added we resort to a Bayesian estimation, a problem that can be solved recursively by a Kalman filter or any of its variants. The problem can be further expanded by adding other unknowns (for instance, parameters of ionospheric and tropospheric models), sources of information from other systems, mapping information, and even motion models of the receiver. In the design of multiconstellation GNSS receivers, the vector of unknowns can also include the receiver clock offset with respect to each system in order to take advantage of a higher number of in-view satellites and using them jointly in the navigation solution, therefore increasing accuracy.

The analytic representation of a signal received from a GNSS satellite can be generically expressed as

$$r(t) = \alpha(t)s_T(t - \tau(t))e^{-j2\pi f_d(t)}e^{j2\pi f_c t} + n(t) \tag{9.3}$$

where $\alpha(t)$ is the amplitude, $s_T(t)$ is the complex baseband transmitted signal, $\tau(t)$ is the time-varying delay, $f_d(t) = f_c \dot{\tau}(t)$ is the Doppler shift, f_c is the carrier frequency, and $n(t)$ is a noise term. These signals arrive at the Earth's surface at extremely low power (e.g., -158.5 dBW for the GPS L1 C/A code and -157 dBW for Galileo E1), well below the noise floor. In order to estimate its distances to satellites, the receiver must correlate time-aligned replicas of the corresponding pseudorandom code with the incoming signal, in a process called *dispreading*, which provides processing gain only to the signal of interest. After coarse and fine estimation stages of the synchronization parameters (usually known as *acquisition* and *tracking*, respectively), signal processing output is in the form of observables: (1) the pseudorange (code) measurement, equivalent to the difference between the time of reception (expressed in the time frame of the receiver) and the time of transmission (expressed in the time frame of the satellite) of a distinct satellite signal and, optionally, (2) the carrier-phase measurement, actually being a measurement of the beat frequency between the received carrier of the satellite signal and a receiver-generated reference frequency. Then, depending on the required accuracy, the navigation solution can range from pseudorange only, computationally low demanding, and limited accuracy least squares methods to sophisticated combinations of code and phase observables at different frequencies for high-demanding applications such as surveying, geodesy, and geophysics.

The particularizations of Equation (9.3) for each system and frequency link civil signals can be found in the corresponding Interface Control Documents (ICDs). Table 9.1 summarizes those references.

9.3 Challenges in the Acquisition and Tracking of GNSS Signals

It is well known that sources of accuracy degradation due to atmospheric effects can be effectively mitigated by differential systems, even with long baselines [4], but interferences and multipath remain as potential causes of downgraded performance.

Table 9.1 Common GNSS standards

GNSS civil signal	Interface Control Document	Carrier frequency (MHz)	Reference bandwidth (MHz)
GPS L1 and L2C	Interface Specification IS-GPS-200 Revision E. Navstar GPS Space Segment/Navigation User Interfaces, El Segundo, CA, June 2010 Downloadable from gps.gov	1575.42 (L1) and 1227.60 (L2C)	20.46 (L1 and L2C)
GPS L1C (available with first Block III launch, currently scheduled for 2013)	Interface Specification IS-GPS-800 Revision A. Navstar GPS Space Segment/User Segment L1C Interfaces, El Segundo, CA, June 2010 Downloadable from gps.gov	1575.42	30.69
GPS L5 (first Block IIF satellite launched on May 2010)	Interface Specification IS-GPS-705 Revision A. Navstar GPS Space Segment/User Segment L5 Interfaces, El Segundo, CA, June 2010 Downloadable from gps.gov	1176.45	24
GLONASS	Global Navigation Satellite System GLONASS. Interface Control Document. Navigational radiosignal in bands L1, L2. Edition 5.1, Moscow, Russia, 2008	1602.00 (L1) and 1246.00 (L2)	7.875
Galileo	Signal In Space Interface Control Document. Ref: OS SIS ICD, Issue 1.1, European Commission, September 2010 Downloadable from http://ec.europa.eu/enterprise/policies/satnav/galileo/open-service/index_en.htm	1176.45 (E5A), 1207.14 (E5B), 1278.75 (E6), and 1575.42 (E1)	20.46 (E5A), 20.46 (E5B), 40.92 (E6), and 24.552 (E1),

Interferences are one of the most jeopardizing sources of accuracy degradation, and even denial-of-service, of GNSS receivers [5], both if they are used for precise positioning or as a method for distributed synchronization (i.e., time dissemination). This is specially grave in strategic applications such as reference stations, geodesy and surveying, Earth observation, piloted and autonomous aircraft landing, high-precision timing for communication networks and power grid management, machine control, pseudolites and repeaters, GNSS reflectometry, road construction, precision agriculture, and others.

9.3.1 Interferences

Typically, interferences are classified into (1) *continuous waves*, narrowband signals generated by ultrahigh frequency (UHF) and very high frequency (VHF) television broadcasting, by some VHF omnidirectional radio-range (VOR) and instrument landing system (ILS) harmonics, by spurious signals caused by power amplifiers working in the nonlinearity region, or by oscillators present in many electronic devices such as personal computers and mobile phones; (2) *pulsed*, for instance those caused by distance measurement equipment (DME) and tactical air navigation system (TACAN), both emitting pulsed modulations at frequencies in the range 962–1213 MHz, frequencies that include the Galileo E5 and GPS L5 bands, or by wind profile radars (WPR) that operate in the band 1270–1295 MHz, close to the GPS L2C, GLONASS L2, and Galileo E6 bands; (3) *swept interferences*, signals characterized by a narrow instantaneous band at a central frequency that changes over time, usually harmonics of frequency modulated signals, which can be produced by telecommunication systems such as television and radio broadcasting; and (4) *wideband interferences*, understood as signals occupying most of the frequency band of interest. Basically, interferences affect the operation of the low-noise amplifier (LNA) and the automatic gain control (AGC) of the radio frequency (RF) front-end, and consequently have an impact in the performance of the carrier and code tracking loops, which results in deterioration of observables or even in complete loss of lock, thus becoming a disruptive event in the operation of a GNSS receiver. Analysis of RF interference effects in GNSS receivers can be found in Reference [6].

Interferences could also be intentional, like military jamming or hijackers using GPS jammers to prevent a stolen vehicle from being tracked. For instance, it is possible to construct a transmitter of signals nearly identical to those sent by a satellite, with the objective of forging fake navigation messages, transmit them over an area with one or more receivers, and this in way manipulate their PVT solutions. These techniques are known as spoofing, when the attacker synthesizes its transmissions, or meaconing, when (parts of) legitimate GNSS transmissions are reused. Spoofing effects are analyzed in Reference [7].

9.3.2 Multipath Propagation

Multipath propagation occurs when the GNSS signal at the receiver antenna is composed of a mixture of direct line-of-sight signals from the navigation satellite and additionally of one or more signal echoes due to the reflections of the direct signal in the environment of the user. Different to communication systems, where the multipath propagation can be constructively used to improve the overall system performance, in satellite navigation it is a harmful effect. The same principle of operation of a satellite navigation system is based on measuring the propagation time of the navigation signal to the receiver; therefore any additional delay of the signal that is not translated into the line-of-sight distance between the receiver and the navigation satellite results in a measurement error deteriorating positioning and timing performance. Such a measurement error in the receiver code delay loop can vary from a few centimeters to several tens of meters in the worst case depending on the GNSS signal and the code loop architecture used.

The strongest signal echoes occur due to the specula reflection mechanism in the vicinity of the receiver. Such echoes can be characterized by deterministic propagation models where parameters can be assessed by the methods of the geometric optics theory and its extensions. In a more general case, the multipath appears as the result of superposition of a large number of weak echoes produced by the scattering of the GNSS direct signal in the environment of the receiver. Such multipath propagation is modeled by statistical models where parameters can be chosen to fit to some empirical data from channel measurements.

Though quite seldom in the practice, the specula multipath echoes may pose a problem in the applications with tough requirements on the positioning and timing accuracy or/and integrity of the obtained measurements. The following features of the multipath echoes can be used to detect their presence and mitigate their harmful effect:

- A multipath echo is always delayed with respect to the direct signal.
- A multipath echo is, in statistical sense, weaker than the direct signal.
- A strong multipath echo arrives at the receive antenna from the direction that is different to the direction of arrival of the direct signal.

The first and the second properties of the multipath echoes are extensively utilized by the state-of-the-art multipath mitigation techniques operating in the time or frequency signal processing domains. The third feature of the multipath propagation mechanism is of great interest for the antenna array receivers exploiting the spatial domain.

9.4   Design of Antenna Arrays for GNSS

Aside from (or in combination with) time–frequency countermeasures [8, 9] and receiver autonomous integrity monitoring (RAIM) consistency checks, which are proved to be effective against continuous wave interferences, spatial diversity provided by antenna arrays enables a powerful tool for interference mitigation.

Among the number of techniques that can be applied in order to exploit the presence of a plurality of antennas, those based in the digital domain are known to offer superior performance in terms of adaptability to varying scenarios. For instance, Reference [10] proposes a statistical approach based on the generalized likelihood ratio test for deriving a new array-based acquisition test function that exhibits robustness against the presence of interferences when detecting the presence of in-view satellites. However, the most popular approach to the use of several antennas in a GNSS receiver is called *beamforming*, a technique that consists of several antennas whose outputs are controlled in phase and gain, that is, multiplied by complex weights, and then combined to obtain a single output in order to achieve a gain pattern that can be manipulated electronically.

Three approaches might be considered in the architecture of antenna array beamforming: digital, analog, and a mixed strategy. Performance in terms of computational cost, estimation accuracy, hardware complexity, and availability of commercial off-the-shelf (COTS) devices should be analyzed:

- In an analog architecture, weights combining the signal of each antenna are implemented by means of attenuators and phase shifters. A set of different weights needs to be implemented for each satellite to be tracked.
- In digital beamforming, the signal of each antenna is processed digitally. In this case, the weights are simply complex multipliers that are implemented in programmable devices such as FPGAs.

- In the mixed analog/digital strategy, the antenna elements are grouped in subarrays. Each subarray implements a delay-and-sum beamforming with only phase shifters. Antenna outputs are downconverted and digitized, leading to a digital beamforming stage where nulling capabilities and other advanced processing can be performed.

Regardless of the way those weights are implemented, a crucial aspect is how those weights are computed. Although a rough estimation of the receiver position and array attitude would be enough to compute the expected directions of arrival of GNSS signals, and thus point the radiation pattern to those directions, this approach lacks the ability to adapt to varying scenarios and the potential presence of unknown interferers. Thus, the latest developments compute their beamweights based on sophisticated techniques that operate with the received signals in the digital domain.

9.4.1   Hardware Components Design

9.4.1.1   Antenna Elements and Array

The design of the antenna for GNSS receivers should meet the following system-specific requirements:

- *Radiation pattern* of the antenna subsystem should allow for hemispherical coverage in order to receive GNSS signals from all navigation satellites in the sky.
- *Polarization*: direct-path GNSS signals have right-hand circular polarization (RHCP). RHCP purity is important for mitigating a strong multipath tending to have prevailing left-hand circular polarization (LHCP).
- *Frequency response*: good out-of-band interference rejection is desired.
- *Low cross-talk* between antenna RF front-ends is desired since it determines the maximum depths of the spatial zeros produced by the adaptive antenna.
- *Phase responses* of the antenna elements and RF front-ends should be precisely calibrated for a stable antenna phase center position.
- *Constant group delay* over different antenna processing channels and different satellites (i.e., different directions of arrival) is desired.
- *Size*: the size of the antenna may become an issue in some applications, for example, mobile units. In other applications the antenna has to fit in a standardized installation footprint, for example, in aviation.

Due to the growing use of GPS receivers in past years, a variety of antennas suitable for different GNSS applications exist now. The mostly used antenna designs are helical, patch, and planar spiral antennas.

Helical antennas, single or quadrifilar (see Figure 9.1), are well known for their excellent circular polarization (CP) properties. They exhibit good CP for a broad range of angles. In addition, they have broad radiation patterns (as shown in Figure 9.2) suitable for the omnidirectional signal reception. Generally, quadrifilar helix antennas (QHA) are mounted vertically and have good radiation characteristics along the horizon. As a consequence, they are suitable for keeping line-of-sight (LOS) of satellites positioned near the horizon. One of the main drawbacks of helical antennas for many GNSS applications is that they are not low profile. They are also more likely to

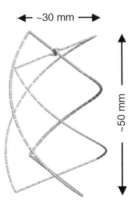

Figure 9.1 Quadrifilar helix antenna. © 2006 Cobham. Reprinted, with permission, from [11].

receive the multipath echoes at low elevations that occur due to the signal scattering/reflection from the ground.

Patch antennas (see Figure 9.3) are planar structures that inherently radiate in the forward hemisphere and have a directivity value of 5 dBiC or more. They are less able to keep the line-of-sight signal of near horizon satellites (see Figure 9.4) but are more immune to multipath from low angles. Although not as good as helical antennas in terms of polarization, patch antennas can achieve suitable CP for a broad range of directions. The frequency response of a typical patch antenna is usually a single band. Circular polarization can be achieved with a single feed by intentionally destroying the symmetry of the patch structure. Alternatively, a symmetrical structure (square, circular patch, or annular ring) is excited with two or four orthogonal feeds.

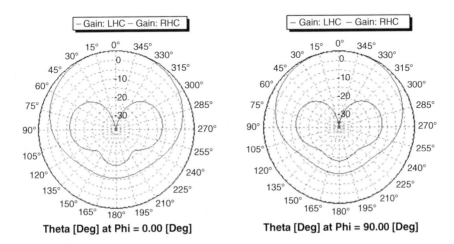

Figure 9.2 Radiation pattern of the air-filled QHA without ground plane. © 2006 Cobham. Reprinted, with permission, from [11].

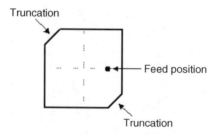

Figure 9.3 Truncated patch geometry. © 2006 Cobham. Reprinted, with permission, from [11].

The extension of the patch antenna design principle to the multiple frequency band signal reception results in *the stacked patches*. These are also planar structures. They have similar properties to the single patch antennas except that they are arranged and fed in such a way as to provide a multiband response. Typically, the GPS/Galileo frequency requirements may be satisfied with two levels of stacked patches. The first level of patch provides a resonance at the L1 band, while the second level provides a resonance at the E5 band, covering both the E5a and the E5b bands. Figure 9.5 shows a prototype antenna consisting of two stacked patches excited by four probes. The antenna is made of low-cost epoxy and glass loaded

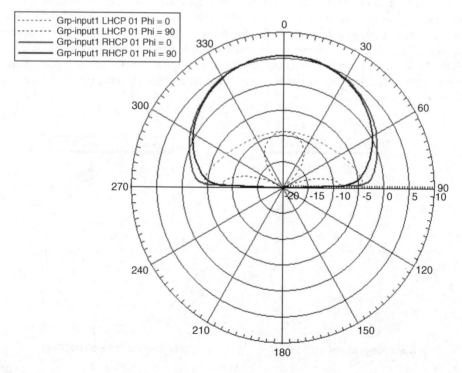

Figure 9.4 Radiation pattern of the truncated patch. © 2006 Cobham. Reprinted, with permission, from [11].

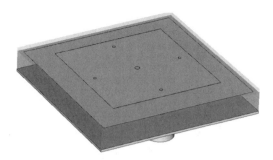

Figure 9.5 Full size probe fed stacked patches antenna. © 2006 Cobham. Reprinted, with permission, from [11].

substrates. The antenna size is 63 mm × 63 mm × 11 mm. The frequency response of this antenna covers the GPS/Galileo bands with a VSWR (voltage standing wave ratio) better than 1.5 (see Figure 9.6). The antenna possesses a good CP characteristic where the axial ratio on the axis is better than 2 dB in the frequency bands of interest. The antenna also provides good coverage over a broad range of angles. The pattern presents omnidirectional properties in azimuth.

The structure of a *spiral antenna* introduced in Reference [12] is shown in Figure 9.7. It has spiral slots of equal length and hence operates in a single frequency band. However, slots of different sizes can be used with corresponding changes in the feeding arrangement in order to achieve dual band operation. The spiral slots are surrounded by a system or concentric slot ring and vias in the outer periphery to form an electromagnetic band (high-impedance) structure. The purpose of such a structure is to reduce the sensitivity of the

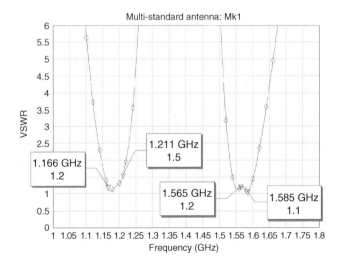

Figure 9.6 VSWR over GPS/Galileo frequency bands. © 2006 Cobham. Reprinted, with permission, from [11].

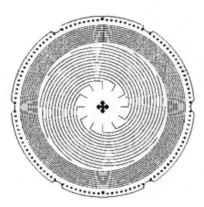

Figure 9.7 Spiral GPS L1 spiral pinwheel antenna. © 2000 NovAtel Inc. Reprinted, with permission, from [12].

antenna to spurious signals arriving at near horizon directions and so increase the robustness of the antenna to the multipath propagation effect.

Anti-jam antenna arrays are still primarily used in military GNSS applications as a means to combat interference. They are often referred to as controlled reception pattern antennas (CRPAs), which mitigate radio interference by shaping the antenna beam and placing nulls in the direction of the interference source. An all-digital approach offers the greatest flexibility and robustness in a wide range of interference scenarios.

Because of technological reasons, the patch antennas are mostly used for constructing an antenna array. An example of an antenna array of seven patch antenna elements in presented in Figure 9.8, where a high dielectric lens is used to reduce the overall size of the array.

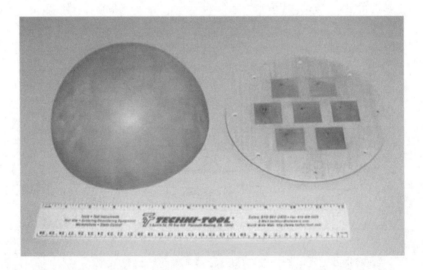

Figure 9.8 Miniaturized GPS antenna array with patch element. © 2001 NAVSYS Corporation. Reprinted, with permission, from [13].

9.4.1.2   RF Front-End and A/D Conversion

The signals of GNSS satellites are transmitted in the L-band (1164 MHz and 1610 MHz) in a distance of more than 20 000 km to the user. Due to free space loss the received signal power on the Earth's surface is extremely low. For GPS L1 signals it is typically -157 dBW (2×10^{-13} W). Thus the received GNSS signals are buried below the noise floor. These small signal power levels are the main reason for the vulnerability of GNSS receivers against any kind of radio frequency interference (RFI). For GNSS receivers, superheterodyne architectures with low intermediate frequencies are widely used and also recommended for interference robust multiantenna receivers. This front-end topology translates the radio frequency to a low intermediate frequency in one or more mixing stages. Nowadays ADCs digitize signals with a resolution of 12 and even more bits and a dynamic range of more than 70 dB. The next subsections provide guidelines for the design of suitable RF front-ends for GNSS receivers.

9.4.2   Array Signal Processing in the Digital Domain

In the literature there can be found some digital beamforming implementations aimed at the GNSS. One of the first reported platforms is the NAVSYS high-gain advanced GPS receiver (HAGR), which is composed of a 16 antenna–element array receiver and uses dedicated hardware to create up to 12 independent and parallel beamformings to simultaneously point the antenna array beam to 12 GPS satellites. HAGR also includes a software-defined GPS receiver to control the beamforming algorithm. However, few technical details about the implementation are available. Other achievements in hardware-based digital beamforming platforms are focused on specific beamforming algorithms. It is worth mentioning the QR decomposition (QRD)-based beamforming engine [14] made by Xilinx engineers and the minimum variance distortionless response (MVDR) beamformer implementation reported in Reference [15]. The trend in digital beamforming implementations is to tightly combine hardware and software designs in order to improve the execution speed of the adaptive beamforming algorithms, which usually require matrix inversion operations. In that sense, it is important to mention the recent advances achieved by the efficient implementation of the QRD-RLS algorithm [16], which exploits the hardware/software co-design strategy.

9.4.2.1   Digital Beamforming: The Array as a Spatial Filter

In the concept of beamforming the antenna array is considered as a spatial filter. In fact each array element can be treated as an element of a spatial tap delay line that delivers the measurement of arriving signals at the corresponding spatial delay. At a given instant of time, the measurements observed at all spatial taps can be combined together in a way that improves the reception of the signal(s) of interest and minimizes the contribution of unwanted signals and noise. As elegantly shown in Reference [17], the antenna array in this concept is a spatial-domain counterpart of a tapped delay line time-domain filter. The term beamforming in this context refers to the process of shaping the reception/radiation pattern of an antenna array that is achieved by adjusting the amplitude and phase responses of individual antenna channels and summing up their outputs in order to produce a single signal. In the simplest case, so called narrowband beamforming, the shaping of the array reception pattern is

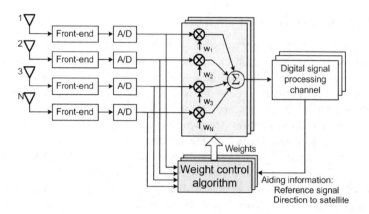

Figure 9.9 Generic block diagram of narrowband beamforming.

achieved through the adjustment of the array complex weights as shown in Figure 9.9. With wideband signals or large array apertures, a wideband or space–time beamforming should be used (see Figure 9.10). Here, in order to shape the antenna channel response over the signal spectrum content a tapped-delay line is utilized.

In both narrowband and wideband beamforming cases the array weights are generated by some control algorithm. The narrowband assumption can be used with beamforming as long as the effective width of the signal's autocorrelation function stays much larger than the maximum signal propagation time over the array aperture:

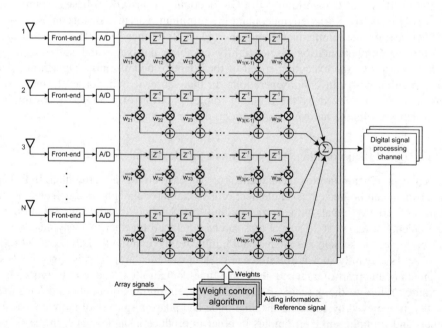

Figure 9.10 Generic block diagram of wideband beamforming.

$$\frac{\int\limits_{-\infty}^{\infty} |R(\tau)|^2 \mathrm{d}\tau}{|R(0)|^2} = \tau_{eff} >> \Delta t_{max} \tag{9.4}$$

or, alternatively,

$$BW << 1/\Delta t_{max} \tag{9.5}$$

where $R(\tau)$ is the signal autocorrelation function, τ_{eff} is the effective width of the peak of the autocorrelation function, Δt_{max} is the maximum signal propagation time over the array aperture, and BW denotes the signal bandwidth.

For a reasonable antenna array size, for example, below 0.5 m, the narrowband beamforming assumption is valid with all Galileo signals. The assumption comes, however, at a limit with the AltBOC E5 signal when assuming the use of the entire transmitted bandwidth of 92 MHz. For this reason we will further focus on the narrowband beamforming when describing the approaches to the generation of the array weights.

Given the array weights, the resulting array reception pattern can be calculated as follows:

$$AP(\theta, \varphi) = \sum_{i=1}^{N} w_i \exp\left(j\frac{2\pi}{\lambda}(x_i \cos\varphi\sin\theta + y_i\sin\varphi\sin\theta + z_i\cos\theta)\right) AP_i(\theta, \varphi) \tag{9.6}$$

where w_i denotes the complex weight of the signal from the ith array element, λ is the wavelength, the term $\exp(\ldots)$ describes the array manifold function, and $AP_i(\theta, \phi)$ denotes the reception pattern of the ith array element with the coordinates $[x_i, y_i, z_i]$ at the given polarization.

The weight control algorithm adjusts the array reception pattern using a priori information and/or one of the performance optimization criterion. The most frequently used techniques to control the array weights are:

- Deterministic beamforming.
- Linearly constrained minimum variance (LCMV) beamforming.
- Minimum mean squared error (MMSE) beamforming.

With *deterministic beamforming*, a deterministic set of array weights is used in order to steer the main lobe of the array reception pattern in the direction of arrival of a desired signal. For this purpose, the array weights are chosen to match the antenna array response in the desired steering direction (θ, ϕ) as follows:

$$w_i = \left(\exp\left(j\frac{2\pi}{\lambda}(x_i \cos\varphi\sin\theta + y_i\sin\varphi\sin\theta + z_i\cos\theta)\right) AP_i(\theta, \varphi)\right)^* \tag{9.7}$$

where $(\cdot)^*$ denotes complex conjugate.

The *LCMV algorithm* is used on the premise that the direction of arrival of the desired signal is known. The array weights are adjusted by the algorithm in a way to preserve the predefined antenna array response for the GNSS signal and to maximize the output SINR or, equivalently, to minimize the noise and interference power at the beamformer output, which is formulated as follows:

$$\min_{\mathbf{w}} E\left\{|y(t)|^2\right\} = \min_{\mathbf{w}} \mathbf{w}^H E\left\{\mathbf{x}(t)\mathbf{x}^H(t)\right\}\mathbf{w}, \qquad \text{s.t.} \mathbf{a}(\theta,\phi)\mathbf{w}^H = 1 \qquad (9.8)$$
$$= \min_{\mathbf{w}} \mathbf{w}^H \mathbf{R}_{xx}(t)\mathbf{w},$$

with an optimum steady-state solution

$$w_{LCMV} = \mathbf{R}_{xx}^{-1}\mathbf{C}\left(\mathbf{C}^H\mathbf{R}_{xx}^{-1}\mathbf{C}\right)^{-1}g, \quad \text{where } \mathbf{C} = \mathbf{a}(\theta,\phi) \text{ and } g = 1 \qquad (9.9)$$

where $\mathbf{x}(t)$ and $y(t)$ denote the beamforming input and output, correspondingly, $\mathbf{R}_{xx}$ is the array covariance matrix, $E\{\cdot\}$ stands for the expectation operator, and $\mathbf{a}(\theta,\phi)$ is the array manifold vector with the elements

$$a_i(\theta,\varphi) = \exp\left(j\frac{2\pi}{\lambda}\left(x_i\cos\varphi\sin\theta + y_i\sin\varphi\sin\theta + z_i\cos\theta\right)\right)AP_i(\theta,\varphi) \qquad (9.10)$$

The *MMSE algorithm* uses a reference signal, $r(t)$, for the weight adaptation process. The array weights are adjusted in such a way as to minimize the difference between the beamformer output and the reference signal in the least-mean squares sense, so that

$$\min_{\mathbf{w}} E\left\{|r(t) - y(t)|^2\right\} = \min_{\mathbf{w}} E\left\{|r(t) - \mathbf{w}^H\mathbf{x}(t)|^2\right\} \qquad (9.11)$$

with an optimum steady-state solution as follows:

$$\mathbf{w}_{MMSE} = \mathbf{R}_{xx}^{-1}\mathbf{p}, \quad \text{where } \mathbf{p} = E\{\mathbf{x}(t)r^*(t)\} \qquad (9.12)$$

In a GNSS receiver the reference signal for the MMSE control algorithm can be generated by the signal tracking module in the form of either a PRN code replica with the delay estimated by the code tracking loop or a current data bit sample estimated by the carrier tracking loop.

It can be shown [18] that both LCMV and LMS algorithms provide an optimum solution for the antenna weight vector in terms of the maximum SINR. The adaptation process in these algorithms can be performed in a block-wise manner or sample-per-sample way. In the first case, a block of array output samples collected over some time is used to obtain the estimation of the covariance and cross-correlation matrices, for example, for the array covariance matrix $\mathbf{R}_{xx}$ as follows:

$$\hat{\mathbf{R}}_{xx}[k+1] = \gamma\hat{\mathbf{R}}_{xx}[k] + \mathbf{x}[k]\mathbf{x}^H[k] \qquad (9.13)$$

where k is the index of the update epoch and γ is a memory factor defining how fast the past data are 'forgotten', $0 < \gamma \leq 1$. With the sample-per-sample adaptation, the array weight vector is presented in the recursive form as follows:

$$\mathbf{w}[k + 1] = \mathbf{w}[k] + \Delta\mathbf{w}[k] \tag{9.14}$$

and the correction term $\Delta\mathbf{w}[k]$ is obtained at each adaptation epoch from the instantaneous values of the array outputs and reference signal [19]:

$$\Delta\mathbf{w}[k] = \gamma(\mathbf{a}^*(\theta, \phi) - \mathbf{x}[k]\mathbf{x}^H[k]\mathbf{w}[k]) \text{ with LCMV algorithm}$$

$$\Delta\mathbf{w}[k] = \gamma\mathbf{x}[k](r^*[k] - \mathbf{x}^H[k]\mathbf{w}[k]) \text{ with MMSE algorithm} \tag{9.15}$$

9.4.2.2 Statistical Array Processing: The Array as a Spatial Sensor

Instead of considering the antenna array as a spatial filter it can be treated as the spatial sensor of the arriving signals. Following that concept, the desired improvement of the system performance with the use of the navigation signal can be formulated as the following parameter estimation problem: the time delay and carrier phase of the signal of interest have to be estimated in the presence of unwanted signals and noise. The corresponding signal model describing the measured array outputs can be introduced as follows (for example, see Reference [20]):

$$\mathbf{y}(t) = \sum_{\ell=1}^{L} \mathbf{s}_\ell(t) + \mathbf{n}(t) \tag{9.16}$$

where L is the number of arriving signals $\mathbf{s}_\ell(t)$ and $\mathbf{n}(t)$ accounts for the receiver noise assuming an additional complex white Gaussian noise. Further, each arriving signal is given by

$$\mathbf{s}_\ell(t) = \mathbf{a}(\varphi_\ell(t), \vartheta_\ell(t))\, \alpha_\ell(t)\, c(t - \tau_\ell) \tag{9.17}$$

where α_ℓ stands for the complex amplitude of the ith signal, τ_ℓ stands for the time delay, $c(t)$ denotes a pseudorandom noise (PRN) code used by the navigation signal, and φ_ℓ and ϑ_ℓ stand for the azimuth and elevation angles of arrival, respectively. Here, $\mathbf{a}(\phi, \vartheta)$ denotes the array manifold vector. Vectors $\mathbf{y}(t)$, $\mathbf{s}_\ell(t)$, and $\mathbf{s}(t)$ in the above equations have the length that is equal to the number of array elements M, for example, $\mathbf{y}(t) \in \mathbb{C}^{M \times 1}$.

The spatial observations $y[n]$ of the antenna array outputs are collected at N time instances, whereas $y[n] = y(nT_s)$ with $n = 1, 2, \ldots, N$. The channel parameters are assumed constant during the observation interval NT_s, which means that the coherence time of the assumed channel is $T_{coh} \geq T_s N$. Collecting the snapshots of the observation interval leads to

$$\begin{aligned}
\mathbf{Y} &= [\mathbf{y}[1], \quad \mathbf{y}[2], \quad \ldots, \quad \mathbf{y}[N]] \in \mathbb{C}^{M \times N} \\
\mathbf{N} &= [\mathbf{n}[1], \quad \mathbf{n}[2], \quad \ldots, \quad \mathbf{n}[N]] \in \mathbb{C}^{M \times N} \\
\mathbf{S}_\ell &= [\mathbf{s}_\ell[1], \quad \mathbf{s}_\ell[2], \quad \ldots, \quad \mathbf{s}_\ell[N]] \in \mathbb{C}^{M \times N} \\
\boldsymbol{\theta} &= \left[\boldsymbol{\theta}_1^T, \quad \boldsymbol{\theta}_2^T, \quad \ldots, \quad B_L^T\right]^T \\
\boldsymbol{\theta}_\ell &= [\alpha_\ell, \quad \tau_\ell, \quad \varphi_\ell, \quad \vartheta_\ell]^T
\end{aligned} \tag{9.18}$$

Using matrix notation, the signal model can be written as

$$\mathbf{Y} = \mathbf{S}(\boldsymbol{\theta}) + \mathbf{N} = \sum_{\ell=1}^{L} \mathbf{S}_\ell(\boldsymbol{\theta}_\ell) + \mathbf{N} \tag{9.19}$$

Assuming spatially and temporally uncorrelated noise contribution elements in $\mathbf{N}$, the covariance of the noise is then given by $\mathbf{I} \cdot \sigma_n^2$. The likelihood function with the introduced signal model for the problem of estimating the parameter vector $\boldsymbol{\theta}_\ell = [\alpha_\ell,\ \tau_\ell,\ \varphi_\ell,\ \vartheta_\ell]^T$, $\ell = 1, 2, \ldots, L$, is then represented by the conditional probability density function (PDF) as follows:

$$p(\mathbf{Y}; \boldsymbol{\theta}) = \frac{1}{(\pi \sigma_n^2)^{MN}} \exp\left(-\frac{\|\mathbf{Y} - \mathbf{S}(\boldsymbol{\theta})\|_F^2}{\sigma_n^2} \right) \tag{9.20}$$

Here, $\| \cdot \|_F$ denotes the Frobenius norm of a matrix. The maximum likelihood estimator (MLE) is then given by

$$\hat{\boldsymbol{\theta}} = \arg \max_{\boldsymbol{\theta}} \{p(\mathbf{Y}; \boldsymbol{\theta})\} \tag{9.21}$$

Estimation of $\boldsymbol{\theta}$ is a computationally extensive task since there is no analytical solution for the global maximum. The conditional PDF $p(\mathbf{Y}; \boldsymbol{\theta})$ is generally not a concave function of $\boldsymbol{\theta}$. Since the values for maximization of the complex amplitude α_ℓ can be given in closed form as a function of the other parameters, the computation of the MLE is a three-L-dimensional nonlinear optimization procedure.

9.5 Receiver Implementation Trade-Offs

9.5.1 Computational Resources Required

In the structure of the GNSS receiver, the beamforming operation can be placed before or after the PRN code correlation process. The corresponding block diagrams are shown in Figures 9.11 and 9.12. As can be seen from these figures, the placement of the beamforming has a large effect on the recourse partitioning between the high-rate and low-rate processing

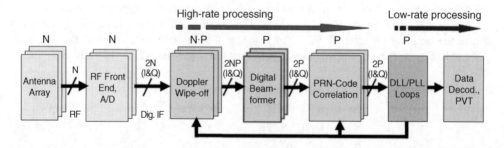

Figure 9.11 Block diagram of a GNSS array receiver with beamforming before PRN code correlation.

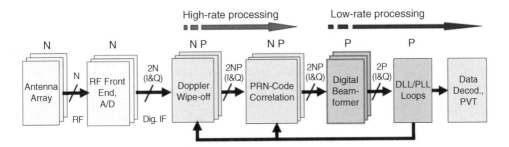

Figure 9.12 Block diagram of a GNSS array receiver with beamforming after PRN code correlation.

parts of the receiver. Normally, implementation of the beamforming after the PRN-code correlation in the low-rate processing part allows for implementation of more sophisticated algorithms and the addition of some aiding functions like direction finding and coupling with inertial sensors. The option before PRN code correlation may potentially provide lower overall complexity and therefore lower power consumption.

The number of bits to be used for sample quantization at each of the receiver's digital stages are important design parameters, since they have a direct impact on the arithmetic computation architecture and the achieved performance, mostly in terms of robustness against the presence of interferences. In order to show this effect, we simulated the Galileo E1B and E1C Open Service signals described in the Galileo's Signal in Space Interface Control Document. This band, centered at $f_{GalE1} = 1575.420$ MHz and with a reference passband bandwidth of 24.5520 MHz, uses the composite binary offset carrier (CBOC) modulation, defined in the baseband as

$$s(t) = \frac{1}{\sqrt{2}} (e_{E1B}(t)(\alpha sc_A(t) + \beta sc_B(t)) - e_{E1C}(t)(\alpha sc_A(t) - \beta sc_B(t))) \tag{9.22}$$

where the subcarriers $sc(t)$ are defined as

$$sc_A(t) = \text{sign}\left(\sin\left(2\pi f_{s,E1A} t\right)\right) \tag{9.23}$$

$$sc_B(t) = \text{sign}\left(\sin\left(2\pi f_{s,E1B} t\right)\right) \tag{9.24}$$

and $f_{s,E1A} = 1.023$ MHz and $f_{s,E1B} = 6.138$ MHz are the subcarrier rates, $\alpha = \sqrt{\frac{10}{11}}$<use solidus>, and $\beta = \sqrt{\frac{1}{11}}$<use solidus>. Channel B contains the I/NAV type of navigation message, $D_{I/NAV}$ intended for safety-of-life (SoL) services:

$$e_{E1B}(t) = \sum_{l=-\infty}^{+\infty} D_{I/NAV}\left[[l]_{4092}\right] \oplus C_{E1B}\left[|l|_{4092}\right] p\left(t - lT_{c,E1B}\right) \tag{9.25}$$

where $\oplus$ is the exclusive–or operation (modulo–2 addition), $|l|_L$ means l modulo L, $[l]_L$ is the integer part of l/L, and $p(t)$ is a rectangular pulse of a chip period duration centered at $t = 0$

and filtered at the transmitter. In the case of channel C, it is a pilot (dataless) channel with a secondary code, forming a tiered code:

$$e_{E1C}(t) = \sum_{m=-\infty}^{+\infty} C_{E1Cs}\left[|m|_{25}\right] \oplus \sum_{l=1}^{4092} C_{E1Cp}[l]p\left(t - mT_{c,E1Cs} - lT_{c,E1Cp}\right) \qquad (9.26)$$

with $T_{c,E1B} = T_{c,E1Cp} = 1/1.023$ μs and $T_{c,E1Cs} = 4$ ms. The C_{E1B} and C_{E1Cp} primary codes are pseudorandom memory code sequences defined in the Interface Control Document (Annex C.7 and C.8). The binary sequence of the secondary code C_{E1Cs} is 0011100000001010110110010.

For the numerical simulations, we considered a circular, $N = 8$, omnidirectional element antenna array, with the corresponding RF front-ends delivering complex signals in baseband (I&Q components), although adapting the beamformer to a suitable intermediate frequency is straightforward. The receiver baseband bandwidth was set to 6.5 MHz (which contains 92.88% of the total signal power), and the sampling frequency to $f_s = 13$ Msps. The simulated interference was a wideband, LTE-like signal centered at the same frequency of the Galileo signal and occupying the same bandwidth. Figure 9.13 shows the performance of four different beamformers considering different resolutions of the ADC and applying a

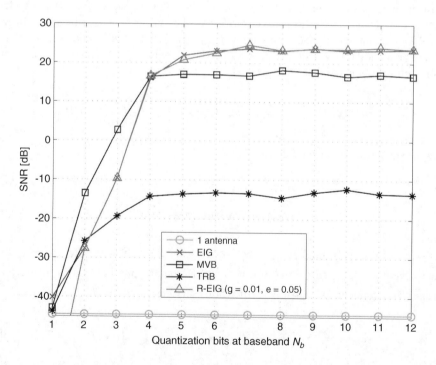

Figure 9.13 Wideband interference mitigation versus quantification resolution performance for different beamformers. Initial (prebeamformer) SNIR was set to −50 dB. Average results of 1000 independent realizations.

quatification of 8 bits in the beamweights. In addition to the beamformers presented above (deterministic, LCMV, and MMSE beamformers), we tested the eigenbeamformer presented in Reference [21] and defined as

$$w_{EIG} = PE\left\{ \left(\hat{\mathbf{R}}_{\mathbf{XX}} - \hat{\mathbf{a}}^H \hat{P}_{d_0} \hat{\mathbf{a}} \right)^{-1} \left(\hat{\mathbf{a}}^H \hat{P}_{d_0} \hat{\mathbf{a}} + \hat{\sigma}_n^2 \mathbf{1} \right) \right\} \quad (9.27)$$

where $PE\{\cdot\}s$ is the operator that yields the principal eigenvector of a matrix, that is, the eigenvector that corresponds to the maximal eigenvalue, $\hat{P}_{d_0}$ is the estimation of the receiving signal power, and $\hat{\sigma}_n^2$ is the estimation of the noise power.

Results show that an ADC providing four bits per sample provides a high level of protection against interferences. This performance can be enhanced by an automatic gain control (AGC) that avoids saturation of the ADC and provides an extra margin of protection.

9.5.2 Clock Domain Crossing in FPGAs/Synchronization Issues

In order to define the system low-level architecture, an important design principle is the synchronous methodology, in which all storage components (registers composed by a set of flip-flops[1]) are controlled by a common clock signal. Implicitly, digital signal processing techniques and the associated mathematical tools assume a synchronized system. Indeed, design and analysis so far are based on an ideal clocking scenario, assuming that the entire system can be driven by a single clock signal and that the sampling edge of this clock signal can reach all the components at the same time. In reality, this is hardly possible, since each input port of a gate and each wire introduce small values of resistance and capacitance. Thus, the design must consider a nonideal clock signal and data distribution. The clock distribution network is the circuit that delivers the clock signal to all the flip-flops in the system. FPGA devices usually have one or more prerouted and prefabricated clock distribution networks, and if the VHDL code is developed properly, the synthesis software can recognize the existence of the clock signal and automatically map it to one (or more) of such networks. The effect of the clock distribution network can be modeled by propagation delays from the clock source to various registers. Because of the variation in buffering and routing, the propagation delays may be different. The key characteristic is the difference between the arrival times of the sampling edges, which is known as the *clock skew*. For multiple registers, we consider the worst-case scenario and define the clock skew as the difference between the arrival times of the earliest and latest sampling edges.

Multiple clocks may exist or become necessary for several reasons, such as inherent multiple clock sources (interaction with external systems), circuit size (clock skew increases with the size of the circuit and the number of flip-flops), design complexity (since a large digital system is frequently composed of several small subsystems of different performance and power requirements, applying pure synchronous design methodology may introduce unnecessary constraints), and power considerations (the dynamic power of a CMOS device is proportional to the switching frequency of transistors, which is correlated to the system clock frequency). These considerations have a direct impact on the synchronous design

---

[1] A flip-flop is an electronic circuit that has two stable states and thereby is capable of serving as one bit of memory.

methodology, since they imply the need to divide the system into multiple synchronous subsystems and to design special interfaces between those subsystems.

In such a multiple clock system, the term *clock domain* is used to describe a block of circuitry in which the flip-flops are controlled by the same clock signal, even in the case of using a derived clock signal with different (but known) clock frequencies or phases. Note that these derived clock signals need their own individual clock distribution networks even though they are in the same clock domain.

The proper operation of a clocked flip-flop depends on the input being stable for a certain period of time before and after the clock edge. If the setup and hold-time requirements are met, the correct output will appear at a valid output level at the flip-flop output (i.e., the output voltage will be either above the 'high' or below the 'low' thresholds) after a certain clock-to-output delay. However, if these setup and hold-time requirements are not met, the output of the flip-flop may take much longer than the expected maximum clock-to-output delay to reach a valid logic level. This is called unstable behavior, or *metastability* [22]. When a flip-flop enters a metastable state, its output voltage is somewhere between the low and the high thresholds, and cannot be interpreted as either logic '0' or logic '1'. If the output of the flip-flop is used to drive other logic cells, the in-between value may propagate to downstream logic cells and lead the entire digital system into an unknown state, and thus a functional failure. Since the error patterns and failures are not easily repeatable (a typical setup consists of an external clock source with a random phase), we should adopt good design practices in systems with multiple clock domains, especially in asynchronous clock-domain crossing paths.

The problem of metastability can be prevented by adding special devices between the source and the destination domains [23]. These devices, known as synchronizers, isolate metastability, delivering a clean signal to the downstream logic. Such devices commonly consist of multiple flip-flop stages with a handshake protocol. In the proposed design, we have used asynchronous FIFOs as a way to make the clocks independent at the write and the read ports. This solution, when implemented using a RAM memory block in combination with Gray encoding/decoding of input/output pointers in order to minimize bit transitions, is also known to mitigate glitches and latency-related problems. An FIFO memory used as a synchronizer has independent read and write clock ports, dividing the FIFO into two clock regions and providing a suitable data path between both clock domains. Each of the FIFO clock regions has their own set of control signals, synchronous to the respective clock signal. This feature allows the use of FIFOs as data buffers when both clocks have a different frequency. FPGA vendors provide such Intellectual Property (IP) cores, which could even include correction checking [24].

9.6 Practical Examples of Experimentation Systems

9.6.1 *L1 Array Receiver of CTTC, Spain*

The first example of implementation of an antenna array-based receiver in this chapter was developed in the Center Tecnològic de Telecomunicacions de Catalunya (CTTC), Spain. It consists of an eight-element array disposed in a circular shape, a radio frequency front-end that downconverts and digitizes the signals received by each antenna, and a digital engine implemented in FPGA technology that is able to process the signals, compute the weights,

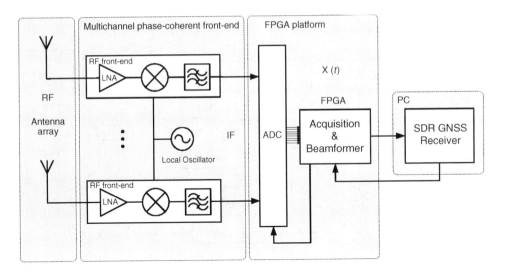

Figure 9.14 Receiver block diagram [25].

and perform the beamforming in real-time. In the following, we provide an overview of its design, implementation, and performance.

Figure 9.14 presents the overall system block diagram of the multiantenna receiver. From left to right, the first block is the antenna array, which is composed of eight antenna elements distributed in a circular shape with half-wavelength separation between them. Each of the antenna terminal elements is connected to a dedicated RF front-end channel. The multi-channel RF front-end is in charge of selecting the GNSS band and adapting the received antenna signal both in frequency and amplitude to be digitized by the ADC. It is composed of an RF amplification and filtering stage, a frequency downconversion stage, and an IF amplification and filtering stage. A common local oscillator (LO) is used for all the hetero-dyne stages, and the sample clock keeps phase coherence with the LO signal. Table 9.2 shows the electrical measurements performed on the CTTCs RF front-end.

Table 9.2 RF front-end electrical measurements [25]

Parameter	Measurement
RF frequency	1575.4200 MHz
IF frequency	69.9988 MHz
Passband bandwidth	17.50 MHz
Stopband bandwidth	22.00 MHz
Gain per channel	58.00 dB
Noise figure	2.18 dB
1 dB compression point	−65.30 dBm
Third-order intercept point (IP3)	53.70 dB
Image rejection at 1435.42	≥ 57 dB
Phase noise (10 kHz)	−82.00 dBc

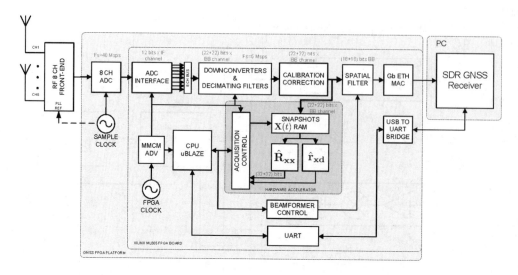

Figure 9.15 Detailed block diagram of the receiver's digital part [25].

The front-end output is then fed to a multichannel ADC based on the Texas Instruments' ADS5273 chipset, an eight-channel ADC integrated circuit that supports a sampling frequency up to 70 Msps with 12 bits per sample. In this implementation, the sampling frequency was set to $f_s = 40$ Msps, implying that each digital output transmits at a bit rate of $40 \times 12 = 480$ Mbps. This data flow is injected into a Xilinx Virtex-6 FPGA via FMC (FPGA Mezzanine Card), the ANSI/VITA 57.1-2008 standard, which defines I/O mezzanine modules with connection to an FPGA or other device with reconfigurable I/O capability.

The FPGA logic is in charge of downconverting the front-end signal from IF to baseband, calibrating the array (compensating phase mismatch between RF channels, as well as mutual coupling effects), and implementing an acquisition algorithm and a digital beamforming. In order to compute the calibration parameters, GNSS-like signals can be injected to the antennas in order to measure the effect of each front-end channel and compensate them digitally [26].

The high-level block diagram of the proposed digital processing architecture is shown in Figure 9.15. From left to right, each of the f_{IF} output channels of the antenna array front-end is sampled synchronously by the ADC board. The sample clock provides the 10 MHz reference for the front-end LO frequency synthesizer. The FPGA logic circuitry reads the digital samples coming from the ADC and performs a digital downconversion from f_{IF} to the baseband. In this step, in order to reduce the sample rate, the digital downconverter stage implements a decimation of factor 8 and thus the baseband sample rate is 5 Msps.

The baseband samples are fed to a calibration correction block, which compensates the differences both in phase and in amplitude between channels. At this point, the calibrated snapshots are fed both to the real-time spatial-filter processing block and to the array acquisition hardware accelerator. The spatial-filter block can implement a wide range of beamforming algorithms. The resulting spatially filtered single output sample stream is fed to the gigabit Ethernet module, which is in charge of grouping the signal samples into Ethernet packets. A commodity PC receives the packets, where the samples can be stored for postprocessing, or processed in real-time by a software receiver.

Most array-based acquisition algorithms and digital beamforming techniques require the computation of the autocorrelation and the cross-correlation vectors of the received signal snapshots, as well as inversions of matrices and other sophisticated matrix algebra operations that require floating-point precision. The computation can be split into logic circuitry blocks implemented in very high speed integrated circuit hardware description language (VHDL) or in an embedded processor that provides floating-point capabilities and higher-level programming. For this prototype, the embedded software processor is implemented using the Xilinx Microblaze core, which is used both for the acquisition and for the spatial-filter algorithms.

The acquisition hardware accelerator module is directly connected to the software processor by means of a dedicated FIFO-style connection called the fast Simplex link (FSL) bus, which provides the processor with access to the portion of the acquisition algorithm implemented in VHDL using a specifically developed peripheral.

The flexibility of the design allows experimentation with different digital signal processing strategies without the need for hardware modifications, only requiring changes in the FPGA programming. As a example, the performance of several acquisition algorithms was measured in harsh interference environment conditions. The following test functions were executed using 1 ms of captured snapshots:

- **GLRT colored:** is the acquisition algorithm presented in Reference [27], whose test function is defined as

$$T_1(\mathbf{X}) = \max_{f_d, \tau} \left\{ \hat{\mathbf{r}}_{\mathbf{xd}}^H(f_d, \tau) \hat{\mathbf{R}}_{\mathbf{XX}}^{-1} \hat{\mathbf{r}}_{\mathbf{xd}}(f_d, \tau) \right\} > \gamma \qquad (9.28)$$

where $\mathbf{X}$, referred to as the space–time data matrix, is defined as the antenna array baseband snapshot, where each row corresponds to one antenna and each column stores the output of the ADC during the integration time (K samples), $\hat{\mathbf{R}}_{\mathbf{XX}} = \frac{1}{K} \mathbf{X} \mathbf{X}^H$ is the estimation of the autocorrelation matrix of the array snapshots, also known as the sample covariance matrix (SCM), $\hat{\mathbf{r}}_{\mathbf{xd}}(f_d, \tau) = \frac{1}{K} \mathbf{X} \mathbf{d}^H$ is the cross-correlation vector between the array snapshot matrix and the local replica, generated with a delay τ and a Doppler shift f_d, and γ is the detectionthreshold.

- **GLRT white:** is the white noise version of the GLRT for the unstructured array signal model, whose test function is defined as

$$T_2(\mathbf{X}) = \max_{f_d, \tau} \left\{ \frac{\hat{\mathbf{r}}_{\mathbf{xd}}^H(f_d, \tau) \hat{\mathbf{r}}_{\mathbf{xd}}(f_d, \tau)}{\mathrm{Tr}(\hat{\mathbf{R}}_{\mathbf{XX}})} \right\} \qquad (9.29)$$

where $\mathrm{Tr}(\cdot)$ stands for the trace operator.

- **Single antenna MF:** is the noncoherent matched filter commonly used in single-antenna receivers [28]. The test function is defined for the ith element as

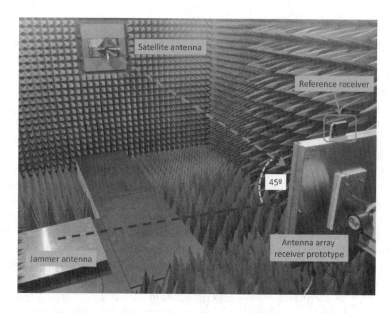

Figure 9.16   Measurements in the anechoic chamber [25].

$$T_3(s, i) = \max_{f_d, \tau} \left\{ \sum_{t=0}^{K-1} \left\| x_i(t) e^{-j2\pi f_d t} d^*(t, \tau) \right\|^2 \right\} \tag{9.30}$$

The performance of the implemented acquisition algorithms was tested in harsh interference environment conditions. The test functions defined above were executed using 1 ms of captured snapshots. Figure 9.16 is a picture of the anechoic chamber setup. The satellite signal is transmitted using the horn antenna, with $\theta = 0°$ and $\varphi = 90°$, simulating a realistic situation where a high-elevation satellite is received. The interference is transmitted using an auxiliary monopole antenna with an approximated DOA of $\theta = 45°$ and $\varphi = 45°$, which simulates a moderate elevation jammer or a communication signal coming from a nearby communication tower. The satellite signal power and the interference (or jammer) power is given in terms of the carrier-to-noise density ratio (C/N0) and the interference-to-noise density ratio (I/N0), respectively, measured at the IF output of the front-end. The AGC function was turned off and the front-end was configured at maximum gain. In both scenarios, the interference and the GPS L1 C/A signal were generated by a couple of Agilent E4438C vector signal generators. The satellite signal's C/N0 was set at 44 dB Hz, which was measured using a reference receiver based on the Sirf Star III chipset by Cambridge Silicon Radio (CSR). Two in-band interference scenarios were tested:

- **Continuous wave interference:** In this scenario, a continuous wave (CW) jammer impinges into the array with $f_{int} = 1545.43$ MHz and I/N0 = 133 dB Hz.

- **4G: /LTE interference** This experiment simulates a situation where an LTE-like signal coming from a nearby base station is interfering with the GPS L1 C/A signal. It is known that there is concern about the interferences that the deployment of a 4G network could cause, especially when it uses the 1552.7 MHz band [29]. Since the RF front-end prototype is equipped with highly selective SAW filters, in our setup the interference generator is unable to reach the interference power levels required to interfere with the GPS L1 band with out-of-band spurious emissions. In order to test the protection against a possible wideband interference, a simulated LTE base station downlink signal [30] with 5 MHz of channel bandwidth impinges into the array with $f_{int} = 1545.42$ MHz and I/N0 $= 133$ dB Hz. In this situation, the interference is superposed over the entire acquisition bandwidth.

In order to define an interference protection metric offered by the proposed acquisition with respect to other acquisition algorithms, we used the so-called *generalized signal-to-noise ratio* or *deflection coefficient* to measure the detectors performance. The deflection coefficient is defined as

$$
\delta = \frac{(\mathrm{E}\{T(\mathbf{X}, \mathrm{H_1})\} - \mathrm{E}\{T(\mathbf{X}, \mathrm{H_0})\})^2}{var\{\mathrm{E}\{T(\mathbf{X}, \mathrm{H_0})\}\}}
\tag{9.31}
$$

where $\mathrm{H_0}$ refers to the *null hypothesis*, when the desired satellite signal is absent, and $\mathrm{H_1}$ refers to the *alternative hypothesis*, when the desired signal is present. $\mathrm{E}\{T(\mathbf{X}, \mathrm{H_1})\}$ and $\mathrm{E}\{T(\mathbf{X}, \mathrm{H_0})\}$ stand for the test statistic expectation in $\mathrm{H_1}$ and $\mathrm{H_0}$, respectively, and $var\{\mathrm{E}\{T(\mathbf{X}, \mathrm{H_0})\}\}$ stands for the variance of the test statistic in $\mathrm{H_0}$. This quantity measures the effectiveness of the quadratic statistic in separating the two hypotheses. In the measurements, the expectation operators and the variance operator were substituted by their sample mean and sample variance, respectively. Figure 9.17 shows the evolution of the estimated deflection coefficients in the CW interference scenario for an I/N0 from 90 to 150 dB Hz sweep. From the results we can see that the acquisition based on the GLRT for colored noise offers a jammer or interference protection of 33 dB with respect to the single antenna MF and the white version of the GLRT.

The deflection coefficient measurement was also done for the LTE interference scenario. Figure 9.18 shows the results. The GLRT colored offers 25 dB of protection for wideband inband interferers with respect to the single antenna MF and the white version of the GLRT. It is important to take into account the fact that the front-end reached the compression point at the highest power region of the interference power sweep (I/N0 from 135 to 150 dB Hz). This situation degrades the performance of the acquisition algorithms and thus a better performance is expected, enabling the AGC to avoid the front-end saturation.

9.6.2 GALANT, a Multifrequency GPS/Galileo Array Receiver of DLR, Germany

The second example of an experimentation system is the Galileo antenna array demonstrator (GALANT) that is being developed by the Institute of Communications and Navigation of the German Aerospace Center (DLR) with the main focus on improving robustness of a Galileo/GPS receiver against radio frequency interference and multipath

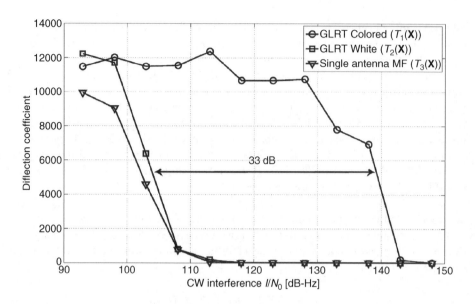

Figure 9.17 Deflection coefficient evolution in the CW interference scenario [25].

effects in safety-of-life (SoL) Galileo applications or any other applications with a safety critical context. The aim of the GALANT initiative is to perform the necessary predevelopment research as well as to develop, build, and demonstrate an integrated SoL receiver system. The integration of an adaptive array antenna was considered by the receiver design from the early beginning of the project.

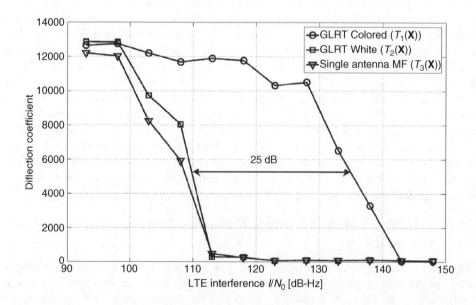

Figure 9.18 Deflection coefficient evolution in the LTE interference scenario [25].

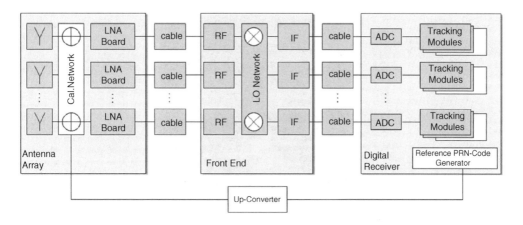

Figure 9.19 Block diagram of online calibration in the GALANT receiver.

At the time of this writing (October 2011) the receiver demonstrator was capable of receiving and processing Galileo and GPS signals in E1/L1 and E5a/L5 frequency bands employing an antenna array. In order to allow for controlling the reception pattern of the antenna array in given direction(s) and performing an accurate estimation of directions of arrival of incoming signals, the antenna channels of the receiver are calibrated for each operating frequency band using an online procedure. This procedure makes use of a specially generated signal that is added to the received GNSS signals directly at the antenna outputs of array elements. The phases and amplitudes of the calibration signal measured in the antenna channels of the receiver enable an assessment to be made of the difference in the channel transfer characteristics and the corresponding corrections to the array signals to be produced. The block diagram of the implementation of the online calibration procedure in the GALANT receiver is shown in Figure 9.19. The specific features of this implementation are that reserved PRN codes of GPS/Galileo are used to produce GNSS-like calibration signals without any data modulation having the spectrum content and the power level that are representative for the received GNSS signals. Such a calibration signal can be acquired and tracked in the receiver digital signal processing part as any other GNSS signal without any specific modification required. The estimations of the power and carrier phase of the calibration signal are obtained by the code and carrier tracking loops in the different antenna channels. The corrections for the array signals are produced based on these estimations as well as taking into account, when appropriate, for example, by deterministic beam and null steering, the reception patterns of the array antenna elements. The reception patterns of the array elements are measured during the receiver integration and further assumed to be time invariant. More information about the operation of the online calibration in the GALANT receiver will be given when describing individual blocks of the receiver hardware and software.

The hardware of the GALANT demonstrator mainly consists of four components (see Figure 9.20):

- An active antenna array [1] with calibration network. The antenna array consists of four single antenna elements arranged in a 2×2 rectangular uniform array with the center-to-

Figure 9.20 Main components of the GALANT receiver.

center interelement spacing of 90 mm. The design of the antenna array allows for mixing a calibration signal to the receiver GNSS signals at each operating frequency band.

- An RF front-end [2] with four antenna channels for each frequency band (an E1/E5a double-band RF front-end of the later modification is shown in Figure 9.21). The RF front-end channels convert each of the received antenna signals down to an intermediate frequency and perform signal amplification and filtering. The RF front-end box [2] also contains two upconverters for producing E1 and E5a calibration signals transported over cables to the antenna array.
- An FPGA board with integrated analog-to-digital (A/D) and digital-to-analog (D/A) converters. The signal processing on the FPGA board includes signal conditioning after A/D conversion, digital downconversion as well as Doppler and PRN-code correlation. A part of the FPGA is also allocated to the generation of calibration signals in the digital form on an intermediate frequency. The D/A converters available on the FPGA board are used to obtain the calibration signals in the analog form.
- An embedded Intel 2 GHz dual core PC card where the software for controlling the PFGA board and processing the correlator outputs is running.

In conventional dual frequency navigation antennas, a combined output is used for both frequency bands. In the case of radio interference, for example, of the aviation distance

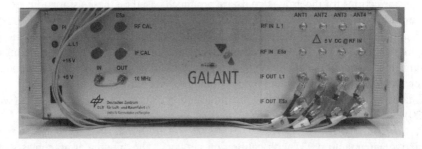

Figure 9.21 E1/E5a RF front-end.

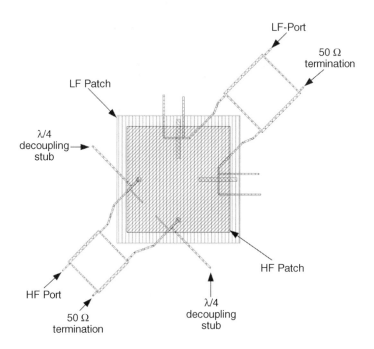

Figure 9.22   Schematic top view of the dual-band (E5 and E1) antenna element with highly isolated outputs. © 2011 IEEE. Reprinted, with permission, from [31].

measurement equipment (DME), it is not desirable to share one analog signal path for both frequency bands. Due to the potentially high power of such radio interference a signal distortion caused by saturation effects in the LNA may harm both bands. In the GALANT system a novel array element has been designed, which allows a frequency separation providing individual highly isolated outputs for E1 and E5 bands. The resulting antenna geometry is sketched in Figure 9.22. The main parameters for the optimization of this antenna are also indicated in the picture. The signals received in the E1 band are coupled to the high-frequency (HF) ports, whereas those in the E5 band are delivered to the low-frequency (LF) ports. Circular polarization operation is obtained by combining both HF and LF ports using 90° hybrids, which can be optimized for operation in the E1 and E5 bands, respectively.

Four of the single antenna elements are combined to a 2×2 antenna array with center-to-center interelement spacing of 90 mm, as shown in Figure 9.23. Measured curves of the gain for this array when pointing the main beam to boresight are given in Figure 9.24. It can be observed that the gains of the array in the lower and higher frequency bands are different. In fact the interelement spacing at the central frequencies are given by $d = 0.35 \; \lambda \text{E5}$ at 1189 MHz and $d = 0.47 \; \lambda \text{E1}$ at 1575 MHz. The resulting theoretical values for the array factor gain are then 3.7 dBi at 1189 MHz and 6.3 dBi at 1575 MHz.

The passive antenna array is connected to an RF board via SMP connectors. The RF board is a six-layer FR4 printed circuit board that contains the directional couplers and distribution networks for the E1 and E5 calibration signals, as well as two LNAs and a bandpass filter for each antenna element and each frequency band. These components are shown in the block diagram of the signal propagation path for a single array element in Figure 9.25 in the part

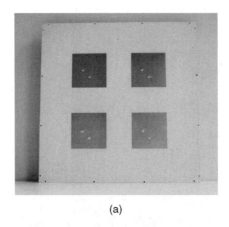

(a)

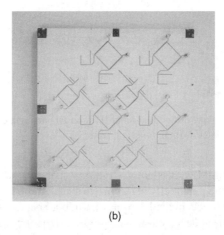

(b)

Figure 9.23 Top and bottom view of the dual-band 2×2 array. © 2011 IEEE Reprinted, with permission, from [31].

showing the dual-band active antenna. Though the design of the antenna array supports reception of the Galileo signals with the full bandwidth of the E5 band (1.164–1.215 GHz), only the E5a part component of the E5 signal is currently used by the GALANT receiver. This is because of higher robustness against radio interference provided by the separate processing of E5a and E5b signals, which is an important aspect in safety critical applications. The RF front-end hardware can, however, be easily modified in order to process E5a and E5b coherently if a higher ranging accuracy due to the larger bandwidth of approximately 50 MHz is desired.

The designs of the E1 and E5a paths in the RF front-end (see Figure 9.25) differ basically only in the RF filtering. The E1 front-end is designed for the reception of the signals of Galileo safety-of-life service, which have the 1 dB bandwidth of 14 MHz while the corresponding navigation signal on E5a has the bandwidth of 20 MHz. As mentioned above, the first part of the analog signal processing path is integrated in the active antenna array. Some filtering of the signal already occurs due to the frequency selective properties of the array

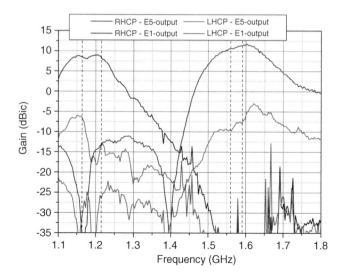

Figure 9.24 Gain curves at the LF and HF outputs of the 2 × 2 array. © 2011 IEEE Reprinted, with permission, from [31].

element. The required rejection of out-of-band signals is completed further by two ceramic RF filters coming next in the signal processing chain. The order of the filters are second and third, correspondingly. In order to limit the effect of the signal losses in the RF filters and obtain an acceptable noise figure the first LNA is placed directly after the directional coupler. Thanks to the relatively low gain of 13 dB and a high IP3, the LNA is very robust against

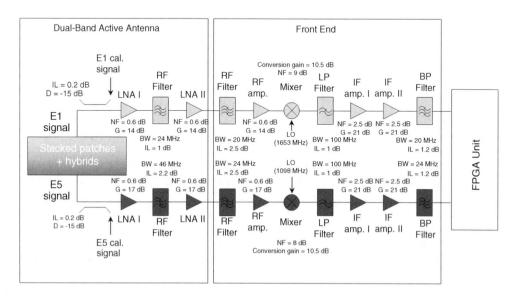

Figure 9.25 Block diagram of signal path for a single antenna element. © 2011 IEEE Reprinted, with permission, from [31].

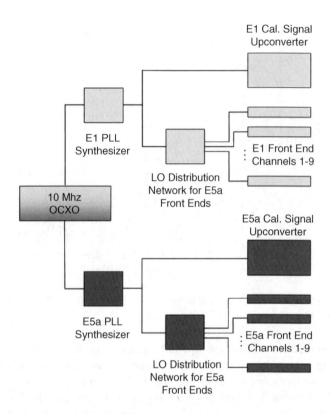

Figure 9.26   Local oscillator distribution network.

high power interference. The second LNA is identical to the first one. It boosts the signal level again by 13 dB and ensures a system noise figure of less than 1.6 dB. Splitting the LNA gain in multiple amplification stages with the filters in between helps to minimize the risk of saturation of the LNAs by strong out-of-band interference.

The downconversion stages in E5a and E1 paths bring the incoming signal to an intermediate frequency of 78 MHz. The corresponding local oscillator frequencies give enough spectral separation between the RF signal and the unwanted alias, which helps to relax the requirements on the anti-aliasing performance of RF filters. Due to this, a polyphase filtering at IF becomes obsolete. A third-order Chebychev lowpass filter is utilized to mitigate the RF breakthrough effects of the mixer. This is necessary to avoid a further amplification of the RF signal at the IF stage. This would harm the crosstalk isolation and, as a result, reduce the maximum depth of the spatial nulls in the array beam pattern. The total gain of two IF stage amplifiers is 40 dB. The last component in the analog signal processing chain is the Type I Chebychev anti-aliasing bandpass filter of the seventh order.

The mixers in the downconversion stage of the RF front-ends obtain the reference signals with the frequencies of 1.65342 GHz in the E1 path and 1.09845 GHz in the E5a path over a local oscillator (LO) distribution network, shown in Figure 9.26. A 10 MHz oven-controlled crystal oscillator is used as the reference frequency for two PLL synthesizers, which generate the LO reference signals for the E1 and E5a bands. These reference signals are also utilized in the upconverters when generating the array calibration signals. The upconversion of the calibration signal is performed by passive double balanced mixers. These mixers use the same local oscillator frequency as the multichannel front-ends. Before the signal is fed to the calibration network of the antenna array it is attenuated by approximately 80 dB.

For the distribution of the synthesizer outputs within the multichannel front-ends, two 1:9 distribution networks were developed and optimized for each frequency of interest. They are integrated in the six-layer FR4 printed circuit board of the front-ends. Because of a high LO power, which needs to be about 5 dBm, it is necessary to avoid unwanted radiation of the LO signals. To minimize this radiation the LO distribution network was designed using strip-line technology and integrated in the lower three layers of the front-end board. Moreover, the E1 and E5a distribution networks have been implemented in different layers and are, therefore, isolated from each other.

The outputs of the RF front-ends are further processed in the FPGA board, a Lyrtech VHS-ADC FPGA card. A single VHS-ADC board is populated with a powerful DSP-FPGA (Virtex-4 SX55) and provides eight high-rate A/D converters (maximum 105 Msps at 14 bit) together with variable gain amplifiers at inputs. Multiple VHS-ADC modules can also be cascaded in the case of larger antenna arrays. The incoming IF signals are digitized at the rate of 104 MHz, which causes an intentional aliasing down to a new digital IF of 26 MHz. The 14 bit resolution of the ADCs together with the digital steerable variable gain amplifiers results in a very high dynamic range, which is ideally suited for the use of digital signal processing algorithms for radio interference detection and mitigation.

The digitized signals are received by the Xilinx FPGA where custom logic for the baseband processing is implemented. In detail, the following custom built entities are instantiated for each array element:

- A digital lowpass filter (LPF);
- a digital automatic gain control (AGC) that reduces the bit width of the input samples based on a dynamic analysis of the signal amplitude;
- a downsampler to provide samples at reasonable rates;
- several modules for correlating the input data with a locally generated reference code and carrier signals.

Furthermore, each VHS-ADC module requires address logic and cPCI transfer. The cPCI transfer logic is provided by Lyrtech whereas the address logic is custom built. Last-mentioned logic prepares the tracking module register contents to be transferred between the Host-PC and the VHS-ADC module as a continuous data block.

The modules for the signal correlation are based on a generic VHDL description supporting GPS, GIOVE, and Galileo signals in the E1 and E5a bands. All PRN codes used by each module are memory based and can be updated via the host PC during operation. With a single antenna, each tracking module for one GNSS signal consists of five correlator modules plus one code and one carrier numerically controlled oscillator (NCO). With four antenna elements in the array the beamforming tracking module consists of 20 correlators plus two NCOs for the code and carrier. Up to 16 beamforming tracking modules fit into the utilized Virtex-4 SX55 FPGA, which then holds in total 320 correlators. Further key features of the tracking modules are:

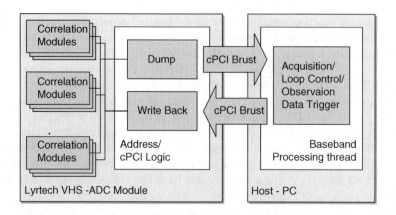

Figure 9.27   cPCI bus transfer schematic.

- variable resolution of the carrier and code NCOs;
- generation of BPSK- and BOC-modulated codes;
- the correlator spacing is adjustable during runtime:
- NCOs support the bump jumping algorithm used to track BOC signals.

Whereas the computationally extensive correlation process is realized in FPGA, the control of signal acquisition and tracking loops is performed by the software on the host PC. This makes the data transfer between the host PC and the FPGA critical for real-time operation due to associated latencies. In order to minimize the effect of latencies, the register contents of the tracking modules are transferred in continuous blocks (see Figure 9.27) instead of individually addressing the tracking module registers. The practice shows that the real-time operation can be surely achieved due to the high throughput of burst transfers of the cPCI bus. The host PC periodically acquires tracking module data that is copied into the PC's memory and can be accessed by the baseband processing software. After updating the tracking loops the software writes register contents data with new NCO control values back to the FPGA.

The address map of a single correlation module covers about 40 bytes for the correlator control values and output data. Accordingly, for a real-time operation, a 12 channel receiver supporting a 2×2 antenna array requires about 4 kbyte to be transferred via the cPCI bus in less than 1 ms (700 μs are typically assumed). In total, transfers of 5.3 Mbyte/s have to be guaranteed. Respectively, the 3×3 antenna array requires 11.8 Mbyte/s to be transferred. Both transfer rates are in the scope of the VHS-ADC specification, which allows a throughput of up to 20 Mbyte/s.

The entire range of functions supported by the software of the host PC includes:

- real-time operation of the data transfer with the correlation modules in FPGA over cPCI bus;
- real-time updating of tracking loops for all active multiantenna satellite channels with optional beamforming;
- direction of arrival estimation;

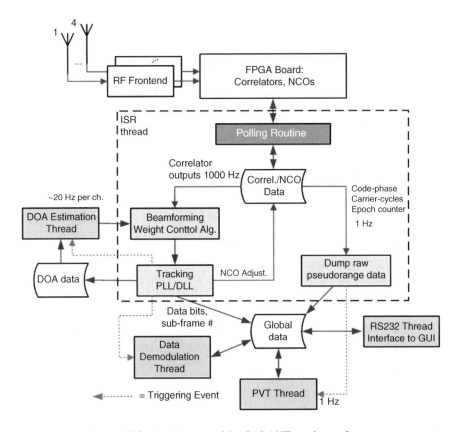

Figure 9.28 Architecture of the GALANT receiver software.

- navigation data demodulation;
- pseudorange calculation;
- PVT estimation;
- data transfer of receiver status information and PVT results to a graphical user interface (GUI) over serial RS232 port.

The Microsoft Windows XP operating system running on the host PC allows the signal processing software to be executed in multiple threads. On the one hand, this feature is very helpful in achieving real-time operation of the baseband processing since different priorities can be assigned to the threads. On the other hand, higher computational performance can be achieved on the Intel Dual Core architecture since the software modules can be executed in parallel on both CPU cores. Figure 9.28 shows the software architecture of the host PC.

The beamforming algorithms supported by the the signal processing software on the host PC are:

- The deterministic beamforming, by using the direction of arrival information obtained with the use of the GNSS system almanac or the direction of arrival estimation technique.

- Minimum mean squared error (MMSE) beamforming with the least mean squares (LMS) iterative procedure used for the array weight adaptation. The update rate of the adaptation process is the same as the rate of the PRN code correlators. This technique also maximizes the signal-to-noise plus interference ratio and has the advantage that it does not require the knowledge of the directions of arrival of the signal of interest.

The estimation of the directions of arrival of the GNSS incoming signal is performed using the two-dimensional unitary ESPRIT technique [32]. Since the estimation is performed after the PRN code correlation, it allows the detection of the direct GNSS satellite signal as well as strong multipath echoes. The DOA algorithm is implemented in an individual thread of the receiver software and is called by a tracking channel about 10–20 times per second, which depends on the length of the array data block used by the ESPRIT technique. A longer block length allows better DOA estimation, increasing, however, the reaction time of the receiver on the changes in the signal environment. The real-time functioning of the receiver has been tested with simultaneous and continuous DOA estimation for 12 tracked satellites with the update rate of 20 times per second.

The performance of the GALANT antenna array receiver has been evaluated under realistic signal conditions during two measurement campaigns in the German Galileo Test Environment (GATE) in Berchtesgaden in 2010 [33] and 2011 [34]. In the following, a short overview of the results of the last campaign of 2011 will be given.

First, the results for radio frequency scenario tests will be presented. The geometry of the measurement setup is shown in Figure 9.29. The interferer, like a pulsed DME, continuous wave, or broadband noise signal, was generated with a programmable signal generator at the E5a and E1 frequencies. A horn antenna, which was mounted on a mast and steered into the direction of the GALANT antenna array, was used for the signal transmission. The line-of-sight distance between the antennas was 11.3 m. At this short distance the far-field assumption is still fulfilled while strong radio interference can still be generated for the

Figure 9.29   Measurement setup.

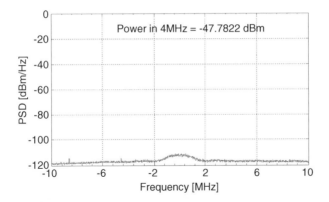

Figure 9.30 Power spectral density of the front-end output, interference-free case.

receivers under test, even with relatively low transmitted powers. Furthermore, the impact of the interferer on the environment can be kept at a minimum. For example, the transmitted interference power of 0 dBm at the output of the signal generator corresponds to the incoming power of −45.2 dBm at L1 and of −42.2 dBm at E5a at the GALANT antenna aperture. With a nominal received signal power of −125 dBm for GATE and GPS, the transmitted interference power of 0 dBm corresponds approximately to the interference-to-signal ratio (I/S) of 80 dB.

Broadband noise interferer. In this scenario a 4 MHz broad noise signal centered at 1575.42 MHz was transmitted. The GALANT receiver operated in the GPS/Galileo L1/E1. The interferer power was increased stepwise. Between each step the interferer was switched off for a short time to allow for the reacquisition of lost signals. Figure 9.30 shows the measured power spectral density (PSD) at the IF output of the front-end of antenna 1. The small rise in the measured PSD around the center frequency is due to the superposition of all received Galileo and GPS signals. Figure 9.31 shows the measured

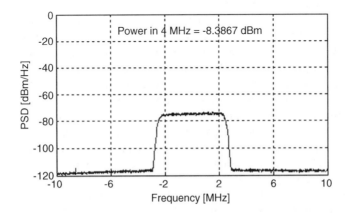

Figure 9.31 Power spectral density of the front-end output, noise interferer with 4 MHz bandwidth (I/S ≈ 65 dB).

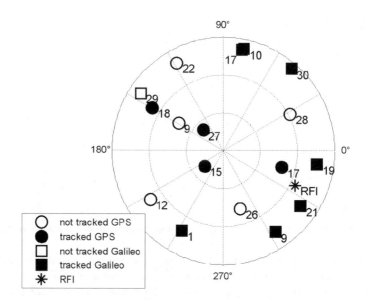

Figure 9.32 Skyplot of visible, tracked GNSS signals and RFI source, corrected for antenna attitude.

PSD of the same front-end output while the noise interferer with -15 dBm transmit power corresponding to the I/S ratio of approximately 65 dB was present. A skyplot of the tracked satellites and the interference source is shown in Figure 9.32. Because of the availability of 15 beamforming tracking channels in the GALANT receiver, where one channel was reserved for tracking the array calibration signal, not all visible satellites could be tracked simultaneously.

The representative reception patterns obtained with the help of MMSE adaptive beamforming for three of the tracked signals are shown in Figure 9.33. (a) shows the pattern for Galileo PRN 30. It can be observed that in the interference-free case the main beam points in the arrival direction of the signal (see Figure 9.32). In the case of interference (see Figure 9.33b), the reception pattern changes but the main beam still points in the direction of the satellite and a spatial null is steered towards the interferer. Figure 9.33(c) shows the array reception pattern for Galileo PRN 19, where the direction of arrival is close to that of the interferer. As in the case of PRN 30, the main beam points in the direction of the satellite signal, but as soon as the radio interference appears, the main beam moves away while a spatial null is placed into the direction of the interferer. It is obviously not possible to nullout the interference strongly and still preserve a high gain to the satellite signal. Figure 9.33 (d) and (e) show the beam patterns for GPS PRN 18 in the nominal and interfered case. Since PRN 18 has a large spatial separation to the interference source, the main beam is almost certainly not influenced by the RFI.

It is obvious that the spatial separation of the interferer and a satellite signal has a strong effect on the achievable jammer robustness. Figure 9.34 shows the C/N_0 ratios of the tracked satellites plotted for different transmitted jammer powers. Comparing Figure 9.34 with Figure 9.32 clearly shows the dependence between the maximum tolerable I/S ratio and the

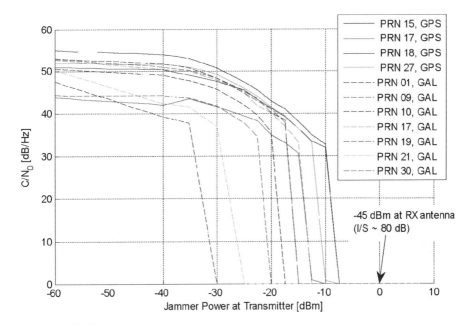

Figure 9.33   Antenna array reception patterns with MMSE digital beamforming.

corresponding spatial separation between the RFI and the navigation signal. The effect of the spatial separation is illustrated by Figure 9.35, where the maximum I/S levels at which the GATE signals are still tracked (in gray) and the interference levels at which all satellites are lost (in black) are plotted versus angular separation between the navigation and interferer signals. The data for this plot were collected from three different measurement sets. When switching to the next set, the measurement van was moved by a few meters so that the interferer was observed by the GALANT antenna array at the azimuth angles of 106°, 89°, and 69°. As the result, 24 different separation angles were available for eight GATE transmitters. The interference suppression increases up to a separation angle of about 40° and then

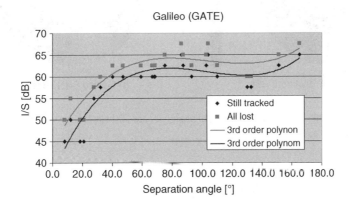

Figure 9.34 C/N_0 versus interference power of broadband noise interference on L1/E1.

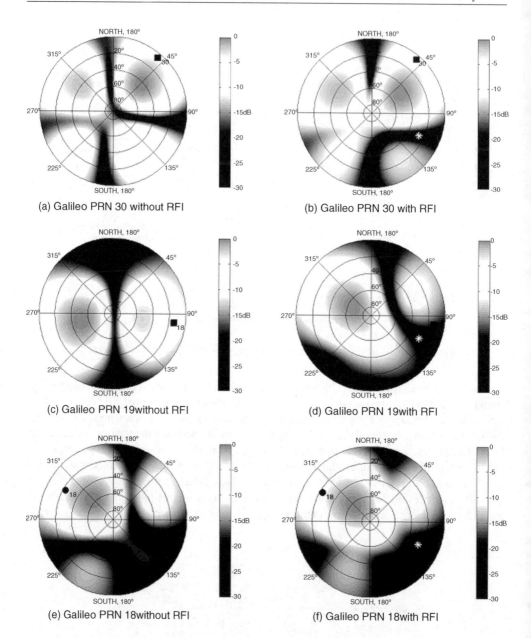

(a) Galileo PRN 30 without RFI

(b) Galileo PRN 30 with RFI

(c) Galileo PRN 19without RFI

(d) Galileo PRN 19with RFI

(e) Galileo PRN 18without RFI

(f) Galileo PRN 18with RFI

Figure 9.35 I/S at the receive antenna at which signals are still tracked or lost versus corresponding spatial separation to the interferer.

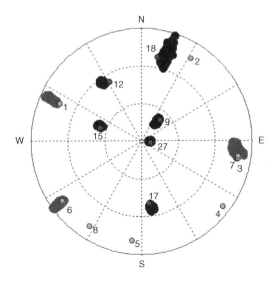

Figure 9.36 Skyplot with results of the direction of arrival estimation, Location 1.

remains more or less constant. This behavior can be well explained by the width of the array main beam. Additional measurements with CW and DME interferers showed similar results and are therefore not presented here.

Direction of arrival estimation. The obtained direction of arrival estimation results of the field tests of the GALANT demonstrator in GATE are summarized in Figures 9.36 and 9.37 and Table 9.3. In order to allow high update rates of the estimations, the

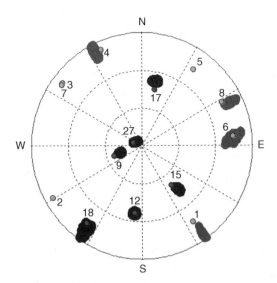

Figure 9.37 Skyplot with results of the direction of arrival estimation, Location 2.

Table 9.3 Estimation error characteristics, Location 1

Signal	Azimuth		Elevation		Number of observations
	Mean	Std	Mean	Std	
GPS, PRN 9	−2.26°	1.23°	3.07°	1.16°	285
GPS, PRN 12	−5.31°	0.71°	−2.55°	1.10°	285
GPS, PRN 15	2.38°	0.69°	−1.02°	0.56°	285
GPS, PRN 17	−2.16°	0.61°	−3.25°	0.74°	285
GPS, PRN 18	5.26°	1.56°	−3.82°	4.92°	284
GPS, PRN 27	−2.61°	4.95°	0.32°	0.48°	285
GATE, Tx 1	0.19°	0.37°	−6.75°	2.12°	285
GATE, Tx 3	−7.35°	0.77°	3.86°	0.97°	285
GATE, Tx 6	0.09°	0.41°	−3.62°	2.22°	202
GATE, Tx 7	−2.73°	0.85°	1.79°	1.50°	285

length of the data blocks processed by the direction-finding algorithm was set to 50 I&Q samples. Because of the spreading effect of the PRN code correlation on all signals different to the signal of interest, only the useful signal is usually assumed to dominate at the prompt correlator output in typical operational conditions. By using this assumption, only a single direction was estimated in each beamforming tracking channel.

Because of the computational complexity, the current version of the real-time direction-finding software does not take actual reception patterns of the array elements into account by assuming isotropic sensors in the underlying signal model. As shown in Reference [35] with results of offline postprocessing, this can lead to increased estimation errors. Moreover, at very low elevations, where the actual reception patterns of the array elements differ most from the isotropic sensor assumption, invalid estimation results may be obtained. This effect can be observed in Figures 9.36 and 9.37 where the direction estimations, which are plotted in gray for Galileo and in black for GPS signals, are not available for all GATE transmitters. It can also be observed that the DOA estimations become available for GATE transmitters 4 and 8 while transmitters 7 and 3 are lost as the measurement van changed from Location 1 (Figure 9.36) with the heading of −263.5° to Location 2 (Figure 9.37) with the heading of −54.5°. This effect is significant only for the signals coming from low elevation angles and is caused, as mentioned above, by the mismatch between the signal model used by the ESPRIT algorithm for the DOA estimation and the actual patterns of the array elements. The direction-finding software for GPS satellites was affected much less because of the much higher elevation angles at which they were received by the antenna array.

In Table 9.3 the mean and standard deviations of the DOA estimation errors in Location 1 are summarized. Since the DOA estimation results were collected over a time period of approximately 5 minutes, where the changes of the angles of arrival of the GPS signals become notable, the error analysis was carried out taking the movement of the satellites into

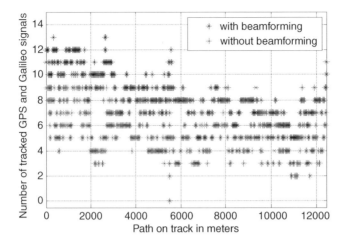

Figure 9.38   Number of tracked signals on a 12 km track.

account. In Figures 9.36 and 9.37 the positions of the tracked GPS satellites are shown for the middle of the 5 minute long tests.

Availability measurements. The enhancement of the signal reception provided by the digital beamforming technology also improves the availability of the navigation signals. This effect is demonstrated in Figure 9.38, where the number of tracked satellites with (black asterisks) and without (gray asterisks) digital beamforming during two consecutive test drives along a 12 km route in the GATE environment are plotted. The mean values of the satellites tracked in both cases are shown in Table 9.4, early indicating the gain of array processing with respect to the availability of the navigation signals. As mentioned earlier, the elevations of GATE transmitters are low compared to the satellites in a nominal GPS constellation. That explains the relatively low number of tracked Galileo signals in Table 9.4. The effect of the improved availability of the navigation signals can be further characterized on the level of positioning. On nearly 100% of the track, more than four satellite signals are tracked with digital beamforming. Without beamforming, more than four satellite signals are tracked only on 85.82% of the test track.

Table 9.4   Mean of tracked GPS and Galileo signals on a 12 km track

	Number of tracked signals	
	With DBF	Without DBF
GPS	4.96	4.02
Galileo	3.23	1.11
GPS + Galileo	8.19	5.13

References

1. Stone, J. (inventor) (December 16, 1902) Method of determining the direction of space telegraph signals. U.S. patent 716,134.
2. Bancroft, S. (1985) An algebraic solution of the GPS equations. *IEEE Transactions on Aerospace and Electronic Systems*, **21** (AES-1), 56–59.
3. Borre, K. and Strang, G. (1997) *Linear Algebra, Geodesy, and GPS Wellesley*, Wellesley-Cambridge Press.
4. Hernández–Pajares, M., Miguel Juan, J., Sanz, J. *et al.* (2010) Wide-Area RTK. High precision positioning on a continental scale. *Inside GNSS*, **5** (2), 35–46.
5. Thomas, M. (2011) *Global Navigation Space Systems: Reliance and Vulnerabilities*, The Royal Academy of Engineering, London, UK.
6. Kaplan, E.D. and Hegarty, C.J. (2006) Interference, multipath, and scintillation, understanding GPS, in *Principles and Applications*, 2nd edn (ed. A House), Artech House, Norwood, pp. 243–299.
7. Juang, J.C. (2009) Analysis of global navigation satellite system position deviation under spoofing. *IET Radar, Sonar and Navigation*, **3** (1), 1–7.
8. Balaei, A.T., Motella, B. and Dempster, A.G. (2008) A preventive approach to mitigating CW interference in GPS receivers. *GPS Solutions*, **12** (3), 199–209.
9. Borio, D., Camoriano, L., Savasta, S. and Lo Presti, L. (2008) Time–frequency excision for GNSS applications. *IEEE Systems Journal*, **2** (1), 27– 27.
10. Arribas, J., Fernández–Prades, C. and Closas, P. (2012) Antenna array based GNSS signal acquisition for interference mitigation. *IEEE Transactions on Aerospace and Electronic Systems (accepted)*.
11. Esbri Rodriguez, O., Antreich, F., Konovaltsev, A. *et al.* (2006) Antenna-based multipath and interference mitigation for aeronautical applications: present and future. Proceedings of the 19th International Technical Meeting of the Satellite Division of the Institute of Navigation (ION GNSS'06), Fort Worth, TX.
12. Kunysz, W. (2000) High performance GPS pinwheel antenna. Proceedings of the 13th International Technical Meeting of the Satellite Division of The Institute of Navigation (ION GNSS'00), Salt Lake City, UT.
13. Brown, A. and Morley, D. (2001) Test results of a 7-element small controlled reception pattern antenna. Proceedings of the 14th International Technical Meeting of the Satellite Division of The Institute of Navigation (ION GNSS'01), Salt Lake City, UT.
14. Dick, C., Harris, F., Pajic, M. and Vuletic, D. (2006) Real-time QRD-based beamforming on an FPGA platform. Proceedings of the Fortieth Asilomar Conference on Signals, Systems and Computers (ACSSC'06), Pacific Grove, CA, pp. 1200–1204.
15. Liu, J., Weaver, B., Zakharov, Y. and White, G. (2007) An FPGA-based MVDR beamformer using dichotomous coordinate descent iterations. Proceedings of the IEEE International Conference on Communications (ICC'07), Glasgow, Scotland, pp. 2551–2556.
16. Lodha, N., Rai, N., Krishnamurthy, A. and Venkataraman, H. (2009) Efficient implementation of QRD-RLS algorithm using hardware–software co-design. Proceedings of the IEEE International Parallel and Distributed Processing Symposium (IPDPS'09), Rome, Italy.
17. Krim, H. and Viberg, M. (1996) Two decades of array signal processing research: the parametric approach. *IEEE Signal Processing Magazine*, **13** (4), 67–94.
18. Litva, J. and Lo, K.-Y.T. (1996) *Digital Beamforming in Wireless Communications*, Artech House, Norwood, MA.
19. De Lorenzo, D. (2007) Navigation Accuracy and Interference Rejection for GPS Adaptive Antenna Arrays. PhD Dissertation, Department of Aeronautics and Astronautics, Stanford University.
20. Antreich, F., Nossek, J. and Utschick, W. (2007) Maximum likelihood delay estimation in a navigation receiver for aeronautical applications. *Aerospace Science and Technology*, **12** (3), 256–267.
21. Fernández-Prades, C., Closas, P. and Arribas, J. (2011) Eigenbeamforming for interference mitigation in GNSS receivers. Proceedings of the International Conference on Localization and GNSS (ICL-GNSS), Tampere, Finland, pp. 93–97.
22. Chu, P.P. (2006) *RTL Hardware Design Using VHDL. Coding for Efficiency, Portability, and Scalability*, John Wiley & Sons, Inc., Hoboken, NJ.
23. Crews, M. and Yong, Y. (February 2003) Practical Design for Transferring Signals between Clock Domains: A Simple Circuit Addresses the Errors and Limitations of Asynchronous Design. EDN, pp. 65–71.
24. Xilinx, Inc. (March 2011) FIFO Generator v8.1, DS317 Product Specification. San José, CA.

25. Arribas, J., Fernández-Prades, C. and Closas, P. (2011). Antenna array based GNSS signal acquisition: real-time implementation and results. Proceedings of the ION GNSS, September 2011.
26. Fernández-Prades, C. (2006) Advanced Signal Processing Techniques for Global Navigation Satellite System Receivers. PhD Dissertation, Department of Signal Theory and Communications, Universitat Politècnica de Catalunya (UPC), Barcelona, Spain.
27. Arribas, J., Fernández-Prades, C. and Closas, P. (2011) Array-based GNSS acquisition in the presence of colored noise. Proceedings of the 36th IEEE International Conference on Acoustics, Speech and Signal Processing (ICASSP'11), Prague, Czech Republic, pp. 2728–2731.
28. Kaplan, E.D. and Hegarty, J. (2005) *Understanding GPS. Principles and Applications*, Artech House Publishers, Norwood, MA.
29. Boulton, P., Borsato, R., Butler, B., and Judge, K. (2011) GPS interference testing lab, live, and lightsquared. *Inside GNSS*, **6** (4), 32–45.
30. LTE (2011) 3GPP. 3GPP TS 36.104: Base Station (BS) Radio Transmission and Reception v.8.12.0, June.
31. Heckler, M.V.T., Cuntz, M., Konovaltsev, A. *et al.* (2011) Development of robust safety-of-life navigation receivers. *IEEE Transactions on Microwave Theory and Techniques*, **4** (59), 998–1005.
32. Haardt, M., Zoltowski, M. and Mathews, C. (1998) ESPRIT and closed-form 2D angle estimation with planar arrays, in *Digital Signal Processing Handbook* (eds V. Madisetti and D. Williams), CRC Press, Boca Raton, FL.
33. Cuntz, M., Greda, L.H., Konovaltsev, A. *et al.* (2010) Lessons learnt: the development of a robust multi-antenna GNSS receiver. Proceedings of the 23rd International Technical Meeting of the Satellite Division of the Institute of Navigation (ION GNSS'10), Portland, OR, pp. 2852–2859.
34. Cuntz, M.K.A., Sgammini, M., Hättich, C. *et al.* (2011) Field test: jamming the DLR adaptive antenna receiver. Proceedings of the 24th International Technical Meeting of the Satellite Division of the Institute of Navigation (ION GNSS'11), Portland, OR.
35. Konovaltsev, A., Cuntz, M. and Meurer, M. (2010) Novel calibration of adaptive antennas. Proceedings of the 23rd International Technical Meeting of the Satellite Division of the Institute of Navigation (ION GNSS'10), Portland, OR.

10

Multiband RF Front-Ends for Radar and Communications Applications

Roberto Gómez-García, José-María Muñoz-Ferreras and
Manuel Sánchez-Renedo
Department of Signal Theory and Communications, University of Alcalá, Madrid, Spain

10.1 Introduction

The development of software-defined radios (SDRs) should pave the way for achieving compact equipments capable of acquiring multistandard/multiband services [1]. An analogue-to-digital converter (ADC) just after the antenna would directly implement the concept [2]. Unfortunately, available ADC sampling rates and resolutions do not currently permit large radio frequency (RF) bandwidths to be handled. As a result, novel and revisited RF front-end configurations for SDR applications are now under research to overcome this bottleneck. Note that these schemes may be used not only for telecommunications purposes but also for other modern systems such as multifrequency radars. This introduction presents a brief summing-up of previously devised solutions for the realization of RF front-ends with emphasis on the SDR framework.

10.1.1 Standard Approaches for RF Front-Ends

From the receiver perspective, some alternatives to RF front-ends are interesting. One example is the superheterodyne architecture incorporating several mixing stages [3]. Figure 10.1(a) shows its RF receiver chain. Here, the incoming signal is down-converted to a low frequency after filtering, where the acquisition is done. The first local oscillator (LO) usually varies its frequency for system reconfiguration in order to enable the acquisition of different channels, whereas the second conversion translates the signal from the intermediate frequency (IF) to

Microwave and Millimeter Wave Circuits and Systems: Emerging Design, Technologies, and Applications,
First Edition. Edited by Apostolos Georgiadis, Hendrik Rogier, Luca Roselli, and Paolo Arcioni.
© 2013 John Wiley & Sons, Ltd. Published 2013 by John Wiley & Sons, Ltd.

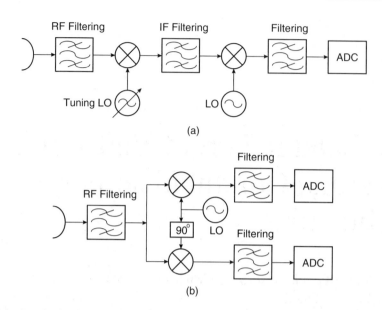

(a)

(b)

Figure 10.1 Standard architectures for RF front-ends: (a) superheterodyne approach, (b) zero-IF and low-IF alternatives.

baseband or to a low frequency. Furthermore, this second conversion usually acquires the in-phase and quadrature (I-Q) channels.

The superheterodyne solution has some drawbacks, such as its great size or the need for careful designs regarding the image bands and spurious signals. Other approaches exploit the zero-IF and low-IF receivers [4–7]. Figure 10.1(b) shows their common architecture. In the zero-IF solution, I-Q imbalances and direct-current (DC) offsets may be critical. In the low-IF alternative, where the I-Q conversion is no longer necessary, the image frequency problem is reinstated. All the aforementioned architectures may also find serious difficulties for simultaneously acquiring several bands.

10.1.2 Acquisition of Multiband Signals

The previous schemes do not easily circumvent the problem of acquiring multiband signals, which is the interest here. Fortunately, a simple choice to this aim which avoids mixers is direct sampling [8–11]. In this scheme, shown in Figure 10.2, a basic RF front-end, shaped by the cascade of an amplifying stage and a filtering section, supplies the signal to the ADC.

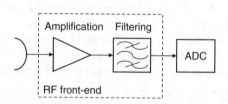

Figure 10.2 Multiband direct-sampling architecture to implement the SDR concept.

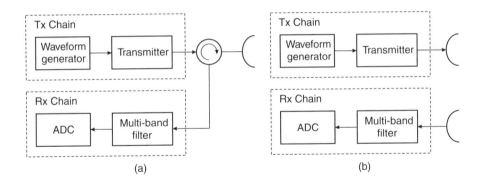

Figure 10.3 Direct-sampling architectures for multifrequency radars: (a) circulator + single-antenna solution, (b) transmitter and receiver antenna solution.

The idea behind this approach is that a sparse spectrum does not need to be sampled at the Nyquist rate to be properly acquired (i.e. without aliasing). The key element in Figure 10.2 is the RF filter, which should be of multipassband type.

This architecture is also appropriate for multifrequency radar systems, which should simultaneously collect target echoes coming from different bands. In the radar counterpart, the direct-sampling solution, depicted in Figure 10.3, incorporates a multiband bandpass filter in the receiver chain before the ADC. Depending on the waveform type, the scheme of Figure 10.3(a) may be more appropriate than the one of Figure 10.3(b). Indeed, the architecture of Figure 10.3(b) is useful for multiband continuous-wave radars, because of the usual low power isolation levels of circulators. Note that amplification stages, not depicted here for clarity, should obviously be included in the RF chains.

10.1.3   The Direct-Sampling Architecture

As explained, the idea behind the direct-sampling approach is that a sparse spectrum does not need to be sampled at the Nyquist rate to be properly acquired (i.e. without aliasing). Indeed, sampling frequencies below that imposed by the Nyquist criterion may be used here. For instance, if a single-band signal is to be acquired, sub-Nyquist sampling frequencies that avoid aliasing may exist. In addition, as detailed later, these lower frequencies are analytically computable [12]. In contrast, if several bands are to be sampled, no formulas to obtain the permissible sampling frequencies are available. In this case, the best that can be done is to derive spectral intervals within which these frequencies can be found and to apply search algorithms for their calculation [8, 13].

10.1.3.1   The Theorem of Bandpass Sampling

Acquiring a baseband real-valued analogue signal requires the sampling frequency to be at least twice the highest frequency component of its spectrum [14]. Otherwise, undesired aliasing generally emerges. Nevertheless, when the real-valued signal has a sparse spectrum, such as a bandpass or a multichannel signal, a sampling frequency lower than that corresponding to the Nyquist rate may be used without aliasing [12, 15].

As an example, it is widely known that a real-valued bandpass signal can be properly acquired by a sub-Nyquist sampling rate f_s if its positive and negative frequency bands do not overlap in the spectral interval $[-f_s/2, f_s/2]$. In this case, an analytical expression for the sampling frequencies f_s avoiding aliasing is available as follows [12]:

$$\frac{2f_H}{n} \le f_s \le \frac{2f_L}{n-1} \tag{10.1}$$

where

$$2 \le n \le \frac{f_H}{f_H - f_L} \tag{10.2}$$

Here $n \in \{0, 1, 2, \ldots\}$ and f_L and f_H refer to the lowest and highest frequency components of the bandpass signal, respectively.

10.1.3.2 Acquisition of Complex Frequency-Sparse Signals

If the spectrum sparsity becomes complex, as for a multichannel acquisition system, no equation enabling the calculation of the permissible sub-Nyquist sampling frequencies exists. Fortunately, the derivation of some constraints for these frequencies is quite obvious and the search for them may also be accomplished [8, 13].

For a real-valued multichannel signal, the minimum sampling frequency $f_{s,\min}$ that may avoid aliasing is twice the sum of the bandwidths of its channels [8]. In this case, the spectrum of the sampled multiband signal occupies all the available bandwidth $[-f_{s,\min}/2, f_{s,\min}/2]$ without band overlapping. It is worth noting that a sampling rate lower than $f_{s,\min}$ will necessarily give rise to aliasing.

A multiband acquisition system using this minimum sampling frequency alleviates the need for expensive high sampling rate ADCs [16]. Moreover, it can pave the way for implementing the direct-sampling solution to the SDR concept [1, 2]. In this direct-sampling solution, if the minimum sub-Nyquist frequency is used, less stringent demands are imposed on the ADC. In this chapter, the analytical conditions that a multiband signal should meet to ensure that the minimum sampling frequency does not imply aliasing are derived. Moreover, design rules and prototypes of novel signal-interference multipassband filters suitable for these direct-sampling RF front-ends are also shown.

10.2 Minimum Sub-Nyquist Sampling

Given a multiband real-valued signal, the main purpose addressed here is to acquire it by using the minimum sub-Nyquist sampling frequency.

10.2.1 Mathematical Approach

As earlier mentioned, the minimum rate $f_{s,\min}$ that may avoid aliasing is twice the sum of the bandwidths of the channels, meaning that the channels do not overlap within the margin $[-f_{s,\min}/2, f_{s,\min}/2]$. If the number of channels is N and their bandwidths are set to be B_i, with $i \in \{1, 2, \ldots, N\}$, then this minimum sub-Nyquist sampling rate $f_{s,min}$ is

$$f_{s,min} = 2 \sum_{i=1}^{N} B_i \tag{10.3}$$

The next two sections concentrate on the acquisition of multichannel signals for the cases of arbitrary dual-band and evenly spaced equal-bandwidth multiband signals.

10.2.2 Acquisition of Dual-Band Signals

Let it assume a dual-band acquisition system. The spectral allocation of the two channels is depicted in Figure 10.4. The lower band has a bandwidth B_L, whereas B_U refers to the upper-channel bandwidth. The separation between channels is given by Δ and f_L represents the lowest signal frequency component.

As commented, the theory of bandpass sampling analytically provides the sub-Nyquist frequencies that do not give rise to aliasing for a single-band signal [12]. Unfortunately, no direct equations exist for more complex spectra, such as the one shown in Figure 10.4.

The interest is to sample the incoming signal at the minimum sampling frequency $f_{s,min}$. Indeed, this considerably alleviates the requirements imposed on the receiver ADC. Therefore, given the bandwidths B_L and B_U, the conditions that the Δ and f_L parameters should satisfy to guarantee that the minimum sub-Nyquist sampling frequency can be employed without producing aliasing must be found.

In order to acquire real-valued dual-band signals, it comes as trivial that the minimum sampling frequency $f_{s,min}$ that may circumvent aliasing is given by

$$f_{s,min} = 2(B_L + B_U) \tag{10.4}$$

Note that $f_{s,min}$ can be much lower than the Nyquist frequency, written as

$$f_{Nyq} = 2(f_L + B_L + \Delta + B_U) \tag{10.5}$$

for the generic dual-band signal shown in Figure 10.4.

Hence, if the lowest frequency f_L and the band separation Δ are adequately chosen, it will be possible to acquire the dual-band signal properly at the minimum sampling frequency, thus avoiding undesired aliasing. The rules to set f_L and Δ correctly thus become a valuable system design tool to conveniently allocate the signal channels.

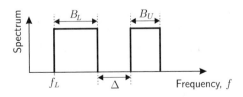

Figure 10.4 Spectral allocation for the two channels of a dual-band signal.

A first case implying that the bands of the sampled signal do not overlap within the frequency interval $\left[-f_{s,min}/2, f_{s,min}/2\right]$ is the one in which, after sampling, the lower channel situates without spectral inversion in the interval $[0, B_L]$ and the upper channel adjacently appears in the margin $\left[B_L, f_{s,min}/2\right]$, also without spectral inversion. Both conditions are simultaneously satisfied if the relationships below are met:

$$f_L - mf_{s,min} = 0 \tag{10.6}$$

$$f_L + B_L + \Delta - nf_{s,min} = B_L \tag{10.7}$$

with $m, n \in \{0, 1, 2, \ldots\}$. Consequently, by using Equations (10.4), (10.6) and (10.7) and by noting that $f_L \geq 0$ and $\Delta \geq 0$, the following design equations for f_L and Δ are finally deduced:

$$f_L = p(B_L + B_U) \tag{10.8}$$

$$\Delta = q(B_L + B_U) \tag{10.9}$$

where $p \in \{0, 2, 4, 6, \ldots\}$ and $q \in \{0, 2, 4, 6, \ldots\}$. A dual-band acquisition system fulfilling Equations (10.8) and (10.9) may use an ADC driven by the minimum sampling frequency, because no aliasing will arise. Furthermore, no spectral inversion is to be observed for the acquired bands in this first case. Note that spectral inversion is not a problem, as it can easily be counteracted at the digital processing level.

A total of eight cases for a dual-band signal can be mathematically analyzed. These are as follows:

- **Case 1:** lower channel not inverted in margin $[0, B_L]$; upper channel not inverted in margin $[B_L, B_L + B_U]$.
- **Case 2:** lower channel not inverted in margin $[0, B_L]$; upper channel inverted in margin $[B_L, B_L + B_U]$.
- **Case 3:** lower channel inverted in margin $[0, B_L]$; upper channel not inverted in margin $[B_L, B_L + B_U]$.
- **Case 4:** lower channel inverted in margin $[0, B_L]$; upper channel inverted in margin $[B_L, B_L + B_U]$.
- **Case 5:** lower channel not inverted in margin $[B_U, B_L + B_U]$; upper channel not inverted in margin $[0, B_U]$.
- **Case 6:** lower channel not inverted in margin $[B_U, B_L + B_U]$; upper channel inverted in margin $[0, B_U]$.
- **Case 7:** lower channel inverted in margin $[B_U, B_L + B_U]$; upper channel not inverted in margin $[0, B_U]$.
- **Case 8:** lower channel inverted in margin $[B_U, B_L + B_U]$; upper channel inverted in margin $[0, B_U]$.

Analogously to Equations (10.8) and (10.9), two equations for f_L and Δ can be obtained for each case. Table 10.1 details, for completeness, all the equations that enable the setting of

Table 10.1 Selection of the f_L and Δ parameter values to avoid aliasing when using the minimum sampling frequency in a dual-band acquisition system

Case 1	Case 2
$f_L = p(B_L + B_U), p \in \{0, 2, 4, 6, \ldots\}$	$f_L = p(B_L + B_U), p \in \{0, 2, 4, 6, \ldots\}$
$\Delta = q(B_L + B_U), q \in \{0, 2, 4, 6, \ldots\}$	$\Delta = (r-1)B_L + rB_U, r \in \{1, 3, 5, 7, \ldots\}$
Case 3	**Case 4**
$f_L = (p-1)B_L + pB_U, p \in \{2, 4, 6, 8, \ldots\}$	$f_L = (p-1)B_L + pB_U, p \in \{2, 4, 6, 8, \ldots\}$
$\Delta = rB_L + (r-1)B_U, r \in \{1, 3, 5, 7, \ldots\}$	$\Delta = r(B_L + B_U), r \in \{1, 3, 5, 7, \ldots\}$
Case 5	**Case 6**
$f_L = (r-1)B_L + rB_U, r \in \{1, 3, 5, 7, \ldots\}$	$f_L = (r-1)B_L + rB_U, r \in \{1, 3, 5, 7, \ldots\}$
$\Delta = s(B_L + B_U), s \in \{1, 3, 5, 7, \ldots\}$	$\Delta = sB_L + (s-1)B_U, s \in \{1, 3, 5, 7, \ldots\}$
Case 7	**Case 8**
$f_L = r(B_L + B_U), r \in \{1, 3, 5, 7, \ldots\}$	$f_L = r(B_L + B_U), r \in \{1, 3, 5, 7, \ldots\}$
$\Delta = (s-1)B_L + sB_U, s \in \{1, 3, 5, 7, \ldots\}$	$\Delta = p(B_L + B_U), p \in \{0, 2, 4, 6, \ldots\}$

the lowest frequency parameter f_L and the interband separation Δ, for prefixed values B_L and B_U for the channel bandwidths.

The sets of equations detailed in Table 10.1 can be compacted. In fact, if the parameters f_L and Δ satisfy any of the equivalent sets of equations provided in Table 10.2, then the minimum sampling frequency $f_{s,min}$ does not give rise to aliasing.

Finally, it is interesting to highlight the fact that the described procedure for the dual-band scenario may be easily extended to more than two channels. Indeed, if N is the number of bands and B_i, $i \in \{1, 2, \ldots, N\}$, are the prescribed channel bandwidths, then the lowest frequency parameter f_L and the interband channel separations Δ_n, $n \in \{1, 2, \ldots, N-1\}$, should be properly selected to assure that the minimum sampling frequency (10.3) avoids aliasing. It can be easily inferred that, for a general multiband situation, there is a total of $N! \times 2^N$ cases, each of them having N independent equations to be satisfied.

As an example of the latter, in a multichannel acquisition system with N bands, a first case could be the one where the bands consecutively appear in the margin $\left[-f_{s,min}/2, f_{s,min}/2\right]$

Table 10.2 Compact equations for the f_L and Δ parameter values to avoid aliasing when using the minimum sampling frequency in a dual-band acquisition system

$f_L = k(B_L + B_U), k \in \{0, 1, 2, 3, \ldots\}$	$f_L = (k-1)B_L + kB_U, k \in \{1, 2, 3, 4, \ldots\}$
$\Delta = (r-1)B_L + rB_U, r \in \{1, 3, 5, 7, \ldots\}$	$\Delta = rB_L + (r-1)B_U, r \in \{1, 3, 5, 7, \ldots\}$
or	or
$\Delta = p(B_L + B_U), p \in \{0, 2, 4, 6, \ldots\}$	$\Delta = s(B_L + B_U), s \in \{1, 3, 5, 7, \ldots\}$

without spectral inversion. In such a situation, it can be shown that, in order to use the minimum sub-Nyquist sampling frequency, the lowest frequency parameter f_L and the interband channel separations Δ_n, $n \in \{1, 2, \ldots, N - 1\}$, should meet

$$f_L = p \sum_{i=1}^{N} B_i \qquad (10.10)$$

$$\Delta_n = q \sum_{i=1}^{N} B_i \qquad (10.11)$$

where $p \in \{0, 2, 4, 6, \ldots\}$ and $q \in \{0, 2, 4, 6, \ldots\}$.

It becomes trivial that the number of cases rapidly increases with the number of channels. A complete rigorous analysis for more than two channels is simple, but tedious. However, in multiband acquisition scenarios, evenly spaced equal-bandwidth channels are usual. The next subsection emphasizes this relevant type of multiband signal.

10.2.3 Acquisition of Evenly Spaced Equal-Bandwidth Multiband Signals

Let us suppose there is a multiband signal having N bands, with channels evenly separated by Δf and showing identical bandwidths equal to B. Figure 10.5 depicts the spectrum of a typical multiband signal, $S(f)$, where f_{L1} is the lowest frequency component of the first channel. Obviously, the condition $\Delta f \geq B$ must be satisfied. In Figure 10.5, note that only the spectrum amplitude for positive frequencies has been represented. Nevertheless, energy at negative frequencies will also be found since the signals at the receiver antenna are real-valued (i.e. $|S(f)| = |S(-f)|$ and $\angle\{S(f)\} = -\angle\{S(-f)\}$).

For the case under study of evenly spaced equal-bandwidth multiband systems, the possible minimum sampling frequency $f_{s,\min}$, according to Equation (10.3), is

$$f_{s,\min} = 2NB \qquad (10.12)$$

This section details the conditions that the signal parameters, as shown in Figure 10.5, should meet to guarantee that the minimum sub-Nyquist sampling frequency $f_{s,\min}$ does not generate aliasing – in other words, which values should the f_{L1}, B and Δf parameters take to ensure that $f_{s,\min}$ is a valid sampling frequency.

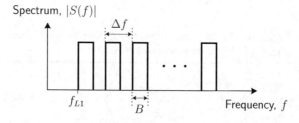

Figure 10.5 Spectrum amplitude of an evenly spaced equal-bandwidth multiband signal. © 2011 IEEE. Reprinted, with permission, from [17].

Table 10.3 Selection of the f_{L1}, Δf and B parameter values to avoid aliasing when using the minimum sampling frequency

---

Condition on f_{L1}

$$f_{L1} = k\frac{f_{s,\min}}{2} = kNB, k \in \{0,1,2,3,\ldots\}$$

Condition on Δf

N even	$N = 2^l, l \in \{1,2,3,4,\ldots\}$	$\Delta f = 2(1+2m)B, m \in \{0,1,2,3,\ldots\}$
		or
		$\Delta f = (1+2nN)B, n \in \{0,1,2,3,\ldots\}$
	$N \neq 2^l, l \in \{1,2,3,4,\ldots\}$	$\Delta f = 2(1+2m)B$, with $m \in \{0,1,2,3,\ldots\}$ and $\mathrm{rem}(2(1+2m),p) \neq 0$ where p is any odd integer so that $(p \neq 1)$ and $\mathrm{rem}(N,p) = 0$
		or
		$\Delta f = (1+2nN)B, n \in \{0,1,2,3,\ldots\}$
N odd	$N = 1$	Δf not applicable
	$N \neq 1$	$\Delta f = 2mB$, with $m \in \{1,2,3,4,\ldots\}$ and $\mathrm{rem}(m,p) \neq 0$ where p is any odd integer so that $(p \neq 1)$ and $\mathrm{rem}(N,p) = 0$
		or
		$\Delta f = (1+2nN)B, n \in \{0,1,2,3,\ldots\}$

The aforementioned design conditions have been obtained and are given in Table 10.3 [17]. It should be remarked that, although an exhaustive inspection process was used for their derivation, they are valid regardless of the number of bands in the evenly spaced equal-bandwidth multiband acquisition system under consideration. Note that, in Table 10.3, $\mathrm{rem}(a,b)$ is the remainder after division of a by b.

Table 10.4 details, as an illustrative example, some possible values for the $\Delta f/B$ quotient, ensuring that $f_{s,\min}$ avoids aliasing once the first requisite f_{L1} in Table 10.3 is also fulfilled [17]. The missing values for $\Delta f/B$ inevitably cause aliasing, since they do not satisfy the second condition in Table 10.3.

Table 10.4 Some possible values for the $\Delta f/B$ quotient, not implying aliasing when $f_{s,\min}$ is used

N	$\Delta f/B$
2	1, 2, 5, 6, 9, 10, 13, 14, 17, 18, 21, 22, 25, 26, 29, 30, . . .
3	1, 2, 4, 7, 8, 10, 13, 14, 16, 19, 20, 22, 25, 26, 28, 31, 32, . . .
4	1, 2, 6, 9, 10, 14, 17, 18, 22, 25, 26, 30, . . .
5	1, 2, 4, 6, 8, 11, 12, 14, 16, 18, 21, 22, 24, 26, 28, 31, 32, . . .
6	1, 2, 10, 13, 14, 22, 25, 26, . . .
7	1, 2, 4, 6, 8, 10, 12, 15, 16, 18, 20, 22, 24, 26, 29, 30, 32, . . .
8	1, 2, 6, 10, 14, 17, 18, 22, 26, 30, . . .
9	1, 2, 4, 8, 10, 14, 16, 19, 20, 22, 26, 28, 32, . . .
10	1, 2, 6, 14, 18, 21, 22, 26, . . .
11	1, 2, 4, 6, 8, 10, 12, 14, 16, 18, 20, 23, 24, 26, 28, 30, 32, . . .
12	1, 2, 10, 14, 22, 25, 26, . . .

10.3 Simulation Results

In this section, some simulation results are reported. These results are provided to demonstrate the correctness of the equations derived in the previous section.

10.3.1 Symmetrical and Asymmetrical Cases

To validate the suitability of the obtained design equations, it is necessary to simulate multiband signals and verify the effects of the subsampling process on them. To this aim, two different types of multiband signal are considered here, as follows:

- A multiband signal $s_1(t)$ with each channel having spectral symmetry with respect to the centre frequency. Specifically, a rectangular shape for its spectral bands is assumed.
- A multiband signal $s_2(t)$ with each channel exhibiting spectral asymmetry with regard to the centre frequency. Concretely, an increasing frequency ramp is analysed in this case.

The spectrum corresponding to one channel of these real-valued signals $s_1(t)$ and $s_2(t)$, respectively, can be written as

$$S_1(f) = A \operatorname{rect}\left[\frac{f-f_c}{B}\right] + A \operatorname{rect}\left[\frac{f+f_c}{B}\right] \tag{10.13}$$

$$S_2(f) = \left[\frac{A}{B}(f-f_c) + \frac{A}{2}\right] \operatorname{rect}\left[\frac{f-f_c}{B}\right] + \left[-\frac{A}{B}(f+f_c) + \frac{A}{2}\right] \operatorname{rect}\left[\frac{f+f_c}{B}\right] \tag{10.14}$$

where A is the spectrum amplitude, B is the occupied bandwidth, f_c is the centre frequency and $\operatorname{rect}[f]$ is a frequency rectangular-shape pulse of unity amplitude and width, centred at the zero frequency.

By resolving the inverse Fourier transform of Equations (10.13) and (10.14) analytically, the expressions for one channel of the multiband signals in the time domain can be obtained as

$$s_1(t) = A \operatorname{sinc}(Bt)\cos(2\pi f_c t) \tag{10.15}$$

$$s_2(t) = \frac{A}{\pi t}\left(\cos(\pi Bt) - \frac{1}{\pi Bt}\sin(\pi Bt)\right)\sin(2\pi f_c t) + \frac{A}{\pi t}\sin(\pi Bt)\cos(2\pi f_c t) \tag{10.16}$$

where $\operatorname{sinc}(x) = \sin(\pi x)/(\pi x)$.

Different channels of the multiband signals will have different values for the A, B and f_c parameters in Equations (10.15) and (10.16). Indeed, the sampling of these signals gives rise to the basis to visualize possible aliasing effects and, hence, to the capability of verifying the mathematical framework previously given. Note also that the sampling of Equation (10.16) enables the observation of possible spectral inversion.

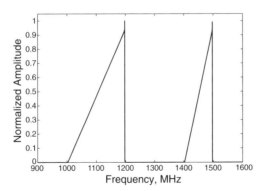

Figure 10.6 Spectrum magnitude of the frequency-asymmetrical dual-band signal (simulation parameters: $B_L = 200\,\text{MHz}$, $B_U = 100\,\text{MHz}$, $f_L = 1000\,\text{MHz}$ and $\Delta = 200\,\text{MHz}$).

10.3.2 Verification of the Mathematical Framework

To check the equations detailed in Section 10.2.2, let us consider an asymmetrical dual-band signal with a 200 MHz bandwidth lower channel centred at 1100 MHz and a 100 MHz bandwidth upper channel around 1450 MHz. Figure 10.6 shows the simulated spectrum magnitude of this dual-band signal, after assuming a ramp-like shape for each channel (Equation (10.16)). Note also that only positive frequencies have been represented here, as well as the appearance of the well-known Gibbs phenomenon [14].

The dual-band signal of Figure 10.6 has a lowest frequency $f_L = 1000\,\text{MHz}$ and an inter-band spectral separation $\Delta = 200\,\text{MHz}$. Noting that $B_L = 200\,\text{MHz}$ and $B_U = 100\,\text{MHz}$, the minimum sub-Nyquist frequency that may avoid aliasing is $f_{s,\min} = 600\,\text{MHz}$, according to Equation (10.4). This minimum frequency is much lower than the Nyquist frequency ($f_{Nyq} = 3000\,\text{MHz}$), given by Equation (10.5). On the other hand, it can easily be shown that the f_L and Δ values meet the equations given in Table 10.1 (case 3). In particular, the aforementioned rules are satisfied for $p = 4$ and $r = 1$. Consequently, taking into account the results expounded in Section 10.2.2, the minimum sampling frequency should avoid aliasing. This is confirmed in Figure 10.7, which has been obtained by subsampling the dual-band signal of Figure 10.6 at the rate $f_{s,\min} = 600\,\text{MHz}$. As can be seen, no aliasing arises. Furthermore, as stated by case 3, it can be observed that the lower channel appears inverted in the interval $[0, B_L]$, whereas the upper channel is not inverted and adjacently emerges in the margin $[B_L, B_L + B_U]$. Note also that, for representation clarity, a little guard band has been considered in the simulation process.

The design rules provided in Table 10.1 have been validated with many examples for all eight cases. If the f_L and Δ parameters do not fulfil the design equations, then the minimum sub-Nyquist sampling frequency $f_{s,\min}$ irremediably causes aliasing. Figure 10.8 shows an illustrative example of this situation. The only difference in relation to Figure 10.7 is that now $\Delta = 400\,\text{MHz}$, implying that the equations are not met.

Additionally, it is worth remarking that the fact that the minimum sub-Nyquist frequency avoids aliasing does not necessarily imply that a higher sampling frequency should also circumvent this band overlapping. As a proof of that, let us consider again the signal assumed in

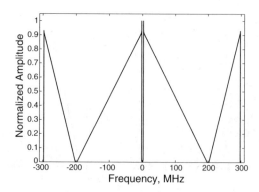

Figure 10.7 Spectrum magnitude of the frequency-asymmetrical dual-band signal after subsampling at the minimum sub-Nyquist frequency: aliasing is avoided (simulation parameters: $B_L = 200\,\text{MHz}$, $B_U = 100\,\text{MHz}$, $f_L = 1000\,\text{MHz}$, $\Delta = 200\,\text{MHz}$ and $f_s = f_{s,\text{min}} = 600\,\text{MHz}$).

Figure 10.6, but now sampled at a higher sampling frequency $f_s = 700\,\text{MHz}$. Figure 10.9 shows that aliasing arises for this example.

Subsequently, in order to verify Equations (10.10) and (10.11), the acquisition of a triple-band signal is addressed. The channel bandwidths are $B_1 = 100\,\text{MHz}$, $B_2 = 50\,\text{MHz}$ and $B_3 = 25\,\text{MHz}$, respectively; the lowest frequency is $f_L = 1050\,\text{MHz}$, whereas the interband separations are $\Delta_1 = 350\,\text{MHz}$ and $\Delta_2 = 700\,\text{MHz}$. Figure 10.10 shows the spectrum magnitude of this signal after sampling at the minimum rate $f_{s,\text{min}} = 350\,\text{MHz}$. In this case, aliasing and spectral inversion are not observed since the signal parameters fulfil Equations (10.10) and (10.11).

In order to verify the equations provided in Section 10.2.3, let us consider Figure 10.11, which represents, in normalized terms, the amplitude spectrum of a four-channel signal. Here, the normalized frequency $f_{Nz} = f/f_0$ has been used, where f_0 is the centre frequency of the quad-band signal. Thus, $B_{Nz} = B/f_0$, $\Delta f_{Nz} = \Delta f/f_0$ and $f_{L1,Nz} = f_{L1}/f_0$.

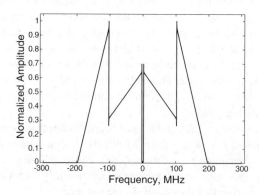

Figure 10.8 Spectrum magnitude of the frequency-asymmetrical dual-band signal after subsampling at the minimum sub-Nyquist frequency: aliasing exists (simulation parameters: $B_L = 200\,\text{MHz}$, $B_U = 100\,\text{MHz}$, $f_L = 1000\,\text{MHz}$, $\Delta = 400\,\text{MHz}$ and $f_s = f_{s,\text{min}} = 600\,\text{MHz}$).

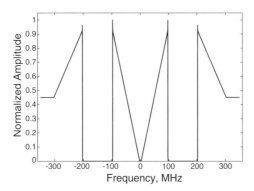

Figure 10.9 Spectrum magnitude of the frequency-asymmetrical dual-band after subsampling: undesired aliasing emerges (simulation parameters: $B_L = 200\,\text{MHz}$, $B_U = 100\,\text{MHz}$, $f_L = 1000\,\text{MHz}$, $\Delta = 200\,\text{MHz}$ and $f_s = 700\,\text{MHz}$).

For the particular example of $B_{Nz} = 2/15$, $\Delta f_{Nz} = 4/15$ and $f_{L1,Nz} = 8/15$, it can be easily verified that the conditions listed in Table 10.3 are met. This means that the minimum sub-Nyquist sampling frequency, $f_{s,\min,Nz} = f_{s,\min}/f_0 = 16/15$, does not give rise to aliasing in this case. Figure 10.12 shows the normalized representation of the magnitude spectrum of the sampled signal $s_1(t)$, according to Equation (10.15). Effectively, it is proven that no aliasing emerges in this signal when sampled at the minimum sub-Nyquist rate.

10.4   Design of Signal-Interference Multiband Bandpass Filters

According to Figures 10.2 and 10.3, the key element in the multiband RF front-end architecture based on the direct-sampling paradigm is the input RF multichannel filter. Traditionally, microwave bandpass filters have been designed through circuit networks shaped by electromagnetically coupled resonators. However, this solution can hardly be used to design

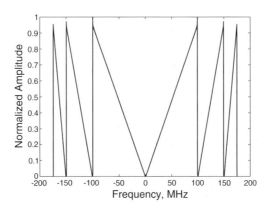

Figure 10.10 Spectrum magnitude of the frequency-asymmetrical triple-band signal after subsampling: aliasing is circumvented (simulation parameters: $B_1 = 100\,\text{MHz}$, $B_2 = 50\,\text{MHz}$, $B_3 = 25\,\text{MHz}$, $f_L = 1050\,\text{MHz}$, $\Delta_1 = 350\,\text{MHz}$, $\Delta_2 = 700\,\text{MHz}$ and $f_s = f_{s,\min} = 350\,\text{MHz}$).

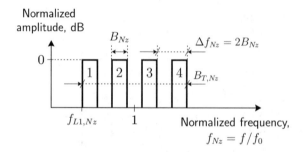

Figure 10.11 Normalized representation of the amplitude spectrum of the quad-band signal. © 2011 IEEE. Reprinted, with permission, from [17].

multiband bandpass filters with four or more bands, particularly when the multiband filtering action must be created in an ultra-wideband (UWB) spectral range.

Fortunately, the signal-interference approach for the development of high-frequency filters turns out to be correct when multiband features are imposed on their transfer functions. In this section, the basic theory to synthesize signal-interference multiband bandpass filters is described. The interest here is not only in evenly spaced equal-bandwidth multiband filters but also those with spectral asymmetry between the bands.

10.4.1 Evenly Spaced Equal-Bandwidth Multiband Bandpass Filters

The circuit detail of the proposed multiband bandpass filter with evenly spaced equal-bandwidth channels is reported in Figure 10.13 [17, 18]. The fundamental element in its structure is the signal-interference transversal filtering section, shaped by the parallel connection of two different transmission-line segments. The Z and θ variables refer to the characteristic impedance and electrical length of each line, respectively.

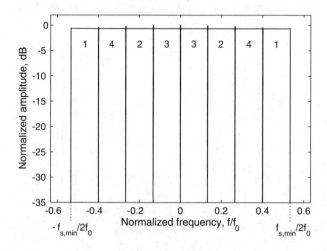

Figure 10.12 Spectrum magnitude of the sampled quad-band signal: no aliasing arises (simulation parameters: $B_{Nz} = 2/15$, $\Delta f_{Nz} = 4/15$, $f_{L1,Nz} = 8/15$ and $f_{s,Nz} = f_{s,\min,Nz} = 16/15$). © 2011 IEEE. Reprinted, with permission, from [17].

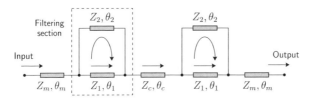

Figure 10.13 Detail of the proposed multiband bandpass filter with symmetrical evenly spaced equal-bandwidth channels. © 2011 IEEE. Reprinted, with permission, from [17].

The transversal filtering section, by means of feedforward signal-interference principles, allows multiband bandpass filtering transfer functions with N one-pole passbands and inter-band power transmission zeros within the spectral range $[0, 2f_0)$ to be synthesized. To this aim, the following design equations for their parameters must be used:

$$\frac{1}{Z_1} - \frac{1}{Z_2} = \frac{1}{Z_0} \tag{10.17}$$

$$\theta_1(f_0) = \frac{N\pi}{2}, \quad \theta_2(f_0) = \frac{(N+2)\pi}{2} \tag{10.18}$$

where Z_0 is the reference impedance. Formulas for the resulting N transmission-pole frequencies are as follows:

$$\{f_1, f_2, \ldots, f_N\} = \left\{ \frac{2kf_0}{N+1}, k = 1, 2, \ldots, N \right\} \tag{10.19}$$

Note that, in Figure 10.13, two identical filtering sections have been cascaded. This is done to an increase in the out-of-band power rejection levels as required in real applications. Input/output transmission line segments have also been added to enhance in-band power matching for all the passbands. Obviously, more selectivity can be attained by cascading more transversal sections, but at the expense of higher power insertion losses.

10.4.2 Stepped-Impedance Line Asymmetrical Multiband Bandpass Filters

Spectral asymmetry can be incorporated into this class of microwave filter by using stepped-impedance lines in the transversal filtering section [19]. This feature can be exploited to develop multipassband filters with bands having different bandwidths.

The main concept behind this idea is inspired by the well-known equivalence between a conventional half-wavelength open-ended transmission line resonator and its stepped-impedance counterpart, as indicated in Figure 10.14, where the relationship shown must provide a perfect correspondence at the design frequency f_d.

Indeed, by using the equivalence outlined in Figure 10.14, transversal filtering sections with strongly asymmetrical passbands can be synthesized. As an example, Figure 10.15 shows two different stepped-impedance line realizations of the network formed by two in-parallel lines of π and 2π electrical length at $f_d = f_0$, respectively (i.e. $N = 2$ in Equations (10.18) and (10.19)). In this case, an asymmetrical dual-band transfer function is attained.

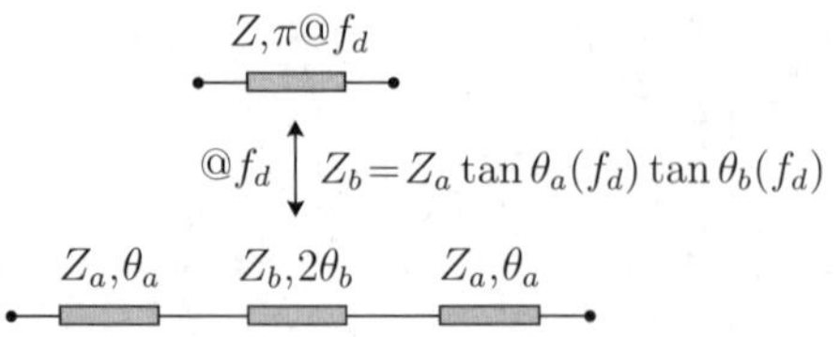

$Z,\pi@f_d$
$@f_d$
$Z_b=Z_a\tan\theta_a(f_d)\tan\theta_b(f_d)$
Z_a,θ_a
$Z_b,2\theta_b$
Z_a,θ_a

Figure 10.14 Conventional half-wavelength open-ended transmission line resonator and its equivalent stepped-impedance counterpart at f_d.

10.5 Building and Testing of Direct-Sampling RF Front-Ends

In a direct-sampling approach to simultaneously acquire several bands, the use of the minimum sub-Nyquist sampling frequency relaxes the demands imposed on the ADC. If the multiband signal to be acquired meets the conditions indicated in this chapter, then this minimum sampling rate $f_{s,\min}$ circumvents undesired aliasing. As previously mentioned, the key element in this direct-sampling-based RF front-end is the multiband bandpass filter carrying out the multichannel signal preselection. In this section, two different microstrip prototypes of a multiband bandpass filter intended for a minimum-sampling rate RF front-end are manufactured and characterized. The design process for these microwave filters is based on the theoretical principles expounded in Section 10.4.

10.5.1 Quad-Band Bandpass Filter

Let us assume a quad-band acquisition system, whose four channels have the following values for the centre frequencies: $f_{c1} = 1.2$ GHz, $f_{c2} = 1.73$ GHz, $f_{c3} = 2.27$ GHz and $f_{c4} = 2.8$ GHz. The desired bandwidth for each channel is $B = 267$ MHz. Under this assumption, the corresponding evenly spaced equal-bandwidth quad-band acquisition system has the parameters: $f_{L1} = 1.07$ GHz, $B = 267$ MHz and $\Delta f = 533$ MHz. Hence, according to Table 10.3, the minimum sampling frequency avoids aliasing in this system.

By using the design rules reported in Section 10.4.1, a quad-band bandpass filter has been designed, constructed and measured for this application. A photograph of the manufactured microstrip prototype is shown in Figure 10.16.

The simulated (circuit model and electromagnetic predictions) and measured power transmission and reflection responses for this circuit are plotted in Figure 10.17. Simulations have

$Z_b,2\theta_b$
Z_a,θ_a
Z_a,θ_a
Input
$Z_1,2\pi@f_d$
Output
(a)
$Z_b,2\theta_b$ $Z_a,2\theta_a$ $Z_b,2\theta_b$
Z_a,θ_a
Z_a,θ_a
Input
$Z_2,\pi@f_d$
Output
(b)

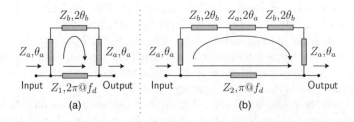

Figure 10.15 Stepped-impedance line versions of the transversal filtering section shaped by two in-parallel transmission lines of π and 2π electrical lengths at f_d. © 2011 IEEE. Reprinted, with permission, from [19].

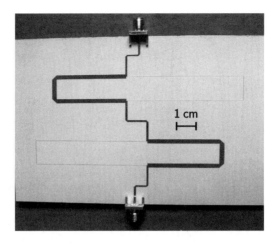

Figure 10.16 Photograph of the manufactured microstrip quad-band bandpass filter prototype (substrate: CER-10 of Taconic). © 2011 IEEE. Reprinted, with permission, from [17].

been performed with the commercial packages HP-Ees of Libra and HFSS v10.0 of Ansoft. Measurements have been carried out by means of an HP-8720C network analyser of Agilent. As shown, the desired bands are correctly approximated. To this end, the generation of the out-of-band transmission zeros is essential.

10.5.2 Asymmetrical Dual-Band Bandpass Filter

Let us assume a dual-band acquisition system, with channels exhibiting the following centre frequencies: $f_{c1} = 1.53$ GHz and $f_{c2} = 3$ GHz. The lower-channel bandwidth is specified as

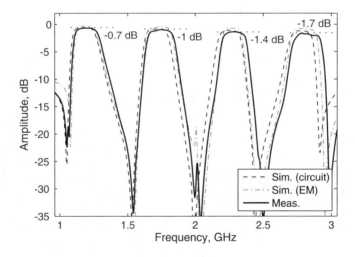

Figure 10.17 Simulated and measured power transmission ($|S_{21}|$) and reflection ($|S_{11}|$) responses for the manufactured microstrip quad-band bandpass filter prototype: (a) $|S_{21}|$, (b) $|S_{11}|$. © 2011 IEEE. Reprinted, with permission, from [17].

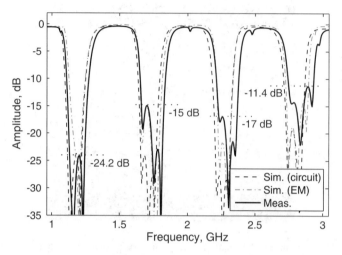

Figure 10.17   (*Continued*)

$B_L = 133$ MHz, whereas the bandwidth for the upper channel is $B_U = 400$ MHz. Note the high asymmetry imposed on this dual-band signal. Consequently, the lowest frequency is $f_L = 1.47$ MHz and the interband separation is $\Delta = 1.2$ GHz. This signal meets the conditions given in Table 10.1 (case 6, $r = 3$ and $s = 3$) and, hence, the minimum sampling frequency ($f_{s,\min} = 2(B_L + B_U)$) does not give rise to aliasing.

By using the theory indicated in Section 10.4.2, a frequency-asymmetrical dual-band bandpass filter has been built and tested for this system. A photograph of the developed microstrip prototype is provided in Figure 10.18.

The simulated (circuit model and electromagnetic results) and measured power transmission and reflection responses for this prototype are drawn in Figure 10.19. For the simulation

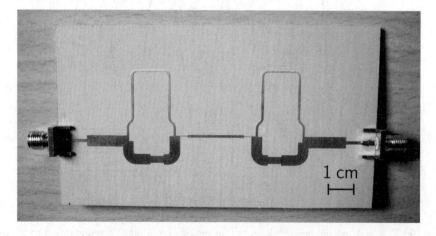

Figure 10.18   Photograph of the frequency-asymmetrical microstrip dual-band bandpass filter prototype (substrate: CER-10 of Taconic).

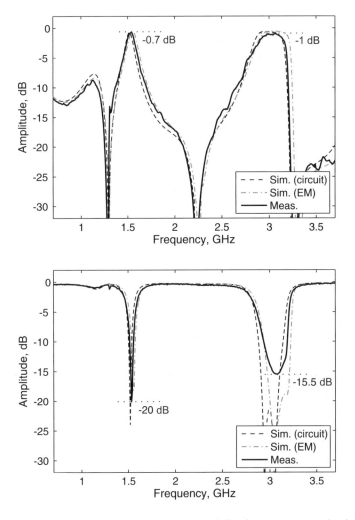

Figure 10.19 Simulated and measured power transmission ($|S_{21}|$) and reflection ($|S_{11}|$) responses for the microstrip frequency-asymmetrical dual-band bandpass filter prototype: (a) $|S_{21}|$, (b) $|S_{11}|$.

and characterization processes, the same tools and equipment as in the previous example were employed. As shown, again, the desired bands are fairly approximated, taking this benefit from the creation of the stopband transmission nulls.

10.6   Conclusions

The direct-sampling approach is an interesting alternative to implement the SDR paradigm. Indeed, if a sub-Nyquist sampling frequency is employed, less stringent demands are imposed on the ADC. In this chapter, the conditions that a multiband signal should meet to guarantee that the minimum sub-Nyquist sampling frequency circumvents aliasing have been explored. The analysis has been made for a dual-band signal and for an evenly spaced equal-bandwidth multiband signal. Simulations have confirmed the validity of the derived design rules. On the other hand, it has been shown that the key element in the multiband direct-sampling RF front-end is the input RF multiband bandpass filter, which can be implemented by using novel signal-interference techniques. Moreover, the employment of stepped-impedance lines in this kind of high-frequency filter concept enables frequency-asymmetrical filtering transfer functions to be attained. For experimental validation, two microstrip multiband bandpass filter prototypes for two different examples of a multichannel sub-Nyquist acquisition system have also been designed, manufactured and characterized.

References

1. Cruz, P. *et al.* (2010) Designing and testing software-defined radios. *IEEE Microwave Magazine*, **11** (4), 83–94.
2. Mitola, J. (1995) The software radio architecture. *IEEE Communications Magazine*, **33** (5), 26–38.
3. Lockie, D. and Peck, D. (2009) High-data-rate millimeter-wave radios. *IEEE Microwave Magazine*, **10** (5), 75–83.
4. Bagheri, R. *et al.* (2006) Software-defined radio receiver: dream to reality. *IEEE Communications Magazine*, **44** (8), 111–118.
5. Chan, S.C. *et al.* (2007) Design and complexity optimization of a new digital IF for software radio receivers with prescribed output accuracy. *IEEE Transactions on Circuits and Systems – I: Regular Papers*, **54** (2), 351–366.
6. Sanduleanu, M.A.T. *et al.* (2008) Receiver front-end circuits for future generations of wireless communications. *IEEE Transactions on Circuits and Systems – II: Express Briefs*, **55** (4), 299–303.
7. Hueber, G. *et al.* (2008) An adaptive digital front-end for multimode wireless receivers. *IEEE Transactions on Circuits and Systems – II: Express Briefs*, **55** (4), 349–353.
8. Akos, D.M. *et al.* (1999) Direct bandpass sampling of multiple distinct RF signals. *IEEE Transactions on Communications*, **47** (7), 983–988.
9. DeVries, C.A. and Mason, R.D. (2008) Subsampling architecture for low power receivers. *IEEE Transactions on Circuits and Systems – II: Express Briefs*, **55** (4), 304–308.
10. Rivet, F. *et al.* (2008) A disruptive receiver architecture dedicated to software-defined radio. *IEEE Transactions on Circuits and Systems – II: Express Briefs*, **55** (4), 344–348.
11. Barrak, R. *et al.* (2009) Optimized multistandard RF subsampling receiver architecture. *IEEE Transactions on Wireless Communications*, **8** (6), 2901–2909.
12. Vaughan, R.G. *et al.* (1991) The theory of bandpass sampling. *IEEE Transactions on Signal Processing*, **39** (9), 1973–1984.
13. Tseng, C.H. and Chou, S.C. (2003) Direct downconversion of multiple RF signals using bandpass sampling. Proceedings of the IEEE International Conference Communications, Anchorage, AK, 11–15 May 2003, pp. 2003–2007.
14. Oppenheim, A.V., Schafer, R.W. and Buck, J.R. (1989) *Discrete-Time Signal Processing* (ed. E Cliffs), Prentice-Hall, NJ.
15. Mishali, M. and Eldar, Y.C. (2010) From theory to practice: sub-Nyquist sampling of sparse wideband analog signals. *IEEE Journal of Selected Topics in Signal Processing*, **4** (2), 375–391.
16. Lewyn, L.L. *et al.* (2009) Analog circuit design in nanoscale CMOS technologies. *Proceedings of the IEEE*, **97** (10), 1687–1714.
17. Muñoz-Ferreras, J.M. *et al.* (2011) RF front-end concept and implementation for direct sampling of multiband signals. *IEEE Transactions on Circuits and Systems – II: Express Briefs*, **58** (3), 129–133.
18. Gómez-García, R. *et al.* (2011) Microwave transversal six-band bandpass planar filter for multi-standard wireless applications. Proceedings of the 2011 IEEE Radio and Wireless Symposium, Phoenix, AZ, 16–19 January 2011, pp. 166–169.
19. Gómez-García, R. *et al.* (2011) Signal-interference stepped-impedance-line microstrip filters and application to duplexers. *IEEE Microwave and Wireless Components Letters*, **21** (8), 421–423.

11

Mm-Wave Broadband Wireless Systems and Enabling MMIC Technologies

Jian Zhang, Mury Thian, Guochi Huang, George Goussetis and Vincent F. Fusco

Institute of Electronics and Communications and Information Technology (ECIT), Queens University of Belfast, Belfast, UK

11.1 Introduction

Over the past decade there has been an increasing interest in the development of milli-meter (mm)-wave broadband wireless systems and enabling technologies [1]. Consistent effort in this direction has been largely motivated by the foreseen opportunities for V-band technology in the broader consumer electronics market. The benefits of utilizing the spectrum in the range 57–66 GHz are numerous. The available bandwidth within this range of frequencies is capable of very large data throughput transmission, currently pro-jected in excess of 5 Gb/s [2]. Given the rapid increase in bandwidth demand and coupled with the saturation of available bandwidth at lower frequencies, mm-wave systems are emerging as the natural choice for emerging high-throughput communication applica-tions. Significantly, propagation of signals at 60 GHz is inherently limited to a short range due to an oxygen absorption peak, which introduces typical free-space propagation losses of 10–15 dB/km [1]. Consequently, near co-sited systems operating on the same frequency can be inherently noninterfering and frequency reuse becomes naturally available across distances of a few tens of meters. Driven by these opportunities and forecasted applications, spectrum regulators globally are converging towards the availa-bility of this band for unlicensed short-range data transfer applications.

Microwave and Millimeter Wave Circuits and Systems: Emerging Design, Technologies, and Applications,
First Edition. Edited by Apostolos Georgiadis, Hendrik Rogier, Luca Roselli, and Paolo Arcioni.
© 2013 John Wiley & Sons, Ltd. Published 2013 by John Wiley & Sons, Ltd.

The vibrant environment set by this landscape has driven rapid developments in the standardization of V-band short-range communication systems. Industry groupings and standardization bodies have been issuing sets of standards for 60 GHz communication systems; at the time when this chapter was written the latest version by a major body was issued in 2010. Pre-commercial demonstrators have already been produced by several groups within the academia [2–4] and industry [2–6], while the first commercial V-band chipset capable of transmitting up to 480 Mb/s was launched in 2007 [7]. Yet, despite the rapid developments in this field, significant engineering challenges remain to be addressed in order to deliver fully the envisaged capability for V-band wireless systems.

Since from the outset the majority of envisaged applications for V-band wireless systems have been in the consumer electronics sector, price has been the key factor in making the technology commercially viable. Hence, despite the performance advantages of III–V semiconductor technology, such as GaAs, at mm-wave frequencies, a more feasible choice in order to realize an RF front-end at a minimized price level for widespread integration in consumer electronic products is required. Lower cost solutions (e.g., mm-wave centered CMOS (typical values for a 65 nm CMOS transistor are $f_T \sim 225$ GHz, $f_{max} \sim 300$ GHz) [8] or BiCMOS (typical values for a 130 nm SiGe BiCMOS transistor $f_T \sim 200$ GHz, $f_{max} \sim 285$ GHz) [9] can offer a significantly improved performance-to-cost ratio, while also allowing for integration of the baseband circuitry with the RF front-end [2]. Within the context of a rapidly changing landscape in the RF semiconductor industry, the technology of choice for mass-production V-band broadband systems are likely to be determined as the lowest cost solution that meets the minimum performance requirements. At the time this chapter is written, and despite the rapid developments in mm-wave CMOS technology, SiGe BiCMOS appears to be a dominant solution.

The synthesis of circuits and subsystems in V-band RF front-ends necessitates an accurate microwave monolithic integrated circuit (MMIC) design library consisting of well-validated passive, active small signal as well as large signal component models, with realistic parasitics included, which are compatible with modern circuit simulation tools such as ADS [10] or CADENCE [11]. Circuit design also has to be validated for layout using numerical electromagnetic simulation tools, such as CST [12] and HFSS [13]. The development of the component library should, in turn, be informed by the overall system architecture. The system architecture, and the associated performance requirements on individual components, should be carefully selected to be commensurate with price-to-performance ratio requirements. Although there are detailed aspects of MMIC design that are dependent on the selected semiconductor technology, the development of a design library linked with a study of the system architecture is a significant step toward establishing an enabling technology for V-band wireless communication systems regardless of the chosen technology.

This chapter aims to provide an introduction to the design of V-band wireless systems for short-range communications. Section 1.2 presents an overview of the standardization activities globally while Section 1.3 looks into some potential system architectures for 60 GHz communication systems. The design of MMIC components in SiGe BiCMOS technology, selected to emphasize different system insertion aspects, is presented in Section 1.4 and an outlook is given in Section 1.4.4.

11.2 V-Band Standards and Applications

Spectrum regulators globally have converged in allocating an unlicensed band around 60 GHz band (frequencies of 57–64 GHz are available in North America and Korea, 59–66 GHz in Europe and Japan), as shown in the Figure 11.1. The main drive for this allocation is short-range broadband wireless communications. Compared with the microwave band (2.5 GHz and 5 GHz), the amount of available bandwidth at 60 GHz is plentiful and provides the ability to support high-rate wireless communications.

Standardization in 60 GHz wireless communication systems is currently under development by several industry consortia and international standard organizations. IEEE 802.15 Task Group 3c [14] has considered a millimeter-wave alternate physical layer for the IEEE 802.15.3 standard for WPANs (IEEE 802.15c completed in October 2009). The European Computer Manufacturer Association (ECMA version 2.0 in December 2010) [15] has developed a 60 GHz technology standard for very high data rate short-range unlicensed communications to support bulk data transfer such as downloading data from a kiosk and HD multimedia streaming. WirelessHD [16] is an industry-led effort to define a next-generation wireless HD interface specification for consumer electronics products. The consortium completed the WirelessHD specification version 2.0 in May 2010. In addition, some WiFi companies also exploit 60 GHz spectrum through the development of the IEEE 802.11. WiGig [17] (Wireless Gigabit Alliance) is privately developed through an industry consortium and has already published version 1.0 (backward compatible with the IEEE 802.11 standard) in May 2010, with consumer products expected to roll-out in late 2011.

11.2.1 IEEE 802.15.3c Standard

The IEEE 802.15.3c standard provides an internationally defined physical layer framework to support consumer 60 GHz WPANs. The specification was issued in October of 2009 by Task Group 3c IEEE 802.15 working group. According to this standard, the frequency range

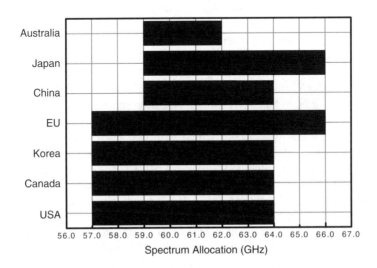

Figure 11.1 International unlicensed spectrum around 60 GHz.

Table 11.1   Millimeter-wave PHY channelization according to IEEE
802.15.3c standard [14]

Channel	Central frequency, f_c (GHz)	Bandwidth (GHz)
1	58.320	2.1600
2	60.480	2.1600
3	62.640	2.1600
4	64.800	2.1600

between 57 024 and 65.88 GHz is divided into four channels, each of 2.16 GHz, shown in
Table 11.1.

The IEEE 802.15.3c describes three types of transmission modes: single carrier mode (SC
PHY), high speed interface mode (HIS PHY), and audio/visual mode (AV PHY). In addition,
to promote coexistence and interoperability, a common mode signaling (CMS) is defined for
channel scanning as well as providing a low rate communication mode.

The *SC PHY* is specified with a high degree of flexibility in order to allow implementers
the ability to optimize for different applications. This standard provides three kinds of modulation schemes according to different wireless connectivity applications. Low rate (Type 1) is
specified for data rates up to 1.5 Gb/s. Medium rate (Type 2) is specified for data rates up to
3 Gb/s. High rate (Type 3) is specified for data rates in excess of 5 Gb/s. The modulation
scheme selection for each type is different. For the low rate SC transmission mode, the
preferred modulation scheme is $\pi/2$ BPSK; for the medium rate SC transmission mode, the
preferred modulation scheme is $\pi/2$ QPSK; for the high rate SC transmission mode, the preferred modulation scheme is $\pi/2$ 8-PSK and $\pi/2$ 16-QAM.

The *HSI PHY* is designed for devices with low-latency, bidirectional high-speed data
wireless connectivity applications. It uses OFDM. The HSI PHY supports a variety of modulation and coding schemes (MCSs), which can be further classified according to the
throughput delivered. For the coding mode EEP, MCS0–MCS3 are specified for data rates
up to 2.695 Gb/s with modulation scheme QPSK; MCS4–MCS6 are specified for data rates
up to 5.390 Gb/s with modulation scheme 16-Q AM; MCS7 is specified for data rates up to
5.775 Gb/s with modulation scheme 63-Q AM. For the coding mode UEP, MCS8–9 are
specified for data rates up to 2.503 Gb/s with modulation scheme QPSK; MCS10–11 are
specified for data rates up to 5.005 Gb/s with modulation scheme 16-QAM.

The *AV PHY* is designed for NLOS operation and the transport of uncompressed, high-
definition video and audio. It is implemented with two PHY modes: the high-rate PHY
(HRP) and the low-rate PHY (LRP), both of which use OFDM. The HRP mode supports
seven kinds of MCs. For the coding mode EEP, MCS0–MCS1 are specified for data rates up
to 1.904 Gb/s with modulation scheme QPSK and MCS2 is specified for data rates up to
3.807 Gb/s with modulation scheme 16-QAM. For the coding mode UEP, MCS3 is specified
for data rates up to 1.904 Gb/s with modulation scheme QPSK and MCS4 is specified for
data rates up to 3.807 Gb/s with modulation scheme 16-QAM. For the coding mode MSB,
MCS5–MCS6 is specified for data rates up to 1.904 Gb/s with modulation scheme QPSK.
The LRP mode supports four kinds of MCs. MCS0–MCS3 are specified for data rates up to
10.2 Mb/s with modulation scheme BPSK.

Table 11.2 Receiver specification according to IEEE 802.15.3c standard [14]

Mode		Reference sensitivity (dBm)	Frame (bit) error rate	Receiver maximum input level (dBm)
SC		−46–70	FER < 8%	> −10
HSI	MSC0	< −50	BER < 10^{-6}	> −25
	MSC1	< −70		
AVI	LRP	< −70	BER < 10^{-7}	> −30
	HRP	< −50		> −24

The transmitter and receiver specifications of all the three transmission modes are summarized in Tables 11.2 and 11.3, respectively.

11.2.2 ECMA-387 Standard

The second version of the ECMA-387 specification for the high-rate 60 GHz PHY, MAC, and HDMI PALs was issued in December of 2010 [15]. This standard specifies a PHY (physical layer), MAC (medium access control) sublayer, and an HDMI (high-definition multimedia interface) protocol adaptation layer for 60 GHz wireless networks. The PHY operates in the 57–66 GHz frequency band. Table 11.4 depicts the channel-band allocation. Channel bonding allows the system to use two or more channels at the same time.

Two types of devices are defined in the standard: type A and type B. Both of the device types can work independently or coexist and communicate with each other.

Type A. There are 22 kinds of mode in type A devices. They operate at an SCBT mandatory mode (A0) at 0.397 Gb/s with other optional SCBT modes (A1–A13) at data rates of

Table 11.3 Transmitter specification according to IEEE 802.15.3c standard [14]

Mode			EVM (dB)	Central frequency tolerance	Nominal used bandwidth	TX power on/off ramp	Subcarrier frequency spacing
SC	1		−7	< ±25 μHz/Hz	N/A	< 9.3 ns	N/A
	2		−14				
	3		−21				
HSI	< 1.5 Gb/s		−7	< ±20 μHz/Hz	1815 MHz	N/A	5.15 625 MHz
	2.1–2.7 Gb/s		−14				
	2.8–5.3 Gb/s		−21				
	> 5.4 Gb/s		−23				
AVI	LRP	LRP0	−10	< ±20 μHz/Hz	1.76 GHz	N/A	4.96 MHz
		LRP1	−10				
		LRP2	−12				
		LRP3	−12				
	HRP	HRP0	−10	< ±20 μHz/Hz	92 MHz	N/A	2.48 MHz
		HRP1	−14				
		HRP2	−19				

Table 11.4 Band allocation according to the ECMA-387 standard [15]

Channel	Central frequency, f_c (GHz)	Bandwidth (GHz)
1	58.320	2.16
2	60.480	2.16
3	62.640	2.16
4	64.800	2.16
5	59.400	4.32
6	61.560	4.32
7	63.720	4.32
8	60.480	6.48
9	62.640	6.48
10	61.560	8.64

0.794 Gb/s to 6.350 Gb/s (without channel bonding) and optional OFDM modes at data rates of 1.008 Gb/s to 4.032 Gb/s (A14–A21).

Type B. There are four kinds of mode in type B devices. They operate using DBPSK at data rates of 0.794 Gb/s to 1.588 Gb/s (without channel bonding), with optional modes of DQPSK and UEP-QPSK at data rates of 3.175 Gb/s.

The transmitter and receiver specification of both types of devices are categorized in Tables 11.5 and 11.6, respectively.

11.2.3 WirelessHD

WirelessHD is an industry-led effort to define a next-generation wireless HD interface specification for consumer electronics products. The first edition of the WirelessHD specification was completed in January 2008 and the second one was issued in May of 2010. It is a wireless high-definition digital interface technology band and represents the first consumer application of 60 GHz technology. It primarily delivers Gb/s streamed video and audio. It is suitable for a wide range of devices including televisions, Blu-ray players, and a variety of

Table 11.5 Receiver specification according to the ECMA-387 standard [15]

Type		Minimum receiver sensitivity (dBm)	Receiver clear channel assessment performance
Type A	SCBT (A0–A13)	$-43.5 \sim -60$	CCA will be launched to indicate a busy channel when transmissions at a receiver level are equal to or greater than -85.0 dBm
	OFDM (A14–A21)	$-50.2 \sim -60.6$	
Type B	B0	-60.7	
	B1	-57.7	
	B2	-54.6	
	B3	-54.6	

Table 11.6 Transmitter specification according to the ECMA-387 standard [15]

Type		EVM (dB) RMS (dB)	Transmitter center frequency tolerance	Symbol clock frequency tolerance	Transmitter power control	Clock synchronization
Type A	SCBT (A0–A13)	EVM −4.8 ~ −12	<±20 ppm	<±20 ppm	Changed with 2 dB step size within ±1 dB or 20% of the change	Should be derived from the same oscillator
	OFDM (A14–A21)	RMS −7.3 ~ −20.9				
Type B	B0	RMS −7				
	B1	RMS −7				
	B2	RMS −8.2				
	B3	RMS −8.2				

other source devices. Two types of physical layers are defined in the standard: the high-rate PHY (HRP) and the low-rate PHY (LRP).

The *high-rate PHY* (HRP) is used for data transfer with a capability in excess of 3 Gb/s. It uses OFDM. Multiple data rates are supported by the HRP with various modulations and coding schemes. The HRP supports a variety of modulation and coding schemes (MCSs), which can be further classified according to the throughput delivered. For the coding mode EEP, MCS0–MCS1 are specified for data rates up to 1.904 Gb/s with modulation scheme QPSK; MCS2 is specified for data rates up to 3.807 Gb/s with modulation scheme 16-QAM. For the coding mode UEP, MCS3 is specified for data rates up to 1.904 Gb/s with modulation scheme QPSK; MCS4 is specified for data rates up to 3.907 Gb/s with modulation scheme 16-QAM. For the coding mode MSB, MCS5–MCS6 is specified for data rates up to 1.904 Gb/s with modulation scheme QPSK.

The *low-rate PHY* (LRP) is used for carrying data with a capability of less than 40 Mb/s. It can also be used for low-rate data streaming of audio. The LRP uses BPSK modulation, with data rates up to 40 Mb/s for all four schemes.

A total of four channels in the frequency range of 57–66 GHz are defined for HRP and LRP, as shown in Table 11.7.

11.2.4 WiGig Standard

The wireless gigabit (WiGig) specification is established by the Wireless Gigabit Alliance and developed based on the existing IEEE 802.11 standard for wireless communication at multigigabit. The first edition of the WiGig specification was issued in May of 2010 [17]. The WiGig standard is quite close to its IEEE 802.11 counterpart and there is speculation that the two groups could merge into a single standard [18]. According to the WiGig standard, the spectrum is divided into four channels, each 2.16 GHz wide. WiGig supports triband devices, 2.4, 5, and 60 GHz bands. The delivered data speed could be up to 7 Gb/s. The standard's application would be, but not limited to, fast data communication applications, such as uncompressed video transmission. The specification supports

Table 11.7   HRP/LRP frequency plan according to the WirelessHD standard [16]

Channel		Central frequency, f_c (GHz)	Bandwidth
HRP	1	58.320	2.16 GHz
	2	60.480	2.16 GHz
	3	62.640	2.16 GHz
	4	64.800	2.16 GHz
LRP	1_1	58.161 375	98 MHz
	1_2	58.320	98 MHz
	1_3	58.478 215	98 MHz
	2_1	60.321 375	98 MHz
	2_2	60.480	98 MHz
	2_3	60.638 215	98 MHz
	3_1	62.481 375	98 MHz
	3_2	62.640	98 MHz
	3_3	62.798 215	98 MHz
	4_1	64.641 375	98 MHz
	4_2	64.800	98 MHz
	4_3	64.958 215	98 MHz

two types of modulation and coding schemes: OFDM with speeds up to 7 Gb/s and SC with speeds up to 4.6 Gb/s.

11.3   V-Band System Architectures

Several system architectures suitable for V-band broadband wireless communication systems have been proposed in the literature [2]. In this section, the traditional heterodyne and homodyne architectures are reviewed for the case of V-band receivers. The large available bandwidth at V-band further allows simpler architectures with reduced spectral efficiencies in order to provide the required data throughput, while offering opportunities for reduced cost and improved system power efficiency. This is discussed in Section 1.3.3.

11.3.1   Super-Heterodyne Architecture

The super-heterodyne architecture has been a standard choice for wireless receivers for a long time [18]. A super-heterodyne system converts received RF signals into baseband in several steps. The RF signals are first down-converted into the intermediate frequency (IF) with a high-frequency local oscillator (LO). After channel selection filtering, these IF signals are down-converted once again into the baseband with an IF LO. Generally, this approach provides not only high channel selectivity and sensitivity, but also immunity to DC offset, LO leakage, and I/Q mismatch at the cost of the image frequency and half IF problem [19]. Figure 11.2 shows a block diagram of a typical super-heterodyne receiver. An RF band-pass filter (BPF) lies between the antenna and low-noise amplifier (LNA) to reject the out-of-band signal. The filtered in-band signals are then amplified by a following LNA in order to keep the system noise floor as low as possible. The image signal can be rejected by either an

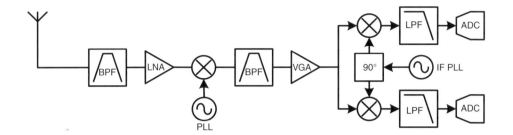

Figure 11.2   Super-heterodyne receiver.

image-rejection mixer or by a following image-rejection bandpass filter. The first mixer translates the RF signal down to the first IF frequency. An automatic gain control (AGC) amplifier adjusts signal levels for optimum I/Q demodulator operation. After splitting into two branches of I/Q quadrature signals, the first IF signals are down-converted into baseband with two mixers and the second quadrature IF LOs, which come from either quadrature LO or with the help of a 90 degrees coupler.

11.3.2   Direct Conversion Architecture

Direct conversion, also called zero-IF or homodyne conversion, down-converts the received RF signals directly to baseband without any intermediate stage [20–23]. This architecture is simpler and requires fewer RF components than the super-heterodyne counterpart. Therefore it leads to lower DC consumption and the possibility of easier monolithic one-chip integration since the IF saw filter and subsequent stages are now replaced with a lowpass filter. Figure 11.3 shows a block diagram of a typical direct conversion receiver. An RF bandpass filter (BPF) is employed between the antenna and the low-noise amplifier (LNA) to reject the out-of-band signal. The filtered in-band signals are then amplified by a following LNA in order to keep the system noise floor as low as possible. Then, signal levels are adjusted by an automatic gain control (AGC) amplifier for optimum I/Q demodulator operation. After splitting into two branches of I/Q quadrature signals, the first IF signals are down-converted into

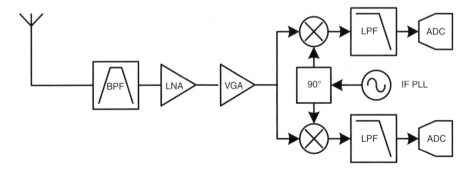

Figure 11.3   Direct-conversion receiver.

baseband with two mixers and the second quadrature IF LOs, which come from either quadrature LO or with the help of a 90° coupler.

One of the major drawbacks of direct-conversion receivers is LO self-mixing. This results in DC offset and degrades the receiver dynamic range. I/Q mismatch is another issue that has to be taken into account in direct-conversion receivers. It mainly comes from the immature RF technology, the tolerance of which becomes larger as the operating frequency increases.

11.3.2.1 Example: Single-Carrier Direct Conversion Architecture

In the remaining part of this section, direct-conversion architecture is chosen as an example to analyze the system performance. This is motivated by the following:

- It is the simplest architecture, which facilitates the implementation.
- No highly selective bandpass filters are necessary and no image problems exist.
- A low sampling rate requirement of ADC.

The receiver architecture is used as an example, since the dual properties mostly apply to the transmitter. The simplified direction conversion block diagram shown in Figure 11.4 is employed.

A primary characteristic for any wireless system is a link budget. A link budget is a method of accounting for all the gains and losses in a communication system, from the power output to the received signal at the receiver.

Consider a high-capacity link operating with the following real-world performance characteristics:

- Operating frequency band: 60 GHz
- LNA: 15 dB gain; 6 dB NF; −5 dBm IIP3
- Mixer: 3 dB gain; 10 dB NF; 5 dBm IIP3
- LPF: 8 dB gain; 10 dB NF; −10 dBm IIP3
- AGC: 10–40 dB gain; 15 dB NF; 0 dBm IIP3

The calculated system characteristics including cumulative gain, noise figure IIP3, and signal noise rate can be summarized in Table 11.8.

The calculated RF-end characteristics are then added to the baseband system for system simulation. Here a differential binary phase shift keying (BPSK) is chosen as modulation/demodulation scheme and a 2 Gb/s data stream as input. Figure 11.5 shows a block diagram of a DBPSK-based modulator/demodulator in Matlab Simulink.

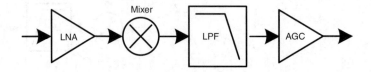

Figure 11.4 Simplified direct-conversion receiver.

Table 11.8 Calculated system characteristics

	LNA	Mixer	LPF	AGC	
				Low	High
Gain(dB)	15	3	8	10	40
NF(dB)	6	10	10	15	
IIP3(dBm)	−5	5	−10	0	
Cumulative gain(dB)	4	36–66			
Cumulative NF(dB)	1	6.75			
Cumulative IIP3(dBm)	2	−35			
SNR(dB)	3	9.23			

Figure 11.6 gives the simulated bit error rate performance of the proposed DBPSK-based modulator/demodulator. With a 9.2 dB SNR input and 2 Gb/s data stream, the simulated bit error is around 1×10^{-4}. From the curve, an improved 12 dB SNR performance is shown to be required to obtain a level of 1×10^{-7} BER.

11.3.3 Bits to RF and RF to Bits Radio Architectures

Due to large bandwidth unlicensed spectrum availability at V-band, high-speed mediocre spectral efficiencies as low as 0.25 bit/s/Hz can be considered and hence simple radio architectures become a viable proposition. Such radio architectures offer significant cost and power reduction possibilities. Of this class of radio the impulse radio architecture, shown in Figure 11.7, is probably the simplest [24]. Here pulse-position modulation (PPM) is generated using pulse widths from 2 ns to 200 ps (equivalent to bit rates of 500 Mb/s to 5 Gb/s). The key element here is a single-pole single-throw (SPST) switch; see Section 11.4.4 below. With this type of wireless, direct baseband to mm-wave signal up-conversion to 60 GHz

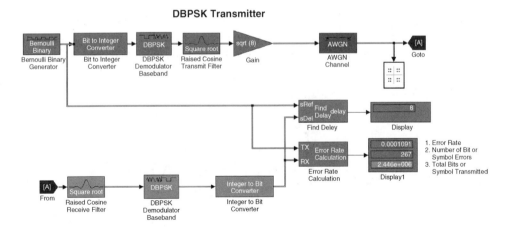

Figure 11.5 Block diagram of the DBPSK-based modulator/demodulator.

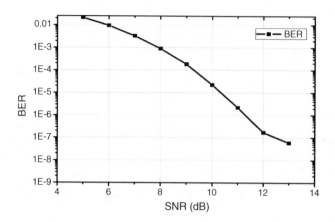

Figure 11.6 Simulated BER performances.

occurs. A Gaussian filter is typically used in order to ensure compliance with the 60 GHz spectral transmission mask.

11.4   SiGeV-Band MMIC

For the 60 GHz applications, III–V semiconductor technologies such as GaAs and InP have mostly been used in the past since compound semiconductors have possessed the best overall performance for mm-wave MMICs. However, silicon-based mm-wave MMIC technology has recently developed largely due to advances in process technology and lithographic techniques. There are two main driving forces for the development of the silicon technologies toward mm-wave MMICs: the lower cost compared to compound semiconductors and the possibility of integrating both digital and multifunction analog designs on the same chip. The lower cost is not only due to lower production costs, which originate from cheaper epitaxial material, but also from the large number of silicon foundries that offer mature and stable processing together with high yield and large wafer sizes.

Among the emerging silicon technologies, SiGe-based HBT and CMOS silicon technologies have been used, where the latter is often highlighted as the most interesting due to its low cost and the former due to its better, close to III–Vs, high-frequency performance.

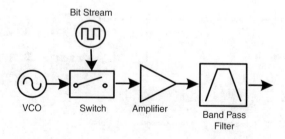

Figure 11.7 Impulse radio architecture.

Therefore, throughout this section, the emphasis is placed on SiGe-HBT as the most promising solution for the 60 GHz wireless communication system in terms of the cost/performance ratio.

11.4.1 Voltage Controlled Oscillator

In the transceiver system, one of the most important and complex blocks is the phase-locked loop (PLL) which serves as a local oscillator (LO) to the mixing circuit to up/down-convert the modulated carrier signal. In addition to the requirement for a wide operation frequency tuning range, the PLL should exhibit a low spurious tone response so as not to deteriorate the signal-to-noise ratio (SNR) in wideband systems [25, 27]. Obviously, a wideband low-gain voltage controlled oscillator (VCO) is a prerequisite for the construction of such a demanding PLL system [25, 27].

However, as the frequency increases, the parasitic parameters within the resonating tank elements seriously constrain the frequency tuning range. A well-known problem is that many harmonics coexist with the fundamental signal available at the VCO output. Thus harmonics can be taken advantage of in order to decrease the design complexity at the high operating frequency. The technique utilizing the second harmonic of the VCO core is very attractive for deployment in the mm-wave frequency band, because the VCO core needs only to oscillate at half of the desired frequency so that the trade-off among phase noise, frequency tuning range, and power dissipation is alleviated. In this section, the second harmonic signal is utilized in order to generate the desired frequency band using a summing circuit [26]. The switched varactor array is introduced in order to obtain a low-gain wide tuning range and an active single-ended to differential (STD) converter is also adopted in order to create a differential pair.

11.4.1.1   Design of the 60 GHz Wide Tuning Range SiGeVCO

A simple block diagram of the VCO is shown in Figure 11.8(a). The entire circuit consists of four blocks: a VCO core, a summing circuit, a single-ended to differential converter, and a differential amplifier. The VCO core is designed to oscillate at fundamental frequency f_0. The output signals of the VCO core are summed together using two capacitors C9 and C10. Thus the even harmonics are extracted and all the odd harmonics are canceled. As a result the second harmonic emerges at the output port of the summing circuit. The extracted single-ended even harmonics are then converted to a differential pair by the following single-ended to differential (STD) circuit since in many transceiver systems a differential signal is preferred. After that, a differential cascode amplifier is adopted to amplify the desired signal further at the $2f_0$ frequency and improve the phase and amplitude balances of the STD output signals.

There are three varactor banks, Cv1 to Cv3, among which two are used for coarse frequency tuning and the other, Cv1, is used for continuous fine tuning. C5 and C6 are functioned as DC blocking capacitors. R2 and R3 provide a DC feed-through path for the PN junction varactor diodes. Customized spiral inductors, L1-2, are simulated under three-dimensional EM EDA and modeled using a double-π equivalent RLC circuit [28]. The simulated inductance and quality factor (Q-factor) at 30 GHz are about 142 pH and 24 respectively. The variable capacitor banks include three varactor groups CV1–3, two DC blocking capacitors C5–6, and two DC feeding resistors R2–3. The simulated

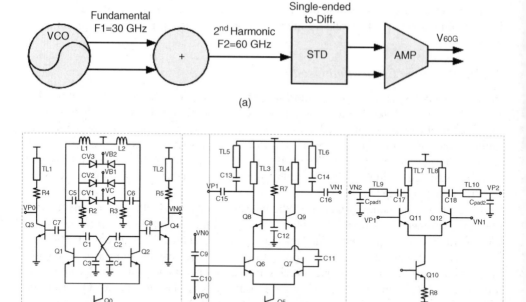

(a)

(b)

Figure 11.8 (a) Block diagram of the wide tuning range VCO, (b) schematic of the wide tuning range VCO. © 2011 IET. Reprinted, with permission, from [26].

Q-factor is about 6.5 at 30 GHz. Transistors Q1–2 and capacitors C1–4 form a negative resistance in order to compensate for the loss of the LC tank. A common emitter amplifier with series RL load functions as a buffer amplifier stage to alleviate the capacitive effect of the VCO core. The second harmonic signal is extracted single-ended by the following summing circuit, which is constructed by a pair of capacitors, C9 and C10, and then converted to a differential pair by a single-ended to differential converter circuit [29]. The output of the summing stage is then injected into the base of Q6. A part of the output signal of Q6 is fed back to the base of Q7 by using C11 to connect the Q7 base node with the Q6 collector node. Transistors, Q8–9, are introduced to alleviate the Miller effect in order to achieve an enhanced bandwidth. A differential common emitter amplifier with high common mode rejection is used to amplify the desired signal and reduce the phase and amplitude imbalance of the differential signal [30]. An impedance peaking LC network is used as the load of the STD circuit. The high impedance of the LC network is achieved by setting the resonant frequency of the LC network to about 60 GHz. Therefore the gain of the STD circuit can be enhanced. The output 50 Ω impedance matching network is included for testing purposes. It is constructed by placing a series CLC network at the output path. Transmission line (T-line) inductors TL9–10 in the

CLC network are constructed using the top and second metallization layers. In order to provide better impedance matching, the line impedance of TL9–10 is designed to be 50 Ω at 60 GHz. For accurate simulation, the parasitic capacitance from the I/O pad is also modeled by a 40 fF capacitor and included in the circuit simulation. All bias networks in the proposed VCO are omitted in the schematic shown in Figure 11.1(b). Resistors are added to the emitter node of tail current sources to increase the output impedance to provide better common mode noise rejection [30]. The transmission line inductors used in the circuit are those available in the standard process design kit.

11.4.1.2 Measurement Results

The proposed wideband VCO was fabricated using 0.35 μm SiGe technology. The microphotograph of the proposed VCO is shown in Figure 11.9.

The VCO was measured using an Agilent 11 974 V preselected millimeter mixer and an Agilent E4407B spectrum analyzer. The VCO has four discrete frequency bands, which improve the frequency tuning range and decrease the VCO gain. The VCO operates from 58.85 to 70.85 GHz, as shown in Figure 11.10, with about 20% frequency tuning range. The measured VCO gain for the four frequency bands increases from 0.97 GHz/V to 1.6 GHz/V, which is due to the decrease of the varactor size. The output power of the VCO ranges from −40 to −35 dBm, as shown in Figure 11.11. The measured phase noise at 3 MHz offset, shown in Figure 11.12, ranges from −78 dBc/Hz to −89.5 dBc/Hz.

The output power spectrum at 69.5 GHz is shown in Figure 11.13. The entire circuit occupies the $770 \times 550\ \mu\text{m}^2$ die area and consumes 62 mA under a 3.5 V DC supply.

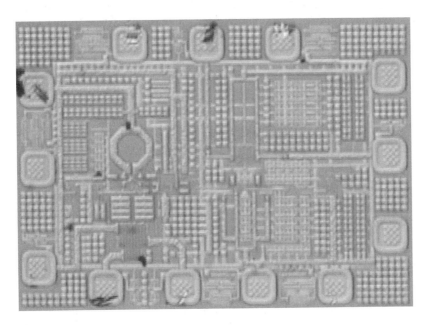

Figure 11.9 Microphotograph of the wide tuning range VCO. © 2011 IET. Reprinted, with permission, from [26].

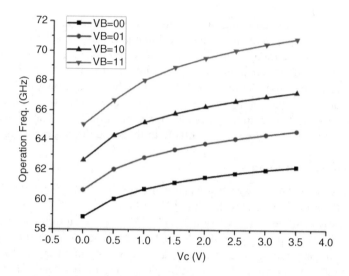

Figure 11.10 Measured operating frequency of the VCO. © 2011 IET. Reprinted, with permission, from [26].

11.4.2 Active Receive Balun

A balun can convert signals that are unbalanced (single-ended) about ground to signals that are balanced (differential) or the reverse [31]. It is a key component in a feeding network wherein differential transmission lines (CPS lines) are used to feed the two arms of a dipole antenna. Compared with the traditional passive balun, such as transformer or Marchand

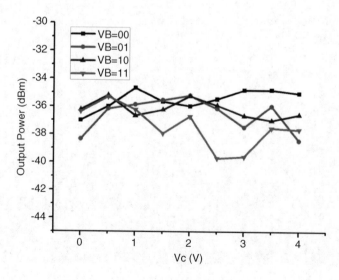

Figure 11.11 Measured output power of the proposed VCO. © 2011 IET. Reprinted, with permission, from [26].

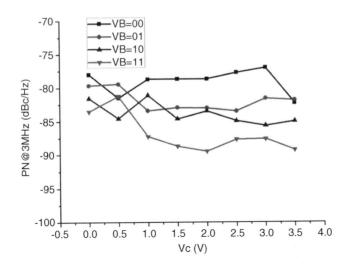

Figure 11.12   Measured phase noise at 3 MHz offset of the wide tuning range VCO. © 2011 IET.
Reprinted, with permission, from [26].

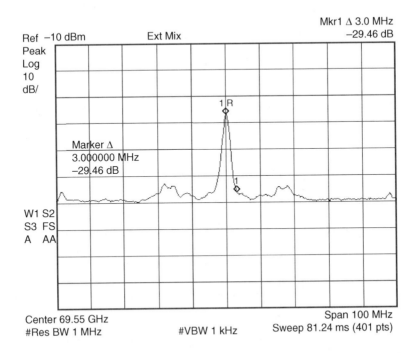

Figure 11.13   Measured output power spectrum at 69.5 GHz. © 2011 IET. Reprinted, with
permission, from [26].

balun, the active balun presented here shows positive conversion gain and occupies a compact physical size [32].

11.4.2.1 Circuit Design

Figure 11.14 shows the designed V-band active balun, which is composed of one single-ended to differential converter (STD) and one emitter coupled differential amplifier [33, 34].

An LC impedance peaking network is adopted for the load of the STD circuit. If the series resistance of the transmission line inductors is ignored, the impedance ZL seen in the network from the collector node of Q3 or Q4 can be expressed as

$$Z_L = \frac{sL_4(1 + s^2 L_5 C_4)}{1 + S^2 C_4 (L_4 + L_5)} \tag{11.1}$$

where L_4 and L_5 are the inductance of transmission lines TL4 and TL5, respectively. By properly choosing the values for L_4, L_5, and C_4, ZL can be made very large in the desired frequency band in order to force the output current to flow into the following stage. However, in practice the series resistance in the transmission line inductor will limit the impedance peaking effect. In this design, for the purpose of input stage impedance matching, the transmission lines, L1 to L3, shown in Figure 11.1, are selected with 50 Ω characteristic impedance at the desired frequencies. Note that, at millimeter-wave frequencies, for example, 60 GHz, the base impedance of a bipolar transistor, is relatively

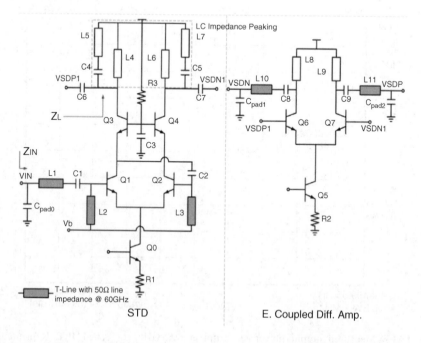

Figure 11.14 Schematic of the active balun. © IET. Reprinted, with permission, from [32].

small compared with a 50 Ω transmission line. Therefore, the input impedance can be simplified as

$$Z_{in} = \frac{1}{sC_1 C_{pad0}} \times \frac{1 + s^2 C_1 (L_1 + L_2) + sC_1 (R_{s1} + R_{s2})}{s(R_{s1} + R_{s2}) + [1 + s(L_1 + L_2)]} \tag{11.2}$$

where C_1 and C_{pad0} are the capacitance of the AC coupling capacitor C1 and bonding pad, respectively. ZB is the impedance of the transistor Q1 looking into the base node and is much less than ZL1. ZL1 is the characteristic impedance of the transmission line L1, which is around 50 Ω at 60 GHz. As can be seen from Equation (11.2), the second term in the denominator and C_{pad0} in the first term can be ignored when C_1 is much larger than C_{pad0}. As a result, the input impedance is primarily determined by the characteristic impedance of the transmission line L1. In the design, C_1 is set to 1 pF and the bond pad capacitance C_{pad0} is measured at around 40 fF. Thus, a good 50 Ω impedance matching can be achieved over a wide frequency band.

For the active part, a part of output signal at the collector node of Q1 is coupled to the base of Q2 through capacitor C2, which helps alleviate the unbalanced distribution of the input signal between Q1 and Q2 due to the fact that the output impedance of the tail current source is not sufficiently high. Thus the differential mode current flowing into the output node VSDN1 is increased. Transistors Q3 and Q4 are introduced to alleviate the Miller effect in order to achieve an enhanced bandwidth. A differential common emitter amplifier with high common mode rejection is used to amplify the desired signal and reduce the phase and amplitude imbalance of the differential signal through further compression of the common mode signal generated by the first stage. The output 50 Ω impedance matching network is included for the testing purpose. It is constructed by placing a series CLC network at the output path.

11.4.2.2 Measurement Results

The presented active balun was fabricated using 0.35 μm SiGe technology. The microphotograph of the circuit is shown in Figure 11.15. The circuit was measured on wafer using GSG probes. An Agilent E8361A network analyzer was used to measure the S-parameters.

The circuit worked under a 17.5 mA and 3.3 V DC bias. The measured S-parameters of the circuit are shown in Figure 11.16. Ideally the signal incident from the input port is distributed to both differential output ports with equal amplification, resulting in an equal conversion gain and zero phase error. However, due to the accuracy limitation of the model, fabrication, and simulation tolerance, the measured conversion gain of the two differential ports are 8 dB and 6.4 dB at 60 GHz, respectively, which exhibits a magnitude imbalance around 1.6 dB. The measured phase error is less than 5° over the frequency range of 55 to 65 GHz.

The noise figure of the active balun was also measured, as shown in Figure 11.17. It exhibits a 7.5 dB noise performance at 60 GHz and is less than 8 dB from 55 GHz to 65 GHz.

11.4.3 On-Chip Butler Matrix

The Butler matrix is a key component in the switched beam array systems [35–37]. The main beam of the radiation pattern of the antenna can point to the optimal direction and consolidate the communication link by judging the power levels of detected EM waves.

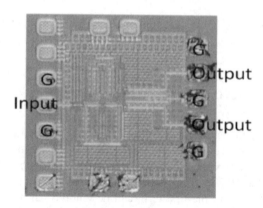

Figure 11.15 Microphotograph of the active balun chip. © IET. Reprinted, with permission, from [32].

A simple block diagram of the Butler matrix is shown in Figure 11.18 [38]. When the different input ports are excited, the Butler matrix functions as a beamforming network to provide four output signals with equal power levels and the progressive phases of +45°, −45°, +135°, and −135°, respectively. Hence, one can switch the direction of the radiation main beam by exciting the designated input port. Previously 60 GHz Butler matrix designs have been given in substrate intergraded waveguide [39] and microstrip [38] technology, but these occupy a large area and are thus unsuitable for MMIC realization. In this section, an example of a miniaturized SiGe-based V-band 4 × 4 Butler matrix MMIC is presented. It exhibits an average insertion loss of 2 dB with an amplitude variation less than 0.5 dB and

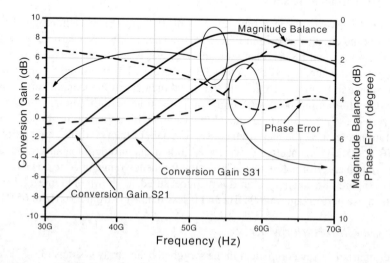

Figure 11.16 The measured conversion gain, magnitude balance, and phase error. © IET. Reprinted, with permission, from [32].

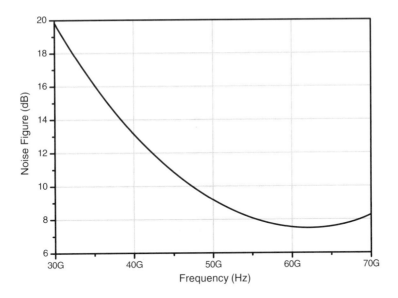

Figure 11.17 Measured noise figure performance of the active balun. © IET. Reprinted, with permission, from [32].

phase imbalance of less than 5° from 55 to 65 GHz. The chip area is only $0.5 \times 0.9\,\text{mm}^2$ including all pads.

11.4.3.1 Coupler Design

The 3 dB 90° coupler is the most common key component in the Butler matrix circuit. Usually it is designed with branch-line couplers based on a quarter-wavelength transmission line

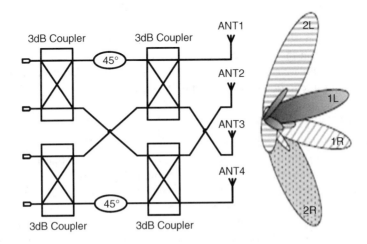

Figure 11.18 Butler matrix network.

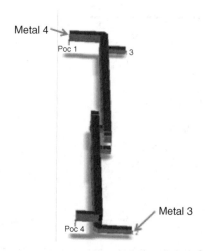

Figure 11.19   Layout of the transformer coupler.

or lumped component. However, such a design suffers either from a large chip size or high process tolerance/parasitic effect on component fabrication, especially at high frequencies. For example, a microstrip line with length λ/4 equals 0.5 mm at 60 GHz in a typical Si process. As a consequence, a physically compact transformer coupler is used to solve this problem [38]. This is implemented with a pair of mutually coupled transmission lines in a broadside coupling configuration using the SiGe multimetal layer structure. Figure 11.19 shows a layout example, where the first winding starting from port-1 is placed on the thicker metal-4 layer (2.8 μm thicknesses) and then is routed down on the metal-3 (1.5 μm thickness) to port-2. At the same time, the second winding starting from port-3 is placed on the metal-3 layer and then is raised up on metal-4 to port-4. These windings are completely overlapped and the ground plane beneath the stacked windings is taken away to ensure a high coupling value. Thus only 250 μm winding length is needed, which is approximately equal to λ/10.

The EM simulated results of the proposed transformer coupler have shown that insertion loss is 3.5 ± 0.5 dB and the output phase difference is $90° \pm 0.5°$ in 55–65 GHz, and that return loss and isolation are greater than -23 dB. The performance of the quadrature coupler is hence suitable for integration in an on-chip Butler matrix.

11.4.3.2   Measurement Results

The proposed Butler matrix was fabricated using 0.35 μm SiGe technology. The microphotograph of the circuit is shown in Figure 11.20. The Butler matrix was measured on-wafer using GSG probes. During the measurement, port-1 is used for the excitation input port and all the other input ports are terminated with 50 Ω. Port-5 to port-8 are used as output ports and are unterminated when not being probed.

The measured insertion loss and phase shift are shown in Figure 11.21. Ideally, the signal incident from one of the input ports is distributed to all four output ports with equal amplitude, resulting in a theoretical 6 dB insertion loss. However, due to the metallization and

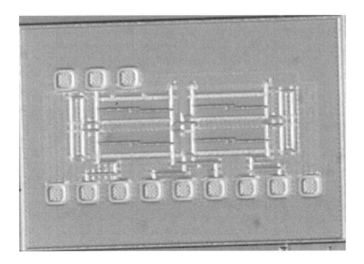

Figure 11.20 Micrograph of the fabricated Butler matrix.

substrate losses, fabrication, and simulation tolerance, an extra insertion loss of 2 dB is measured and the amplitude imbalance is around 0.5 dB at 60 GHz. The ideal relative phase shift between the adjacent output ports is 45°. The measured phase error is about 3° at 60 GHz.

The measured amplitude and phase characteristics of the Butler matrix MMIC are substituted into a linear 1×4 phased array with $\lambda/2$ antenna spacing for the array factor calculation [40]. These results are shown in Figure 11.22 and confirm the beamforming operation over a 10 GHz bandwidth from 55 to 65 GHz.

11.4.4 High GBPsSiGeV-Band SPST Switch Design Considerations

From the discussions in Section 11.3.3 of direct bits to mm-wave radio it can be seen that speed of the switch used to perform the data encoding plays a fundamentally important role. High-speed switches have to date not received much attention, particularly at 60 GHz. Switches implemented using CMOS technology [41] are not as fast as the SiGe HBT switches reported in Reference [42] for use in 24 GHz automotive radar systems. In the remainder of this section, the design and performance of a low-loss broadband V-band SiGe HBT single-pole single-throw (SPST) switch with sub-100 ps switching time is discussed [43].

The SPST switch circuit topology is depicted in Figure 11.23. To enable high-speed switching operation, the current-steering technique is adopted [44]. The tail current (I_{BIAS}) can be steered either to the Q_1–Q_2 or Q_3–Q_4 pair using the differential pair Q_5–Q_6. The dummy pair Q_1–Q_2 makes the switch absorptive at its input as required to reduce unwanted transient-related effects such as pulse reflection and frequency pulling of a signal source. The switch is designed to be reflective at the output, allowing it to directly drive a nonlinear PA. The switching time is determined by (a) the transistor rise time and (b) the output matching network bandwidth, which in fact dominates switching time performance. Broadband output matching networks (L_2–L_3–C_2) are used while simple L_1–C_1 networks are sufficient to match the input to 50 Ω. A current mirror with emitter resistive degeneration prevents sudden increases of current caused by small V_{BE} fluctuations.

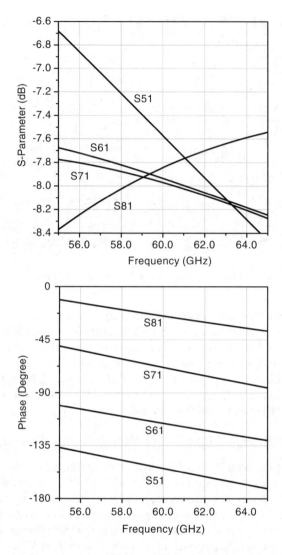

Figure 11.21   Measured insertion loss and phase shift for the fabricated Butler matrix prototype.

The switch in Figure 11.23 is designed and implemented using 0.35 μm SiGe technology with a transistor peak f_T at 170 GHz and f_{max} at 250 GHz at a collector current density of 5 mA/μm² [45]. The chip occupies a 780×600 μm² die and is switched ON and OFF by applying 0.5 V DC at 23.5 mA and 0 V DC at 0 mA, respectively. The circuit operates from ± 1.8 V DC supply voltage with 1.05 V DC applied to the bases of Q_1–Q_4 while the bases of Q_5–Q_6 are 0 V biased. The measured S-parameters are illustrated in Figure 11.24. Here insertion loss is less than 1 dB from 40 GHz to 70 GHz, while isolation greater than 16 dB up to 70 GHz was measured. The measured input referred a 1 dB compression point at 50, 55, and 60 GHz ranges from 1 to 2 dBm. The power handling of

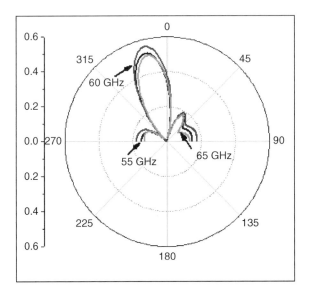

Figure 11.22 Calculated radiation patterns based on the measured Butler matrix S-parameters with isotropic antenna elements.

SiGe switches is inferior to the PIN diode or FET switches, but an impulse radio requires an input signal as low as −5 dBm.

Figure 11.25 shows the modulated output signal when a 3.3 Gb/s pulse train is applied to the switch. The rise and fall times of the output signal are, respectively, 25 ps and 50 ps. With the contribution of the rise/fall time of the pulse train coming from the generator being negligibly small, the actual total rise and fall time of the switch itself is about 75 ps, suggesting

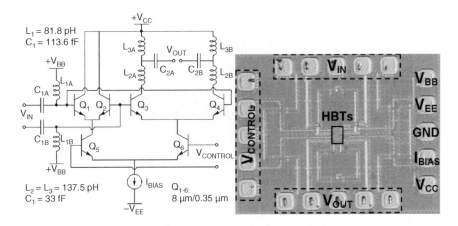

Figure 11.23 High-speed SPST switch circuit with current steering: circuit layout and microphotograph. © 2011 IET. Reprinted, with permission, from [43].

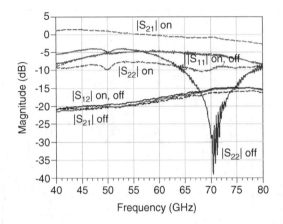

Figure 11.24 Measured S-parameters for SPST prototype (solid lines for OFF and dash lines for ON). © 2011 IET. Reprinted, with permission, from [43].

that switching rates up to 13 Gb/s are achievable. The simulated noise figure across the 30 GHz bandwidth is lower than 6.6 dB. Table 11.9 shows a performance comparison with recently published switches.

11.5 Outlook

The area of 60 GHz systems for short-range communications is a rapidly evolving field. This chapter has provided an overview of the recent and ongoing developments on V-band wireless systems in regard to relevant standards, system architectures, and enabling MMIC

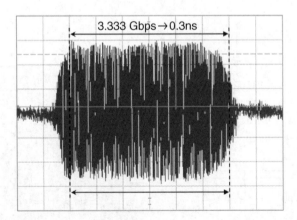

Figure 11.25 Measured modulated output signal with a carrier frequency of 55 GHz. Vertical and horizontal axes are 17.5 mV/div and 50 ps/div, respectively. © 2011 IET. Reprinted, with permission, from [43].

Table 11.9 Comparison of the switch described here with recently published SPST switches © 2011 IET. Reprinted, with permission, from [43]

Reference	[43]	[47]	[42]	[44]	[46]
Technology	0.35 μm SiGe	GaAs	0.8 μm SiGe	0.8 μm SiGe	InGaP/GaAs
Frequency (GHz)	40–70	0–80	14–26	20–26	20–26
Rise/fall time (ps)	25/50	<1000	60/60	100	100^a
IL (dB)	-1.5^b-0.9	<5	-2.5^b-1	0.3–3.3	-8^b–12
ISO (dB)	>16	>17	>30	>42	>24
RL (dB)	>5/>5	>5/>1	>5/>5	>10/>10	>3/>3
Input on/off	>8/>4	—	>5/>5	>5/>5	>3/>5
Output on/off					
P1dB (dBm)	2 at 60 GHz	—	—	0.5 at 23 GHz	0.5^a at 24 GHz
Noise figure (dB)^a	<6.6	—	—	13.3 at 24 GHz	—

^{a}Simulation results.
^{b}The minus (−) sign indicates gain.

technologies. The emphasis was placed on SiGe BiCMOS technology, which is the most mature and cost-effective solution during the period when this chapter was written.

The future of V-band broadband wireless technology will strongly depend on its adoption in a market segment of sufficient size to justify the required investments and in this respect the identification of a driving application will be a significant development. Among the advantages of 60 GHz is the small form factor of the antenna (particularly significant for handheld and portable devices), the large available unlicensed bandwidth, and the inherent interference suppression due to atmospheric absorption. A significant challenge remains the reduction of the RF front-end cost and its packaging. The development of viable directive and potentially steerable antenna systems remains a further engineering challenge that stems from the power budget requirements of 60 GHz systems [48]. From a packaging perspective, the system-on-chip versus system-in-package options will eventually be determined by the performance-to-cost ratio.

To date, the envisaged applications include the transmission and reception of uncompressed high-definition television (HDTV) signals between home appliances (monitor, blue ray, game consoles), which will remove time delays associated with image compression algorithms. The technology is also suitable for the wireless universal serial bus (USB) connection for potential future desktop and portable devices, including smartphones as well as wireless kiosk applications. Consumer electronic manufacturers have already started integrating 60 GHz technology in commercial products; this is likely to expand as the challenge for robust design and packaging of mm-wave parts and systems to consumer domain cost and performance becomes more mature. The significant technological potential and the ever-increasing need for yet larger bandwidths and information throughput promise that V-band and mm-wave wireless broadband systems will remain an attractive option for the future.

From a technology perspective, the rapid trends and developments in mm-wave CMOS RFIC are likely to provide within the near future competitive options for the implementation of V-band front-ends [2]. Nevertheless, it is the authors' aspiration that some of the concepts, techniques, and architectures described in this chapter will remain valuable guidelines regardless of the semiconductor technology. Meanwhile, within the broader semiconductor technology roadmap, SiGe BiCMOS technology could prove more relevant to other and potentially more niche applications. Beyond automotive radar in the 77 GHz band, these could include the FCC licensed E-bands (71–76 GHz, 81–86 GHz, and 92–95 GHz), which are available for wireless backhaul applications [49] as well as active and passive mm-wave and submm-wave imaging applications [50, 51].

References

1. Zhouyue, P. and Khan, F. (2011) An introduction to millimeter-wave mobile broadband systems. *IEEE Communications Magazine*, **49** (6), 101–107.
2. Rappaport, T.S., Murdock, J.N. and Gutierrez, F. (2011) State of the art in 60-GHz integrated circuits and systems for wireless communications. *Proceedings of the IEEE*, **99** (8), 1390–1436.
3. Fusco, V. and Wang, C. (2010) V-band 57–65 GHz receiver. *IET Proceedings, Microwaves, Antennas and Propagation*, **4** (1), 1–7.
4. Wang, C. and Fusco, V. (2010) Reduced architecture V-band transmitter. *IET Proceedings, Microwaves, Antennas and Propagation*, **4** (11), 1948–1954.
5. Forstner, H., Ortner, M. and Verweyen, L. (2011) A fully integrated homodyne upconverter MMIC in SiGe:C for 60 GHz wireless applications. Proceedings of the 11th Silicon Monolithic Integrated Circuits in RF Systems (SiRF), IEEE, pp. 129–132.
6. Ortner, M., Forstner, H., Verweyen, L. and Ostermann, T. (2011) A fully integrated homodyne downconverter MMIC in SiGe:C for 60 GHz wireless applications. Proceedings of the 11th Silicon Monolithic Integrated Circuits in RF Systems (SiRF), IEEE, pp. 145–148.
7. SiBeam Takes Wraps off Wireless, HD, Technology, EE, Times (2007) [cited October 25, 2011], available from: http://www.electronics-eetimes.com/en/sibeam-takes-wraps-off-wireless-hd-technology.html?cmp_id=7&news_id=200000922.
8. Tomkins, A., Aroca, R.A., Yamamoto, T. *et al.* (2009) A zero-IF 60 GHz 65 nm CMOS transceiver with direct BPSK modulation demonstrating up to 6 Gb/s data rates over 2 m. *IEEE Journal of Wireless Link Solid-State Circuits*, **44** (8), 2085–2099.
9. Natarajan, A., Reynolds, S.K., Tsai, M.-D. *et al.* (2011) A fully-integrated 16-element phased-array receiver in SiGe BiCMOS for 60-GHz communications. *IEEE Journal of Solid-State Circuits*, **46** (5), 1059–1075.
10. Advanced Design System (ADS), Agilent Technologies, http://www.home.agilent.com/agilent/product.jspx?cc=GB&lc=eng&ckey=1297113&nid=-34346.0.00&id=1297113.
11. Cadence Virtuoso, http://www.cadence.com/products/cic/analog_design_environment/pages/default.aspx.
12. CST Microwave Studio, http://www.cst.com/content/products/mws/overview.aspx.
13. Ansoft High Frequency Structure Simulator, accessed 6 June 2012, http://www.ansoft.com/products/hf/hfss/.
14. IEEE Standard for Information Technology: Telecommunications and Information Exchange between Systems – Local and Metropolitan Area Networks.
15. Standard ECMA-387 [cited December 21, 2011], available from: http://www.ecma-international.org/publications/standards/Ecma-387.htm.
16. WirelessHD [cited December 21, 2011], available from: http://www.wirelesshd.org/.
17. WiGig [cited December 21, 2011], available from: http://wirelessgigabitalliance.org/specifications/.
18. Merritt, R. (2010) 60 GHz Groups Face Off in Beijing over Wi-Fi's Future, *EE Times* [cited December 21, 2011], available from: http://www.eetimes.com/electronics-news/4199522/60-GHz-groups-face-off-in-Beijing-over-Wi-Fi-s-future.
19. Razavi, B. (1998) *RF Microelectronics*, Prentice Hall.
20. Reynolds, S., Floyd, B., Pfeiffer, U. and Zwick, T. (2004) 60 GHz transceiver circuits in SiGe bipolar technology. Digest of Technical Papers, International Solid-State Circuits Conference (ISSCC), IEEE, pp. 442–538.
21. Floyd, B.A., Reynolds, S., Pfeiffer, U. *et al.* (2005) SiGe bipolar transceiver circuits operating at 60 GHz. *IEEE Journal of Solid-State Circuits*, **40** (1), 156–167.

22. Gunnarsson, S.E., Karnfelt, C., Zirath, H. *et al.* (2005) Highly integrated 60 GHz transmitter and receiver MMICs in a GaAs pHEMT technology. *IEEE Journal of Solid-State Circuits*, **40** (11), 2174 –2186.

23. Gunnarsson, S.E., Karnfelt, C., Zirath, H. *et al.* (2007) 60 GHz single-chip front-end MMICs and systems for multi-Gb/s wireless communication. *IEEE Journal of Solid-State Circuits*, **42** (5), 1143 –1157.

24. Lee, J., Huang, Y., Chen, Y. *et al.* (2000) A low-power fully integrated 60 GHz transceiver system with OOK modulation and on-board antenna assembly. Digest of Technical Papers, International Solid-State Circuits Conference (ISSCC), IEEE, pp. 316–317.

25. Rogers, J., Plett, C. and Dai, F. (2006) *Integrated Circuit Design for High-Speed Frequency Synthesizer*, Artech House.

26. Huang, G. and Fusco, V. (2011) 60 GHz wide tuning range SiGe bipolar voltage controlled oscillator for high definition multimedia interface and wireless docking applications. *IET Proceedings, Microwaves Antennas and Propagation*, **5** (8), 934–939.

27. Dyadyuk, V., Bunton, J., Kendall, R. *et al.* (2007) Improved spectral efficiency for a multi-gigabit mm-wave communication system. IEEE European Microwave Conference, pp. 810–813.

28. Ahn, Y.-G., Kim, S.-K., Chun, J.-H. *et al.* (2009) Efficient scalable modeling of double-π equivalent circuit for on-chip spiral inductors. *IEEE Transactions on Microwave Theory and Techniques*, **57** (10), 2289–2300.

29. Huang, G., Kim, S.-K. and Kim, B.S. (2009) A wideband LNA with active balun for DVB-T application. IEEE International Symposium Circuits and Systems, pp. 421–424.

30. Gray, P.R., Hurst, P.J., Lewis, S.H. *et al.* (2010) *Analysis and Design of Analog Integrated Circuits*, 5th edn, John Wiley & Sons, Ltd.

31. Ferndahl, M. and Vickes, H. (2011) The combiner matrix balun: a transistor-based differential to single-ended module for broadband applications. Digest IEEE International Microwave Symposium.

32. Zhang, J., Huang, G. and Fusco, V. A Miniaturized V-band SiGe active receive balun MMIC with gain. *Electronics Letters* (submitted).

33. Huang, G., Kim, S.-K. and Kim, B.-S. (2009) A wideband LNA with active balun for DVB-T application. Circuits and systems. IEEE International Symposium ISCAS, pp. 421–424.

34. Ma, S., Fang, J. and Lin, F. (dated September 19, 2000) Accurate and tunable active differential phase splitters in RFIC applications. U.S. Patent 6,121,809.

35. Cetinoneri, B., Atesal, Y.A. and Rebeiz, G.M. (2011) An 8X8 Butler matrix in 0.13 μm CMOS for 5–6-GHz multibeam applications. *IEEE Transactions on Microwave Theory and Techniques*, **59** (2), 295–301.

36. Chin, T.Y., Wu, J.C., Chang, S.-F. and Chang, C.C. (2010) A V-band 8X8 CMOS Butler matrix MMIC. *IEEE Transactions on Microwave Theory and Techniques*, **58** (12), Part 1, 3538–3546.

37. Chang, C.C., Lee, R.H. and Shih, T.-Y. (2010) Design of a beam switching steering Butler matrix for phased array system. *IEEE Transactions on Antennas and Propagation*, **58** (2), 367–374.

38. Chin, T.Y., Chang, S.F., Chang, C.C. and Wu, J.C. (2008) A 24-GHz CMOS Butler matrix MMIC for multi-beam smart antenna systems. IEEE Radio Frequency Integrated Circuits Symposium.

39. Chen, C.J. and Chu, T.-H. (2010) Design of a 60-GHz substrate integrated waveguide Butler matrix – a systematic approach. *IEEE Transactions on Microwave Theory and Techniques*, **58** (7), Part 1, 1724–1733.

40. Balanis, C.A. (2005) *Antenna Theory: Analysis and Design*, John Wiley & Sons, Ltd.

41. Uzunkol, M. and Rebeiz, G.M. (2010) A low-loss 50–70 GHz SPDT switch in 90 nm CMOS. *IEEE Journal on Solid-State Circuits*, **45** (10), 2003–2007.

42. Gresham, I. and Jenkins, A. (2003) A fast switching, high isolation absorptive SPST SiGe switch for 24 GHz automotive applications. Proceedings of the European Microwave Conference, Munich, Germany, pp. 903–906.

43. Thian, M., Buchanan, N.B. and Fusco, V.F. (2011) Ultrafast low-loss 40–70 GHz SPST switch. *IEEE Microwave and Wireless Components Letters*, **21** (12), 682–684.

44. Hancock, T.M. and Rebeiz, G.M. (2005) Design and analysis of a 70-ps SiGe differential RF switch. *IEEE Transactions on Microwave Theory Techniques*, **53** (7), 2403–2410.

45. Böck, J., Schäfer, H., Aufinger, K. *et al.* (2004) SiGe bipolar technology for automotive radar applications. Proceedings of the IEEE Bipolar/BiCMOS Circuits Technology Meeting, pp. 84–87.

46. Yang, J.-R., Kim, D.-W. and Hong, S. (2008) A 24-GHz active SPST MMIC switch with InGaP/GaAs HBTs. *Microwave Optical Technology Letters*, **50** (8), 2155–2158.

47. gotMIC, SSS090A01 SPST switch [Internet accessed October 31, 2011], available from: http://www.gotmic.se/switches.html.

48. (2009) Special Issue on 60–90 GHz Communications. *IEEE Transactions on Antennas and Propagation*, **57** (10)
49. Pi, Z. and Khan, F. (2011) An introduction to millimeter-wave mobile broadband systems. *IEEE Communications Magazine*, **49** (6), 101–107.
50. Cooper, K.B., Dengler, R.J., Llombart, N. *et al.* (2011) THz imaging radar for standoff personnel screening. *IEEE Transactions on Terahertz Science Technology*, **1** (1), 169–182.
51. Appleby, R. and Anderton, R.N. (2007) Millimeter-wave and submillimeter-wave imaging for security and surveillance. *IEEE Proceedings*, **95** (8), 1683–1693.

12

Reconfigurable RF Circuits and RF-MEMS

Robert Malmqvist[1,2], Aziz Ouacha[1], Mehmet Kaynak[3], Naveed Ahsan[4] and Joachim Oberhammer[5]

[1]Swedish Defence Research Agency (FOI), Stockholm, Sweden
[2]Uppsala University, Uppsala, Sweden
[3]IHP GmbH, Frankfurt (Oder), Germany
[4]Linköping University, Linköping, Sweden
[5]KTH Royal Institute of Technology, Stockholm, Sweden

12.1 Introduction

RF communication and remote sensing (e.g. radar/radiometric) systems are facing the demands of increasing complexity, number of frequency bands and standards, as well as increased bandwidths and higher frequencies for higher data throughput, while at the same time the power consumption, the form factor of the systems and the overall cost need to be reduced. Many of today's RF, microwave and millimetre-wave circuit and system solutions are adding more and more complexity (and thus cost) to the analogue RF front-end and for certain applications it needs to meet stringent demands of high linearity/large-signal handling capability (robustness), high sensitivity and frequency selectivity. The current solution of using multiple RF front-end circuits for different frequency bands may then no longer be cost-effective for the increasing number of bands and frequency standards (see Figure 12.1a). Especially when considering the increased demands on size and cost reduction, reconfigurable and tunable (e.g. switched) circuits and systems are perhaps the only viable solution with respect to achieving further miniaturization, higher functionalities and affordability. For such reconfigurable systems, low-loss, low-power and high-linearity tuning components and strategies for realizing highly integrated (e.g. single-chip) adaptive RF circuits and front-ends are thus very much needed. Reconfigurable systems may be required, for example, when the operating frequency band and/or bandwidth need to be tuned and/or the antenna

Microwave and Millimeter Wave Circuits and Systems: Emerging Design, Technologies, and Applications,
First Edition. Edited by Apostolos Georgiadis, Hendrik Rogier, Luca Roselli, and Paolo Arcioni.
© 2013 John Wiley & Sons, Ltd. Published 2013 by John Wiley & Sons, Ltd.

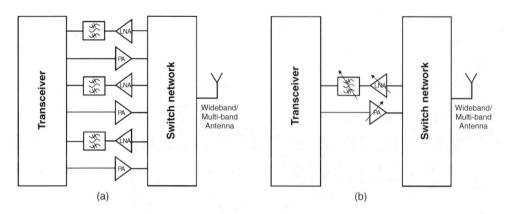

(a)   (b)

Figure 12.1 Two wideband/multiband RF front-end architectures based on using either (a) multiple (parallel) or (b) single (frequency-agile/tunable) LNAs, PAs and filters.

has to be reconfigured. Reconfigurable subcircuits (e.g. switching, matching and phase shifting RF components) may be utilized to implement tunable (multiband) LNAs, PAs and filters that can be commercially very attractive since such devices can be useful for many different frequency bands and applications (see Figure 12.1b).

To exemplify, today's wireless RF systems for point-to-point communication (radio links) can operate at many different frequencies (sub-bands) within the 5 to 40 GHz range and the use of such highly adaptive (frequency-agile) front-end components could result in a reduced system complexity and cost savings due to component reuse. Whereas most of today's RF circuits are designed for a specific (fixed) function and frequency range, a much higher degree of flexibility would be possible using highly reconfigurable circuit implementations and front-ends architectures. This chapter presents examples of reconfigurable RF circuits that have been realized using either fully transistor-based solutions or by employing RF microelectromechanical systems (RF-MEMS). In this chapter we will first address a novel approach for implementing reconfigurable RF circuitry based on the concept of programmable microwave function arrays (PROMFAs). We will then review various reconfigurable circuit designs based on the emergence of high-performance RF-MEMS switches being developed in GaAs, GaN and SiGe RFIC/MMIC process technologies. In the final section, we will present an overview of various state-of-the-art RF-MEMS-based phase shifter designs intended for electronic beamsteering antennas and phased array systems.

12.2 Reconfigurable RF Circuits – Transistor-Based Solutions

12.2.1 Programmable Microwave Function Arrays

Most of today's microwave circuits are designed for specific functions and special needs. This conventional way of designing these high-frequency circuits limits their ability to adapt to new demands and consequently it limits the flexibility of the whole system. There is a growing trend to have flexible and reconfigurable circuits –circuits that can be digitally programmed to achieve various functions based on specific needs. With the recent advances in technology, these demands can now be met. Some efforts have been made with evolvable

hardware, such as in Reference [1], where a mixer is automatically adjusted for best performance. The idea of having flexible RF systems with reconfigurable circuit blocks is becoming
more and more popular. Various solutions have already been proposed for example: a flexible
VCO based on tunable active inductor in Reference [2] and a reconfigurable bandpass filter
for multifunctional systems in Reference [3]. Some approaches to reconfiguration are based
on MEMSs (microelectromechanical systems), such as a reconfigurable power amplifier and
a flexible low-noise amplifier design reported in References [4] and [5], respectively. The
FPAA (field programmable analogue array) is another approach for the implementation of
reconfigurable circuits [6–11]. In this approach, a two-dimensional array of CABs (configurable analogue blocks) is utilized for the implementation of flexible filters. These CABs
include digitally configurable transconductance amplifiers connected through a hexagonal
interconnect network. Various filter configurations can be implemented by means of reconfigurable signal routing inside the array. This section provides the details of the so called
PROMFA circuit concept, which consists of a matrix of analogue building blocks that can be
dynamically reconfigured [12, 13]. Either each matrix element can be programmed independently or several elements can be programmed collectively to achieve a specific function.
The PROMFA circuit can therefore realize more complex functions, such as filters or
oscillators.

12.2.2   PROMFA Concept

The PROMFA concept was proposed by Ahsan *et al.* [13] and is based on an array of
generic cells. The idea is similar to an FPAA approach, but for analogue signals in the
microwave region. A block schematic overview of such a system is illustrated in Figure
12.2(a). Here the analogue building blocks (PROMFA cells) are connected together, and to
control the behaviour of each analogue cell a digital control logic is placed between them.
The individual PROMFA cells can be configured in a number of ways, for example as an
amplifier, power splitter, power combiner, router, and so on. The array can therefore realize
more complex functions, such as filters or oscillators. The PROMFA system utilizes a two-
dimensional mesh network topology, which is also a common routing topology used by
digital network routers. The PROMFA cell has several different possibilities that can be
used to connect the four different ports to each other. Because of the symmetry of the cell,
any port can be either an input or an output. The signal path can be either of pass-transistor
type (simple switch) or amplifying, depending on the biasing of the cell. This gives a
number of different configurations. A single cell is described in the block schematic in Figure 12.2(b). At each of the four ports, a transistor switch is placed to open or close the port.
The transistor size is adjusted to achieve good matching both to other cells and to a 50 Ω
system. The eight grey blocks are common source amplifier stages with transistor arrays.
The white blocks are switching transistors, used for bidirectional signal paths. By activating
different stages, the different functions can be realized (see Figure 12.2c and d).

Figure 12.3 illustrates the PROMFA subcell that includes a bidirectional amplifier stage
and a bypass (pass-transistor switch) [13]. The input signal can be routed either through a
direct path (pass-transistor) or through a bidirectional amplifier stage. The bidirectional
amplifier stage allows the possibility to amplify the signal in both directions depending on
requirement. The circuit is designed in such a way that it has symmetry that makes it a
reciprocal network.

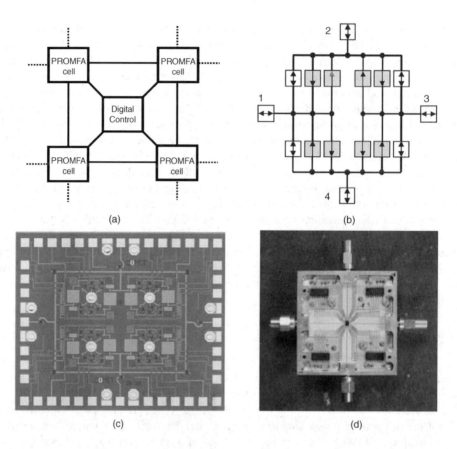

Figure 12.2 (a) Block schematic showing PROMFA cells connected in an array configuration, (b) block schematic of a single PROMFA cell, (c) a chip photograph and (d) a photograph of such a chip when mounted in a test fixture. © 2007 IEEE. Reprinted, with permission, from [13].

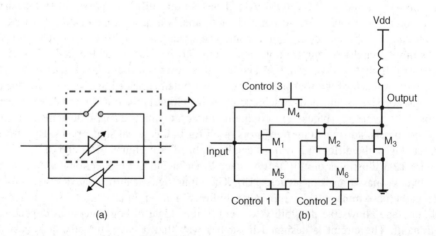

Figure 12.3 Illustration of a PROMFA subcell: (a) block schematic and (b) circuit schematic (transistor array with bypass switch). © 2007 IEEE. Reprinted, with permission, from [13].

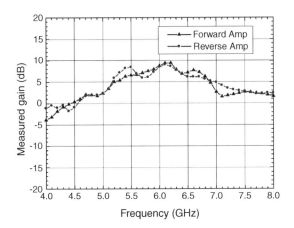

Figure 12.4 Measured gain of a reconfigurable (bidirectional) amplifier. © 2007 IEEE. Reprinted, with permission, from [13].

Figure 12.4 shows the measured gain of a bidirectional amplifier demonstrating the reciprocity of a PROMFA cell [13]. The forward and reverse amplifiers are not allowed to operate concurrently and based on requirement one of them is switched on at a time. Figure 12.5(a) shows the comparison of simulated and measured results for amplifier gain (s_{31}) and isolation (s_{13}). Figure 12.5(b) shows the variation in input and output reflection coefficients (s_{11} and s_{33}) versus frequency. The amplifier transistor array also provides three possibilities of phases. The delay through an amplifier corresponds to a phase shift Φ, which can be expressed as

$$\Phi = \pi + 2\pi n = \omega_c \tau = 2\pi f_c \tau \tag{12.1}$$

where τ is the amplifier time constant and is given by

$$\tau \propto \frac{C_L}{g_m} \tag{12.2}$$

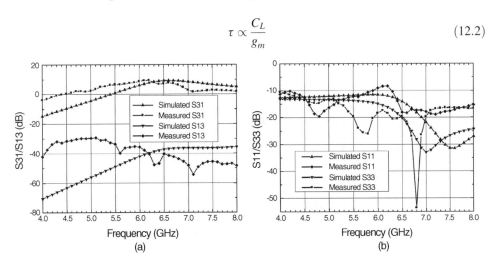

Figure 12.5 Measured and simulated response of a PROMFA amplifier circuit: (a) gain s_{31} and reverse isolation s_{13} and (b) input/output match s_{11}, s_{33}. © 2007 IEEE. Reprinted, with permission, from [13].

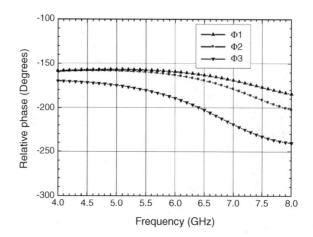

Figure 12.6 Relative phase shift of a transistor array amplifier. © 2007 IEEE. Reprinted, with permission, from [13].

This means that τ can be controlled by the variation of g_m (for a given load capacitance). In addition, g_m is dependent on the transistor width (W_T):

$$g_m \propto W_T \tag{12.3}$$

Therefore, different delays can be achieved by changing the width of the transistor. The control transistors are used to get three phase possibilities, Φ_1, Φ_2 and Φ_3. Figure 12.6 shows the relative phase shift of the transistor array amplifier with three possibilities [13]. Figure 12.7 shows the comparison of the measured and simulated phase response for the phase possibility Φ_3 [13]. A single PROMFA subcell only provides three possibilities. The results indicate that the relative phase shift has a reasonable

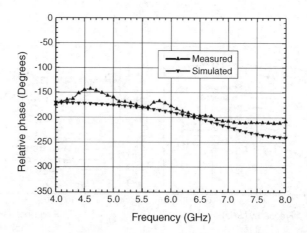

Figure 12.7 Measured and simulated phase response. © 2007 IEEE. Reprinted, with permission, from [13].

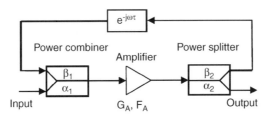

Figure 12.8 Block schematic of a recursive bandpass filter. © 2007 IEEE. Reprinted, with permission, from [13].

flat response. The concept can be extended to obtain other phase possibilities by connecting additional PROMFA cells. Each PROMFA subcell requires seven control lines that include three control lines for each transistor array and one for the bypass. Four control lines are also needed for the port switches. Therefore, in total, 32 control lines are needed for a single PROMFA cell. For the first prototype, four off-chip serial to parallel converters each controlling one subcell were used in Reference [13]. The test chip was controlled through four serial lines using a computer-based data acquisition card and Lab-View© software.

12.2.3   Design Example: Tunable Band Passfilter

Active recursive bandpass filters are interesting for microwave applications [14, 15]. A traditional recursive bandpass filter is accomplished according to the model in Figure 12.8. This model is based on positive feedback, where a certain part of the signal going through is fed back to the input through a time delay (τ). The input coupling factor is denoted by α_1 while β_1 is the feedback coupling factor. Similarly α_2 and β_2 are coupling factors of the output power splitter. The amplifier gain and noise figure are denoted by G_A and F_A respectively. By varying the delay in the feedback loop, a phase shift is accomplished that changes the centre frequency of the filter. Figure 12.9 shows the illustration of

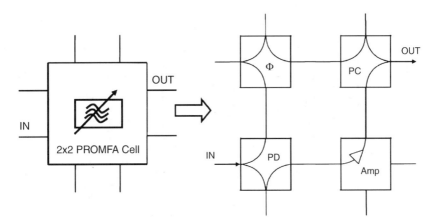

Figure 12.9 Illustration of a tunable filter using 2×2 PROMFA cells. Reproduced from [16] with permission from LiU-Tryck Linköping University Sweden.

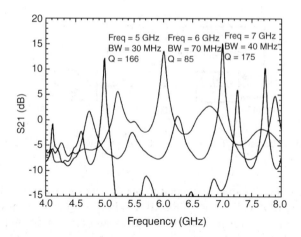

Figure 12.10 Simulated response of a tunable band pass filter based on measured results of a generic PROMFA cell. © 2007 IEEE. Reprinted, with permission, from [13].

a digitally tuned bandpass filter using a 2×2 PROMFA cell and Figure 12.10 shows the simulated response of such a filter at 5, 6 and 7 GHz [13]. The simulated results are based on the measured data of a single PROMFA cell. The dynamic reconfiguration of PROMFA cells provides flexibility. Figure 12.11 shows the configuration in which two T/R modules (transceivers) can share the same tunable filter based on a 2×2 PROMFA cell [16].

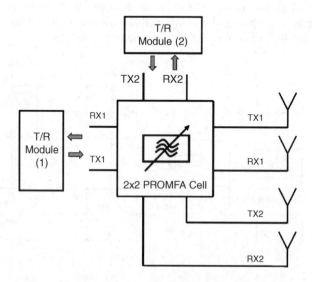

Figure 12.11 Illustration of two T/R modules sharing the same tunable filter. Reproduced from [16] with permission from LiU-Tryck Linköping University Sweden.

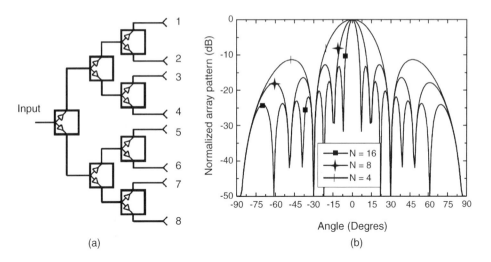

(a) (b)

Figure 12.12 (a) Block schematic of an active beam-forming network ($N = 8$) and (b) normalized array pattern for different numbers of antenna elements. © 2007 IEEE. Reprinted, with permission, from [13].

12.2.4 Design Examples: Beamforming Network, LNA and VCO

Another suitable application of the PROMFA cell is in the implementation of beamforming networks. The reciprocity of the PROMFA cell allows it to be an appropriate choice for this application. The corporate-fed arrays are mostly used for beamforming applications due to their versatility. Conventional networks make use of passive power dividers that have significant feedline losses. An active corporate feed network using PROMFA cells can be a suitable choice as it can reduce the feedline losses. Figure 12.12(a) shows the block diagram of an eight-element active corporate feed network using active PROMFA power dividers [13]. The possibility of having a bidirectional amplifier makes it possible to use the same network as a receiver with active power combiners. Another advantage of having generic cells in the beamforming network is that we can choose antenna elements depending on requirement. Especially in the case of large arrays, it might be useful to have different possibilities of active antenna elements. Figure 12.12(b) shows the normalized linear array pattern with different antenna elements. As an example, normalized array patterns at 6 GHz with $N = 16, 8, 4$ are shown in Figure 12.12(b) [13].

The PROMFA approach provides one viable solution for flexible RF systems and its CMOS implementation has certain advantages. For example, a highly integrated solution with both analogue and digital parts on the same chip may realize a low-cost PROMFA circuit. The high level of integration in modern CMOS technology also provides the possibility of large PROMFA arrays, but they are rather difficult to implement in GaAs due to its limited yield. The block diagram of a PROMFA cell utilized for CMOS implementation is shown in Figure 12.13(a) [16]. This single cell architecture is similar to Figure 12.2(b) but without any port switches; instead it has isolation switches both at the input and output of each amplifier. The single cell consists of four identical unit cells with low-noise amplifiers in each unit cell. The circuit schematic of the unit cell amplifier is shown in Figure 12.13(b) [16].

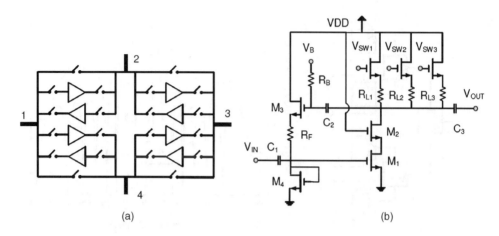

Figure 12.13 (a) Block schematic of a PROMFA cell and (b) circuit schematic of a unit cell low-noise amplifier. Reproduced from [16] with permission from LiU-Tryck Linköping University Sweden.

The forward and reverse amplifiers of a unit cell can be activated simultaneously to realize a feedback oscillator. A simple model of a feedback oscillator is shown in Figure 12.14 [16]. The oscillation frequency can be controlled by the variation in g_m. As g_m is a function of $(V_{GS} - V_{TH})$, this mechanism can be used to tune the oscillator frequency. In the proposed architecture this can be achieved by changing the bias voltage (V_B) of the feedback transistor M_3. Along with analogue tuning, the proposed CMOS PROMFA unit cell oscillator also has the possibility of step tuning. Step tuning can be achieved by changing the state of 3-bit control switches in each amplifier. With load switching, it is also possible to introduce different phase shifts that will result in different oscillation frequencies. The PROMFA cell architecture and its dynamic reconfiguration capability provide the possibility of having a series of cascaded unit oscillators (Figure 12.15) [16]. With a harmonic combination of multiple oscillators the phase noise can also be reduced [17]. In a four-unit cell oscillator, differential output signals can be obtained from adjacent ports (i.e. 1 and 2 or 3 and 4, respectively).

Figure 12.16 shows the photograph of a CMOS PROMFA chip mounted on FR4 PCB [16]. A single-unit cell has an active chip area of $0.09\,\text{mm}^2$ including coupling capacitors. A summary of chip test results for all three configurations (LNA, tunable filter and tunable oscillator) is given in Table 12.1 [16]. The chip test results can be considered promising as the tested reconfigurable RF circuit shows good performance in all three configurations.

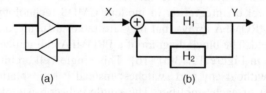

Figure 12.14 Block schematics (a and b) of a PROMFA unit cell oscillator. Reproduced from [16] with permission from LiU-Tryck Linköping University Sweden.

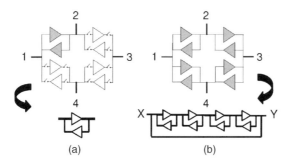

Figure 12.15 Different oscillator configurations in a PROMFA cell: (a) a single unit cell oscillator and (b) using four oscillators in a ring. Reproduced from [16] with permission from LiU-Tryck Linköping University Sweden.

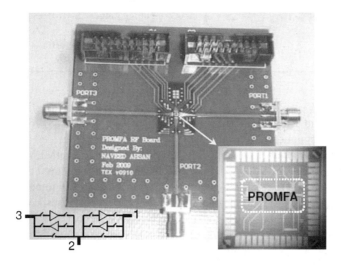

Figure 12.16 Photograph of a test fixture with a PROMFA oscillator chip mounted on a printed circuit board. Reproduced from [16] with permission from LiU-Tryck Linköping University Sweden.

12.3   Reconfigurable RF Circuits Using RF-MEMS

RF switches and switching circuits are needed in almost all wireless applications ranging from, for example, telecommunication (base stations, radio links, handsets, etc.) to RF sensing systems (e.g. radar and radiometric sensors) [18]. Switching devices are used, for example, for band selection in multiband systems, as building blocks in filter and capacitance banks, as redundancy switches in receivers and switch matrices, and as Dicke switches in radiometers. Nowadays, solid-state semiconductor switches are commonly used in such circuits but they are rather narrow band, nonlinear and have relatively high losses, especially at microwave and millimetre-wave frequencies. RF-MEMS switches are attractive options for replacing them in switching circuits for certain applications because of their superior performance for wideband operation (e.g., lower losses and DC power consumption, higher linearity and isolation). In fact, RF-MEMS-based switches and switching circuits have

Table 12.1 Summary of measured results [16] Reproduced from [16] with permission from LiU-Tryck Linköping University Sweden

Configuration	Measured results (single-unit cell)
LNA	$Gain_{typical} = 4\,dB$ $NF_{min} = 2.65$ $BW_{3\,dB} = 2.7\,GHz$ $P_{1\,dB@1\,GHz} = -8\,dBm$ $IIP_{3@1\,GHz} = +1.1\,dBm$ $P_{DC\,max} = 8.8\,mW$
Tunable filter	$Tuning\ range = 600\,MHz–1.2\,GHz$ $s_{21@1.1\,GHz} = 5.6\,dB$ $Q_{@1\,GHz} = 1.6$ $P_{DC@1\,GHz} = 13\,mW$
Tunable oscillator	$Tuning\ range = 600\,MHz–1.8\,GHz$ $Phase\ noise_{@(1.2\,GHz,\ \Delta f\,=\,1\,MHz)} = -94.7\,dBc/Hz$ $P_{OUT@1.2\,GHz} = -8\,dBm$ $P_{DC@1.2\,GHz} = 18\,mW$

attracted great interest from academia and industry, with potential applications in wireless communication and RF sensing (see, for example, References [18] to [20]). RF-MEMS switches are also commercially available as discrete packaged devices (specified from DC up to 40 GHz) from a number of vendors (e.g. switches from RadantMEMS, OMRON, Delf-MEMS, among others) [21–23]. As RF-MEMS technology has matured over the last years it has reached a level where MEMS switches have been successfully integrated into practical RF systems with proven long-term reliability. In 2007, US companies Lockheed Martin and Radant successfully demonstrated the use of a proof-of-concept X-band MEMS-based steerable antenna containing 25 000 MEMS switches integrated within a commercial air-borne radar system [24]. An ongoing trend is also with respect to the monolithic integration of RF-MEMS switches within RFIC/MMIC processes as some semiconductor IC foundries (Rockwell, OMMIC, IHP and Selex) have integrated MEMS switches on top of GaAs, SiGe and GaN substrates [25–30]. Next, we will review the current status with respect to the monolithic/hybrid integration of RF-MEMS and active devices with some examples of RF-MEMS reconfigurable amplifiers. Finally, within the last section we will review some state-of-the-art RF-MEMS phase shifter designs for electronic beamsteering.

12.3.1 Integration of RF-MEMS and Active RF Devices

Reconfigurable high-performance (low-loss/DC power and high isolation/linearity) front-ends are needed in RF systems for wireless communication, space, aerospace, defence and security applications within the microwave/mm-wave range. Martinez proposed that the realization of more 'Intelligent Microsystems' could result in a paradigm shift in RF component design since such highly flexible microsystems will have the ability to optimize their performance over broad bandwidth, overcoming the trade-offs that current RF components must endure [31]. For end-users who may have different and also changing demands, following such an approach will allow a higher degree of flexibility (compared with today's existing solutions in where the overall system performance is largely controlled by the manufacturer). Such a unique ability is expected to lead to very efficient RF systems with reduced size, weight, power and cost (due to facilitating component reuse and achieving further miniaturization compared with using conventional technologies). Due to its well-known superior RF properties the RF-MEMS has been proposed by many as an enabling technology to achieve those highly attractive benefits based on reconfigurable (adaptive) front-end architectures [18–20]. With respect to integration of RF-MEMS and active RF devices most of the successful implementations demonstrated so far have been realized mainly using hybrid integration schemes and relatively few examples of active MEMS circuit designs have been presented above 5–10 GHz [32–38]. Intelligently controlled L-band and X-band GaAs PAs using off-chip (wire-bonded) reconfigurable RF-MEMS-based output tuners were reported in References [32] and [33]. However, hybrid integration using bond-wires is not optimal for many demanding mm-wave applications (such as, for example, 60, 77, 94 and 120–140 GHz systems) requiring monolithic integration to improve on RF performance and cost-effectiveness. The X-band and Ka-band reconfigurable (switched) PAs and LNAs reported by the US foundry Rockwell Scientific remain to this date as the first real examples of a successful monolithic integration of active devices with RF-MEMS switches in a GaAs MMIC foundry process also demonstrating enhanced circuit functionalities [25–27]. Next, we will review the current status with respect to the monolithic integration of RF-MEMS in GaAs/GaN and SiGe processes exemplifying also with some realized reconfigurable MEMS-MMIC designs.

12.3.2 Monolithic Integration of RF-MEMS in GaAs/GaN MMIC Processes

RF-MEMS together with active RF circuitry have so far with a few notable exceptions [25–27] mostly been realized as hybrid circuits and mainly up to 5–10 GHz, which still leaves room for significant improvements to be made with respect to RF performance, frequency range and functionality, as well as to achieve reduced complexity and lower costs. By monolithic integration of RF-MEMS and MMICs, a higher degree of functionality would be possible, for example in active reconfigurable (multiband or switched wideband) antenna front-ends. The first examples of RF-MEMS MMIC-based Ka-band phase shifters and single-chip C-band/X-band/Ka-band MEMS reconfigurable LNAs and PAs were demonstrated in 1999–2004 using the Rockwell Scientific GaAs process [25–27] (see Table 12.2). These GaAs amplifiers showed no performance degradation after MEMS switch fabrication. Albeit, packaging aspects were not addressed in References [25] to [27]. In Europe, the MMIC foundry OMMIC (France) has recently developed an ohmic contact type of RF-MEMS switch using a GaAs process technology [28, 39] (see Figure 12.17). Figure 12.17(b) and (c) show photographs and measured s-parameter data of such a GaAs MEMS SPST switch with a circuit area of $400 \times 680 \, \mu m^2$ (including RF and DC pads). The GaAs MMIC-based MEMS switch made by OMMIC contains several parallel so-called flex slots to ensure a good contact is made between the four switch beams used and the corresponding contact bumps when the switch is pulled down (the actuation voltage of this type of MEMS switch was found to be around 30–70 V). Such a wideband microstrip-based GaAs MEMS switch circuit was reported by Malmqvist et al. [39], having less than 1 dB of transmission loss up

Table 12.2   RF-MEMS reconfigurable LNA data at X-band to K-band

Frequency (GHz)	Gain (dB)	Noise figure (dB)	$P_{1\,dB}$ (dBm)	MEMS integration	Reference
4–12 GHz	25 (at 8–12 GHz) 23 (at 4–8 GHz)	1.1	N/A	Monolithic (GaAs)	[25]
26–30 GHz	22	N/A	N/A	Monolithic (GaAs)	[27]
24 GHz	16.0/8.2 (on) $-11.5/-20.8$ (off)	2.8/4.9	15/20	Hybrid (GaAs/quartz)	[38]
20 GHz	13.9 (on) -3.4 (off)	4.1	16	Hybrid (GaAs)	[37]
20 GHz	13.7 (on) 5.9 (off)	4.2	16	Hybrid (GaAs/quartz)	[37]

to 75 GHz and better than 10 dB of isolation up to 67 GHz. A CPW-based GaAs MEMS switch design was reported by Rantakari *et al.* to have a loss of less than 1 dB with more than 12 dB of isolation up to 95 GHz, respectively [28].

To achieve multiband/multistandard operation using a single active RF device a frequency-agile low-noise (or high-power) amplifier circuit topology may be implemented utilizing reconfigurable impedance matching networks as illustrated in Figure 12.18. Some examples of RF-MEMS-based reconfigurable LNA matching networks made on silicon and quartz substrates can be found in Reference [41]. Figure 12.19 shows a chip photo (a) and measured small-signal results (b and c) of a 1-bit RF-MEMS matching network (top) and a wideband unmatched LNA break-out circuit (bottom) that both have been fabricated on the same GaAs wafer by the MMIC foundry OMMIC (the chip area is 3 mm^2) [40]. The 1-bit GaAs MEMS matching network shown in Figure 12.19 was tunable between 16 and 24 GHz (40% tuning range) with minimum 0.4 dB and 4.5 dB of in-band losses when both switches were in up-state and down-state, respectively. The highly linear RF properties of such a reconfigurable MEMS matching network were demonstrated as the *s*-parameter results remained virtually the same when measured at two different RF input power levels (-25 dBm and 9 dBm, respectively). The higher in-band loss of the GaAs MEMS matching network in the down-state was explained by a higher than expected switch contact resistance ($R_{on} = 4\,\Omega$). Simulated results of a 15–20 GHz tunable LNA, also presented in Reference [40], indicated that it should be possible to obtain a similar high in-band gain and low-noise figure at the two different centre frequencies (sub-bands), assuming that the MEMS switch contact resistance could be made sufficiently small ($R_{on} = 1$–$2\,\Omega$). Such frequency-agile RF-MEMS-based MMICs (with the active RF circuit part was made on the same chip) may be looked upon as initial steps towards realizing highly integrated (single-chip) reconfigurable active microwave/mm-wave circuits and front-ends.

In order to characterize the RF power handling capability of some OMMIC GaAs MEMS switches, both hot and cold switching tests done at 1 and 4 GHz, respectively, were reported by Rantakari *et al.* [28]. Figure 12.20(a) shows a photograph of a GaAs MEMS switch after such RF breakdown tests were carried out at FOI (Swedish Defence Research Agency). Within the presented power stress tests the RF power was inserted at either the anchor end

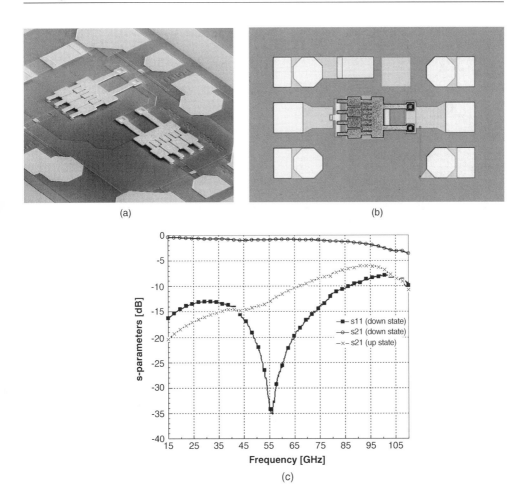

Figure 12.17 (a) Close-up photograph of two ohmic contact-based RF-MEMS switches fabricated by the GaAs MMIC foundry OMMIC. Reproduced with permission from Ommic SAS and (b, c) chip photograph and measured *s*-parameter data of a GaAs MMIC-based RF-MEMS SPST switch, respectively. © 2011 IEEE. Reprinted, with permission, from [39].

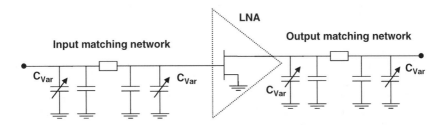

Figure 12.18 Circuit schematic of a frequency-agile LNA using reconfigurable impedance matching networks. Reproduced from [40] with permission from EuMA.

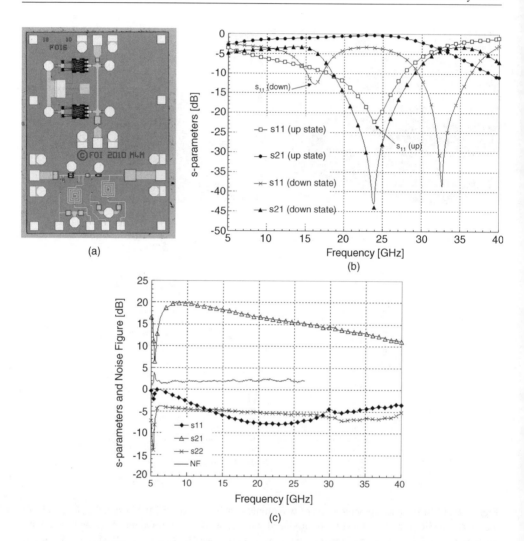

Figure 12.19 (a) One-bit RF-MEMS reconfigurable matching network (top) and a wideband unmatched LNA circuit (bottom) fabricated on the same GaAs wafer and (b, c) measured small-signal data of a GaAs MEMS matching network and unmatched LNA, respectively. Reproduced from [40] with permission from EuMA.

of the MEMS switch (to the left) or from the opposite side. In cold-switching conditions, the RF power was only applied when the MEMS SPST switch was fixed in either the 'ON' or 'OFF' states, and in hot switching conditions the MEMS SPST switch was switched once between the on and off states with RF power applied at the same time. By gradually increasing the RF power in 1 dB steps starting from the 25 dBm level, the gain compression and power handling capabilities were measured. The normalized loss compression curve of one such MEMS switch tested in the on state is shown in Figure 12.20(b). The measured loss compression was also very uniform for all switches tested in the on state (i.e. less than

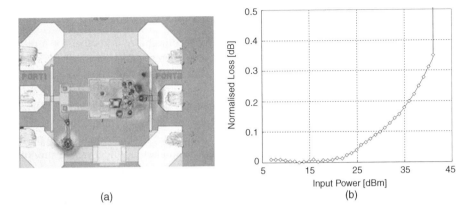

(a) (b)

Figure 12.20 (a) Photograph of a GaAs RF-MEMS switch after an RF power breakdown test and (b) measured normalized switch (on state) transmission loss as function of input power. Based on Reference [28].

0.4 dB of increased insertion loss when the RF input power level was varied from 10 to 41 dBm). The cold switching characteristics of 10 GaAs-based RF MEMS SPST switches were tested and all the switches were able to handle more than 40 dBm (10 W) of RF input power [28]. Five of the switches were tested in the on state and five in the off state. The breakdown power level was roughly 41 dBm (12.5 W) in all cases (as also is shown in Figure 12.20). Rantakari *et al.* [28] also reported the hot-switching test results of seven GaAs MEMS SPST switches and for four of those the breakdown power level (i.e. when the switches were physically destroyed) varied from 36 to 41 dBm of RF input power (4–12 W). The remaining three tested GaAs MEMS switches were stuck in the on state at an input power level of 32–34 dBm (1–2 W).

In spite of the recent remarkable progress in microwave and millimetre-wave applications of wideband-gap (WBG) devices, no results have been reported so far on a monolithically integrated reconfigurable RF-MEMS-based switched amplifier (LNA/PA) on GaN/SiC substrates. This is also in spite of the acknowledged superior RF properties of such enabling technologies (e.g., in terms of very high power efficiency and low-loss low-power switching) that should make them ideal candidates for use in future intelligent microsystem (IMS) applications. Crispoldi *et al.* reported on a successful demonstration of RF MEMS switches and active RF circuit function blocks (i.e. transistors) fabricated on the same GaN wafer [30]. Experimental data of such GaN-based RF MEMS switches and fixed (i.e. not reconfigurable) amplifier circuits showed capabilities for monolithic integration with promising results. It is evident that a successful combination of GaN MMIC amplifiers and high-power MEMS switches would enable new types of RF components with unprecedented performance. An RF-MEMS based triband (1.4, 2.5 and 3.6 GHz) GaN PA realized as a hybrid circuit was recently reported by Liu *et al.* exhibiting a high efficiency of more than 46% while delivering an output power of more than 4.3 W [42]. Zine-El-Abidine *et al.* presented a monolithically integrated GaN-based tunable MEMS capacitor with a tuning range of 60% at 5 GHz that may be employed for the implementation of GaN on-chip tunable matching networks and reconfigurable amplifiers [43].

12.3.3   Monolithic Integration of RF-MEMS in SiGeBiCMOS Process

The latest developments in SiGe technologies have become more attractive for mm-wave frequency applications due to the high performance of hetero-junction bipolar transistors (HBTs). Meanwhile, SiGe HBTs show f_t and f_{max} values of 300 and 500 GHz, respectively [44]. There is a growing interest in providing fully integrated solutions for 60 GHz WLAN, 77 GHz radar, 71–86 GHz point-to-point communication, 94 GHz imaging and 122 GHz sensors applications. Such applications may benefit from using highly adaptive systems (e.g. switched) and can be realized using RF-MEMS switches due to their superior RF performance and high linearity. Furthermore, for such high-frequency applications, a monolithic integration of the switch with a high-performance CMOS or BiCMOS platform would be advantageous over any heterogeneous integration with the basic IC process because it provides the shortest connection paths between switch and circuitry resulting in the lowest parasitics. Figure 12.21(a) illustrates the embedded RF-MEMS switch integration in IHPs 0.25 μm SiGe:C BiCMOS process [29]. The capacitive switch is built between the Metal2 and Metal3 layers. High-voltage electrodes are formed using Metal1 while Metal2 is used as the RF signal line. A thin Si_3N_4/TiN layer stack, which is part of the BiCMOS metal–insulator–metal (MIM) capacitor, forms the switch contact region. This configuration creates a height difference between the high-voltage electrode and the signal line, which serves simultaneously as a stopping layer for the switch membrane. The membrane is realized using a stress compensated Ti/TiN/AlCu/Ti/TiN Metal3 layer stack. An SEM view of the switch is shown in Figure 12.21(b) [29]. Figure 12.22 shows the laser Doppler vibrometer (LDV) measurement results of the switch as the total z displacement of the contact region versus time [29]. The LDV measurements provide the real-time speed and displacement of a single node on the movable membrane, which helps to characterize the mechanical behaviour of the RF-MEMS switch.

A z displacement of the switch membrane (gap between the signal line and membrane) of 1.2 μm can be observed from the same figure. Furthermore, the LDV measurement also provides the switch-on and switch-off times of the embedded RF-MEMS switch. The switch-on time is 5 μs under 30 V electrode voltage and the switch-off time (release

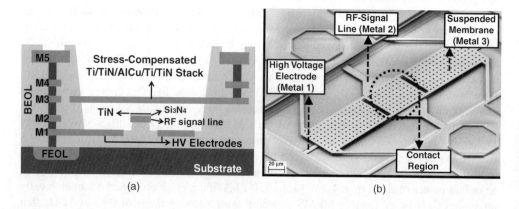

(a)   (b)

Figure 12.21   Embedded RF-MEMS switch within a SiGe foundry process technology: (a) schematic cross-section view and (b) SEM view. © 2009 IEEE. Reprinted, with permission, from [29].

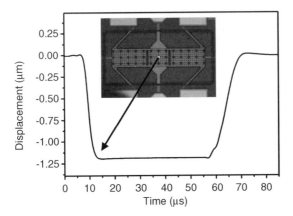

Figure 12.22 Vertical displacement of one selected node (specified by the green point) at 30 V pulse actuation measured by a laser Doppler vibrometer. © 2009 IEEE. Reprinted, with permission, from [29].

time) is less than 10 µs (Figure 12.22) [29]. The pull-down voltage of the switch varies between 18 V and 22 V over the wafer due to a remaining Metal3 layer stress profile, resulting in a different initial gap between the high-voltage electrode and the membrane. However, 30 V is applied to establish a strong and homogeneous contact over the wafer. The off state capacitance (membrane up) varies between 25 fF and 30 fF, while the on state (membrane down) ranges between 210 fF and 250 fF, providing an off/on state capacitance ratio of at least 1:8, but typically 1:10. The *s*-parameter measurement results are given in Figure 12.23 for different sites of the wafer at 30 V bias voltage. The

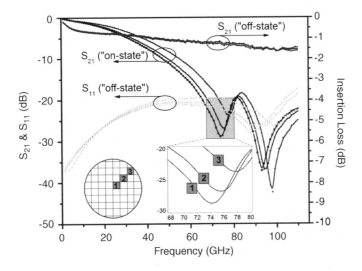

Figure 12.23 Measured *s*-parameter data for switches made at three different locations on a wafer. Insertion shows that the resonance frequency shifts from 74 GHz to 78 GHz due to 'on-state' capacitance variation. © 2009 IEEE. Reprinted, with permission, from [29].

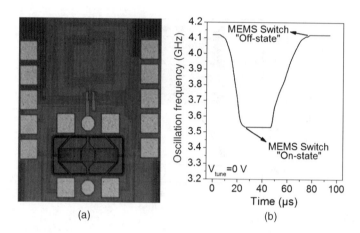

(a)   (b)

Figure 12.24   (a) Photograph of a fabricated dual-band MEMS switched LC-VCO and (b) variation of the oscillator frequency at a 30 V step excitation with a 50 μs pulse width. The peak-to-peak RF signal amplitude on the signal line is 5 V. © 2009 IEEE. Reprinted, with permission, from [29].

insertion loss of the switch in the 1–110 GHz frequency range is below 1.65 dB. The isolation for frequencies between 60 GHz and 100 GHz is better than 15 dB. S_{11} of the switch, given in Figure 12.23 for the off state, demonstrates a good input matching to 50 Ω. The resonance characteristic of the switch differs only by ±2 GHz over an 8 inch wafer, as can also be seen from Figure 12.23 [29].

An SiGe embedded RF-MEMS switch is very promising for reconfigurable RFICs. Figure 12.24(a) shows a die photo of a dual-band BiCMOS VCO where frequency switching is achieved using such an RF-MEMS switch [29]. The switch is connected to the resonance node of the oscillator, therefore changing the effective load on the resonance node. The high-capacitance ratio of 1:10 allows the oscillation frequency to be switched at 0 V tuning voltage from 3.55 GHz to 4.15 GHz (see Figure 12.24b).

Figure 12.25 shows the VCO tuning behaviour [29]. The VCO frequency can be switched to two different frequencies using the embedded RF-MEMS switch. The output frequency can also be tuned around the switched frequencies using a varactor, available in the BiCMOS process. Figure 12.26 shows the output response of the VCO picked up by a signal source analyser [29]. It shows the output response after different numbers of hot-switching cycles under a 5 V peak-to-peak RF signal on the signal line of the switch. As can be seen from Figure 12.26, no significant performance degradation can be observed, even after 10 billion switching cycles.

12.3.4   Design Example: RF-MEMS Reconfigurable LNA

RF-MEMS-based LNAs and PAs have, so far, with a few notable exceptions, mainly been realized up to 5–10 GHz (see. for example, References [25] to [27], [33], [35], [37], and [38]). Table 12.2 presents an overview of some published RF-MEMS reconfigurable low-noise amplifier data within the 10 to 30 GHz frequency range. GaAs based X-band and Ka-band reconfigurable LNA and PA MMICs with RF-MEMS switches integrated on-chip were reported by Hacker *et al.* [27]. Albeit, the noise figure and linearity of the

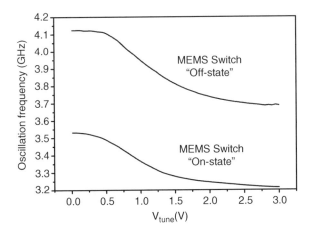

Figure 12.25 Tuning characteristics of a dual-band SiGe MEMS LC-VCO. Continuous frequency tuning is achieved by using an MOS varactor for both states. © 2009 IEEE. Reprinted, with permission, from [29].

single-chip GaAs MEMS switched LNA circuits were not presented in this case. A K-band (18–26 GHz) RF-MEMS reconfigurable and multifunctional (low-NF and high-linearity) LNA hybrid circuit has been reported by Malmqvist *et al.* [38] and consists of a dual-LNA MMIC (optimized for low noise and high linearity, respectively) wire-bonded to a MEMS SPDT switch network made on a quartz substrate. A circuit schematic and a photograph of the K-band MEMS LNA are shown in Figure 12.27(a) and (b), respectively (total circuit size is $3 \times 5\,\text{mm}^2$) [38]. Such reconfigurable MEMS-based LNA circuitry may be integrated with, for example, different down-converter stages (selected for the lowest possible NF or the highest possible dynamic range) in adaptive wideband or multiband front-ends.

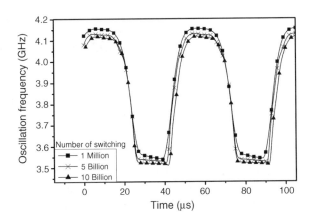

Figure 12.26 Reliability test of the switch. Output frequency response of SiGe MEMS LC-VCO after a certain number of switching cycles. © 2009 IEEE. Reprinted, with permission, from [29].

Reconfigurable multifunctional LNA realised using a dual-LNA
MMIC combined with an RF-MEMS SPDT switch network

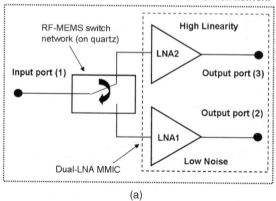

(a)

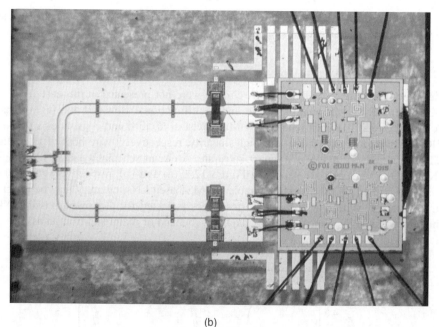

(b)

Figure 12.27 A K-band RF-MEMS switched reconfigurable dual-LNA hybrid circuit: (a) block circuit schematic and (b) photograph. Reproduced from [38]. With permission from Hindawi Publishing Corporation.

Figures 12.28 and 12.29 show measured (three-port) *s*-parameter, noise figure and 1 dB compression point of the K-band reconfigurable dual-path LNA hybrid circuit, shown in Figure 12.27(b), when the RF-MEMS switches used within the two branches were switched on and off, respectively [38]. When the high-gain and low-NF (LNA1) path was activated (MEMS switched on) the other low-gain and high-linearity (LNA2) path was deactivated (MEMS switched off) and vice versa. Here, only one of the two individual (fixed) LNA circuits (LNA1 or LNA2) was DC biased at the time to minimize the overall power

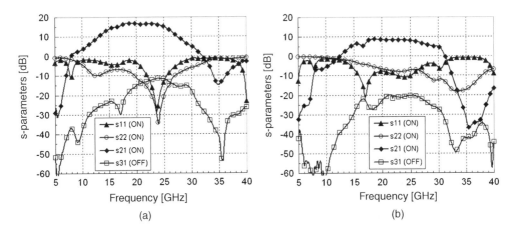

Figure 12.28 Measured *s*-parameters of a K-band RF-MEMS reconfigurable dual-LNA circuit: (a) high-gain LNA path switched on (low-gain off) and (b) low-gain LNA path switched on (high-gain off). Reproduced from [38]. With permission from Hindawi Publishing Corporation.

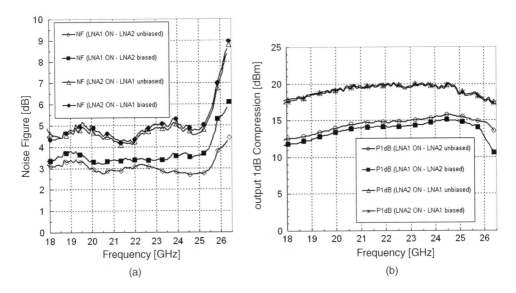

Figure 12.29 (a) Measured noise figure and (b) 1 dB compression point of a K-band RF-MEMS reconfigurable dual-LNA circuit. Reproduced from [38]. With permission from Hindawi Publishing Corporation.

consumption. Compared with the two (fixed) LNA circuits (i.e. LNA1 and LNA2 data), the MEMS switched LNA paths showed a minimum 0.6 dB higher noise figure and 1 dB lower gain, respectively, as a result of the combined losses of the MEMS switch network and bond-wires used as interconnects. The in-band gain of one of the two MEMS LNA paths was reduced by 25–30 dB when the other path was deactivated. As expected, the linearity of the tested MEMS switched dual-LNA circuit was found to be limited by the active devices used and not by the RF-MEMS switching elements. The K-band reconfigurable dual-LNA circuit presented in Reference [38] was believed to be the first reported RF-MEMS-based multifunctional (low-noise and high-linearity) amplifier with $NF = 3$–$5\,dB$ and $P1dB = 12$–$20\,dBm$ at 18–25 GHz.

12.3.5 RF-MEMS-Based Phase Shifters for Electronic Beam Steering

RF components based on microelectromechanical systems (MEMSs) have attracted significant attention since the first RF MEMS switch was introduced in 1991 [45]. Since then, RF-MEMS components have shown much lower insertion loss, higher linearity over a large bandwidth and lower power consumption as compared to solid-state technology [18, 46]. Especially in radar systems based on phased antenna arrays, phase shifters are widely employed. Ferrite-based phase shifters have good performance, but cannot be easily integrated and are more expensive in fabrication. RF-MEMS devices, mainly switches, switched or tuned capacitive loading of a transmission line, have been extensively researched, as they have near-ideal performance, are very compact, can be integrated on microwave-grade substrate, can be fabricated by standard, highly parallel semiconductor fabrication processes and are characterized by relative ease of circuit design [18]. So far, three main types of MEMS phase shifters have been introduced: (a) MEMS-switched true-time delay-line (TTD) phase shifter networks, (b) distributed MEMS transmission-line (DMTL) phase shifters [47] and (c) loading of a three-dimensional micromachined transmission line by a movable dielectric block [48]. In the following, these types are discussed and examples are given, and a table summarizing the key performance of selected phase shifters of these types is presented.

(a) RF-MEMS switched true-time delay-line phase shifter networks
  For switched line (SL) phase shifters, low-loss and high-isolation RF-MEMS switches are employed for routing the signal over transmission lines of different lengths, thus resulting in a true-time delay between the input and the output depending on the selected signal path. Different arrangements are possible:
  (i) Various phase-shifting sections in a sequential arrangement, as shown in Figure 12.30(a), realized by single-pole double-throw (SPDT) switches.
  (ii) Various phase-shifting sections in a parallel arrangement, as shown in Figure 12.30(b), requiring single-pole multiple-throw (SPxT) switches.

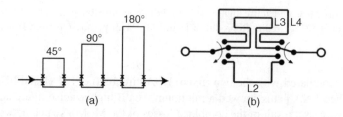

Figure 12.30 RF-MEMS-switched true-time delay-line networks implemented using (a) sequentially switched branches or (b) parallel-switched branches, respectively.

(iii) A combination of (i) and (ii).

Since RF-MEMS switches are used to switch between different paths, this kind of phase shifter inherits all advantages of RF-MEMS switches, resulting in excellent performance. However, they are less suitable for millimetre-wave frequencies, including W-band, because the performance of multiple switches and the necessary lengths of the transmission line degrade the performance and also because it is difficult to design SPxT switches at such frequencies [48]. Stehle *et al.* [49] presented a phase shifter consisting of a 45° loaded-line (LL) phase shifter element and a 90° as well as a 180° switched-line phase shifter element resulting in a 3-bit RF MEMS phase shifter. For the complete device, the return loss was better than $-12\,dB$ and the insertion loss was $-5.7\,dB$ at 76.5 GHz. Design equations for an RF-MEMS switched-line phase shifter comprised of impedance-matched slow-wave unit cells with the optimization goal to maximize the figure-of-merit $\Delta\varphi/dB$ are presented by Lakshminarayanan *et al.* [50]. To verify the equations, 1-bit phase shifters were implemented by cascading a number of unit cells corresponding to various maximum design frequencies. The phase shifter with the maximum frequency of 110 GHz showed an insertion loss of about 2.65 dB, measured $\Delta\varphi/dB$ of 150°/dB and return loss below $-19\,dB$.

(b) Distributed MEMS transmission-line phase shifters (DMTL)

The distributed MEMS transmission-line phase shifter concept is based on a periodically loaded transmission line with capacitive MEMS bridges to vary the line capacitance. Thus, the propagation coefficient of the line is altered and the signal phase between the input and the output of the phase shifter network is changed. However, the line impedance is also changed in addition to varying the capacitance of the line. This imposes a limitation on the maximum usable capacitance ratio, in addition to the ones required by the switch design parameters, in a compromise to the total acceptable mismatch. The capacitance ratio is reported to be limited to approximately 1.3–1.6 [47]. In general, DMTL phase shifters have excellent performance in the millimetre-wave regime in comparison to the TTD phase shifter and are therefore the MEMS phase shifter technology of preferred choice at these high frequencies.

In addition to analogue-mode MEMS-tunable distributed capacitor elements, it is also possible to load the transmission line digitally by employing fixed capacitances, which can be switched by, typically capacitive, MEMS switches to change the load of the line. This alternative method allows for better reproducibility, even though fewer phase shifted states are achieved, and provides the capability of increasing the capacitance ratio, which is more suitable in a millimetre-wave regime [47]. Already in 2000, Barker and Rebeiz [51] presented a design and optimization of distributed MEMS transmission-line (DMTL) phase shifters for the U-band and W-band with analogue tuning capability of the individual stages, fabricated on quartz substrate. The W-band DMTL phase shifter consisted of 48-bridge DMTL elements with pull-down voltage of just over 26 V and a corresponding capacitance ratio of 1.15. Measured phase shift per decibel loss was 70°/dB from 75 to 110 GHz. The average measured insertion loss was $-2.5\,dB$ and return loss was $-11\,dB$ at 94 GHz. In 2003, Hung, Dussopt and Rebeiz [52] demonstrated a low-loss distributed 2-bit W-band MEMS phase shifter on glass substrate, as shown in Figure 12.31. Each unit cell of this phase shifter consisted of a MEMS bridge and the sum of the two MAM capacitors, which were fabricated using the cross-over

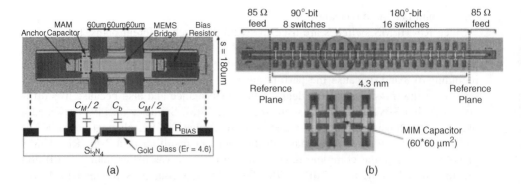

Figure 12.31 A low-loss distributed 2-bit W-band MEMS phase shifter: (a) a single cell with its corresponding profile and (b) the complete 2-bit W-band DMTL phase shifter implemented using 24 MEMS switches. Reproduced from [51]. With permission from EuMA.

between the MEMS bridge and the CPW ground plane. Since the simulated phase shift of the unit cell was about 1.2° at 80 GHz, a 90° section with 8 switches and a 180° section with 16 switches were cascaded. Phase shifts of 0°, 89.3°, 180.1° and 272° were measured at 81 GHz, which were close to the designed frequency and phase shifts. The return loss was better than −11 dB, the average insertion loss was −2.2 dB and the phase error was equal to ±2°. The highest measured in-band insertion loss was equal to −2.9 dB at 94 GHz.

(c) MEMS tunable dielectric-block loaded-line phase shifter

Both the TTD network and the DMTL MEMS phase shifter concepts have the disadvantage that the capacitive MEMS switches or tunable capacitor, employed in both concepts, are composed of thin metal bridges that cannot handle large induced current densities at high RF power because of limited heat conductivity to the substrate due to their suspension above the substrate [53–55]. The thin bridges, typically gold membranes, are subject to losing their elastic behaviour drastically, even at slightly elevated temperatures of just around 80 °C, resulting in decreased reliability. These issues result in decreased repeatability of the device performance, and in the worst case even in permanent plastic deformation, buckling or even melting of the thin metal layer. In addition, dielectric charging of the isolation layers of capacitive switches, typically employed both for the TTD and the DMTL concept, has been extensively investigated but is still an unresolved problem of capacitive MEMS switches [56].

A new type of MEMS phase shifters, overcoming these power-handling and reliability limitations, has been introduced by Somjit *et al.* in 2009 [48]. This concept is based on tuning the loading of a three-dimensional micromachined transmission line by a dielectric block placed on top of the line, which is moved by MEMS actuators, as shown in Figure 12.32. The blocks of the single stages of this phase shifter concept consist of a high-resistivity monocrystalline silicon block placed upon and loading a three-dimensional high-impedance micromachined CPW line. The relative phase shift is achieved by moving the dielectric block vertically above the transmission line by electrostatic actuation, which results in different propagation constants

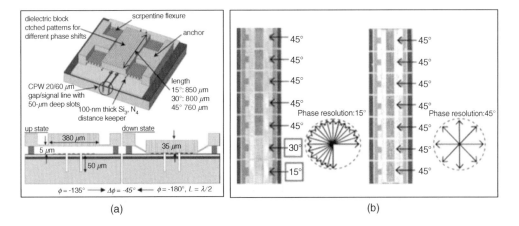

(a)   (b)

Figure 12.32 Binary-coded 4.25-bit W-band monocrystalline silicon MEMS multistage dielectric-block phase shifters: (a) working principle of a single stage of the phase shifter and (b) microscopic pictures of fabricated seven-stage phase shifters. © 2009 IEEE. Reprinted, with permission, from [48].

of the microwave signal depending on the vertical position of the dielectric block. Sequential multistages of these movable dielectric blocks result in a linearly coded phase shifter. As the periodic pattern etched into the dielectric block for releasing the movable structures after the fabrication can be varied by mask design, it is possible to tailor-make the macroscopically effective dielectric constant of each individual block. Thus, each stage can be designed for an individual phase shift. This allows a pseudo-binary coded phase shifter circuit to be designed.

Figure 12.32(b) shows such a binary-coded phase shifter consisting of $15°$, $30°$ and $5 \times 45°$ phase shifting stages and resulting in an effective phase shifter of 4.25 bit for a total of seven stages, along with a linearly coded phase shifter with $7 \times 45°$ phase-shifting stages, the latter resulting in 3 bit for the same number of stages. For the binary-coded phase shifter, at the design frequency of 75 GHz, the maximum return and insertion loss were -17 and -3.5 dB, respectively, which corresponded to a loss of -0.82 dB/bit and a phase shift efficiency of $71.1°$/dB and $490.02°$/cm. This phase-shifter type, in contrast to MEMS TTD phase shifters based on optimizing the switching network for a specific band, are extremely wideband, as the maximum insertion loss and return loss are -4.0 and -12 dB, respectively, for the whole W-band (75–110 GHz). This phase-shifter type was found to be very linear with a third-order input intercept point IIP3 of 49.3 dB [48]. Also, as this concept does not employ any thin metallic bridges, which limit the current and thus the power handling of conventional MEMS TTD and DMTL phase shifters, the power handling is effectively only limited by the heat-sink capability of the transmission line itself, that is not by the MEMS part. A comparable study found that even at 40 dBm of RF input power at 75 GHz, the hottest spot on this phase-shifter design has only increased by 30 °C, which is 10–20 times less than for conventional MEMS TTD and DMTL phase-shifter designs [57]. To summarize this section on RF-MEMS phase shifters, a comparison of several selected W-band phase shifters is provided in Table 12.3. Most of the data is reproduced from Reference [48], with some additions.

Table 12.3 Comparison of W-band RF MEMS phase shifter results Based on Reference [48]

Reference	[49]	[57]	[58]	[50]	[59]	[51]	[52]	[48]
Type	cLL	SL	SL (reflect line)	TTD	DMTL	DMTL	DMTL	LL (dielectric block)
Substrate	Si	Quartz	Quartz	Quartz	Glass	Quartz	Glass	Si
Frequency f_n (GHz)	76	90	80	110	78	94	81	75
Number of bits	3	1	2	1	3	Analogue	2	4.25
Configuration possibilities	8	2	4	2	8	Analogue	4	19
Maximum $\Delta\varphi$ at f_n (°)	315	180	282	410	315	170	272	270
Maximum IL at f_n (dB)	5.8	2.5	6.1	2.6	3.2	2.5	2.2	3.5
Maximum IL/bit at f_n (dB/bit)	1.9	2.5	3.1	2.6	1.1	—	1.1	0.8
Maximum $\Delta\varphi$/loss at f_n (°/dB)	55.3	85.7	70.5	150.0	95.8	70.0	—	71.0
Maximum $\Delta\varphi$/loss (°/dB)	—	32.8	—	—	83.3	—	—	98.3
Minimum RL at f_n (dB)	−12	−12	−9	−19	−12	−11	−11	−17
Maximum IL at W-band (dB)	−8	−6	−7	—	−6	−3	−3	−4
Minimum RL at W-band (dB)	−7	−3	−9	−19	−9	−10	−11	−12

12.4 Conclusions

This chapter has presented some examples of realized reconfigurable RF/microwave circuits that could enable more cost-effective RF system solutions with flexible performance. A novel approach for implementing reconfigurable RF circuitry based on the concept of programmable microwave function arrays (PROMFAs) was reviewed with some included design examples (tunable filter, LNA and VCO). Due to its superior RF properties RF-MEMS is being regarded as one of the key enabling technologies for realizing highly adaptive wideband/multiband front-ends, especially at higher microwave and millimetre-wave frequencies. A comparison of various RF-MEMS-based phase shifter circuits was presented with respect to demonstrated state-of-the-art performance and reliability issues (including small-signal data, power handling and temperature sensitivity).

Furthermore, the emergence of high-performance RF-MEMS switches being developed in GaAs, GaN and SiGe RFIC/MMIC process technologies can result in better performance/lower cost via monolithic integration of active RF devices and RF-MEMS switch circuits. A comparison was presented between reconfigurable RF-MEMS LNAs (within the 10 to 30 GHz range) implemented as single-chip or multichip circuits, respectively. Such highly integrated reconfigurable active RF circuitry could open up new possibilities/approaches for how wideband/multiband front-end architectures may be implemented in future low-cost adaptive RF systems.

References

1. Kasai, Y., Sakanashi, H., Murakawa, M. *et al.* (2000) Initial evaluation of an evolvable microwave circuit. Evolvable Systems: From Biology to Hardware, Lecture Notes in Computer Science 1801 (Proceedings of ICES 2000), Springer Verlag, pp. 103–112.
2. Mukhopadhyay, R., Park, Y., Sen, P. *et al.* (2005) Reconfigurable RFICs in Si-based technologies for a compact intelligent RF front end. *IEEE Transactions on Microwave Theory and Techniques,* **53** (1), 81–93.
3. Fathelbab, W.M. and Steer, M.B. (2005) A reconfigurable bandpass filter for RF/microwave multifunctional systems. *IEEE Transactions on Microwave Theory and Techniques,* **53** (3), 1111–1116.
4. Qiao, D., Molfino, R., Lardizabal, S.M. *et al.* (2005) An intelligently controlled RF power amplifier with a reconfigurable MEMS varactor tuner. *IEEE Transactions on Microwave Theory and Techniques,* **53** (3), 1089–1095.
5. Malmqvist, R., Gustafsson, A., Nilsson, T. *et al.* (2006) RF MEMS and GaAs based reconfigurable RF front end components for wide-band multi-functional phased arrays. European Microwave Conference (EuMA 2006), pp. 1798–1801.
6. Becker, J., Henrici, F., Trendelenburg, S. *et al.* (2008) A continuous-time hexagonal field programmable analog array in 0.13 μm CMOS with 186 MHz GBW. International Solid State Circuits Conference (ISSCC), San Francisco, CA.
7. Becker, J., Henrici, F., Trendelenburg, S. *et al.* (2008) A hexagonal field programmable analog array consisting of 55 digitally tunable OTAs. IEEE International Symposium on Circuits and Systems (ISCAS).
8. Becker, J., Trendelenburg, S., Henrici, F. and Manoli, Y. (2008) A rapid prototyping environment for high-speed reconfigurable analog signal processing. *Reconfigurable Architectures Workshop (RAW),* Miami, FL.
9. Henrici, F., Becker, J., Buhmann, A. *et al.* (2007) A continuous-time field programmable analog array using parasitic capacitance Gm-C filters. IEEE International Symposium on Circuits and Systems (ISCAS), New Orleans, LA.
10. Becker, J. and Manoli, Y. (2004) A continuous-time field programmable analog array (FPAA) consisting of digitally reconfigurable GM-cells. IEEE International Symposium on Circuits and Systems (ISCAS), Vancouver.

11. Becker, J., Trendelenburg, S., Henrici, F. and Manoli, Y. (2007) Synthesis of analog filters on an evolvable hardware platform using a genetic algorithm. Genetic and Evolutionary Computation Conference (GECCO), London, UK.

12. Samuelsson, C., Ouacha, A., Ahsan, N. and Boman, T. (2006) Programmable microwave function array, PROMFA. Asia-Pacific Microwave Conference 2006 (APMC 2006), Yokohama, Japan, December 2006, pp. 1787–1790.

13. Ahsan, N., Ouacha, A., Samuelsson, C. and Boman, T. (2007) Applications of programmable microwave function array (PROMFA). European Conference on Circuit Theory and Design (ECCTD 2007), Seville, Spain, August 2007, pp. 167–167.

14. Andersson, S., Caputa, P. and Svensson, C. (2002) A tuned, inductorless, recursive filter LNA in CMOS. 28th European Solid State Circuit Conference(ESSCIRC), Firenze, Italy, September 24–26, 2002.

15. Malmqvist, R., Danestig, M., Rudner, S. and Svensson, C. (1999) Some limiting factors for the noise optimization of recursive active microwave integrated filters. *Microwave and Optical Technology Letters*, **22** (3), 151–157.

16. Ahsan, N. Reconfigurable and Broadband Circuits for Flexible RF Front Ends, ISBN: 978-91-7393-605-7, ISSN: 0345-7524, PhD Thesis, Linköping Studies in Science and Technology.

17. Kim, J.J. and Kim, B. (2000) A low-phase noise CMOS LC oscillator with a ring structure. IEEE International Solid-State Circuits Conference (ISSCC).

18. Rebeiz, G. (2003) *RF MEMS Theory, Design and Technology*, John Wiley & Sons, Inc., New York.

19. Fischer, G. *et al.* (2003) RF-MEMS and SiC/GaN as enabling technologies for a reconfigurable multi-band/multistandard radio. *Bell Labs Technical Journal*, **7** (3), 169–189, Published by Wiley Periodicals, Inc.

20. Rebeiz, G.M. and Muldavin, J.B. (2001) RF MEMS switches and switch circuits. *IEEE Microwave Magazine*, **2** (4), 59–71.

21. http://www.radantmems.com.

22. http://components.omron.eu/en/misc/search/default.html?q=RF+MEMS+switch.

23. http://www.delfmems.com.

24. Maciel, J.J. *et al.* (2007) MEMS electronically steerable antennas for fire control radars. *IEEE Aerospace and Electronic Systems Magazine*, **22** (11), 17–20.

25. Sovero, E.A. *et al.* (1999) Monolithic GaAs PHEMT MMICs integrated with high performance MEMS micro-relays. International Microwave and Optoelectronics Conference, IMOC '99, Rio de Janeiro, Brazil, August 1999, pp. 257–260.

26. Kim, M., Hacker, J.B., Mihailovich, R.E. and DeNatale, J.F. (2001) A monolithic MEMS switched dual-path power amplifier. *IEEE Microwave and Wireless Components Letters*, **11** (7), 285–286.

27. Hacker, J.B., Kim, M., Mihailovich, R.E. and DeNatale, J.F. (2004) Monolithic GaAs PHEMT MMICs integrated with RF MEMS switches. IEEE 2004 CSIC Digest, 229–232.

28. Rantakari, P., Malmqvist, R., Samuelsson, C. *et al.* (2011) Wide-band RF MEMS switches and switching circuits using a gallium arsenide monolithic microwave integrated circuits foundry process technology. *IET Microwaves, Antennas and Propagation*, **5** (8), 948–955.

29. Kaynak, M., Ehwald, K.E., Drews, J. *et al.* (2009) BEOL embedded RF-MEMS switch for mm-wave applications. IEEE 2009 International Electronic Devices Meeting (IEDM), December 2009, pp. 797–800.

30. Crispoldi, F., Pantellini, A., Lavanga, S. *et al.* (2011) New fabrication process to manufacture RF-MEMS and HEMT on GaN/Si substrate. Proceedings of the 4th European Microwave Integrated Circuit Conference, Rome, Italy, September 2011, pp. 1740–1743.

31. Martinez, E.J. (2002) Transforming MMICs. Proceedings of the Gallium Arsenide Integrated Circuits Symposium 2002, 24th Annual Technical Digest, October 2002, pp. 7–10.

32. Fukuda, A. *et al.* (2006) A novel compact reconfigurable quad-band power amplifier employing RF MEMS switches. Proceedings of the 36th European Microwave Conference, Manchester, UK, September 2006.

33. Qiao, D. *et al.* (2005) An intelligently controlled RF power amplifier with a reconfigurable MEMS varactor tuner. *IEEE Transactions on Microwave Theory and Techniques*, **53** (3).

34. Mukherjee, T. *et al.* (2009) RF-CMOS-MEMS based frequency-reconfigurable amplifiers. IEEE 2009 Custom Integrated Circuits Conference, San Jose, CA, September 2009, pp. 81–84.

35. Joshin, K. *et al.* (2010) K-band CMOS-based power amplifier module with MEMS tunable bandpass filter. Proceedings of the 5th European Microwave Integrated Circuit Conference, Paris, France, September 2010, pp. 440–443.

36. Fouladi, S. *et al.* (2009) Reconfigurable amplifier with tunable impedance matching networks based on CMOS-MEMS capacitors in 0.18-μm CMOS technology. Microsystems and Nanoelectronics Research Conference, Ottawa, ON, Canada, October 2009, pp. 33–36.

37. Malmqvist, R., Samuelsson, C., Simon, W. *et al.* (2011) Reconfigurable wideband LNAs using ohmic contact and capacitive RF-MEMS switching circuits. Proceedings of the 6th European Microwave Integrated Circuit Conference, Manchester, UK, October 2011, pp. 160–163.

38. Malmqvist, R., Samuelsson, C., Gustafsson, A. *et al.* (2011) A K-band RF-MEMS-enabled reconfigurable and multifunctional low-noise amplifier hybrid circuit, in *Active and Passive Electronic Components*, Article ID 284767, 7 pages, Hindawi Publishing Corporation.

39. Malmqvist, R., Samuelsson, C., Gustafsson, A. *et al.* (2011) Monolithic integration of millimeter-wave RF-MEMS switch circuits and LNAs using a GaAs MMIC foundry process technology. IEEE International Workshop Series on Millimeter Wave Integration Technologies, Sitges, Spain, September 2011.

40. Malmqvist, R., Samuelsson, C., Gustafsson, A. *et al.* (2011) Reconfigurable RF-MEMS circuits and low noise amplifiers fabricated using a GaAs MMIC foundry process technology. Proceedings of the 12th International Symposium on RF MEMS and RF Microsystems, Athens, June 2011.

41. Malmqvist, R. *et al.* (2009) RF MEMS based impedance matching networks for tunable multi-band microwave low noise amplifiers. Proceedings of the 2009 International Semiconductor Conference (CAS2009), Sinaia, Romania, October 2009.

42. Liu, R., Schreurs, D., De Raedt, W. *et al.* (2011) RF-MEMS based tri-band GaN power amplifier. *Electronics Letters*, **47** (13).

43. Zine-el-Abidine, I. and Dietrich, J. (2011) A monolithic integration of a tunable MEMS capacitor with GaN technology. 2011 24th Canadian Conference on Electrical and Computer Engineering (CCECE), May 2011, pp. 447–449.

44. Heinemann, B. *et al.* (2010) SiGe HBT technology with fT/f_{max} of 300 GHz/500 GHz and 2.0 ps CML gate delay. IEEE 2010 International Electronic Devices Meeting (IEDM), pp. 688–691.

45. Larson, L.E., Hackett, R.H. and Lohr, R.F. (1991) Microactuators for GaAs-based microwave integrated circuits. Proceedings of Transducers 1991, San Franzisco, CA, June 24–27, 1991, pp. 743–746.

46. Brown, E.R. (1998) RF-MEMS switches for reconfigurable integrated circuits. *IEEE Transactions on Microwave Theory and Techniques*, **46** (11), 1868–1880.

47. Ulm, M. *et al.* (2003) Millimeter-wave microelectromechanical (MEMS) switches for automotive surround sensing systems. 2003 Topical Meeting on Silicon Monolithic Integrated Circuits in RF Systems, April 2003. Digest of Papers, pp. 142–149, 9–11.

48. Somjit, N., Stemme, G. and Oberhammer, J. (2009) Binary-coded 4.25-bit W-band monocrystalline–silicon MEMS multistage dielectric-block phase shifters. *IEEE Transactions on Microwave Theory and Techniques*, **57** (11), 2834–2840.

49. Stehle, A. *et al.* (2008) RF-MEMS switch and phase shifter optimized for W-band. 38th European Microwave Conference, 2008. EuMC 2008, 27–31 October 2008, pp. 104–107.

50. Lakshminarayanan, B. and Weller, T.M. (2007) Optimization and implementation of impedance-matched true-time-delay phase shifters on quartz substrate. *IEEE Transactions on Microwave Theory and Techniques*, **55** (2), 335–342.

51. Barker, N.S. and Rebeiz, G.M. (2000) Optimization of distributed MEMS transmission-line phase shifters – U-band and W-band designs. *IEEE Transactions on Microwave Theory and Techniques*, **48** (11), 1957–1966.

52. Hung, J.-J., Dussopt, L. and Rebeiz, G.M. (2003) A low-loss distributed 2-bit W-band MEMS phase shifter. Proceedings of the 33rd European Microwave Conference, October 2003, pp. 983–985.

53. Rizk, J.B., Chaiban, E. and Rebeiz, G.M. (2002) Steady state thermal analysis and high-power reliability considerations of RF MEMS capacitive switches. IEEE MTT-S International Microwave Symposium Digest, vol. 1, pp. 239–242.

54. Coccetti, F., Ducarouge, B., Scheid, E. *et al.* (2005) Thermal analysis of RF-MEMS switches for power handling front-end. Proceedings of the European Microwave Conference, October 4–6, 2005, vol. 3, pp. 1–4.

55. Chow, L.L.W., Wang, Z., Jensen, B.D. *et al.* (2006) Skin-effect self-heating in air-suspended RF MEMS transmission-line structures. *Journal of Microelectromechanical Systems*, **15**, 1622–1631.

56. Zhen, P., Yuan, X., Hwang, J.C.M. *et al.* (2007) Dielectric charging of RF MEMS capacitive switches under bipolar control-voltage waveforms. IEEE/MTT-S International Microwave Symposium, pp. 1817–1820.

57. Somjit, N., Stemme, G. and Oberhammer, J. (2011) Power handling analysis of high-power W-band all-silicon MEMS phase shifters. *IEEE Transactions on Electron Devices*, **58** (5), 1584–1555.
58. Rizk, J.B. and Rebeiz, G.M. (2003) W-band CPW RF MEMS circuits on quartz substrates. *IEEE Transactions on Microwave Theory and Techniques*, **51** (7), 1857–1862.
59. Rizk, J.B. and Rebeiz, G.A. (2003) W-band microstrip RF-MEMS switches and phase shifters. Microwave Symposium Digest, 2003 IEEE MTT-S International, vol. 3, 8–13 June 2003, pp. 1485–1488.

13

MIOS: Millimeter Wave Radiometers for the Space-Based Observation of the Sun

Federico Alimenti[1], Andrea Battistini[1], Valeria Palazzari[1], Luca Roselli[1] and Stephen M. White[2]

[1]*Department of Electronic and Information Engineering, University of Perugia, Perugia, Italy,*
[2]*Department of Astronomy, University of Maryland, College Park, MD, USA*

13.1 Introduction

A millimeter-wave observation of solar flares directly from outside the Earth's atmosphere (i.e., space-based) has never been attempted yet. Nonetheless, this methodology could open several interesting perspectives. First, the Sun could be studied with high time-on-target, high temporal resolution, and good sensitivity. As a consequence, a large sample of millimeter flare could be obtained. These data, equivalent to several years of equivalent observations from ground-based radiotelescopes, could be used in order to address several outstanding scientific questions regarding particle acceleration and transport in the solar corona. Is there more than one population of accelerated electrons, implying two acceleration mechanisms? What is the energy spectrum of the highest energy electrons? Are there very short (subsecond) pulses of acceleration? Second, the same methodology could also be used to study the relationship between solar flares and coronal mass ejections, the latter having effects on space weather, near-Earth conditions, and telecommunications.

This chapter will explore the above possibility by proposing a feasibility study of a full solar disk radiometer operating at the frequency of 90 GHz. Such an apparatus will be referred to as the Millimeter-wave Instrument for the Observation of the Sun (acronym: MIOS). The chapter is organized as follows. First the scientific backgrounds and the

Microwave and Millimeter Wave Circuits and Systems: Emerging Design, Technologies, and Applications,
First Edition. Edited by Apostolos Georgiadis, Hendrik Rogier, Luca Roselli, and Paolo Arcioni.
© 2013 John Wiley & Sons, Ltd. Published 2013 by John Wiley & Sons, Ltd.

motivations will be discussed. Then the characteristics of both quiet-Sun and solar flare radiation at millimeter-wave will be illustrated. Finally, the MIOS system-level design will be addressed.

13.2 Scientific Background

Solar flares are a sudden release of a great amount of energy stored as high magnetic fields in solar active regions. Lasting minutes to hours, they are sources of radiation and particles. The magnetic energy is rapidly converted into thermal, kinetic, and mechanical energies and the consequence is that the local plasma is heated to several tens of million degrees, while particles are accelerated up to high energies. Flare are unique for the diversity of emission mechanisms they exhibit and the broad range of wavelengths at which they radiate, from radio-wave, millimeter-wave, soft and hard X-rays, up to γ-rays, with energies reaching 1 GeV [1].

Millimeter-wave observations are the most sensitive tool to study the highest energy (i.e., >1 MeV) electrons accelerated in solar flare [2, 3] measurements of particle number, energy, spectrum, and transport of particles, which can be obtained from these observations. The millimeter- and submillimeter-wave flare emission is believed to be the result of synchrotron radiation from relativistic electrons, and from thermal bremsstrahlung from the hot plasma in post-flare loops [4]. The results of these observations can be of paramount help to compare with and improve existing particle acceleration models, the latter constrained by limited ground-based observations [5].

Observations from the ground [3–9] have indicated that:

- the energy spectral index of electrons at high energies may differ from the spectral index of nonthermal but low-energy electrons producing hard X-rays;
- there may be a high-energy component [1, 10] not seen at any other wavelengths;
- rapid (subsecond) time structures may be present [11, 12].

In typical large flares, one needs to observe at the millimeter-wave level to ensure that the emission is in the optically thin limit, as needed for quantitative work and in particular comparison with data at other wavelengths such as γ-rays.

Dedicated observations of solar flares at millimeter waves are needed to obtain the peak frequency and the high-frequency slope of radio spectra [3, 13], both of which are crucial to infer the energy distribution of relativistic electrons. These observations are important also to study energy transport in the corona and in the lower solar atmosphere during the extended phase of solar bursts.

During solar flares, large-scale magnetic field structures can be destabilized and propelled into the interplanetary medium, along with the large masses they contain, to form the so-called coronal mass ejections (CMEs). It is now recognized that [3] CMEs are the principal drivers of space weather and the near-Earth conditions. All these phenomena will emit at radio wavelengths in a frequency range covering over seven orders of magnitude, from a few tens of kHz up to a few tens or hundreds of GHz. Although many of the above phenomena are well documented by the available observations, they raise questions that are still unresolved.

Finally, the latest news from SOHO observations, in particular measurements from the VIRGO instrument, showed a strong relationship between helioseismology and flares: global oscillations of our Star seem to be strongly correlated with solar flares [14]. It would also be important to understand what physical process stands behind the heating of the solar corona. Among several theories, one is related with micro- and nanoflares: that is, a great number of extremely small-scale flares would dissipate magnetic energy in the corona (X-ray imaging observations by YOHKOH), heating it as a consequence.

Ground-based measurements [15] are available in the microwave range at 17 and 35 GHz. In this framework a millimeter-wave channel [16] is a good choice since it is well separated from the above frequencies. To improve the quantitative study of flares at these frequencies, instruments with a high detection probability are needed. This can be achieved with good sensitivity, a large field of view (FoV), with a full disk capability being optimal, a high time-on-target, and a high time resolution, the latter to capture fast time structures. From this point of view it can be noted that ground-based single-dish observations at millimeter wavelengths are limited by fluctuations in atmospheric opacity. As a consequence the sensitivity is met by reducing the FoV to a portion of the solar disk. On the other hand, a full disk capability is possible with nulling interferometers, such as the Nobeyama 80 GHz instrument [15], but the practical sensitivity is limited because the dishes must be small. In any case the maximum time-on-target achievable at ground is limited to about 8 hours per day in good weather conditions.

13.3   Quiet-Sun Spectral Flux Density

The emission captured by a radiotelescope is measured in terms of the intensity I_f, which is the power received per square meter and per solid angle unit within a bandwidth of one Hz [17]. Integrating I_f over the solid angle subtended by the emitting radio source we get a quantity S_f called the spectral flux density:

$$S_f^{QS} = \iint_{\Omega_S} I_f \, d\Omega \tag{13.1}$$

In solar radioastronomy S_f^{QS} is expressed in solar flux units (acronym: sfu), 1 sfu being equal to 10^{-22} W/m^2 Hz. The Sun is quite a strong radio source, being a few orders of magnitude stronger that the brightest nonsolar radio source in the microwave. The intensity radiated by the Sun can be evaluated assuming a black-body model with a brightness temperature T_S; from Plank's law one gets

$$I_f = \frac{2hf^3}{c_0^2} \frac{1}{\exp\dfrac{hf}{kT_S} - 1} \approx \frac{2f^2 kT_S}{c_0^2} \tag{13.2}$$

where $h = 6.63 \times 10^{-34}$ J s is Planck's constant, $k = 1.38 \times 10^{-23}$ J/K is Boltzman's constant, $c_0 = 2.99 \times 10^8$ m/s is the speed of light in a vacuum, and f is the frequency of analysis. In the microwave/millimeter-wave frequency range the Rayleigh–Jeans approximation can also be used since $hf \ll kT_S$. The brightness temperature of the Sun at millimeter waves has been estimated by several authors [18]; the values reported in Table 13.1 have been obtained by the Metäshovi Radio Observatory, a separate research institute of the Helsinki University of Technology, Finland [19].

Table 13.1 Quite Sun properties © 2010 IEEE. Reprinted with permission, from [40]

Frequency (GHZ)	Wavelength (mm)	T_S (K)	S_f^{QS} (sfu)
36.8	8.2	7800	2210
77.1	3.9	7250	9010
87	4.4	7200	11 390

In order to compute Equation (13.1), the spatial distribution of the source intensity will be assumed constant over the whole solar disk and equal to zero outside the disk (the small degree of limb brightening at millimeter wavelengths is ignored here). With reference to Figure 13.1 one can write:

$$I_f^{QS}(\rho, \varphi) = \begin{cases} \dfrac{2kT_S}{c_0^2}f^2, & 0 \le \rho \le \rho_S \\ 0, & \text{elsewhere} \end{cases} \tag{13.3}$$

where ρ_S is the angular radius of the Sun. Such a value is equal to $16'$ or, equivalently, to $\pi/675$ radiants. This distribution is based on a spherical coordinate system, the origin of which is at the observer's position. The variables ρ and ϕ are the angular coordinates along the solar radius and around the observer-to-Sun axis respectively. The differential of the solid angle is

$$d\Omega = \sin \rho \, d\rho d\varphi \approx \rho \, d\rho d\varphi \tag{13.4}$$

and the small-angle approximation is well verified for the Sun since $\rho \le \rho_S \ll 1$. As a result the integral (13.1) can be written as

$$S_f^{QS} = \int_0^{2\pi} \int_0^{\rho_S} \frac{2kT_S}{c_0^2} f^2 \rho \, d\rho d\varphi = \frac{2kT_S \rho_S^2}{c_0^2} f^2 \tag{13.5}$$

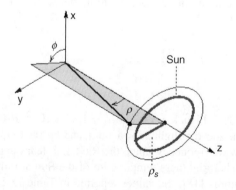

Figure 13.1 Coordinate system used to define the intensity distribution over the solar disk. The antenna is pointed toward the solar center.

The spectral flux density of the quiet Sun can thus be estimated with a rather simple formula. Since T_S has only a weak dependence on frequency in the millimeter range (Table 13.1), this formula shows an increase of S_f^{QS} as the square of the observation frequency. Assuming a brightness temperature of 7200 K from Table 13.1 and evaluating Equation (13.5) at 90 GHz one obtain a spectral flux density of about 12 200 sfu.

13.4   Radiation Mechanism in Flares

Two radiation mechanisms are well established as producing emission at millimeter wavelengths from solar flares. Less relevant for our scientific goals, but probably better known in the wider community, is 'thermal free–free emission', which is produced by the bremsstrahlung mechanism. This mechanism involves collisions of electrons with ions, and the opacity it produces is proportional to $T_e^{-1.5}$, where T_e is the temperature of the electrons. As a consequence the thermal free–free emission is favored in cool dense plasma and is typically only effective for the very high density post-flare loops at temperatures of order 10^7 K that produce the soft X-ray decay phase of flares. The light curve of this component, when present, is very similar to the soft X-ray (0.5–8 keV) light curve. In the very largest flares such a mechanism can produce at most tens of solar flux units. Because the radiating electrons are thermal, this emission component does not tell us much about particle acceleration.

Far more relevant, and generally much brighter, is synchrotron emission from the non-thermal electrons accelerated in solar flares. The synchrotron mechanism is very efficient for nonthermal electrons: electrons moving in a magnetic field exhibit a gyration about the direction of the magnetic field, and the acceleration associated with this gyration leads to radiation by the gyro-magnetic process. The characteristic frequency for this mechanism is the electron gyro-frequency, $f_B = 2.8 \times 10^6 B$, where B is the magnetic field intensity in Gauss. Coronal magnetic field strengths in solar flare sources can be as high as 2000 G but are generally found to be of order 600–800 G, corresponding to a gyro-frequency of 2 GHz. Thus emission at 90 GHz must be at harmonic of at least 15, more typically around 40. Because the typical frequency of gyro-emission of a particle with the Lorentz factor γ is $\gamma^2 f_B$, nonrelativistic particles cannot emit at such high harmonics. As a consequence, the presence of synchrotron emission at millimeter wavelengths requires electrons accelerated to MeV energies. It is this fact that makes millimeter emission such an effective diagnostic of electron acceleration to high energies, in contrast to the microwave range, where electrons with energies of order 300 keV can radiate effectively. This contrast is demonstrated with accurate spectral simulations in Figure 13.2. Synchrotron emission from a large flare can produce a peak flux as large as 10^5 sfu at 100 GHz (many times the total quiet-Sun flux). Nonthermal electrons radiate so efficiently by exploiting this mechanism that it is much easier to detect MeV-energy electrons from their millimeter emission than from their bremsstrahlung γ-rays. This makes a millimeter-wave space-based experiment a very powerful tool for the study of electron acceleration to high energies.

13.5   Open Problems

The list of solar flares that have been detected at millimeter wavelengths from the ground remains small for two very good reasons: firstly, except in the largest events millimeter emission should have a spectrum that falls as frequency increases, limiting flux levels; and,

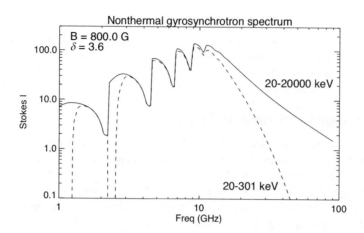

Figure 13.2   Figure showing the effect of MeV-energy electrons on the radio spectrum of a solar flare. The solid line shows the calculated spectrum emitted by electrons in a power-law distribution of energy spectral index -3.6 extending from 20 keV to 20 MeV in a constant magnetic field of 800 G. The dashed curve is the spectrum emitted by the same electron energy distribution but extending only from 20 to 300 keV. When the MeV-energy electrons are absent, emission at millimeter wavelengths is reduced by orders of magnitude whereas microwave emission is not greatly changed. At low frequencies discrete harmonics are seen in the model due the unrealistic assumption of a homogeneous magnetic field in the corona. © 2010 IEEE. Reprinted with permission, from [40].

secondly, observations from the ground are limited by atmospheric opacity fluctuations, which typically impose a detection threshold well above the instrumental noise level. An aperture-synthesis interferometer would get around these problems but no such solar dedicated interferometer exists operating in the 3 mm window: the closest we have is the 9-mm-wavelength 34 GHz system of the Nobeyama Radioheliograph, but in large flares 34 GHz is often optically thick and of limited use for quantitative work. The main solar dedicated source of data in the 3 mm window is the 80 GHz nulling interferometer of the Nobeyama Polarimeter system (a similar 90 GHz nulling interferometer can be deployed at the KOSMA telescope on Gornergrat), whose weakest reliable event detections under optimal weather conditions are around 30 sfu at 80 GHz.

MIOS is designed to reach a sensitivity more than an order of magnitude better than this at 90 GHz. Combining the improved sensitivity with excellent on-target time coverage, it should result in a large increase in event detections and correspondingly a much more complete picture of the millimeter characteristics of flares. In particular, the following major science questions need to be addressed:

- **Accelerated particle populations:** If nonthermal electrons are accelerated by a single mechanism and can be described by a single power-law energy distribution, then the optically thin spectral index of synchrotron emission at millimeter wavelengths and the spectral index of hard X-rays produced by bremsstrahlung should be compatible. This is not the case: in well-observed events, we find that the radio data predict flatter energy spectra than do the hard X-rays (References [20] and [21] give examples). Typically the electrons that produce the radio emission have higher energies than those producing hard

X-rays, so this result could be explained if typical accelerated electron energy spectra have a break upwards by about 2 in the power law index above some energy of order 1 MeV. Such a break is not consistent with any single current candidate for acceleration of the electrons: if this result is confirmed, it implies that one mechanism operates at lower energies (hundreds of keV) and a second mechanism is required at high energies. However, the data indicate that these two mechanisms operate together, providing a strong constraint for explanations. MIOS, with its sensitivity, will provide us with a much more comprehensive picture of the relationship between the millimeter emitting and hard X-ray emitting (seen by the RHESSI satellite) electrons.

- **Rapid fluctuations at submillimeter frequencies:** Subsecond structures have been reported during strong emission levels in observations at submillimeter frequencies [11, 12], but never in the 3 mm window. This may be partly an observational limitation: observations at 3 mm have not generally had sufficient cadence to resolve the structures of duration ~ 0.1 s seen at 210 and 405 GHz. Such rapid phenomena place strong constraints on the acceleration and transport of electrons: they must be accelerated to MeV energies and then lost from the corona in a very short time. The rapid loss suggests either that they are accelerated in very small structures or else they are generated with very small pitch angles and do not undergo pitch-angle scattering in the corona. MIOS will have the time resolution to see such structures at 3 mm: their detection will be important in establishing that the ground-based results are not due to atmospheric or instrumental effects, and in portraying a picture of acceleration that consists of many small episodic events.

- **Bursts with flat or rising spectra:** Anomalous bursts with spectra inconsistent with the normal falling nonthermal synchrotron spectrum have been observed by several instruments. A small burst with a flat radio spectrum out to 90 GHz was reported by White *et al.* in 1992 [22]: its properties were not consistent with any form of thermal emission, so it remains unexplained. SST observations at 210 and 405 GHz have found several flares exhibiting a spectral component rising in the submillimeter range [1, 10, 23] that cannot easily be explained by synchrotron spectra, and are also inconsistent with thermal emission. Such events raise the possibility that yet another emission mechanism is involved, as yet unidentified. Presently there are very few such events and MIOS, with its sensitivity, can be expected to play an important role in the search for further examples.

13.6   Solar Flares Spectral Flux Density

The spectral flux density of the Sun, S_s^f, can be imagined to be composed by two contributions: the first is related to the quiet Sun S_f^{QS} and was estimated in the previous section; the second due to the radiation mechanisms is involved in flares S_f^F:

$$S_f^S = S_f^{QS}(t) + S_f^F(t) \cong S_f^{QS} + S_f^F(t) \tag{13.6}$$

The contribution associated with the quiet Sun is characterized by slow temporal variations. Measurements carried out from an Earth location are primarily influenced by the relative distance between the Earth and the Sun. Such a distance is continuously varying with the orbital position of the Earth and thus has a periodicity of one year. Once corrected for this effect, the measurements show an even slow periodicity associated with the 11 year solar cycle. The contribution due to flares, instead, evolves very quickly in time, the typical duration of a flare event being around 10 minutes.

Because of this great time-scale difference, the quiet Sun spectral flux density can be considered as a constant during a flare event, leading to the approximation of Equation (13.6). As a consequence, the flare contribution can be separated from the total spectral flux density by extracting the only flux variations above the quiet Sun background. At millimeter-wave frequencies the quiet Sun background is around 10 000 sfu while one can be interested in studying small amplitude flares, in the order of 1 sfu. The instrumental sensitivity and stability are thus two critical issues that will be carefully discussed in the present chapter.

13.7   Solar Flares Peak Flux Distribution

In order to estimate the number of detections we expect from MIOS, we investigate the probability distribution of the 90 GHz peak flux. Such a distribution depends on the frequency. At millimeter waves one of the available databases is that of the Nobeyama Radio Observatory, Nobeyama, Japan [24], with two channels at 35 and 80 GHz respectively.

A statistical characterization has been obtained in the following way. First we define the peak flare amplitude as a random variable

$$S = \max\left\{ S_f^F(t) \right\} \tag{13.7}$$

Then, for each flux bin s, the number of observed events with amplitude greater than or equal to s has been determined using the 80 GHz Nobeyama database (see Figure 13.3).

The number of events is proportional to the complementary distribution function, that is, to the probability

$$P(s) = P(S \geq s) \tag{13.8}$$

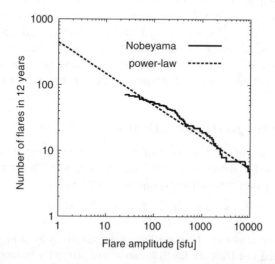

Figure 13.3   Distribution of flare amplitudes from the 80 GHz nulling interferometer at Nobeyama, Japan, in 12 years of observation. For each flare amplitude s on the x axis, the y axis reports the number of observed flares with amplitude greater than or equal to s.

Following the approach in Reference [25] developed for the peak gamma-ray intensity of solar flares, the 80 GHz amplitude is described by a power-law probability distribution function [26]:

$$p(s) = \frac{a-1}{s_{min}} \left(\frac{s}{s_{min}} \right) \tag{13.9}$$

where s_{min} is the minimum flare amplitude to which the power law is maintained (effectively the minimum detectable flare amplitude). The complementary distribution function can easily be evaluated from Equations (13.8) and (13.9):

$$P(s) = \left(\frac{s}{s_{min}} \right)^{-a+1} \tag{13.10}$$

We find that the distribution of flare events in Figure 13.3 can be modeled as follows:

$$NF(s) = N_0 P(S \geq s) \tag{13.11}$$

It is important to recall here that NF is the number events in 12 years of observations by the Nobeyama 80 GHz nulling interferometer. Assuming $s_{min} = 0.1$ sfu for MIOS, the best fitting between the Nobeyama measurements and the power-law model is obtained with $a = 1.48$ and $N_0 = 1350$. This distribution predicts over 100 events above 10 sfu per solar maximum and several hundred events above 1 sfu that MIOS can detect. Similar results have also been reported in Reference [27]. This paper predicts that, in a 12-year solar cycle, there will be 160 events above 10 sfu at 90 GHz and 105 events above 20 sfu.

13.8 Atmospheric Variability

Full-disk millimeter-wave Sun observations are, in principle, possible with both radiometers and nulling interferometers [28]. In ground-based experiments, however, nulling interferometers are the only full-disk instruments providing good sensitivity at frequencies greater than 50 GHz.

The reason behind this performance advantage is that the sensitivity of a ground-based radiometer is mainly limited by the random fluctuation of the tropospheric transmissivity T. A full-disk radiometer measures the total spectral flux density of the Sun S_f^S multiplied by T.

$$S_f^m = s_f^S T \tag{13.12}$$

As a consequence the random fluctuation of the T directly modulates the radiometric measurements. The resulting error is equal to

$$\Delta S_f^m = S_f^S \Delta T \approx S_f^{QS} \Delta T \tag{13.13}$$

with ΔT typically being equal to 10%. At 90 GHz, the quiet Sun spectral flux density is around 10 000 sfu, resulting in an error $\Delta S_f^m \approx 1000$ sfu. The conclusion is that a ground-based full-disk radiometer is unusable for the detection of small flares. Nulling interferometers, on the contrary, are not sensitive to the total spectral flux density of the Sun, so the tropospheric variability does not affect the measured data too much [15]. Their principle of operation is based on the following idea. The radiation coming from a source that is extended with respect to the fringe spacing is canceled-out by interferometry, whereas compact sources like flares are correctly measured [29].

13.9   Ionospheric Variability

In the proposed experiment the Sun will be observed from outside the troposphere. In this case the radio path attenuation is caused only by the ionosphere. This aspect will be analyzed in the present section showing that, with a sufficient orbit height, the ionospheric attenuation and its fluctuation are negligible. As a result, the main factor limiting the application of millimeter-wave radiometers to ground-based observations of flares, is no more an issue for the present space-based mission.

To analyze the ionospheric effect, a model of the radio path between the satellite and the Sun has been set up considering the ionosphere as stratified plasma and treating it according to the classical magnetoionic theory [30].

The adopted scheme is shown in Figure 13.4, where A locates the satellite. The satellite is assumed to be at a certain height h_s from the Earth's surface, along an approximately circular orbit. The Sun is to the right of the figure while the radio path is defined by an x axis originating at the satellite and pointed toward the Sun. In this scheme the satellite orbit (not shown) is orthogonal to the plane of the drawing, making the Sun always visible from the satellite.

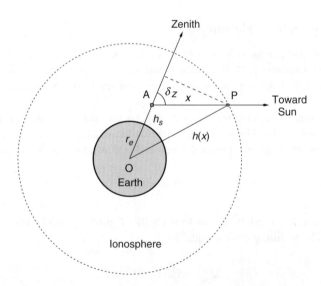

Figure 13.4 Scheme adopted in the computation of the millimeter-wave attenuation of the ionosphere. O is the Earth's center; A is the satellite; P is a generic point within the ionosphere; AP is a segment of length x taken along the direction between the satellite.

The height of a generic point P along the radio path with respect to the Earth's surface (see Figure 13.4) can be determined using simple geometrical considerations:

$$h(x) = -r_e + \sqrt{(r_e + h_s + x \cos \delta_z)^2 + (x \sin \delta_z)^2} \tag{13.14}$$

Modeling the ionosphere as a stratified plasma means that, for each point P along the radio path, the plasma parameters will be different, according to a different height $h(x)$ of P. As a consequence also the radio wave attenuation constant α will be a function of x.

Exploiting the radiative transfer theory, the ionospheric transmissivity is given by

$$T = e^{-\int_0^\infty \alpha(x)dx} \tag{13.15}$$

In order to compute the above integral, the attenuation constant must be expressed in terms of the plasma parameters. It is observed that α is proportional to the plasma conductivity σ:

$$\alpha(x) = \frac{\eta_0}{2}\sigma(x) \tag{13.16}$$

where $\eta_0 = 377\,\Omega$ is the wave impedance of a vacuum. To derive such an expression the relative permittivity of the plasma is approximated with unity, that is, $\varepsilon_r \approx 1$. The plasma conductivity is given by the magnetoionic theory (Reference [30], p. 8):

$$\sigma(x) = \frac{N_e(x)q_e^2 v(x)}{m_e[v^2(x) + \omega^2]} \tag{13.17}$$

In such a relationship $q_e = 1.6 \times 10^{-19}$ C and $m_e = 9.1 \times 10^{-31}$ kg are the electron charge and mass respectively and $\omega = 2\pi f$ is the angular frequency of the radio signal. The plasma parameters are the electron density N_e and the collision frequency v. The latter quantity is defined as the number of collisions per unit time between electrons and heavy particles, that is, ions and neutrals. As stated above, electron density and collision frequency are functions of the height from the Earth's surface and thus from x. In addition they also depend on other variables such as latitude, longitude, month, day, hour and solar activity (sun-spot number). Updated profiles for electronic density and collision frequency are available from several sources [31].

For the present study one is only interested in calculating the lower absolute bound of the ionospheric transmissivity, this being associated with the maximum error due to ionospheric variability. The worst case is, in fact, that of an ionospheric transmissivity varying from unity (ideal case with no attenuation) to the lower absolute bound.

In this perspective, the upper bounds of the electron density and of the collision frequency profiles reported in the tables [31] have been approximated with simple formulas. Since the attenuation constant of radio waves is proportional to the product of the electron density and collision frequency, the upper bound of these quantities guarantee the lower bound of the ionospheric transmissivity. For the electron density a power-law decay has been adopted:

$$N_e(x) = N_0 \left[\frac{h_0}{h(x)}\right]^p \tag{13.18}$$

where h_0 is the height of the F2 ionospheric layer peak, N_0 is the electron density at h_0 and p is the decay exponent. The collision frequency, on the other hand, has been assumed independent of height:

$$\nu(x) = \nu_0 \tag{13.19}$$

ν_0 being the frequency at the F2 layer. The comparison between the two above relationships and the data of tables in Reference [31] is illustrated in Figure 13.5.

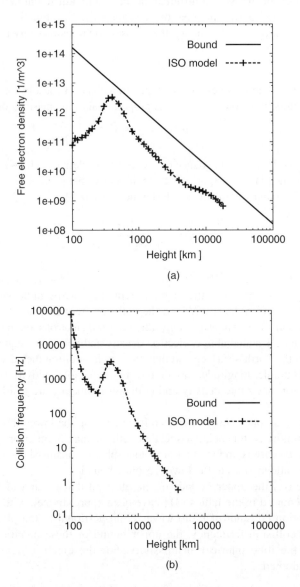

(a)

(b)

Figure 13.5 (a) Upper bounds for ionospheric electron density and (b) collision frequency profiles. The ISO models developed by Chasovitin, table no. 820, available on-line from Reference [31], have been used to set up these bounds.

Table 13.2   Ionospheric profiles

h_0 (km)	N_0 (1/m³)	p	v_0(kHz)
400	1×10^{13}	2	10

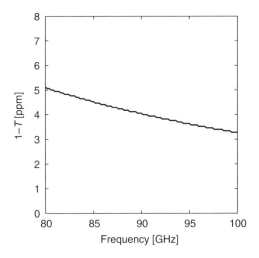

Figure 13.6   Complement to unity $(1 - T)$ of the millimeter-wave ionospheric transmissivity for a radio path oriented toward the Sun. The satellite is at $h_s = 500$ km, while the angle formed by the radio path with the satellite zenith is $\delta_z = 98°$.

The parameters of Equations (13.18) and (13.19) have been selected in such a way as to be more conservative with respect to Reference [31]. The obtained values are reported in Table 13.2

Inserting these profiles in the radio path model and evaluating the integral (13.15) numerically, one finds the ionospheric millimeter-wave transmissivity, the complement to unity of which is depicted in Figure 13.6. At 90 GHz such a transmissivity (absolute worst case) is only 4 ppm below unity. According to Equation (13.13) the impact of ionospheric fluctuations is expected to be less than 0.05 sfu.

13.10   Antenna Design

In the millimeter- and submillimeter wavelength range geometrical optics cannot be applied. One of the main assumptions of geometrical optics is that the size of optical components is larger than the wavelength. At these frequencies, however, propagating waves can have the electric field with a Gaussian distribution. The optics in this domain is therefore often called Gaussian optics, or quasi-optics.

The design of the MIOS antenna is afforded in two steps. First the Gaussian optics [32] is used to obtain an initial guess. Then the design is refined exploiting an antenna design CAD tool, that is, the GRASP 9 Student Edition [33]. The same approach is also used in Reference [34]. The Gaussian modes are obtained as the solution of the Huygens–Fresnel diffraction integral for the E-field distribution in free space. These modes are typically excited by a

Gaussian source of radiation, for instance, a horn antenna. The main properties of Gaussian modes are the following.

The propagation occurs along an axis, say the z axis, with a plane phase front at $z = 0$. The E-field maximum will be on the axis itself and then the amplitude will decay according to a Gaussian distribution. The distance from the z axis to the points where the E-field falls to $1/e$ of the maximum will be defined as the beam size or beam radius. The minimum beam size is called the beam waist w_0 and is localized where the phase front is planar, that is, at $z = 0$ with the above assumptions.

At a certain frequency the beam size w depends on only two parameters, namely the distance from the beam waist z and the beam waist radius:

$$w(z) = w_0 \sqrt{1 + \left(\frac{z}{z_c}\right)^2} \tag{13.20}$$

where z_c is the confocal distance, a function of both w_0 and the wavelength λ:

$$z_c = \pi \frac{w_0^2}{\lambda} \tag{13.21}$$

The half-power beam width Θ_h of the antenna is related to the far field divergence angle θ_0 of the beam:

$$\Theta_h = 1.18\theta_0 \tag{13.22}$$

The latter can be written in terms of both frequency and beam waist radius:

$$\theta_0 = \frac{\lambda}{\pi w_0} \lim_{x \to \infty} \tag{13.23}$$

As a result the fundamental Gaussian mode parameters w_0 and z_c can be found from the knowledge of Θ_h. In the case of MIOS the above parameters are summarized in Table 13.3.

In Reference [29], p. 40, the design of a 45° offset dish antenna operating at 90 GHz is illustrated. The half-power beam width of such an antenna is equal to 1°, while the dish shape has been determined considering the parent paraboloid. The projection of the dish shape on the output plane is a circle of radius $a \simeq 127$ mm. This value is useful for the computation of the output taper edge, that is, the power of the Gaussian beam outside a given radius:

$$T_e = \exp\left[-2\left(\frac{r}{w}\right)^2\right] \tag{13.24}$$

Since, with the parent paraboloid design approach the output plane is also the beam waist location (planar phase front), the above equation can be computed for $r = a$ and $w = w_0$. As a consequence a taper edge of 27 dB is obtained.

Table 13.3 Gaussian optics parameters

Θ_h (deg)	θ_0 (deg)	λ (mm)	w_0 (mm)	z_c (m)
1	0.85	3.3	71.7	4.85

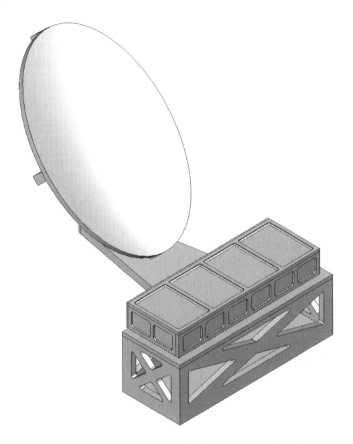

Figure 13.7 Preliminary mechanical drawing of both antenna and millimeter-wave radiometer modules (the box). The latter includes the corrugated feed-horn, the calibration circuitry, and the receiver itself. The main parameters are: offset 45°, $f/D = 0.6$, and dish diameter of 359 mm. © 2010 IEEE. Reprinted with permission, from [40].

The antenna design reported in Reference [29] has thus been considered as the initial guess for a further optimization. This optimization is carried out by exploiting the GRASP 9 Student Edition CAD tool; the final result is shown in Figure 13.7.

13.11 Antenna Noise Temperature

The antenna noise temperature is an important parameter affecting, together with the receiver noise temperature, the system resolution. Such a temperature depends on the scene observed by the antenna and can be evaluated according to Reference [32], p. 144:

$$T_A = \frac{A_e}{\lambda^2} \iint P_n(\rho, \varphi) T_B(\rho, \varphi) \mathrm{d}\Omega \tag{13.25}$$

where λ is the signal wavelength, A_e is the effective area of the antenna, P_n its normalized radiation diagram, and T_B the brightness temperature of the observed source. The variables

ρ and φ are used to describe the spatial variations of the above functions. In the case of the Sun it is particularly convenient to define ρ as the angular coordinate along the solar radius and φ as the angular coordinate around to the antenna-to-Sun axis. Assuming that the brightness temperature is equal to T_S everywhere over the solar disk:

$$T_B(\rho, \varphi) = \begin{cases} T_S, & 0 \le \rho \le \rho_S \\ 0, & \text{elsewhere} \end{cases} \tag{13.26}$$

and using the elementary solid angle $d\Omega = \sin \rho\, d\rho\, d\varphi \approx \rho\, d\rho\, d\varphi$, one obtains

$$T_A = \frac{A_e}{\lambda^2} \int_0^{2\pi} \int_0^{\rho s} P_n(\rho, \varphi) T_{S\rho}\, d\rho d\varphi \approx 2\pi \frac{A_e T_S}{\lambda^2} \int_0^{\rho s} \rho\, d\rho \tag{13.27}$$

The approximation is equivalent to considering the normalized radiation diagram function as equal to unity. Developing the integral (13.27), the following formula is derived:

$$T_A \approx \pi \frac{A_e T_S \rho_s^2}{\lambda^2} \tag{13.28}$$

The antenna theorem (Reference [17], p. 157) states a basic equation between the effective antenna area and the antenna solid angle Ω_a:

$$A_e \Omega_a = \lambda^2 \tag{13.29}$$

Since for a Gaussian beam it is possible to relate the antenna solid angle with the far-field divergence angle (Reference [32], p. 136) and the latter with the half-power beam width [32, 25], one obtains

$$\Omega_a \approx 1.13 \Theta_h^2 \tag{13.30}$$

Θ_h being the above-mentioned half-power beam width. Inserting Equations (13.29) and (13.30) into (13.28) one obtains the antenna noise temperature in terms of system parameters:

$$T_A \approx \frac{\pi}{1.13} \left(\frac{\rho_s}{\Theta_h} \right)^2 T_S \tag{13.31}$$

The formula (13.31) means that the antenna noise temperature can be obtained as the brightness temperature of the source multiplied by the filling factor, that is, the portion of the antenna solid angle filled by the source itself. A numerical example is reported in Table 13.4.

Table 13.4 Antenna noise temperature

T_S (K)	ρ_s (arcmin)	Θ_h (arcmin)	T_A (K)
7200	16	60	1423

13.12 Antenna Pointing and Radiometric Background

The radiometric background generated by the quiet Sun is a function of the antenna pointing direction. In the proposed experiment a heliosynchronous satellite orbit will be assumed. With this choice a virtual 100% time-on-target can be achieved. The maximum radiometric background is given by Equation (13.31) and is obtained for a pointing toward the solar center. The minimum background, instead, will be recorded for a limb pointing at the Sun. In the generic case such a background will be

$$T_A(\rho_0, \varphi_0) = \frac{A_e}{\lambda^2} \iint P_n(\rho - \rho_0, \varphi - \varphi_0) T_B(\rho, \varphi) d\Omega \qquad (13.32)$$

where ρ_0, φ_0 indicate the antenna pointing direction and T is defined as in Equation (13.26). It is apparent from this formulation that the accurate prediction of the MIOS radiometric background is related to the knowledge of P_n, that is, the normalized radiation pattern. For this reason a procedure for the on-flight calibration of the main antenna beam should be considered. This could be attained with a scan of the whole solar disk along the two main heliocentric coordinate directions.

13.13 Instrument Resolution

The resolution of a millimeter-wave radiometer is defined as the standard deviation of the measured antenna noise temperature [35]. Assuming a total-power configuration the well-known resolution formula is

$$\Delta T = \left(T_A + T_R \sqrt{\frac{1}{B\tau} + \left(\frac{\Delta G}{G}\right)^2} \right) \qquad (13.33)$$

The importance of this expression is that it establishes a link between the standard deviation of the measured data and the main instrument parameters, particularly the antenna noise temperature T_A, the receiver equivalent noise temperature T_R, the predetection bandwidth B, the integration time τ, and the gain stability $\Delta G/G$ of the radiometer itself. T_R can be related to the receiver noise figure F_R by

$$T_R = (F_R - 1)T_0 \qquad (13.34)$$

$T_0 = 290$ K being the standard IEEE temperature at which the noise figure is defined. In order to convert a resolution ΔT expressed in terms of antenna noise temperature into a resolution ΔS_f given in terms of solar flux units, the approach proposed in Reference [29], p. 41, is followed:

$$\Delta S_f = \frac{S_f^{QS}}{T_A} \Delta T \qquad (13.35)$$

When the antenna is pointed toward the solar center, S_f^{QS} and T_A can be approximated by Equations (13.5) and (13.31) respectively. Inserting Equation (13.33) into (13.35) one obtains

$$\Delta S_f = S_f^{QS} \left(1 + \frac{T_R}{T_A} \right) \sqrt{\frac{1}{B\tau} + \left(\frac{\Delta G}{G}\right)^2} \qquad (13.36)$$

Table 13.5   Main instrument parameter

f (GHz)	S_f^{QS} (sfu)	Θ_h (arcmin)	T_A (K)	F_R (dB)	T_R (K)	τ (s)	$\Delta G/G$ (ppm)
90	12 200	60	1423	8	1500	0.1–1	100

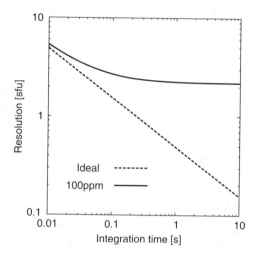

Figure 13.8   Radiometric resolution versus the integration time. The following system parameters have been assumed: noise of the millimeter-wave receiver 8 dB, pre-detection bandwidth 2 GHz, half-power beam width of the antenna 1°. An ideal radiometer is compared with one having a 100 ppm gain stability. © 2010 IEEE. Reprinted with permission, from [40].

    The resolution of the proposed instrument has been analyzed with reference to the system parameters reported in Table 13.5.

    Millimeter-wave receivers for space applications with such a noise figure have already been realized [36]. The resulting receiver noise temperature is $T_R = 1500$ K. In addition the previously estimated 1423 K antenna noise temperature has been assumed. Finally, the integration time has been swept from 10 ms to 10 s. The estimated resolution is reported in Figure 13.8. From the analysis of this figure it is apparent that the specified 3 sfu for 1 s integration time and 10 sfu for 100 ms integration time can be achieved with a 100 ppm gain stability. Such stability must be guaranteed for at least 10 minutes in order to measure a typical flare. The 100 ppm value is 10 times larger than what is presently considered the state of the art [37]. To obtain such stability the radiometer should be periodically calibrated. To this purpose internal calibration standards will be provided, while a thermal stabilization of the instrument could be needed.

13.14   System Overview

The simplified block diagram of the instrument is illustrated in Figure 13.9. It is divided into three main modules, namely, antenna module, millimeter-wave (mmw) radiometer module, and back-end module. The antenna module is constituted by a single parabolic dish and by a circular corrugated feed-horn placed in the focal point of the dish. An offset configuration

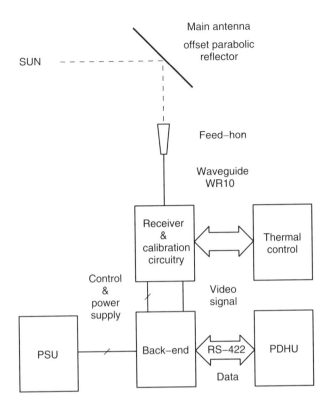

Figure 13.9 Simplified block diagram of MIOS. The instrument is composed of three modules: the antenna module, the mmw radiometer module, and the back-end module.

will be used in order to avoid the obstruction and scattering of the feed system. This will result in an improved antenna efficiency. The main antenna could be placed within the satellite envelope in such a way as to reduce for the temperature variations. The circular corrugated horn will be connected to the receiver input by means of a properly shaped section of WR10 rectangular waveguide.

The mmw radiometer module contains the analog hardware of the instrument, namely, the mmw receiver, the calibration circuitry, the detector diode/diodes, and the video amplifiers. A receiver architecture based on a direct mmw amplification chain is adopted. With this concept, the signal is boosted directly at 90 GHz exploiting advanced InP or GaAs MMIC circuits. In addition, the bandpass filtering and the detection are carried out directly at the operating frequency. Once detected, the signal is adjusted in level-exploiting low-drift instrumentation amplifiers and then passed to the back-end module. Another very important subsystem included in the mmw radiometer module is the calibration circuitry. Since a stability of at least 100 ppm in 10 minutes is required to achieve the scientific goal, it would be necessary to calibrate the instrument periodically for both the receiver gain and the radiometric offset.

These calibrations will be realized exploiting waveguide switches, reference loads, and noise diodes. Temperature sensors will also be used to monitor the receiver components.

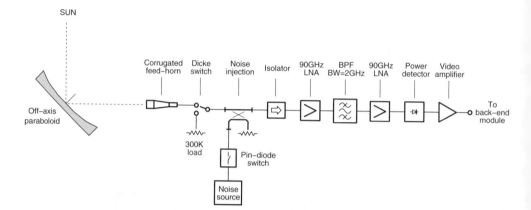

Figure 13.10 Schematic of the millimeter-wave radiometer module: a direct amplification architecture is assumed. The receiver gain is calibrated by noise injection, whereas the radiometer offset is calibrated using a reference load, working as a black-body, and a Dicke switch. © 2010 IEEE. Reprinted with permission, from [40].

The schematic of the millimeter-wave radiometer module is shown in Figure 13.10 illustrating the basic building blocks of the instrument. The back-end module contains 16 bit ADCs. The sampled data will be preprocessed (averaging, etc.) by means of a digital unit. This unit will also act as an instrument controller, accepting instructions from the Processing Data Handling Unit (PDHU) and sending back to the PDHU the corresponding answer. The digital unit could be implemented with an Actel FPGA (RTSX family) ranging from 32 000 to 72 000 gates. With an FPGA of this size, even a simple ALU and/or microprocessor could be easily realized. Both power supply lines and control lines are wired from the back-end to the front-end module. Both the PDHU and the power supply unit (PSU) are linked with the back-end module. A digital bidirectional communication between a PDHU and the instrument is implemented via UART (RS-422 standard) at a standard data rate of 2400 bit/s. In this way the PDHU can fully control the instrument, that is, acquiring the preprocessed measurements, monitor the instrument state, actuating a calibration, or setting a parameter. A suitable housekeeping procedure should be available for this purpose.

13.15 System Design

The resolution analysis shown in Figure 13.8 is based on two main assumptions: receiver noise of less than 8 dB and gain variations below 100 ppm for at least 10 minutes, the latter being the typical duration of a solar flare. In order to complete the first-guess design of MIOS the following steps have been carried out. First the overall receiver gain is evaluated; then gain and noise performances are distributed among the various blocks of Figure 13.10. The gain is determined by selecting a suitable power level P_D at the square-law detector input:

$$P_D = k(T_A + T_R)BG \tag{13.37}$$

where G is the pre-detection gain. Assuming the linearity, the voltage at the detector output will be

$$V_D = \gamma P_D \tag{13.38}$$

γ being the detector sensitivity, usually expressed in mV/W. Combining Equation (13.37) with (13.38), the antenna noise temperature can be related to the output radiometer voltage as

$$V_D = kBG\gamma(T_A + T_R) \tag{13.39}$$

Such a relationship is an important link between the quantity to be measured, T_A, and the quantity effectively treated, V_D. In particular, $kBG\gamma$ plays the role of a temperature-to-voltage conversion constant; T_A is the wanted signal while T_R constitutes the radiometric offset. A more detailed discussion of the retrieval equations will be carried out later on. The derivative of V_D with respect to T_A defines the radiometer sensitivity, that is, the variation of the output voltage associated to a 1 K variation of the antenna noise temperature:

$$\frac{\Delta V_D}{\Delta T_A} = kBG\gamma \tag{13.40}$$

The minimum variation of the output voltage ΔV_D^{min} is obtained by multiplying the radiometer sensitivity (13.30) by the radiometer resolution (13.33). Such a value should be kept well above the input offset and the thermal drift stability of the video amplifier, that is, of the operational amplifier following the square-law detector.

To meet such a criterion it is useful to improve the radiometer sensitivity, which, in turn, can be done by improving both the detector sensitivity and the pre-detection gain. The detector sensitivity is related to the semiconductor material used in the detector diode. Typical values of γ for broadband detectors are 0.5 mV/W for Si Schottky diodes and 1 mV/μW for GaAs Schottky diodes. The pre-detection gain should be determined in such a way as to guarantee the linearity of the detector. To this purpose P_D should be kept below -27 dBm for Si diodes and below -20 dBm for GaAs diodes. Exploiting these data and assuming the instrument parameters defined in Section 13.13, Table 13.6 is obtained. In particular the noise power at the radiometer input $k\,(T_A + T_R)\,B$ is equal to -70.7 dBm, whereas the radiometric resolution is of 0.4 K, corresponding to the specified 3 sfu.

From the analysis of this table emerges the selection of a GaAs detector diode along with an overall pre-detection gain of about 51 dB. The offset and drift parameters of the instrumentation amplifier (IA) have been taken from the AD524C data sheet [38]. To

Table 13.6 Radiometer gain and sensitivity

Type	P_D (dBm)	γ (mV/μW)	G (dB)	$\Delta V_D/\Delta T_A$ (μV/K)	ΔV_D^{min} (μV)	IA offset (μV)	IA drift (μV/°C)
Si	-27	0.5	43.7	0.32	0.13	50	0.5
GaAS	-20	1.0	50.7	3.2	1.3	50	0.5

Table 13.7 MIOS building blocks

Quantity	Unit	Dicke switch	Directional coupler	Isolator	LNA first stage	BPF	LNA second stage
G	dB	−0.7	−1.2	−0.8	28	−3	28
F	dB	0.7	1.2	0.8	5	3	6
T_{eq}	K	51	92	59	627	289	865

distribute gain and noise among the various blocks constituting the radiometer, the performance of off-the-shelf 90 GHz components have been evaluated. A summary of this study is illustrated in Table 13.7. From the Friis formula it is possible to evaluate the noise of the whole receiver as a function of the performances of the single blocks. Developing this well-known relationship one obtains

$$F_R = L_{SW} L_{CP} L_{ISO} F_{LNA}^{(1)} + \frac{L_{SW} L_{CP} L_{ISO}}{G_{LNA}^{(1)}} \left(L_{BPF} F_{LNA}^{(2)} - 1 \right) \tag{13.41}$$

where $L_j = 1/G_j$ is the loss of the generic passive block. For these passive blocks the noise has been assumed equal to the loss itself.

Exploiting the numerical values of the above table and converting the dB quantities into linear units, one obtains

$$F_R = 5.9 + \frac{1.86}{100} (7.95 - 1) = 6.03 \rightarrow 7.8 \, \text{dB} \tag{13.42}$$

As a result it can be concluded that, with the selection of commercial components proposed in Table 13.7, the $F_R < 8$ dB performance (see Section 13.11) can be achieved at the considered operating frequency.

13.16 Calibration Circuitry

Calibration is a fundamental procedure allowing the antenna noise temperature T_A, that is, the quantity under measurement, to be evaluated in terms of the output radiometer voltage V_m, the latter being directly measured by the ADC contained in the back-end unit. To deal with this issue a mathematical model of the output voltage is developed. Considering a video amplifier of voltage gain G_v and input offset voltage V_{OS} the radiometer output can be written as

$$V_m = G_v(V_D + V_{OS}) = kBGG_v\gamma(T_A + T_R) + G_v V_{OS} \tag{13.43}$$

Reversing Equation (13.43) it is possible to derive the relationship between T_A and V_m:

$$T_A = \alpha_r V_m - \beta_r \tag{13.44}$$

where α_r is the radiometric scale and β_r is the radiometric offset. The latter is determined by both the receiver noise temperature and the video amplifier offset:

$$\alpha_r = \frac{1}{kBGG_v\gamma}$$

$$\beta_r = T_r + \frac{V_{OS}}{kbG\lambda}$$

(13.45)

To determine T_A, α_r, and β_r at least three independent measurements are needed: the first one is obtained from the observed scene and the other two are provided by the calibration circuit in Figure 13.10. Such a circuit exploits a noise-adding mechanism to determine the radiometric scale, while a Dicke switch and a load at a known temperature T_{bb} are used to deal with the radiometric offset. In the following such a load will be referred to as the black-body of the instrument. In order to describe the noise source behavior mathematically, the excess noise ratio (ENR) is used. This parameter is defined by

$$ENR = \frac{T_H - T_C}{T_0}$$

(13.46)

where T_H is the hot temperature (noise source switched on), T_C is the cold temperature (noise source switched off), and $T_0 = 290$ K is the IEEE standard temperature. It is worth noting that T_C can be approximated with the physical temperature of the noise source T_{ns}. Indeed, in order to improve the output impedance matching over a wide operational bandwidth, a precision noise source typically exploits an output resistive attenuator and/or an isolator. As a consequence, when the source is switched off, the generated equivalent noise temperature corresponds to the physical temperature of the attenuator/isolator, that is, to that of the source itself. With reference to Figure 13.10, the noise injection circuit can be modeled by

$$T'_A = \begin{cases} T_A + CT_{ns}, & \text{off state} \\ T_A + C(ENR \cdot T_0 + T_{ns}), & \text{on state} \end{cases}$$

(13.47)

where C is the coupling factor of the directional coupler, T_A is the antenna noise temperature, and T'_A is the noise temperature at the receiver input, that is, the quantity that will be measured. Table 13.8 summarizes the four calibration states, obtained by combining the noise-adding mechanism with the Dicke switch operation. In such a table both the Dicke switch and the directional coupler have been considered ideal, that is, lossless. This is because their losses have already been included in the pre-detection gain term, G.

Table 13.8 Calibration states

Dicke switch	Noise source	T'_A	V_m
Scene	Off	$T_A + CT_{ns}$	V_s^{off}
Scene	On	$T_A + C(ENR \cdot T_0 + T_{ns})$	V_s^{on}
Black-body	Off	$T_{bb} + CT_{ns}$	V_{bb}^{off}
Black-body	On	$T_{bb} + C(ENR \cdot T_0 + T_{ns})$	V_{bb}^{on}

The calibration procedure will follow these steps. In the first step the radiometric offset β_r is determined using the black-body. To this purpose the Dicke switch will select the waveguide load and two measurements are carried out, one with the noise source in the on state and another with the noise source in the off state. Repeating the above pair of measurements more times it is possible to reduce the standard deviation of the radiometric offset estimation. The equations are

$$T_{bb} + CT_{ns} = \alpha_r V_{bb}^{off} - \beta_r \tag{13.48}$$

$$T_{bb} + C(ENR \cdot T_0 + T_{ns}) = \alpha_r V_{bb}^{on} - \beta_r \tag{13.49}$$

Solving these equations for β_r gives

$$\beta_r = CT_0 ENR \frac{V_{bb}^{off}}{V_{bb}^{on} - V_{bb}^{off}} - T_{bb} - CT_{ns} \tag{13.50}$$

showing that the radiometric offset can be evaluated from the knowledge of the following calibration parameters: ENR of the noise source, coupling factor of the directional coupler, and physical temperatures of both the black-body and noise source. The latter quantities are directly measured using temperature sensors. In the second step the radiometric scale α_r is determined by observing the scene. To this purpose the Dicke switch will select the scene under observation (i.e., the receiver antenna) and two measurements are carried out, one with the noise source in the on state and another with the noise source in the off state. The equations are

$$T_A + CT_{ns} = \alpha_r V_S^{off} - \beta_r \tag{13.51}$$

$$T_A + C(ENR \cdot T_0 + T_{ns}) = \alpha_r V_S^{on} - \beta_r \tag{13.52}$$

Solving these equations for α_r gives

$$\alpha_r = \frac{CT_0 ENR}{V_S^{on} - V_S^{off}} \tag{13.53}$$

As in the previous case, the radiometric scale is expressed in terms of the calibration circuit parameters. The antenna noise temperature is finally determined by

$$T_A = \alpha_r V_m^{off} - (\beta_r + CT_{ns}) \tag{13.54}$$

where V_m^{off} are the measurements of the scene carried out after the radiometric scale calibration with the noise source in the off state.

The design parameters of the calibration circuit are quoted in Table 13.9; the reported values have been derived assuming $T_{ns} = 300$ K together with a GaAs Schottky detector, a pre-detection gain of 50.7 dB, and a video amplifier detector $G_v = 250$. In these conditions the output radiometer voltage for $T_A = 1423$ K, that is, for a solar spectral flux density of about 12 200 sfu, is $V_m \approx 2.4$ V.

Table 13.9 Calibration circuit parameters

C (dB)	ENR (dB)	$CT_0\,ENR$ (K)	CT_{ns} (K)	α_r (K/V)	β_r (K)
-15	12	145.3	9.5	1232	1547

13.17 Retrieval Equations

The spectral flux density S_f of the Sun can be obtained from the measured antenna noise temperature, the latter determined with the previously described instrument calibration. A closed-form relationship is easily derived assuming the antenna to be pointed toward the Sun center. Combining Equations (13.5) and (13.31) gives

$$S_f \approx 2.26 \frac{kT_A \Theta_h^2}{c_0^2} f^2 \tag{13.55}$$

13.18 Periodicity of the Calibrations

Equations (13.54) and (13.55) allow the spectral flux density of the Sun to be retrieved from the output voltage of MIOS. However, since the physical temperature of the instrument can fluctuate during the orbit time, the calibration of both the radiometric scale α_r and radiometric offset β_r must be periodically repeated. Viewing such an issue from a different perspective, it can be said that the periodic calibration is a tool to keep the gain fluctuations (at least those related to the physical instrument temperature drift) below the 100 ppm threshold stated in Section 13.13. To verify the effectiveness of calibration while determining its periodicity, a thermal drift model of MIOS has first been developed. Such a model starts from the assumption that the physical temperature of the instrument T_p has the following time behavior:

$$T_p(t) = T_p^0 + \Delta T_p^m \sin\left(2\pi \frac{t}{t_{orbit}}\right) \tag{13.56}$$

where T_p^0 is the reference instrument temperature, ΔT_p^m is the maximum allowed temperature variation, and t_{orbit} is the orbital period. Considering an orbital period of about 100 min and exploiting the values $T_p^0 = 300$ K and $\Delta T_p^m = 1$ K, the graph of Figure 13.11 is obtained.

The second step is the description of the temperature impact over pre-detection gain, receiver noise temperature, and video amplifier offset. According to Reference [39], the above function of the temperature has been linearized around $T_0 = 290$ K since T_p^0 is close to T_0 and the temperature variation is limited. For the pre-detection gain a constant slope in dB has been considered:

$$G_{dB} = G_{dB}^0 - \delta_G\left(T_p - T_0\right) \tag{13.57}$$

G_0 being the gain specified at T_0 and δ_G the gain over temperature sensitivity-expressed in dB/K. The pre-detection gain in linear units is easily obtained from the dB value:

$$G = 10^{G_{dB}/10} \tag{13.58}$$

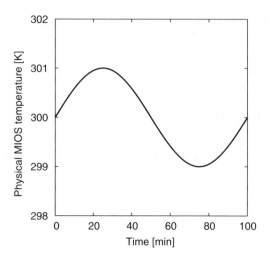

Figure 13.11 Assumed variation of the MIOS physical temperature during the orbit. The reference instrument temperature is equal to 27 °C.

The sign of Equation (13.57) implies that a temperature increase will result in a gain decrease, as typically happens in millimeter-wave amplifiers. The equivalent noise temperature of the receiver has been described by

$$T_R = T_R^0 + \delta_{TR}(T_p - T_0)$$

(13.59)

where T_R^0 is the receiver noise temperature specified at T_0 and δ_{TR} is the sensitivity of T_R with respect to T_p expressed in K/K. To evaluate T^0, Equation (13.34) can be used, provided that the noise figure is specified at T_0, that is, at the IEEE standard temperature. Finally, the video amplifier offset has been modeled by

$$V_{OS} = V_{OS}^0 + \delta_{VOS}(T_p - T_0)$$

(13.60)

where V_{OS}^0 is the offset at T_0 and δ_{VOS} is the offset drift over temperature. The latter value has already been indicated in Table 13.6.

A simulation of MIOS under temperature variations is now feasible. To this purpose the radiometer output voltage V_m has been determined using Equations (13.57) to (13.60) into

$$V_m = kBGG_v\gamma(T_A' + T_R) + G_vV_{OS}$$

(13.61)

where T_A' is described by Table 13.8 as a function of the calibration state. In the simulation an antenna noise temperature constant over time is adopted in such a way as to reproduce the quiet Sun behavior. In particular, $T_A = 1423$ K according to Table 13.4. As expected, the obtained V_m strongly depends on time because of the modeled temperature drifts. At this point a calibration procedure for both the radiometric offset and scale has been simulated. The output radiometer voltage at the calibration instants has been used to evaluate β_r and α_r

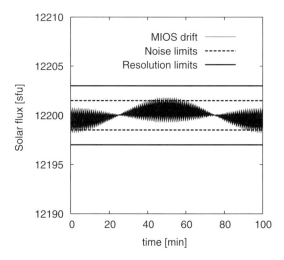

Figure 13.12 Solar spectral flux density retrieved by MIOS in the presence of both temperature drifts and periodic calibrations. In this study a ±1 K variation of the physical temperature is assumed. Such a variation occurs during the orbit, that is in about 100 min, causing both gain and noise fluctuations. Radiometric scale and offset calibrations are simulated every 4 s and 60 s respectively.

with Equations (13.50) and (13.53) respectively. Then the antenna noise temperature has been retrieved with Equation (13.54), while the solar spectral flux density is obtained by Equation (13.55). The results of these simulations are shown in Figure 13.12, illustrating that the drift-induced errors are comparable with the noise limits and well below the resolution limits of the instrument. The resolution limits are equal to ±3 sfu, that is, to the minimum standard deviation of the instrument. The noise limits, indeed, are those computed by Equation (13.35) for $\tau = 1$ s with $\Delta G/G = 0$, that is, the intrinsic noise limits of the radiometer.

The parameters modeling the MIOS thermal drift are quoted in Table 13.10. Then the periodicity of the calibrations has been designed to meet the resolution limits also in the presence of the above ±1 K temperature variation. As a result a scale calibration each of 4 s and a scale calibration each of 60 s are needed.

The property of the simulated calibration procedure is that, after the calibration, the scale/offset constants are applied to the measured data until the next calibration. Interpolation between two consecutive calibrations is a data processing algorithm with superior performances [35]. Other degrees of optimization are related to the adoption of a continuous gain calibration, as in noise-adding radiometers, or to the development of a suitable compensation algorithm [39].

Table 13.10 Thermal Drift Parameters

G^0 (dB)	δ_G (dB/K)	T_R^0 (K)	δ_{TR} (K/K)	V_{OS}^0 (μV)	δ_{VOS} (μV)
50.7	0.1	1540	1.0	25	0.5

13.19   Conclusions

In conclusion, MIOS is an experiment that will improve the knowledge of solar flares by probing the radiation mechanisms associated with high-energy particles. MIOS will operate at 90 GHz with two main goals, namely, obtaining a statistical characterization of flare events at millimeter wavelengths and verifying the existence of fast time structures. A statistical study of solar flares using data from the Nobeyama, Radio Polarimeter (NORP) in Japan has been carried out to verify that MIOS will detect sufficient events to carry out its scientific goals. The probability distribution of solar flare amplitudes has been derived and then the results are scaled appropriately to 90 GHz. Such a distribution predicts that, in a 12-year solar cycle, there will be roughly 160 events above 10 sfu at 90 GHz and 105 events above 20 sfu.

The radiometric resolution shall be less-or-equal than 10 sfu for 100 ms integration time and less-or-equal than 3 sfu for 1 s integration time in at least a 600 ms time scale. Such a radiometric resolution is the standard deviation of the measured data. The design of MIOS is based on a single-dish radiometer. This approach can be pursued since the satellite will fly in the high ionosphere, the variability of which has been found to be very low, that is, about 4 ppm. For this purpose the ionosphere has been modeled as stratified plasma and its variability has been derived considering the difference between the maximum and minimum attenuation.

The system architecture is based on a 35.9 cm offset dish antenna, which will be placed within the satellite body. The antenna offset is equal to $45°$ while a corrugated feed-horn is used to illuminate the dish with $f/D \simeq 0.6$. The receiver is based on a direct amplification configuration, that is, without local oscillator and down-conversion mixer. Such a solution is less expensive, more stable, and more reliable than those using frequency conversion (homodyne or superheterodyne). To attain the specified radiometric sensitivity it is necessary to achieve an overall noise of less than 8 dB and a gain stability better than 100 ppm. The first parameter is achieved by exploiting a 90 GHz low-noise amplifier (LNA) in the receiver chain. To meet the second parameter, instead, a periodic calibration of the radiometer is necessary. In particular, limiting the temperature variations of the instrument to $\pm 1 °C$ around a certain reference temperature, it is found that the radiometric scale must be recalibrated every 4 s while the radiometric offset must be recalibrated every 60 s. The procurement of all the MIOS components, in particular that of the space-qualified LNAs, is possible. The mechanical layout has been defined, while the power and mass budget are available. A first estimation of the MIOS costs indicated that the instrument is feasible within the budget of a small space mission.

References

1. Lüthi, T. (April 2004) Solar Flares at Millimeter and Submillimeter Wavelengths – Instrumental Techniques and Observations. PhD Dissertation, University of Bern, Bern, Switzerland.
2. Raulin, J.P. and Pacini, A.A. (2005) Solar radio emission. *Advances in Space Research*, **35** (5), 739–754.
3. Raulin, J.P., White, S.M., Kundu, M.R. *et al.* (1999) Multiple components in the millimeter emission of a solar flare. *Astrophysical Journal*, **522** (1), 547–558.
4. Trottet, G., Molodij, G. and Klein, K.L. (2005) High-energy electrons in solar flares. *EAS Publication Series*, **14**, 101–106.

5. Miller, J.A., Cargill, P.J., Emslie, A.G. *et al.* (1997) Critical issues for understanding particle acceleration in impulsive solar flare. *Journal of Geophysical Research*, **102** (A7), 14 631–14 659.

6. Pohjolainen, S., Hildebrandt, J., Karlicky, M. *et al.* (2002) Prolonged millimeter-wave radio emission from a solar flare near the limb. *Astronomy and Astrophysics*, **396**, 683–692.

7. Bastian, T.S. (2002) ALMA and the Sun. *Astronomische Nachrichten*, **323** (3/4), 271–276.

8. Kundu, M.R., White, S.M., Shibasaki, K. and Sakurai, T. (2000) Nonthermal flare emission from MeV energy electrons at 17, 34 and 86 GHz. *Astrophysical Journal*, **545** (2), 1084–1088.

9. Ramaty, R., Schwartz, R.A., Enome, S. and Nakajima, H. (1994) Gamma-ray and millimeter-wave emission from the 1991 June X-class solar flares. *Astrophysical Journal*, **436** (2), 941–949.

10. Kaufmann, P., Correia, E., Costa, J.E.R. *et al.* (1985) Solar burst with millimeter-wave emission at high frequency only. *Letters to Nature*, **313**, 380–382.

11. Kaufmann, P., Giménez de Castro, C.G., Makhmutov, V.S. *et al.* (2003) Launch of solar coronal mass ejections and submillimeter pulse bursts. *Journal of Geophysical Research (Space Physics)*, **108**, 1280.

12. Raulin, J.P., Kaufmann, P., Giménez de Castro, C.G. *et al.* (2003) Properties of fast submillimeter time structures during a large solar flare. *Astrophysical Journal*, **592**, 580–589.

13. Giménez de Castro, C.G., Kaufmann, P. and Raulin, J.P. (2005) Recent results on solar activity at sub-millimeter wavelenghts. *Advances in Space Research*, **35** (10), 1769–1773.

14. Karoff, C. and Kjeldsen, H. (2008) Evidence that solar flares drive global oscillations in the sun. *The Astrophysical Journal Letters*, **678** (1), L73–L76.

15. Nakajima, H., Sekiguchi, H., Sawa, M. *et al.* (1985) The radiometer and polarimeter at 80, 35 and 17 GHz for solar observations at Nobeyama. *Publications of the Astronomical Society of Japan*, **37**, 163–170.

16. White, S.M. and Kundu, M.R. (1992) Solar observation with a millimeter-wavelength array. *Solar Physics*, **141**, 347–369.

17. Kraus, J.D. (1966) *Radio Astronomy*, McGraw-Hill.

18. Linsky, J.L. (1973) A recalibration of the quiet sun millimetre spectrum based on the moon as an absolute radiometric standard. *Solar Physics*, **28**, 409–418.

19. Metsähovi Radio Observatory, Finland (2008) Solar radio astronomy at Metsähovi, available online from: HYPERLINK "http://kurp-www.hut.fi/sun/metsahoviaurinko.shtml" http://kurp-www.hut.fi/sun/metsahoviaurinko.shtml.

20. Silva, A.V.R., Wang, H. and Gary, D.E. (2000) Correlation of microwave and hard X-ray spectral parameters. *Astrophysical Journal*, **545**, 1116–1123.

21. White, S.M., Krucker, S., Shibasaki, K. *et al.* (2003) Radio and hard X-ray images of high-energy electrons in an X-class solar flare. *Astrophysical Journal Letters*, **595**, L111–L114.

22. White, S.M., Kundu, M.R., Bastian, T.S. *et al.* (1992) Multifrequency observations of a remarkable solar radio burst. *Astrophysical Journal*, **384**, 656–664.

23. Kaufmann, P., Raulin, J.-P., Giménez de Castro, C.G. *et al.* (2004) A new solar burst spectral component emitting only in the terahertz range. *Astrophysical Journal Letters*, **603**, L121–L124.

24. Nobeyama Radio Observatory (2008) NoRP, The Nobeyama Radio Polarimeter, available online from: HYPERLINK "http://solar.nro.nao.ac.jp" http://solar.nro.nao.ac.jp.

25. Newman, M.E.J. (2005) Peak gamma-ray intensity of solar flares between 1980 and 1989. *Contemporary Physics*, **46**, 323.

26. Clauset, A., Shalizi, C.R. and Newman, M.E.J. (2009) Power-law distributions in empirical data. *SIAM Review*, **51** (4), 661–703.

27. White, S.M. (2008) Solar fare peak fux distributions at 90 GHz, personal communication.

28. Lüthi, T., Magun, A. and Miller, M. (2004) First observation of a solar X-class flare in the submillimeter range with KOSMA. *Astronomy and Astrophysics*, **415**, 1123–1132.

29. Lüthi, T. (December 1999) Nulling-Interferometer zur Beobachtung von Sonneneruptionen bei 90GHz. Master Thesis, University of Bern, Bern, Switzerland.

30. Lawrence, R.S., Little, C.G. and Chivers, H.J.A. (1964) A survey of ionospheric effects upon earth–space radio propagation. *IEEE Proceedings*, **51** (1), 4–27.

31. Chasovitin, Y.K. (2008) Typhoon Obninsk, Kaluga Region, Russia, available online from: HYPERLINK "http://www.wdcb.ru/stp/data/" http://www.wdcb.ru/stp/data/.

32. Goldsmith, P.F. (1998) *Quasioptical Systems: Gaussian Beam, Quasioptical Propagation and Applications*, IEEE Press.

33. GRASP Reference Manual TECC2, available online from: HYPERLINK "http://www.ticra.com" http://www.ticra.com.
34. Vasic, V. (December 2004) An Airborne Millimetre-Wave Radiometer at 183GHz: Receiver Development and Stratospheric Water Vapour Measurements. PhD Dissertation, University of Bern, Bern, Switzerland.
35. Hersman, M.H. and Poe, G.A. (1981) Sensitivity of the total power radiometer with periodic absolute calibration. *IEEE Transactions on Microwave Theory and Techniques*, **29** (1), 32–40.
36. Sholley, M., Barber, G., Raja, R. and Jackson, C. (2000) Advanced MMIC radiometers. Asia Pacific Microwave Conference 2000, pp. 1–5.
37. Tanner, A.B. (2001) A high stability Ka-band radiometer for tropospheric water vapor measurements. IEEE Aerospace Conference 2001, pp. 1849–1863.
38. Analog Devices Inc. AD524 – Precision Instrumentation Amplifier, REV. C, available online from: HYPERLINK "http://www.analog.com" http://www.analog.com.
39. Thompson, D.A., Rogers, R.L. and Davis, J.H. (2003) Temperature compensation of total power radiometers. *IEEE Transactions on Microwave Theory and Techniques*, **51** (10), 2073–2078.
40. Alimenti, F., Palazzari, V., Battistini, A. *et al.* (2012) A system-on-chip millimeter-wave radiometer for the space-based observation of solar flares. Proceedings of the 15th International Conference on Microwave Techniques (COMITE) 2010, Brno, Czech Republic, April 2012, pp. 3–8, 19–21.

14

Active Antennas in Substrate Integrated Waveguide (SIW) Technology

Francesco Giuppi[2], Apostolos Georgiadis[2], Ana Collado[2], Maurizio Bozzi[1] and Luca Perregrini[1]

[1]*University of Pavia, Pavia, Italy*
[2]*Centre Tecnologic de Telecomunicacions de Catalunya (CTTC), Castelldefels, Spain*

14.1 Introduction

In this chapter, the substrate integrated waveguide (SIW) technology is adopted in the development of active low-profile antennas featuring a cavity-backed configuration; this topology consists of a traditional planar antenna where the radiating element is connected to a cavity. In the framework of modern wireless systems, these antenna structures have been extensively studied and implemented, as they offer many advantages, like increased efficiency performance thanks to the suppression of the unwanted surface-wave modes and better heat-dissipation capabilities due to the presence of an adequate metal surface. These features are particularly favourable in large array configurations handling high power levels and make these structures really suitable for wireless sensor networks and space communication applications.

In fact, low-profile antennas find an interesting field of application in the realization of reconfigurable phased-array configurations. Generally, this kind of designs requires compact, active antennas as unit elements, in order to avoid grating lobes when scanning the beam and to minimize spurious radiation. When active antenna elements are used, coupled oscillator arrays (COAs) can be implemented; their dynamical properties, in fact, permit a certain phase distribution to be selected among the elements of the array and hence steer the beam. In particular, the cavity-backed topology can be used in this case for controlling the coupling

Microwave and Millimeter Wave Circuits and Systems: Emerging Design, Technologies, and Applications,
First Edition. Edited by Apostolos Georgiadis, Hendrik Rogier, Luca Roselli, and Paolo Arcioni.
© 2013 John Wiley & Sons, Ltd. Published 2013 by John Wiley & Sons, Ltd.

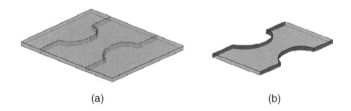

(a)   (b)

Figure 14.1 Comparison between (a) substrate integrated waveguide and (b) traditional metallic waveguide.

among adjacent elements in an array without need of phase shifters; this approach allows for more cost-effective and compact implementations.

In this chapter, the development of cavity-backed active antenna systems based on SIW technology is addressed. Different passive and active topologies are described in this chapter, including the first active SIW antenna ever demonstrated. In particular, the design of SIW antenna oscillators has been carried out by combining full-wave electromagnetic simulation and nonlinear analysis in order to model carefully both the radiating structure and the active circuitry. In addition, coupled oscillator array implementations are studied, based on the proposed antennas, and beam-scanning capabilities are demonstrated.

14.2 Substrate Integrated Waveguide Technology

A promising candidate for the development of a high-performance, low-cost platform for wireless applications is the substrate integrated waveguide (SIW) technology [1–4]; it allows for the fabrication of waveguide-like structures in a dielectric substrate with two ground planes, where the sidewalls of the waveguide are defined by two rows of metal vias (see Figure 14.1). SIW components are therefore intrinsically planar, suitable for straightforward integration with other planar technologies and easily implemented using standard, low-cost fabrication techniques, such as printed circuit board (PCB) and low-temperature co-fired ceramic (LTCC). This technology features similar performance to that achievable using metallic waveguides, such as low losses, good shielding, high power-handling capability, and high quality factor, but it also offers the possibility to integrate any kind of components on the same substrate; passive and active devices, as well as antennas, can be realized on the same dielectric material without the need for transitions between components implemented in different technologies. In addition, several chip-sets can be added and mounted on the same substrate, thus introducing the concept of system-on-substrate (SoS) [5].

SIW technology first appeared in the literature in 1998 [1] and since then a variety of devices have been proposed and demonstrated, including filters, directional couplers, oscillators, power amplifiers, antennas, six-port circuits and circulators. An exhaustive description of the state-of-the-art of SIW technology can be found in Reference [4].

14.3 Passive SIW Cavity-Backed Antennas

Many passive SIW antenna topologies have been proposed in the literature, the majority of them being slotted-waveguide and leaky-wave antennas [4]; these low-profile structures are widely required for many kinds of applications, from space communications (where they are employed in large reflectarray configurations) to radar imaging and wireless sensor networks. In this sense, SIW technology represents a very promising candidate for the realization of planar, cost-effective low-profile antennas, as it will be further demonstrated.

This chapter, in particular, focuses on the realization of active SIW cavity-backed antennas; in this kind of topology, the radiating element, which is usually a slot or a patch, is properly connected to a cavity. This design shows some interesting advantages, first of all the suppression of the undesired surface-wave modes, which increases the antenna efficiency and is particularly useful in array configurations. In addition, the presence of the cavity offers an adequate metal surface for effective heat dissipation, which can help when dealing with active devices in large implementations [6–8]. Moreover, in Reference [9] it is demonstrated how the use of a cavity-backed configuration in an oscillator helps to improve the phase-noise performance of the system. In the literature, passive SIW cavity-backed antennas appeared in References [10] to [12].

The first step in the development of active antenna systems is the careful design of the passive radiating structure; in this section, the dimensioning of passive cavity-backed antennas in SIW technology is addressed.

14.3.1 Passive SIW Patch Cavity-Backed Antenna

Initially, the design of a cavity-backed antenna having a patch as the radiating element has been carried out. The dimensioning was done using an FEM-based full-wave simulator that allows the cavity and the patch to be designed for 10 GHz resonance (Ansys HFSS, used for all the designs proposed in this chapter).

As all the prototypes presented in this chapter, this antenna was realized on an Arlon 25N substrate with a thickness of 0.508 mm and a dielectric constant of 3.38 at 10 GHz. The layout of the prototype is shown in Figure 14.2.

The SIW cavity, resonating on the TM_{120} mode, is created by connecting the two ground planes of the substrate by means of metal vias with diameter $D = 1$ mm and spacing 2 mm, in order to avoid lateral radiation leakage (Figures 14.2 and 14.3). A rectangular ring is etched in the bottom metal plane thus forming a patch antenna. The coplanar feed line in the top plane is extended inside the cavity and a slot is etched perpendicular to the coplanar feed line on the top metal, in order to excite both the patch and the cavity resonance [13]. This antenna topology permits the feeding circuitry and the radiating patch to be isolated, since

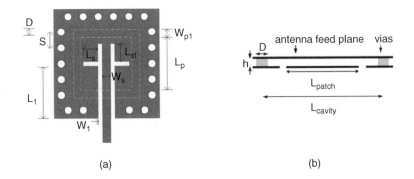

Figure 14.2 Geometry of the passive SIW patch antenna: (a) front view; (b) side view.

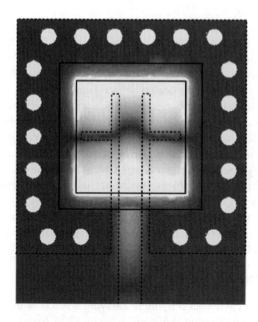

Figure 14.3 Plot of the electric field magnitude for the TM_{120} resonant mode of the cavity.

they are located on opposite sides of the dielectric substrate, thus preventing undesired effects of the feed line on the radiation performance. The dimensions of the cavity-backed antenna oscillator are listed in Table 14.1.

A single-substrate prototype of the presented patch antenna has been realized and measured [14], which is shown in Figure 14.4. In Figure 14.5 the simulated and measured S-parameters of the passive cavity-backed antenna are plotted. The measured structure exhibits a resonance at approximately 10.2 GHz, with a very good agreement between measurements and simulated data.

The simulated gain of the final design is approximately 4.6 dB at broadside, while the measured gain is 3.4 dB (efficiency around 60%), and the E-plane and H-plane radiation patterns are presented in Figure 14.6. The reduced measured broadside gain is attributed to ripple introduced by the small size of the ground plane and additional losses due to the fabrication method.

Table 14.1 Dimensions used in the design

Parameter	Dimension (mm)	Parameter	Dimension (mm)
D	1.00	W_s	1.16
L_{cavity}	11.80	W_{p1}	1.00
L_s	1.92	S	2.00
L_{st}	2.20	L_{patch}	6.60
L_1	6.65	W_1	0.50

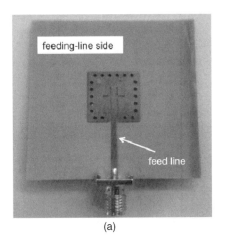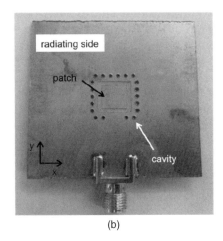

(a)   (b)

Figure 14.4 Photographs of the passive SIW patch antenna prototype: (a) feed-line side, (b) radiating side.

14.3.2   Passive SIW Slot Cavity-Backed Antenna

Another passive cavity-backed SIW antenna topology is presented in this section, which uses a slot as its radiating element. It comprises a slot antenna backed by an SIW cavity, excited by a coplanar line (Figure 14.7). This topology is similar to the structure presented in Reference [10].

The radiating element is a slot etched in the ground plane. An input grounded coplanar line is used to excite both the cavity mode and the slot. As already seen for the antenna presented in Section 14.3.1, this topology also permits the feeding line and the radiating slot to be isolated (Figure 14.7b).

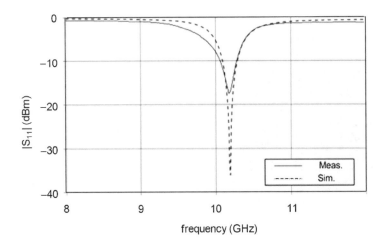

Figure 14.5 Measured (continuous) and simulated (dashed) $|S_{11}|$ of the passive cavity-backed patch antenna.

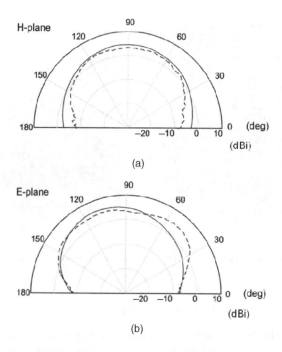

(a)

(b)

Figure 14.6 Simulated (solid) and measured (dashed) radiation pattern of the passive SIW patch antenna at the main resonance: (a) H-plane (xz) and (b) E-plane (yz).

With respect to the antenna proposed in Reference [10], the major differences are the use of a smaller cavity and of a dogbone-shaped slot, which allows for a more compact structure. The particular shape of the slot has been used in order to obtain the desired electrical length within the available cavity space (Figure 14.7a). The design of this passive structure has been carried out with a full-wave simulator.

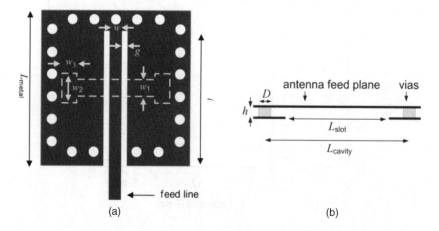

(a)   (b)

Figure 14.7 Geometry of the passive SIW slot antenna: (a) front view; (b) side view. Reproduced from [15]. With permission from ACES.

Table 14.2 Dimensions used in the design

Parameter	Dimension (mm)	Parameter	Dimension (mm)
D	1.00	w_1	1.40
L_{cavity}	11.80	w_2	2.60
L_{slot}	10.00	w_3	1.40
L_{metal}	13.80	w	1.16
i	12.00	o	0.20
g	0.50	h	0.508

The slot and the cavity have been dimensioned for broadside radiation and optimal input matching at 10 GHz. The cavity size has been selected to resonate on the TM_{120} mode. The feed line is a 50 Ω microstrip line outside the SIW cavity, whereas it turns into a coplanar line inside the cavity. The dimensions of the geometrical parameters involved in the design are listed in Table 14.2.

A prototype of the passive cavity-backed SIW antenna has been fabricated and tested, as demonstrated in Reference [15]. Figure 14.8 shows photographs of the feed-line side and of the radiating side of the antenna.

In Figure 14.9 the simulated and measured input matching of the antenna are plotted, which exhibit a very good agreement over the entire frequency band. In particular, the simulation results predict a resonance at 10.14 GHz, whereas the measured results show a minimum of $|S_{11}|$ at 10.17 GHz.

The simulated and measured radiation patterns of the passive antenna are shown in Figure 14.10, for both E-plane and H-plane cuts. The simulated gain of the final design was 4.8 dB at broadside, while the measured gain was 4.6 dB, leading to 70% antenna efficiency. The two patterns were obtained at the frequency of the minimum reflection coefficient, namely 10.14 GHz for the simulated pattern and 10.17 GHz for the measured (Figure 14.10). The asymmetry in the E-plane of the measured gain is potentially attributed to the feed line and connectors in the measurement setup.

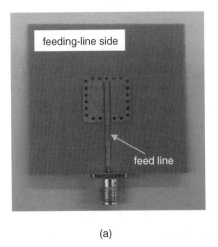

(a)

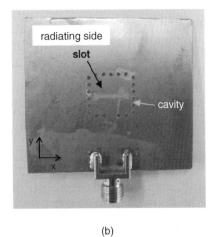

(b)

Figure 14.8 Photographs of the passive SIW slot antenna prototype: (a) feeding-line side; (b) radiating side. Reproduced from [15]. With permission from ACES.

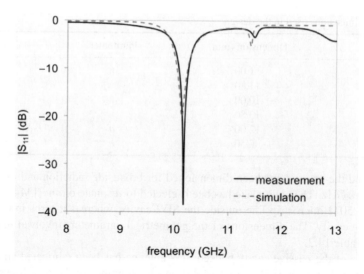

Figure 14.9 Simulated and measured input matching of the passive SIW antenna. Reproduced from [15]. With permission from ACES.

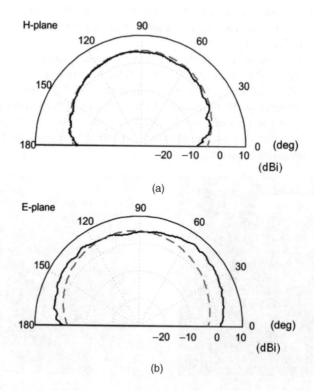

Figure 14.10 Simulated (dashed) and measured (solid) radiation pattern of the passive SIW antenna at the main resonance: (a) H-plane (xz) and (b) E-plane (yz). Reproduced from [15]. With permission from ACES.

14.4   SIW Cavity-Backed Antenna Oscillators

The passive structures proposed in the last section can be used in the development of cavity-backed antenna oscillators; in fact, by properly connecting an active device to the passive antenna, reflection-type oscillators can be obtained. The active antennas can be dimensioned in two steps. First, the passive radiators are designed with full-wave simulation methods and the resulting S-parameters are imported in a circuit simulator. Then, nonlinear Harmonic Balance analysis is used for optimizing the oscillator circuitry, in order to find a stable, steady-state oscillating solution at the desired frequency.

In this section, three different topologies of SIW antenna oscillators are proposed, two of them featuring frequency-tuning capabilities, which are obtained with two different approaches. The possibility to tune the oscillation frequency of the structure may be useful in some coupled oscillator array implementations, as discussed in Section 14.5.

14.4.1   SIW Cavity-Backed Patch Antenna Oscillator

In this section, the first topology of an antenna oscillator in SIW technology is proposed, which uses as the radiating element the passive cavity-backed patch antenna presented in Section 14.2.1.

Once the design of the cavity-backed antenna has been completed, the oscillator circuit is designed by using a commercial Harmonic Balance (HB) simulator available from Agilent ADS. The S-parameters of the passive antenna obtained from the full-wave electromagnetic simulation were used in the HB analysis. Here, in order to avoid the convergence to the trivial DC solution, the HB simulator was made to converge to the oscillating steady state using a properly defined auxiliary generator [16, 17]. The simulation setup is shown in Figure 14.11; the probe consists of an ideal voltage source in series with an ideal band-pass filter. It is connected in parallel to a circuit node, namely to the source of the transistor.

This auxiliary generator helps to force the simulator to find an oscillating solution with nonzero amplitude at the desired frequency. Once the oscillating steady-state solution has been obtained, Transient simulation is used to study its stability.

The active device used is an NE3210S01 HJFET. The gate of the active device is connected to the cavity feed line (Figure 14.12) and the complete circuit is optimized to obtain an oscillation frequency near the measured resonance of the cavity (10.2 GHz). The oscillator is self-biased by placing two resistors from the source terminals to ground. Stubs (Figure 14.12)

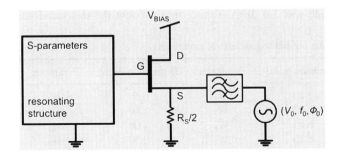

Figure 14.11   Schematic of the setup used for the simulation of the complete oscillator circuit.

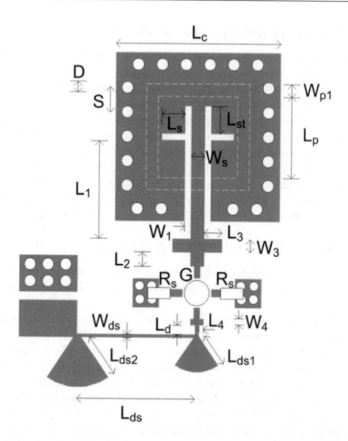

Figure 14.12 X-band active cavity-backed SIW patch antenna oscillator. © 2010 IEEE. Reprinted, with permission, from [14].

are used in the feed line to eliminate unwanted parasitic oscillations and additionally select the desired oscillation frequency. Transient simulation is used to verify the stability of the steady state solutions. The final geometrical dimensions used in the design are listed in Table 14.3.

A prototype of the designed cavity backed antenna oscillator was fabricated and characterized (Figure 14.13) [14]. The measured oscillation frequency was 10.025 GHz and the circuit dissipated 9 mA from a 1.5 V supply. The measured effective radiated power (ERP) at broadside was 3.6 dBm. Taking into account the measured gain of the passive

Table 14.3 Antenna oscillator geometrical parameters

Parameter	Dimension (mm)	Parameter	Dimension (mm)	Parameter	Dimension (mm)
L_c	11.80	W_1	0.50	W_s	1.16
L_s	1.92	W_3	1.10	L_{ds}	9.90
L_p	6.65	W_p	1.00	L_{st}	2.20
L_1	8.10	W_4	0.50	W_{ds}	0.30
L_2	1.20	L_{ds1}	2.90	D	1.00
L_3	1.50	L_d	0.80	s	2.00
L_4	0.40	L_{ds2}	3.90		

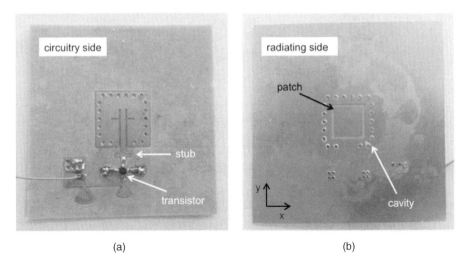

Figure 14.13 Photographs of the active SIW patch antenna prototype: (a) circuit side, and (b) radiating side. © 2010 IEEE. Reprinted, with permission, from [14].

antenna (3.4 dB), the radiated power was found to be 0.2 dBm (1.05 mW), which corresponds to approximately 8% DC-to-RF radiation efficiency. The measured E-plane and H-plane radiation patterns of the active antenna (effective radiated power) are shown in Figure 14.14. Using a probe to capture the radiated signal, the phase noise of the oscillator was measured to be -103 dBc/Hz at 1 MHz offset (Figure 14.15).

14.4.2   SIW Cavity-Backed Slot Antenna Oscillator with Frequency Tuning

The passive slot antenna discussed in Section 14.2.2 has been used as the starting point for the design of an active cavity-backed SIW antenna. The active antenna consists of the combination of the passive antenna and a reflection-type oscillator, which uses the SIW cavity as the resonant element (Figure 14.16). The choice of an SIW resonator leads to single-substrate implementation of both the antenna and the cavity, and the slot is placed on a separate side from the active circuit, thus minimizing the effects of the active components and bias lines in the radiation pattern of the device.

As already shown in Section 14.4.1, the S-parameters of the passive antenna obtained from the full-wave electromagnetic simulation are used in the HB analysis.

The reflection oscillator is obtained by connecting the gate of an active P-HEMT device (NE3210S01) to the cavity feed line (Figure 14.16). Two 16 Ω resistors R_s are then placed between the source terminals of the device and ground, in order to self-bias the circuit. The resulting structure is optimized to obtain an oscillation near the cavity resonance frequency of 10.14 GHz [16, 17].

One of the main parameters involved in the optimization of the active circuit is the stub length L_4 (Figure 14.16). Changing this length allows modification of the input impedance seen by the transistor looking into the cavity; this design parameter can therefore be used to optimize the oscillation frequency and eliminate unwanted parasitic oscillations and was set to $L_4 = 1$ mm.

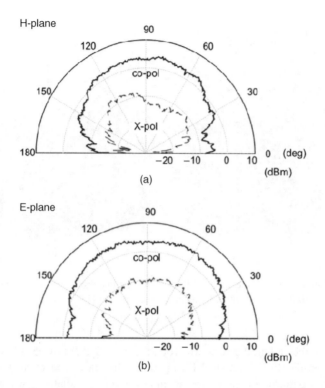

Figure 14.14 (a) H-plane (xz) and (b) E-plane (yz) measured radiation patterns of gain and power product (effective radiated power (dBm)). © 2010 IEEE. Reprinted, with permission, from [14].

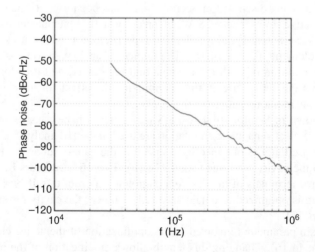

Figure 14.15 Measured phase noise of the active antenna oscillator. © 2010 IEEE. Reprinted, with permission, from [14].

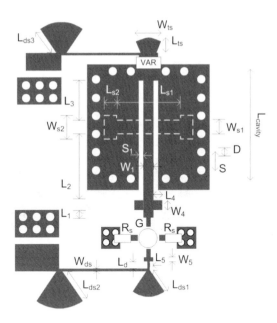

Figure 14.16 Geometry of the active SIW antenna. © 2010 IET. Reprinted, with permission, from [18].

In this antenna topology, frequency-tuning capabilities were introduced. In order to obtain a tunable antenna one via of the cavity wall is removed with respect to the structure used for the passive implementation of Section 14.2.2 (Figure 14.16), and a varactor diode is connected between the top plane and a radial stub with very low input impedance. The tuning of the varactor diode bias voltage V_c produces a change in the loading of the cavity and consequently it modifies its resonance frequency, allowing for the desired tenability to be obtained. In the full-wave simulation the presence of the diode has been taken into account by defining internal ports at the varactor terminal locations; in this way, in the subsequent circuit analysis, it is possible to connect a nonlinear model of the varactor diode to the cavity and predict the achievable tuning range. The selected varactor diode is the MA46H070-1056 with a control voltage V_c ranging from 0 to 20 V. The final dimensions of the circuit used for the design are listed in Table 14.4.

Table 14.4 Geometrical dimensions used in the design

Parameter	Dimension (mm)	Parameter	Dimension (mm)	Parameter (mm)	Dimension (mm)
L_{cavity}	11.80	W_3	0.50	W_{s2}	2.60
L_1	0.80	W_4	1.00	L_{s1}	7.20
L_2	7.30	W_5	0.50	L_{s2}	1.40
L_3	4.40	L_{ds1}	2.85	D	1.00
L_4	1.00	L_{ds2}	3.95	S	2.00
L_5	0.40	L_{ds3}	3.10	W_{ds}	0.30
W_1	1.16	S_1	0.50	L_{ts}	2.00
W_2	1.10	W_{s1}	1.40	W_{ts}	3.00

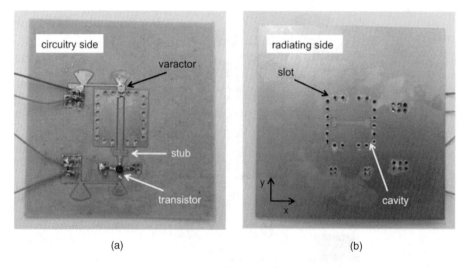

(a)   (b)

Figure 14.17 Photographs of the active SIW slot antenna prototype: (a) circuit side; (b) radiating side. © 2010 IET. Reprinted, with permission, from [18].

A tunable, active antenna prototype was built on Arlon 25N substrate and measured (Figure 14.17); in Figure 14.18, the measured and simulated oscillation frequency of the oscillator versus the biasing voltage V_c are compared. The oscillation frequency varies from 9.82 GHz to 9.98 GHz in simulation and from 9.82 GHz to 10 GHz in measurements, showing a 180 MHz tuning range obtained by tuning the varactor control voltage from 0 to 20 V.

The measured E-plane and H-plane radiation patterns of the antenna oscillator's effective radiated power are shown in Figure 14.19, for V_c equal to 0 and 20 V. The radiation pattern is essentially unchanged within the varactor tuning range. The power–gain product of the SIW cavity-backed slot antenna was evaluated in the anechoic chamber and found to be around 3 dBm for $V_c = 0$ V.

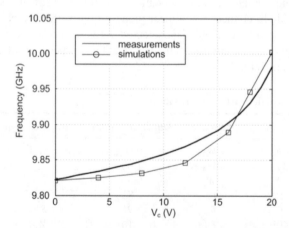

Figure 14.18 Simulated and measured frequency-tuning curves of the active antenna oscillator. © 2010 IET. Reprinted, with permission, from [18].

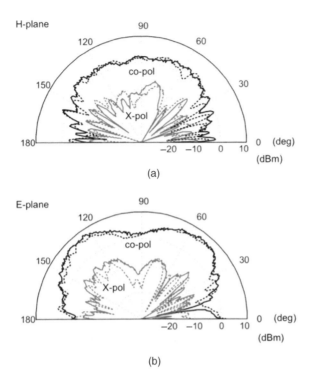

Figure 14.19 Measured radiation patterns (effective radiated power in dBm) for V_c equal to 0 V (solid lines) and 20 V (dashed lines): (a) H-plane (xz), (b) E-plane (yz). © 2010 IET. Reprinted, with permission, from [18].

Based on the measured ERP at broadside and the measured gain of the passive antenna, the radiated power is estimated to be 0.69 mW. The dissipated power of the antenna oscillator is 10.5 mW; thus, its DC-to-RF efficiency is approximately 6.5%. The phase noise performance of the active antenna was measured by capturing the radiated signal using an antenna placed at a short distance from the active antenna oscillator and by processing the received signal using a vector signal analyser (VSA). The measured phase noise was approximately −101 dBc/Hz at 1 MHz offset for control voltages in the range from 0 to 15 V. It monotonically reduced to −105 dBc/Hz as the control voltage was increased to 20 V, as plotted in Figure 14.20.

The active antenna addressed in this section, along with the different issues that arose during the design, is proposed in Reference [18], while a preliminary nontunable version was demonstrated in Reference [15]. The importance of this achievement lies in the fact that these prototypes represent the first active antennas based on substrate integrated technology to be demonstrated and that appeared in the literature.

14.4.3   Compact SIW Patch Antenna Oscillator with Frequency Tuning

As the active antennas presented in this chapter are intended for operation in coupled oscillator array configurations, the compactness is a fundamental issue to be addressed when

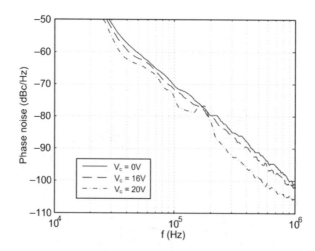

Figure 14.20 Measured phase noise performance of the tunable slot oscillator, for different biasing voltages V_c of the varactor diode.

choosing the suitable topology. In this section, an active SIW patch antenna is presented, whose novel, compact design makes it suitable for use in two-dimensional matrix implementations; another novelty resides in the very large bandwidth and frequency-tuning range, compared to typical patch antennas.

The topology selected for the design and implementation of the cavity-backed patch antenna oscillator is shown in Figure 14.21. The radiating element is a square patch enclosed in an SIW square cavity. On the opposite side with respect to the patch, a short feeding line intended for the excitation of the patch is etched, along with the pads needed for the placement of the active oscillating circuitry. Similarly to the previously presented antennas, this configuration has the advantage of keeping the radiating element and the active circuitry on opposite sides of the substrate. Moreover, its compact shape makes it suitable for application in two-dimensional array implementations.

Electromagnetic analyses were used to design the radiating structure; ports have been defined on the structure in order to model the terminals of the circuitry to be connected (FET, varactor diode, capacitances and resistance), and thus calculate the S-parameters of the antenna with respect to an excitation defined on the gate port (see Figure 14.21). The resonance frequencies of the patch and of the enclosing cavity were at first addressed in this phase of the work. In particular, the patch size was chosen to obtain resonance around 11.8 GHz, while the cavity dimensions were selected in order for the TM$_{120}$ mode to resonate around 12 GHz. The presence of the region etched on the top face where the pads for the circuital elements are placed was considered, as it heavily affects the quality factor of the cavity and it introduces a capacitive effect that modifies the resonance frequency of the desired mode.

After designing the radiating structure, the S-parameters resulting from the full-wave analyses were used in a Harmonic Balance simulator (Agilent ADS) in order to study the oscillating steady-state solutions for the complete active circuitry around 12 GHz.

As shown in Figure 14.21, the gate of the active P-HEMT element (NE3210S01) is connected to the cavity feed line. One of the two source terminals is connected to ground through a 22 Ω resistor, while the other is connected to a capacitance that helps set the

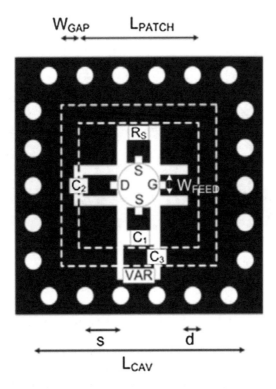

Figure 14.21   X-band compact cavity-backed SIW patch antenna oscillator.

oscillation frequency ($C_1 = 1.5$ pF). The geometrical dimensions of the antenna, as well as the values of the circuit components, are listed in Table 14.5.

Tuning of the oscillation frequency of the device is obtained by properly biasing with voltages from 0 to 20 V a MA46H070-1056 varactor diode, placed at the source terminal of the active device between C_1 and the ground. An additional capacitance $C_3 = 3.3$ pF helps set the width of the tuning range, while $C_2 = 1$ nF is used for biasing purposes.

With respect to the topology presented in Section 14.4.2, in this design the varactor diode is connected in series with the source terminal of the active device, while at the same time it affects the current flowing on the top side of the cavity. In this sense, the diode both affects the load of the resonant structure and the active device itself. As a result, it has a stronger effect on the frequency of oscillation. Using HB simulation, a steady-state, tunable solution around 12 GHz was found and its stability was studied using Transient analysis.

Table 14.5   Antenna oscillator geometrical parameters

Parameter	Dimension	Parameter	Dimension	Parameter	Dimension
L_{CAV}	11.80	s	2.00	L_{PATCH}	6.65
d	1.00	W_{GAP}	1.00	W_{FEED}	1.16

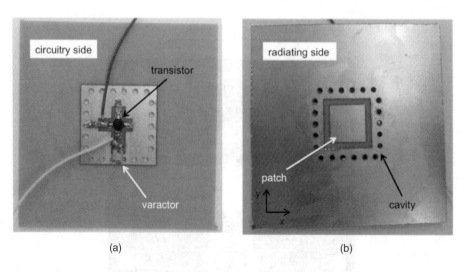

(a)   (b)

Figure 14.22 Photographs of the active SIW antenna prototype: (a) circuit side; (b) radiating side.

A prototype antenna oscillator was implemented and measured (Figure 14.22), whose frequency tuning capability is shown in Figure 14.23. It can be seen that a tuning range of about 500 MHz was achieved.

In Figure 14.24 the ERP radiation patterns for the antenna oscillator are shown for different values of biasing voltages of the varactor diode (0, 10 and 20 V). As it can be seen, as the frequency is increased the value of the power–gain product at broadside remains roughly constant at around 7 dBm. The antenna oscillator draws 12 mA current for 1.5 V supply, constantly over the entire tuning range, with 18 mW dissipated power. Taking into account the simulated gain of the passive radiating structure, which is equal to 5.3 dB, the radiating power was estimated to be 1.48 mW, thus leading to approximately 8% DC-to-RF radiation efficiency. Finally, the noise performance of the circuit was evaluated at 12.3 GHz (13 V control voltage), showing –107 dBc/Hz phase noise at 1 MHz frequency offset (Figure 14.25).

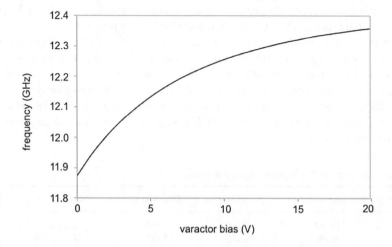

Figure 14.23 Measured frequency tuning of the antenna oscillator.

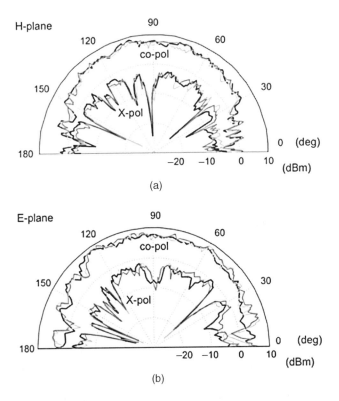

Figure 14.24 ERP co- and cross-polarization radiation patterns for the designed antenna oscillator for varactor bias equal to 0, 10 and 20 V: (a) H-plane (*xz*) and (b) E-plane (*yz*).

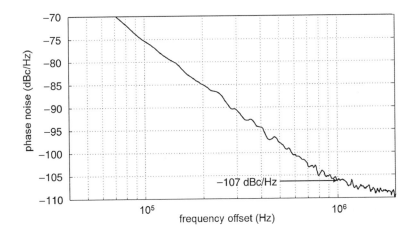

Figure 14.25 Phase noise performance measurement for the designed antenna oscillator.

14.5 SIW-Based Coupled Oscillator Arrays

In the previous section, some SIW active cavity-backed antenna topologies have been proposed, which may be suitable for the realization of coupled oscillator arrays (COAs). A number of applications may require the use of coupled oscillator arrays: among them, the most relevant applications involve coherent distributed sources based on power combining, phased arrays with beam-scanning capabilities without using any phase shifter and advanced reflectarray configurations.

The active SIW antennas can be used in order to combine the output power from each of the elements, thus obtaining a coherent distributed source. The coupling between different elements is achieved by properly connecting their SIW cavities, for instance by opening an aperture in the common wall. In this section, a method intended for establishing the desired coupling among different elements of a phased array is proposed. Thanks to the SIW topology, adjacent cavities can be coupled with proper magnitude and phase, thus allowing for in-phase synchronization of the elements of the array at a common frequency f_s; the output power generated by each oscillator is then combined [19].

One of the most significant features of coupled oscillator arrays relies on the possibility of obtaining beam-steering with no need of phase shifters. The cavity-backed topology was demonstrated to be particularly suitable to this aim [20] and is fully exploited in this section. In fact, since some of the proposed SIW active antennas feature frequency-tuning capabilities, they can be used in a coupled oscillator array configuration in order not only to obtain power combining but also to introduce beam-scanning properties, as shown in Figure 14.26. Once in-phase synchronization is obtained, the free-running frequency of the edge elements of the structure can be slightly detuned in opposite directions, thus establishing different, constant phase-shift distributions along the array and scanning the beam over a certain angle. Generally, this kind of design requires compact, active antennas as unit elements, in order to avoid grating lobes when scanning the beam and minimize spurious radiation, as discussed in Reference [21].

In this section, a preliminary study of the possibility of synchronizing two adjacent elements in an array configuration is presented, with the aim of combining the output power. Then, the study and implementation of a three-element coupled oscillator array with beam-scanning properties is described.

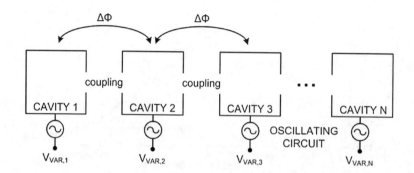

Figure 14.26 Diagram of a coupled oscillator array for the phased array configuration.

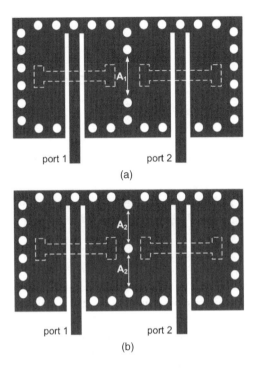

Figure 14.27 Cavity-backed antenna array structure for the study of the coupling magnitude: (a) single-aperture and (b) double-aperture configurations.

14.5.1 Design of Coupled Oscillator Systems for Power Combining

When dealing with cavity-backed antenna oscillators, the easiest way to obtain a certain level of coupling between them is to couple the respective cavities; in order to achieve this, an aperture of the proper shape and width can be used to connect adjacent cavities. For this reason, a preliminary study about the possibility of controlling the coupling between two antenna elements in an array was carried out on the structures shown in Figure 14.27. In this case, the passive array is simply obtained by connecting two cavity-backed slot antennas, such as the one presented in Section 14.2.2, and the geometrical dimensions used are the same.

Two different approaches are adopted when designing an aperture between two SIW cavities. In the first structure, a single window is opened between the two elements, while in the second two apertures are present; in this kind of design, SIW technology appears to be very flexible, as it allows cavities of any shape to be drawn.

Using a full-wave simulator, and defining ports at the feed lines as shown in Figure 14.27, the parameter $|S_{21}|$ can be considered as the strength, or magnitude, of the coupling between the two cavities. Simulated results are presented in Figure 14.28, where the variations of $|S_{21}|$ at resonance and of the resonance frequency itself (where $|S_{11}|$ is minimum) are plotted, versus the aperture width, for the two different configurations.

While the resonance frequency value remains roughly constant, a 9 mm variation in the single aperture configuration produces a change in the coupling strength of approximately 7 dB in both configurations. In particular, in the double-aperture implementation the

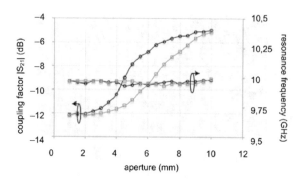

Figure 14.28 Coupling factor magnitude and resonance frequency in the case of a single aperture A_1 = aperture (black lines) or two apertures A_2 = aperture/2 (grey lines). © 2010 IEEE. Reprinted, with permission, from [23].

variation of $|S_{21}|$ is much smoother, thus leading to a more accurate control of the desired coupling factor. This structure was chosen for the further development of the study.

Based on this analysis, two passive structures with a double aperture have been developed, showing different coupling factor magnitudes; the implemented arrays are shown in Figures 14.29 and 14.30, and have been realized. The first one, with one via hole dividing the two cavities, features $A_2 = 4.9$ mm, while the other one has two vias and A_2 = 3.5 mm. The two arrays show simulated coupling magnitudes equal to −7 and −10 dB respectively, evaluated at the resonance frequency of the structures.

A comparison between the measured and simulated $|S_{21}|$ is shown in Figure 14.31. The measured coupling factor curves match very well with simulations; at 10.2 GHz, the measurements show a coupling factor of −7.6 dB for the first array (0.6 dB discrepancy) and of −10.5 dB for the second array (0.5 dB discrepancy).

Based on the first passive array configuration, a coupled antenna oscillator system has been designed, where the active devices are NE32101S01 P-HEMTs, whose gates are connected to the cavity feed lines (Figure 14.32).

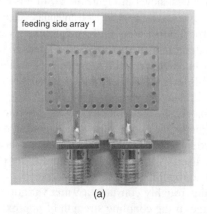
(a)

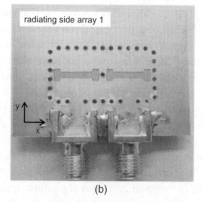
(b)

Figure 14.29 Coupled cavity-backed antenna structure with one via between the adjacent cavities: (a) feeding side and (b) radiating side.

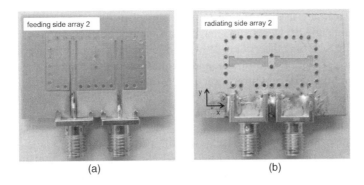

Figure 14.30 Coupled cavity-backed antenna structure with one via between the adjacent cavities: (a) feeding side and (b) radiating side.

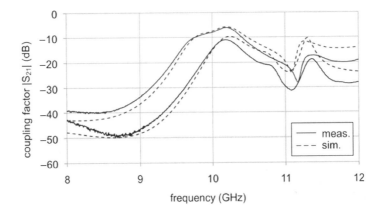

Figure 14.31 Measured (continuous) and simulated (dashed) $|S_{21}|$ for the proposed passive 2×1 arrays.

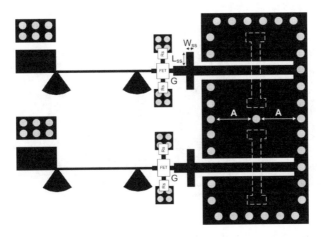

Figure 14.32 Coupled antenna oscillator system layout. © 2010 IEEE. Reprinted, with permission, from [23].

The radiating structure used in the design is identical to the passive one analysed in the previous section, and 16 Ω resistors have been located at the source terminals of the FETs in order to self-bias the circuit. When the individual oscillators are introduced into the array, their oscillation amplitudes and frequencies will change due to their mutual influence through the SIW coupling structure. A further optimization is thus necessary in order to readjust the synchronization frequency of the system, by acting on the stub's dimensions (Figure 14.32); $L_{ss} = 2$ mm and $W_{ss} = 1$ mm have been chosen.

The design of the active array has been carried out using Harmonic Balance simulations (Agilent ADS). As already stated, HB simulation is combined with the use of a properly defined ideal probe [16, 17], in order to ensure its convergence to the oscillating steady-state solution of the system. The stability of the solution is then studied using Transient simulations.

The prototype shown in Figure 14.33 was measured in order to demonstrate the possibility of combining in-phase the output power coming from the two synchronized elements. The system showed a synchronization frequency of 10.09 GHz. Theoretically, when coupling two identical oscillators in-phase one can achieve a 3 dB increase in the total output power with respect to the individual power of each oscillator. In order to determine the gain obtained by combining the two oscillator output powers, the devices were individually biased and their output power and frequencies where measured. The free-running frequencies were found to be $f_1 = 10.038$ GHz and $f_2 = 10.024$ GHz.

Using a probe in order to capture the radiated signal, the output power of the two oscillators was measured. The relative measured output powers of the oscillators were P_1 and $P_2 = (P_1 + 2.7)$ dB, which combined to give a total output power of $P_{out} = (P_1 + 6.7)$ dB $= (P_2 + 4)$ dB. The differences in the individual oscillator output powers are attributed to the limited manufacturing accuracy and component tolerances. Similar considerations can be made about the phase noise performance of the system; the measured phase noises of the individual oscillators are $PN_1 = -97.67$ dBc/Hz and $PN_2 = -97.43$ dBc/Hz at 1 MHz offset from the carrier. As expected, when combining two oscillator outputs the

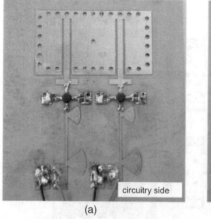 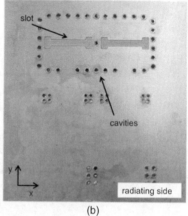

(a)   (b)

Figure 14.33 Photography of the implemented prototype of a coupled antenna oscillator system. © 2010 IEEE. Reprinted, with permission, from [22].

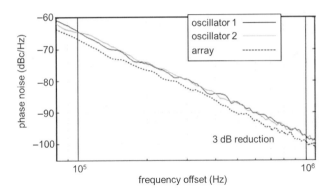

Figure 14.34 Measured phase noise of the individual oscillators and of the 2×1 array of coupled oscillators. © 2010 IEEE. Reprinted, with permission, from [22].

total phase noise of the system was PN_{out}(at 1 MHz) $= -100.8$ dBc/Hz with approximately a 3 dB improvement (Figure 14.34).

Finally, the radiation pattern of the 2×1 active array was evaluated in the anechoic chamber. Figure 14.35 shows the obtained E-plane and H-plane radiation patterns. The product of gain and transmitted power (effective radiated power) is plotted. The E-plane radiation

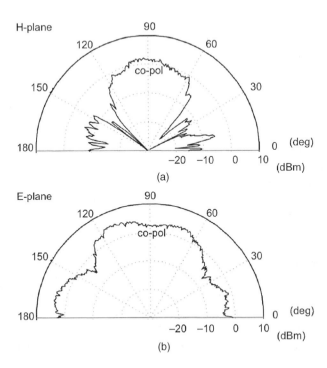

Figure 14.35 (a) H-plane (xz) and (b) E-plane (yz) radiation patterns (effective radiated power) of the 2×1 coupled oscillator array. © 2010 IEEE. Reprinted, with permission, from [22].

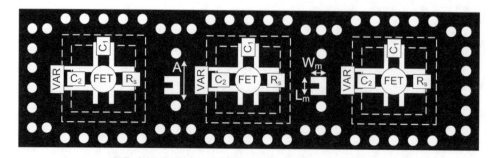

Figure 14.36 Layout of the three-element coupled oscillator array. © 2012 IEEE. Reprinted, with permission, from [25].

pattern shows some ripple, which is potentially attributed to the small ground plane. The measurements verify once more that the two oscillators are locked in-phase, since the radiation peak is located at broadside [22, 23].

14.5.2 Coupled Oscillator Array with Beam-Scanning Capabilities

In this section, a three-element coupled oscillator array based on SIW active antennas is presented. The structure used to develop the study is shown in Figure 14.36; a coupled oscillator array comprising three cavity-backed patch antennas (whose topology has been proposed in Section 14.3.3) is considered. An aperture $A = 5.2$ mm is created between the cavities, whose shape is modified in order for the elements to be half-wavelength spaced approximately (in this case 15 mm). In this way, the magnitude of the coupling between the elements is chosen. The value of the resistors and capacitors mounted on the structure is unchanged with respect to the single-element topology of Section 14.3.3, as well as the other geometrical dimensions.

In order to synchronize the array elements in-phase, the phase of the coupling between them has to satisfy the relation $-90 \leq \Phi_{coupling} \leq 90°$ [24].

With the aim of controlling the parameter $\Phi_{coupling}$, two meander lines were then etched between the cavities on the circuitry side (Figure 14.36); the meander lines act as series capacitors, allowing for modifying the phase of the coupling while leaving the magnitude roughly unchanged. Their dimensions are $W_m = 2.1$ mm and $L_m = 2.2$ mm. The proposed structure was then analysed by means of a full-wave simulator and the resulting S-parameters used in an HB simulation setup that allowed a synchronized solution of the system to be found around 10.5 GHz.

A prototype was then built (Figure 14.37) and measured and a 10.55 GHz synchronization frequency for the elements was found. The coupled oscillator array draws around 40 mA current out of 1.5 V supply. The measured effective radiated power (ERP) at broadside was 15.0 dBm. Taking into account that the simulated gain of the passive antenna was 8.3 dB, the radiated power results were 6.7 dBm (4.7 mW), which corresponds to approximately 7.8% DC-to-RF radiation efficiency.

The control voltages of the varactor diodes of the edge elements $V_{T,1}$ and $V_{T,3}$ have been detuned in opposite directions, in order to establish different phase-shift

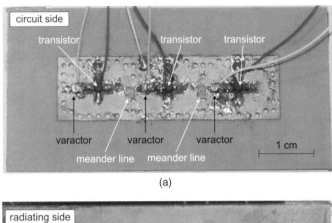

(a)

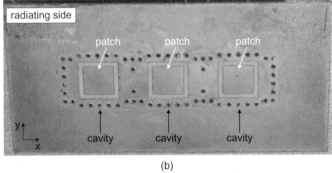

(b)

Figure 14.37 Photography of the three-element coupled oscillator array with meanders: (a) circuitry side and (b) radiating side. © 2012 IEEE. Reprinted, with permission, from [25].

distributions along the array, thus scanning the beam. In addition to that, the control voltage of the central oscillator $V_{T,2}$ has also been adjusted in order to fix the synchronization frequency of the system to the value of 10.55 GHz. The effective radiated power radiation patterns, showing the broadside pointing and the maximum obtained scanning angle, are plotted in Figure 14.38. The maximum scanning angle is around $\pm 20°$.

The relation between the scanning capability and the phase-shift distribution is described by the relation $\Delta\varphi = \pi \sin\Psi_{scan}$, where $\Delta\varphi$ is the phase-shift and Ψ_{scan} is the scanning angle. This means that the maximum obtainable phase-shift distribution along the structure is around $\pm 60°$. Therefore, in the case of this coupled oscillator array, when the phase shift tends to exceed this value, the elements lose their synchronization. Table 14.6 lists the values of the control voltages applied to the varactor diodes in order to get the desired scanning.

The values of the control voltages $V_{T,1}$ and $V_{T,3}$ do not show a symmetrical behaviour, due to the fact that in the practical realization the two edge elements are not exactly identical. Please note also that the voltage of the central oscillator has been changed, in order to maintain the frequency at a fixed value.

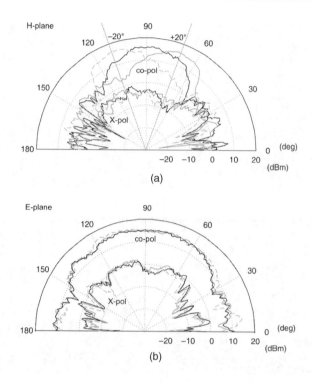

(a)

(b)

Figure 14.38 ERP radiation patterns for a three-element coupled oscillator array with beam-scanning properties: (a) H-plane (xz) and (b) E-plane (yz).

14.6 Conclusions

In this chapter, active antenna systems based on SIW cavity-backed oscillators have been proposed. The demonstration of the first SIW active antenna in the world opens new possibilities for the development of complete, integrated front-ends for wireless sensor networks. Moreover, thanks to the cavity-backed topology, the dynamic properties of the antennas can be used in coupled oscillator arrays, where the beam scan is obtained without the use of any phase shifter.

Possible future perspectives for this work include the realization of larger, two-dimensional arrays, and the study of reflectarray topologies for the introduction of more complex beam-shaping properties.

Table 14.6 Values of the control voltages (in volts) used for scanning the beam

	$V_{T,1}$	$V_{T,2}$	$V_{T,3}$
20°	7	5	15
0°	10.5	8	7.6
−20°	13.8	4.2	9

References

1. Hirokawa, J. and Ando, M. (1998) Single-layer feed waveguide consisting of posts for plane TEM wave excitation in parallel plates. *IEEE Transactions on Antennas and Propagation*, **46** (5), 625–630.
2. Deslandes, D. and Wu, K. (2003) Single-substrate integration technique of planar circuits and waveguide filters. *IEEE Transactions on Microwave Theory and Techniques*, **51** (2), 593–596.
3. Bozzi, M., Perregrini, L., Wu, K. and Arcioni, P. (2009) Current and future research trends in substrate integrated waveguide technology. *Radioengineering*, **18** (2), 201–209.
4. Bozzi, M., Georgiadis, A. and Wu, K. (2011) Review of substrate integrated waveguide (SIW) circuits and antennas. *IET Microwave Antennas and Propagation*, **5** (8), 909–920.
5. Wu, K. (2006) Towards system-on-substrate approach for future millimeter-wave and photonic wireless applications. Proceedings of the Asia Pacific Microwave Conference.
6. Galeis, J. (1963) Admittance of a rectangular slot which is backed by a rectangular cavity. *IEEE Transactions on Antennas and Propagation*, **11** (2), 119–126.
7. Navarro, J.A., Chang, K., Tolleson, J. *et al.* (1991) A 29.3-GHz cavity-enclosed aperture-coupled circular-patch antenna for microwave circuit integration. *IEEE Microwave and Guided Wave Letters*, **1** (7), 170–171.
8. Cheng, J.C., Dib, N.I. and Katehi, L.P.B. (1995) Theoretical modeling of cavity-backed patch antennas using a hybrid technique. *IEEE Transactions on Antennas and Propagation*, **43** (9), 1003–1013.
9. Zheng, M., Gardener, P., Hall, P.S. *et al.* (2001) Cavity control of active integrated antenna oscillators. *IEEE Proceedings on Microwaves, Antennas and Propagation*, **148** (1), 15–20.
10. Luo, G.Q., Hu, Z.F., Dong, L.X. and Sun, L.L. (2008) Planar slot antenna backed by substrate integrated waveguide cavity. *IEEE Antennas and Wireless Propagation Letters*, **7**, 236–239.
11. Bohorquez, J.C., Pedraza, H.A.F., Pinzod, I.CH. *et al.* (2009) Planar substrate integrated waveguide cavity-backed antenna. *IEEE Antennas and Wireless Propagation Letters*, **8**, 1139–1142.
12. Awida, M.H. and Fathi, A.E. (2009) Substrate-integrated waveguide Ku-band cavity-backed 2×2 microstrip patch array antenna. *IEEE Antennas and Wireless Propagation Letters*, **8**, 1054–1056.
13. Hudson, S. and Pozar, D. (2000) Grounded coplanar waveguide-fed aperture-coupled cavity-backed microstrip antenna. *Electronics Letters*, **36** (12), 1003–1005.
14. Giuppi, F., Georgiadis, A., Collado, A. *et al.* (2010) An X-band, compact active cavity backed patch oscillator antenna using a substrate integrated waveguide (SIW) resonator. IEEE International Symposium on Antennas and Propagation (AP-S), Toronto, ON, Canada.
15. Giuppi, F., Georgiadis, A., Bozzi, M. *et al.* (2010) Hybrid electromagnetic and non-linear modeling and design of SIW cavity-backed active antennas. *ACES Journal*, **25** (8), 682–689.
16. Georgiadis, A., Collado, A. and Suarez, A. (2006) New techniques for the analysis and design of coupled-oscillator systems. *IEEE Transactions on Microwave Theory and Techniques*, **54** (11), 3864–3877.
17. Georgiadis, A. and Collado, A. (2008) Nonlinear analysis of a reflectarray cell based on a voltage-controlled oscillator. IEEE International Symposium on Antennas and Propagation (AP-S), San Diego, CA, July 2008, pp. 1–4.
18. Giuppi, F., Georgiadis, A., Collado, A. *et al.* (2010) Tunable SIW cavity backed active antenna oscillator. *Electronics Letters*, **46** (15), 1053–1055.
19. Georgiadis, A., Via, S., Collado, A. and Mira, F. (2009) Push–push oscillator design based on a substrate integrated waveguide (SIW) resonator. European Microwave Conference (EuMC), Rome, October 2009, pp. 1231–1234.
20. Liao, P. and York, R.A. (1993) A new phase-shifterless beam scanning technique using arrays of coupled oscillators. *IEEE Transactions on Microwave Theory and Techniques*, **41** (10), 1.
21. Ip, K.H.Y. and Eleftheriades, A. (2002) A compact CPW-based single-layer injection-locked active antenna for array applications. *IEEE Transactions on Microwave Theory and Techniques*, **50** (2), 481–486.
22. Giuppi, F., Georgiadis, A., Bozzi, M. *et al.* (2010) Active antenna oscillator systems in substrate integrated waveguide (SIW) technology. European Conference on Antennas and Propagation (EuCAP), Barcelona, Spain, 12–16 April 2010.
23. Giuppi, F., Collado, A., Bozzi, M. *et al.* (2010) X-band cavity backed slot antennas and coupled oscillator systems. 40th European Microwave Conference 2010 (EuMC 2010), Paris, 2010.

24. Chang, H.C., Shapiro, E.S. and York, R.A. (1997) Influence of the oscillator equivalent circuit on the stable modes of parallel-coupled oscillators. *IEEE Transactions on Microwave Theory and Techniques*, **45** (8), 1232–1239.
25. F. Giuppi, A. Collado, A. Georgiadis, and M. Bozzi, "Active Substrate Integrated Waveguide (SIW) Antenna with Phase-Shifterless Beam-Scanning Capabilities," to appear at 2012 IEEE MTT-S International Microwave Symposium (IMS2012), Montreal, Canada, June 17–22, 2012.

15

Active Wearable Antenna Modules

Frederick Declercq[1], Hendrik Rogier[1], Apostolos Georgiadis[2]
and Ana Collado[2]
[1]*Department of Information Technology (INTEC), Ghent University, Ghent, Belgium*
[2]*Technological Telecommunications Centre of Catalonia (CTTC), Castelldefels, Spain*

15.1 Introduction

Wearable textile systems are intended to improve the user's awareness of his/her physical conditions such as heart rate, blood pressure, breathing, and so on, and environmental conditions including relative humidity, temperature, and so on, and this by unobtrusive integration of sensing, actuating and computing functionalities into clothing. In addition, unobtrusive integration of wireless communication systems provides real-time monitoring capabilities or it allows communication between different users in an ad hoc network. Typical applications are found in the health-care industry and in rescue operations, during which the deployment of smart wearable systems will help in drastically improving the rescuer's main task, that is, saving human lives. In order to provide a user friendly, inconspicuous, and low-cost intelligent textile system, wearability is an utmost important feature of the system and implies that the system should be lightweight, breathable, and flexible in order not to disturb the user's agility. A typical wearable textile system is composed of six distinct components, being sensors, actuators, power supply, interconnections, communication unit, and a data processing unit.

Body-centric communication requires a body worn antenna that is easy to integrate into a garment and this in an unobtrusive manner. Therefore, antennas composed of textile materials are preferably used since they provide flexibility and conformability to the human body, facilitating the integration into clothing and maintaining breathability of the garment. The vicinity of the human body increases losses due to bulk power absorption and introduces radiation pattern distortion and frequency detuning [1, 2]. Therefore, planar antennas are preferred since the inherent ground plane of the antenna structure minimizes power absorption in the human body and frequency detuning, and radiation is directed away from the

Microwave and Millimeter Wave Circuits and Systems: Emerging Design, Technologies, and Applications,
First Edition. Edited by Apostolos Georgiadis, Hendrik Rogier, Luca Roselli, and Paolo Arcioni.
© 2013 John Wiley & Sons, Ltd. Published 2013 by John Wiley & Sons, Ltd.

human body. Additionally, the low-profile antenna geometry contributes to the ease of integration. Textile planar antennas are constructed from a nonconducting textile substrate together with a patch and ground plane composed of a conductive textile material (electrotextile) [3]. The flexibility of the used materials makes the antennas susceptible to bending and even crumpling. In addition, when applying hydrophilic textile antenna materials, altering atmospheric conditions (e.g., relative humidity) affect the antenna performance [4, 5]. An optimal wearable antenna design, accommodating for the adversities encountered due to the proximity of the human body and the dynamics of the surrounding environment, will improve the overall link budget of the off-body wireless communication link. Several approaches have been studied in wearable antenna design:

1. Sufficient impedance bandwidth is provided to account for frequency detuning caused by the presence of the human body and bending of the antenna [3, 6].
2. A larger groundplane reduces the effect of the human body on the textile antenna's radiation characteristics [6].
3. Textile antennas based on electromagnetic band-gap (EBG) substrates reduce backward radiation, minimizing bulk power absorption in the human body. Moreover, a larger bandwidth and higher radiation efficiency is obtained and the EBG substrate minimizes antenna size [7, 8].
4. Implementation of polarization and pattern diversity in a multiantenna configuration mitigates the shadowing effect of the human body and signal fading, typically observed in an indoor environment [9, 10].
5. Electronic reconfiguration of the radiation pattern by implementing textile arrays allows the radiated beam to be steered in the desired direction and accounts for radiation pattern distortion due to the dynamics of the user [11].
6. The development of active wearable textile antenna modules by integrating electronic circuits underneath the textile antenna reduces losses observed in textile transmission lines or accommodates for the reduced textile antenna gain during operation in the vicinity of the human body [12].

Power autonomy plays a key role in the future proliferation of wearable textile systems. The development of lightweight flexible batteries as proposed in References [13] and [14] tackles the issue of current rigid batteries that harm the ergonomics of the wearable textile system. However, batteries still rely on external energy sources. Energy scavenged from the human body by exploiting sources such as kinetic energy produced by breathing, walking, and vibration can be converted into electrical energy by using piezoelectric energy transformers. Also body heat may be transformed into electric energy by relying on thermoelectric generators [15]. Next to these energy sources, solar energy has proven to be a good candidate for power harvesting and integration of flexible amorphous silicon solar cells into clothing was addressed in Reference [16]. However, the coexistence of solar cells and planar textile antennas results in a *real-estate* problem as the available space for integrating both into the same garment is limited. This problem can be solved by integrating solar cells on to planar textile antennas, that is, by reusing the available area both for RF radiation and for DC-power generation [17].

This chapter focuses on the development and characterization of active wearable antenna modules together with the integration of solar cells on to textile antennas. Design aspects such as material characterization, EM full-wave/circuit co-optimization design techniques and characterization techniques under real test conditions are discussed. First, the electromagnetic property characterization of textile materials is addressed in Section 15.2, discussing two dedicated characterization techniques, being an improved broadband transmission line technique (Section 15.2.3) and a small-band inverse planar antenna characterization method (Section 15.2.4). The latter technique was developed for accurate and fast characterization of antenna performance as a function of relative humidity. Second, the novel concept of active wearable antenna modules is presented in Section 15.3 by means of a design example discussed in Section 15.3.1, being an active integrated wearable receive textile antenna with optimized noise characteristics, operating in the 2.45 GHz ISM band. Finally, in Section 15.3.2, the integration of solar cells on to wearable antennas together with a novel interconnection structure between DC and RF GND is presented by means of a wearable PIFA design operating in the 902–928 MHz UHF band.

15.2 Electromagnetic Characterization of Fabrics and Flexible Foam Materials

15.2.1 Electromagnetic Property Considerations for Wearable Antenna Materials

The electromagnetic properties of interest are the permittivity ε_r and loss tangent $\tan\delta$ of the nonconductive fabric or foam materials and the effective conductivity σ of the electrotextile. For traditional antenna materials such as high-frequency laminates and copper foil, the constitutive parameters are homogeneous and isotropic. When considering materials used in planar textile antenna design, the air cavities present inside the substrate material result in a nonhomogeneous permittivity. Also, the antenna substrate can be composed of multiple fabric layers, contributing to the inhomogeneous nature of the material's permittivity and introducing anisotropic behavior. Moreover, adhesive sheets assembling the different layers contribute to the nonhomogeneous nature and anisotropy of the antenna substrate. The characterization processes presented here take into account all layers and adhesive sheets as used in the final application. Since the inhomogeneities are small compared to the wavelength one can characterize the materials as being homogeneous. Potential anisotropy is accounted for thanks to the fact that the measured dominant field component in both characterization methods presented in this chapter correspond to the dominant component of the mode responsible for radiation by a planar antenna. The conductive fabric used in a wearable textile antenna design typically consists of a copper-plated woven textile material, which is treated as being homogeneous and is modeled by an effective conductivity σ (S/m) and an effective thickness t.

15.2.2 Characterization Techniques Applied to Wearable Antenna Materials

With the introduction of textile fabrics in wearable antenna design, the need for accurate determination of the material's electromagnetic properties emerged. Three different types of microwave characterization techniques are available for the determination of the material's electromagnetic properties. The first technique is the cavity perturbation technique and allows characterization of the anisotropy that is typically found in the electromagnetic properties of woven textile materials [18]. Second, a broadband transmission line technique, relying on scattering parameter measurements of at least two microstrip lines with different lengths, is utilized in order to extract the material's permittivity [19]. In Section 15.2.3 the transmission line method is combined with the matrix-pencil method to minimize deviations in the calculated electromagnetic properties caused by fabrication tolerances and inhomogeneities typically observed in a textile microstrip line structure. The advantage of using the microstrip transmission line method over the cavity perturbation technique is that it allows characterization of the textile material in a setup assembled with the same conductive materials as in the final textile antenna design. This requirement follows from the fact that the effective permittivity of the substrate increases when using electrotextiles as conductive surfaces and that the effect of the glue, used in assembling the different layers, has to be taken into account [20]. Third, techniques based on planar textile resonators are used to characterize the materials. In Reference [21] the real part of the substrate's permittivity is extracted by relying on analytical approximation formulas for the textile antenna's resonance frequency. A similar characterization approach was presented in Reference [4], where the resonance frequency of a textile resonator was estimated by relying on numerical simulations. Fitting the simulated on to the measured resonance frequency allows the dielectric constant of the fabric substrate to be found. The latter approach was accelerated significantly by utilizing surrogate-based optimization techniques that automatically solve the inverse problem, providing a more accurate solution [22, 23]. Furthermore, this method allows the permittivity and loss tangent to be determined. An extension of this method is proposed in Section 15.2.4, where both complex permittivity and effective conductivity of the electrotextile are characterized. The proposed method was used for quantitative characterization of the electromagnetic properties of textile antenna substrates as a function of relative humidity, which allows estimation of the antenna performance in varying climatic conditions.

15.2.3   Matrix-Pencil Two-Line Method

15.2.3.1   Theory

This method relies on the transmission and reflection coefficient measurements of two microstrip lines with different lengths l_2 and l_1, where $l_2 > l_1$ as shown in Figure 15.1. De-embedding of the coax-to-microstrip transitions relies on symmetry, reciprocity, and identical coax-to-microstrip discontinuities of the microstrip line structures [24]. Propagation over the microstrip line is described by a complex propagation factor $\gamma = \alpha + j\beta$ and characteristic impedance Z_0. No prior knowledge of the characteristic impedance Z_0 is required, since the effects of impedance mismatches are included in the de-embedding algorithm [25].

The scattering transfer cascade matrix $\overline{\overline{T}}_D$ of the long microstrip line (length l_2) and the matrix $\overline{\overline{T}}_T$ of the short line (length l_1) are expressed as

$$\overline{\overline{T}}_D = \overline{\overline{T}}_A \cdot \overline{\overline{T}}_L \cdot \overline{\overline{T}}_B \tag{15.1}$$

$$\overline{\overline{T}}_T = \overline{\overline{T}}_A \cdot \overline{\overline{T}}_B \tag{15.2}$$

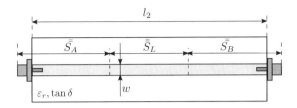

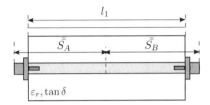

Figure 15.1 Representation of the microstrip lines with lengths l_2 and l_1, consisting of coax-to-microstrip transitions $\bar{\bar{S}}_A$ and $\bar{\bar{S}}_B$ and an ideal lossy transmission line $\bar{\bar{S}}_L$.

where $\bar{\bar{T}}_A$ and $\bar{\bar{T}}_B$ represent the coax-to-microstrip discontinuities and $\bar{\bar{T}}_L$ the lossy transmission line with length $\Delta l = l_2 - l_1$. Solving Equation (15.2) for $\bar{\bar{T}}_B$ and substituting into Equation (15.1) yields

$$\bar{\bar{T}} \cdot \bar{\bar{T}}_A = \bar{\bar{T}}_A \cdot \bar{\bar{T}}_L \tag{15.3}$$

where $\bar{\bar{T}} = \bar{\bar{T}}_D \cdot \bar{\bar{T}}_T^{-1}$, with $\bar{\bar{T}}_D$ and $\bar{\bar{T}}_T$ derived from their scattering parameter counterparts $\bar{\bar{S}}_D$ and $\bar{\bar{S}}_T$. Since $\bar{\bar{T}}_L$ is diagonal, the propagation factor is derived from the eigenvalue equation

$$e^{\pm\gamma\Delta l} = \frac{Tr\left(\bar{\bar{T}}\right) \pm \sqrt{Tr\left(\bar{\bar{T}}\right)^2 - 4\Delta\left(\bar{\bar{T}}\right)}}{2} \tag{15.4}$$

where $Tr\left(\bar{\bar{T}}\right)$ and $\Delta\bar{\bar{T}}$ represent the trace and determinant of $\bar{\bar{T}}$, respectively. Note that $\Delta\bar{\bar{T}} = 1$ because of symmetry and reciprocity. Passivity imposes $\left|e^{-\gamma\Delta l}\right| < 1$, determining the correct sign for the root in Equation (15.4) and thus the correct angular rotation of γ. Substituting $z = 1/2\,Tr\left(\bar{\bar{T}}\right)$ in Equation (15.4) and solving with respect to γ yields $\gamma = (\Delta l)^{-1} \ln\left(z \pm \sqrt{z^2 - 1}\right)$, where a correct definition for the branch cut together with phase unwrapping are required.

Once β is known, the effective permittivity $\varepsilon_{r,eff}$ follows from β/k_0^2. The dielectric constant is then determined by using the transmission line calculator *Linecalc* from *Agilent's Advanced Design System* [26] and then by varying ε_r until the effective permittivity calculated by *Linecalc* equals the measured $\varepsilon_{r,eff}$. Assuming low-loss conductive materials, the conductor losses are negligible compared to the dielectric losses, so the calculated attenuation α_d can be entirely attributed to dielectric loss. Hence, the loss tangent $\tan\delta$ can be determined by [27]

$$\tan \delta \approx 0.0366 \frac{\alpha_d \lambda_0 \sqrt{\varepsilon_{r,eff}} (\varepsilon_r - 1)}{\varepsilon_r (\varepsilon_{r,eff} - 1)} \tag{15.5}$$

with $\alpha_d \approx \alpha$ expressed in dB/m and λ_0 the free-space wavelength. This method allows characterization of the materials as applied in the final textile antenna design. The substrates are characterized as being homogeneous with identical $\varepsilon_{r,eff}$ in each cross-section along the direction of propagation.

15.2.3.2 Reducing Perturbations in the Calculated Propagation Constant

The accuracy of the two-line method becomes poor when the difference in line length $\Delta l = l_2 - l_1$ is electrically small and when connector mounting is not symmetrical. The accuracy of the method can be improved by choosing microstrip lines with a significant electrical length difference. However, a textile transmission line is far from being ideal. The flexible and compressible substrates combined with fabrication inaccuracies are responsible for geometrical deviations in the transversal cross-section of the lines, resulting in varying characteristic impedance along the line. Together with nonidentical coax-to-microstrip transitions, which disturb the symmetry condition, they result in errors on the de-embedded transmission line parameter $e^{-\gamma \Delta l}$. In addition, excitation of evanescent higher-order modes at the coax-to-microstrip discontinuities produces electromagnetic coupling between both transitions. For relatively short microstrip line lengths this coupling can be relatively large and may interfere with the dominant quasi-TEM mode. Hence, de-embedding of transitions yields a propagation constant that does not fully correspond to perfect quasi-TEM mode propagation, resulting in perturbations on the de-embedded transmission line phase and attenuation. This results in a nonphysical behavior of effective permittivity $\varepsilon_{r,eff}$ and $\tan \delta$ as a function of frequency. The use of the matrix-pencil method allows us to minimize these perturbations by applying this technique as an averaging method in which all perturbations present in the calculated $e^{-\gamma \Delta l}$ are interpreted as an unknown noise distribution $n(k)$. The frequency-dependent eigenvalues $y(k)$ are now given by

$$y(k) = e^{\pm \alpha_k \Delta l}[\cos(\beta_k \Delta l) \pm j \sin(\beta_k \Delta l)] + n(k) \tag{15.6}$$

with $k = 1, 2, \ldots, N$ representing the number of frequency points. The frequency-dependent β_k are found by calculating the unwrapped phase of the noiseless part of the data vector $y(k)$, extracted by applying the matrix-pencil method as described in Reference [28]. This method performs optimal fitting of a series of complex exponentials on noise-contaminated data,

$$y(k) = x_1(k) \pm jx_2(k) + n(k) = \sum_{i=1}^{M} R_{1,i} z_{1,i}^k \pm j \sum_{i=1}^{M} R_{2,i} z_{2,i}^k + n(k) \tag{15.7}$$

with M being the number of exponentials. The poles z_i are found as the solution of a generalized eigenvalue problem and the R_i are estimated by solving a linear least-squares problem with all other parameters known. For real-valued data vectors the poles $z_{1,i}, z_{2,i}$ and residues $R_{1,i}, R_{2,i}$ appear in complex conjugate pairs. Therefore, the matrix-pencil method is applied

Table 15.1 Dimensions of the microstrip line configurations in mm

				Copper	Flectron®
		h	w	$l_2 - l_1$	$l_2 - l_1$
Fabric 1	3 layers, plain weave (uneven surface) 98% aramid, 2% antistatic carbon	1.80	5	103-45/104-45	105-43/106-41
Fabric 2	4 layers, twill weave (even surface) 98% aramid, 2% antistatic carbon	1.67	6	98-43/103-45	103-44/104-46
Fabric 3	Nonwoven polyethylene fabric	3.60	15	103-47/102-44	107-46/104-43

separately to the real and imaginary parts of Equation (15.6). Setting the number of exponentials M to two for both real and imaginary parts implements the matrix-pencil method as an averaging technique and estimates the noiseless data vectors $x_1(k)$ and $x_2(k)$.

15.2.3.3 Characterization of Textile Substrates

Description of the Materials to be Characterized

The proposed procedure was used to characterize three different fabrics serving as a textile antenna substrate. *Flectron®*, a copper-plated nylon fabric, is used as the conductive layer, as well as a solid homogeneous copper foil. The composition of the three different substrate materials and the geometrical properties of the microstrip lines are given in Table 15.1. Fabric one and two are both woven fabrics. However, fabric one is a plain weave based on yarns with two different thicknesses, resulting in an uneven surface, whereas fabric two is a twill weave from identical yarns that has an even surface. The substrate of fabric one is composed of an assembly of three layers, whereas for the fabric two substrate four layers are assembled, resulting in an overall substrate thickness of 1.8 and 1.67 mm, respectively, which is sufficient for antenna applications. The third fabric is a single-layered nonwoven polyethylene fabric with a sufficient thickness for antenna applications. For each substrate, two pairs of microstrip lines based on the homogeneous copper foil for the conductive surfaces and two pairs of Flectron®-based microstrip lines were fabricated. The use of the copper foil-based lines allows determination of the loss tangent of the textile substrate when conductor losses are much smaller than dielectric losses. The Flectron®-based microstriplines are used to calculate the permittivity of the substrate while taking into account the effect of the electrotextile. The difference in line length $\Delta l = l_2 - l_1$ was chosen in the vicinity of 60 mm to minimize the low-frequency errors in the extracted effective permittivity.

Experimental Results of the Effective Permittivity and Loss Tangent

Based on the procedure presented in Section 15.2.3.1, the effective permittivity and loss tangent of the three textile substrates in the frequency range from 1 GHz to 10 GHz was extracted. The twill woven aramid fabric 2 is denser than the plain woven aramid fabric 1; hence fabric 2 was expected to have the highest permittivity. Fabric 3 has the smallest density and thus the smallest permittivity. The calculated loss tangent $\tan\delta$ obtained from the copper-based microstrip lines is depicted in Figure 15.2. These values were calculated by substituting the attenuation coefficients calculated by $\alpha_d(k) = \alpha_k = \pm\ln((x_1(k)\Delta l)/(\cos(\beta_k)\Delta l))$, the extracted $\varepsilon_{r,eff}$ obtained from the copper-based microstrip lines, and substituting ε_r in Equation (15.5). The averaged loss tangent

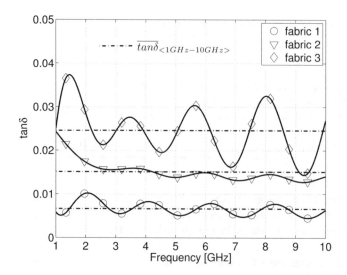

Figure 15.2   Calculated loss tangent in the frequency range from 1 to 10 GHz.

$\overline{\tan\delta}_{\langle 1\ 10\,GHz\rangle}$ obtained by averaging the calculated loss tangent over the measured frequency range is also shown in Figure 15.2. Figure 15.3 shows the measured $\varepsilon_{r,eff}$ ranging from 1 GHz to 10 GHz for the Flectron®-based microstrip lines. The calculated $\varepsilon_{r,eff}$ is the averaged result obtained from the two calculated effective permittivities of the two pairs of microstrip lines.

From the extracted effective permittivities depicted in Figure 15.3 it is concluded that the matrix-pencil technique performs a good frequency-averaging of the calculated effective

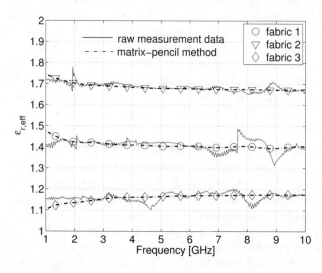

Figure 15.3   Calculated effective permittivity from the Flectron®-based microstrip lines in the frequency range from 1 to 10 GHz.

permittivity. The amplitude of the small periodic oscillations observed in the effective permittivity is minimal when the measured reflection coefficient of one of the lines is minimal. These oscillations are due to multiple reflections and reflections against the edges of the finite substrate or objects in the neighborhood of the test structures. The larger oscillations are caused by the nonidealities in the microstrip line configurations, as discussed in Section 15.2.3.2. Moreover, traction and torsion caused by the measurement cables on the lines and the variable impedance profile along the line due to compressible substrates and production inaccuracies results in additional perturbations of the calculated $\varepsilon_{r,eff}$. The perturbations on $\varepsilon_{r,eff}$ of fabric 2 are the smallest because this substrate is less sensitive to geometrical variations as a result of its higher density/rigidity. The discontinuities observed in the effective permittivity occur when the difference in line lengths is a multiple of half a wavelength. For this condition Equations (15.1) and (15.2) are nearly equal when the substrate exhibits low ohmic losses. Observing the estimated loss tangent as a function of frequency and the averaged loss tangent $\overline{\tan \delta}_{(1-10GHz)}$ of the three substrates in Figure 15.2 tells us that fabric 3 has the highest losses. Fabrics 1 and 2 are made from the same material but fabric 2 has a smaller density than fabric 1, resulting in a lower loss tangent for fabric 1.

Validation

For validation purposes, two planar antennas on each substrate were designed using *Momentum* from *Agilent's Advanced Design System*. The first antenna uses solid copper foil for its conductive layers whereas the second antenna's conductive ground plane and patch were made of Flectron®. The antennas were designed based on the extracted substrate parameters.

The conductor losses were only taken into account for the Flectron®-based antennas using a sheet resistivity of $R_s = 0.1 \, \Omega/\text{sq}$.[3]. The dimensions of the planar antenna are given in Table 15.2 and the topology is depicted in Figure 15.4. The geometrical accuracy of the antennas is about ± 0.5 mm. Reflection measurements of the planar antennas were performed in the range from 2 to 3 GHz and the radiation efficiencies were determined based on full radiation pattern measurements performed in an anechoic chamber. A comparison between simulated and measured resonance frequency/reflection coefficients as depicted in Figure 15.5 allows validation of the extracted permittivity whereas a comparison between measured and simulated antenna radiation efficiency of the copper-based antennas allows evaluation of the accuracy of the extracted loss tangent.

An excellent agreement between measured and simulated reflection coefficients is found for the antennas based on fabric 1 and fabric 2. The resonance frequency error is defined by $|\Delta f_r| = |f_{r,s} - f_{r,m}|$, where $f_{r,s}$ and $f_{r,m}$ represents the simulated and measured resonance frequency, respectively. The resonance frequency errors from the Flectron®-based antennas

Table 15.2 Dimensions of the copper-based and Flectron®-based planar textile antennas in mm

Substrate		L	W	x_f	w_f	s_i	Ground plane size
Fabric 1	Copper	50	50	16	5	2	110×110
	Flectron®	47	51	10	5	2	110×110
Fabric 2	Copper	46	51	11	7	3	110×110
	Flectron®	46	51	11	7	3	110×110
Fabric 3	Copper	52	80	11	21	3	90×140
	Flectron®	53	81	10	20	3	90×140

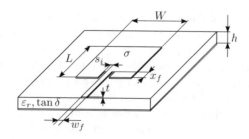

Figure 15.4 Inset fed patch antenna geometry.

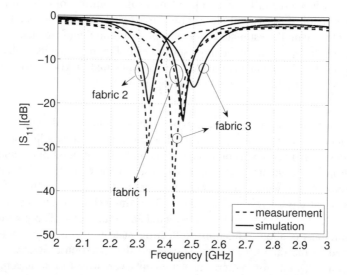

Figure 15.5 Comparison between simulated and measured reflection coefficients of the Flectron®-based textile antennas.

are given in Table 15.3 and show a minimum resonance frequency error of 0 MHz, with a resolution of 10 MHz. The maximum resonance frequency estimation error is 80 MHz and is observed for the Flectron®-based antenna, which is based on fabric 3 as a substrate. To take into account the geometrical accuracy of the patch length L on the calculated dielectric constant, the simulated data were fitted on the measured data by varying ε_r after the patch length was increased/decreased by 0.5 mm. These tolerance values of the dielectric constant are

Table 15.3 Calculated dielectric constant and loss tangent of the substrates

Substrate	$\varepsilon_{r,eff}$ (Flectron®)	ε_r	ε_r tolerance	$\tan \delta_{\langle 1\,GHz-10\,GHz \rangle}$	$\tan\delta_{\langle 1GHz-10GHz \rangle}$ corrected	$|\Delta f_r|$ (GHz)	$|\Delta e_{cd}|$
Fabric 1	1.42	1.57	1.54–1.71	0.007	0.012	0.01	0.11
Fabric 2	1.69	1.91	1.88–1.95	0.015	0.019	0	0.06
Fabric 3	1.14	1.18	1.23–1.28	0.025	0.008	0.08	0.22

given in Table 15.3. The ε_r calculated from the Flectron®-based microstrip lines constructed from fabrics 1 and 2 lie within the tolerance of the geometrical accuracy of the patch antenna. The calculated permittivity of the nonwoven substrate (fabric 3) exhibits the largest error. These inaccuracies can be mainly attributed to the difficulty in realizing identical coax-to-microstrip transitions due to the large thickness of the substrate. The antenna radiation efficiency error is defined by $|\Delta e_{cd}| = \left|e_{cd,f_{r,m}} - e_{cd,f_{r,s}}\right|$, where $e_{cd,f_{r,m}}$ and $e_{cd,f_{r,s}}$ represent the measured and simulated radiation efficiency antennas at their respective resonance frequency. Applying this to the copper-based antennas allows us to evaluate the accuracy of the extracted $\tan\delta$, as described in Table 15.3. It is concluded that the smallest error in antenna efficiency is observed for fabrics 1 and 2, yielding an error of 11% and 6%, respectively. The largest error is observed for fabric 3 (22%). By fitting the radiation efficiencies simulated in ADS-momentum for the copper-based antennas to the corresponding measured efficiencies a corrected loss tangent can be obtained and is presented in Table 15.3.

15.2.4 Small-Band Inverse Planar Antenna Resonator Method

15.2.4.1 Characterization Process

Description of the Inverse Problem
For patch antennas with a fixed and known geometry, antenna performance indicators such as resonance frequency f_r, bandwidth BW, and antenna efficiency e_{cd} are determined by the substrate's complex permittivity and the conductor's conductivity. On the one hand, the resonance frequency of a patch antenna is directly related to the permittivity of the textile substrate by simple analytical approximation formulas. On the other hand, the dependencies of antenna efficiency and bandwidth on the remaining unknown electromagnetic properties $\tan\delta$ and σ are not straightforward and only numerical full-wave techniques can accurately estimate antenna efficiency and bandwidth. The goal is to determine the constitutive parameters of the materials by matching measured performance characteristics to simulated antenna performances extracted from *Momentum* by Agilent Technologies. This comparison is converted into a forward optimization problem by constructing an error function accounting for the simulated and measured antenna performances, given by

$$MSE = a_1 \frac{1}{n} \sum_{i=1}^{n} \left(1 + w_i \left| |S_{11,i}|^{dB} - |\tilde{S}_{11,i}|^{dB} \right| \right)^2 + a_2 \left| e_{cd,f_{r,s}} - \tilde{e}_{cd,f_{r,m}} \right| \qquad (15.8)$$

with n the number of frequency points and w_i the weighting factors for $S_{11,i}$ and $\tilde{S}_{11,i}$, being the simulated and measured reflection coefficients points, respectively, of the antenna at frequency point i; $e_{cd,f_{r,s}}$ and $\tilde{e}_{cd,f_{r,m}}$ represent the simulated and measured antenna efficiencies at their respective resonance frequencies $f_{r,s}$ and $f_{r,m}$. The weighting factors for the two individual error terms in the error function (15.8) are given by a_1 and a_2. By optimizing the electromagnetic properties in the numerical EM software one can minimize Equation (15.8). Hence, the electromagnetic properties yielding a minimum in Equation (15.8) are the solutions of the characterization process. The problem of a nonunique solution is discussed further and is solved by introducing a two-step characterization process.

Note that $\left| e_{cd,f_{r,s}} - \tilde{e}_{cd,f_{r,m}} \right|$ is not squared since the order in magnitude of the efficiency error is similar to the order of magnitude of the mean-squared error function between

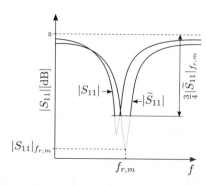

Figure 15.6 Schematic representation of the weighting factor distribution w_i.

measured and simulated antenna reflection coefficients if $a_1 = 1$ and $a_2 = 100$ [29]. Adding 1 to $w_i \left| \left| S_{11,i} \right|^{dB} - \left| \tilde{S}_{11,i} \right|^{dB} \right|$ increases the penalty for values where $\left| \left| S_{11,i} \right|^{dB} - \left| \tilde{S}_{11,i} \right|^{dB} \right| < 1$ and increases the penalty for estimated constitutive parameters near the optimum. To increase the sensitivity of the first part of the error function with respect to ohmic losses, the weighting factors are chosen to be

$$
\begin{cases}
w_i = 0, \left| \tilde{S}_{11,i} \right|^{dB} < \dfrac{3}{4} \left| \tilde{S}_{11} \right|^{dB}_{f_{r,m}} \\[2mm]
w_i = 1, \left| \tilde{S}_{11,i} \right|^{dB} > \dfrac{3}{4} \left| \tilde{S}_{11} \right|^{dB}_{f_{r,m}}
\end{cases}
\tag{15.9}
$$

with $\left| \tilde{S}_{11} \right|^{dB}_{f_{r,m}}$ the measured reflection coefficient at the measured resonance frequency, $f_{r,m}$. A graphical representation of the weighting factor distribution of w_i is depicted in Figure 15.6. Here the focus of the error function is more on a comparison between measured and simulated reflection coefficient points describing the resonance curve's shape and less on the reflection coefficient points describing resonance frequency and peak depth. Ohmic losses and susbtrate permittivity directly affect the shape of the curve whereas the resonance frequency is only affected by the permittivity. From Figure 15.6 one can observe that a match between measured and simulated reflection coefficient points automatically results in a match between measured and simulated resonance frequencies. However, only considering the shape of the resonance frequency does not yields accurate results for extracting an effective loss tangent due to the discrepancies between the simulation model and the real-life antenna. Combining a comparison between measured and simulated antenna efficiencies with a comparison between measured and simulated reflection coefficients as given in Equation (15.8) introduces a higher degree of sensitivity of the error function with respect to ohmic losses. Indeed, antenna efficiency highly depends on ohmic losses, that is, conductor losses and dielectric losses.

Two-Step Characterization Process
The extracted loss tangent is an effective loss tangent since no distinction can be made between conductor losses and dielectric losses in measured bandwidth or measured antenna efficiency, yielding a nonunique solution to the characterization process. To solve the

problem of a nonunique solution in the determination of tan δ and σ, a two-step characterization as depicted in Figure 15.7 is proposed:

1. The inverse characterization process is applied to a textile antenna based on a homogeneous copper foil with known conductivity. A fit between the simulated and measured data yields ε_r and tan δ of the substrate.
2. The characterization process is applied to a textile antenna with identical substrate as in step one, but with the homogeneous copper foil replaced by an electrotextile. The loss tangent extracted from step 1 is used and kept constant whereas an effective conductivity σ and ε_r are optimized in order to realize a fit between measured and simulated data. The extracted permittivity now also accounts for the effect of the electrotextile.

First, an estimated value for the electromagnetic properties is utilized in Agilent Technologies' *Momentum* to design an inset-fed patch antenna as depicted in Figure 15.4, exhibiting a sharp single-mode resonance in the vicinity of 2.45 GHz. The use of a single-mode narrow bandwidth antenna will increase the sensitivity of the resonance frequency with respect to small perturbations in the permittivity, yielding a more accurate determination of ε_r. From this design, a simulation model with the materials' electromagnetic properties as inputs is constructed in *Momentum*. The outputs of the simulation model are the reflection coefficient and antenna efficiency at resonance frequency, which are then compared to the measured data by means of the error function given by Equation (15.8). Then the electromagnetic properties in the simulation model are optimized in order to minimize Equation (15.8), thus utilizing a surrogate-based optimization approach [22].

Materials to be Characterized

Three different materials, exhibiting different sensitivities to moisture, were investigated. The susceptibility of a material to absorb water is characterized by its moisture regain *MR*, which is defined according to the ISO standard 6417 (1–4). A high *MR* implies a higher susceptibility of the material to absorb water. The measured *MR* and the total thickness h of the substrate materials as well as their composition are listed in Table 15.4. From the measured moisture regain one concludes that a substrate built from the woven fabric is more hydrophilic compared to that of the nonwoven fabric and foam substrate. This hydrophilic property of a woven synthetic fabric lies in its open-weave structure. The fleece fabric 2 exhibits the lowest *MR* and is therefore the most water-repellent material in this study. The foam material is an open cell structure and tends to absorb some moisture, as can be seen from its moisture regain. Also the moisture regain of the electrotextile, being a copper-plated polyester, plain woven fabric of thickness $t = 80 \, \mu m$ with a tarnish resistant coating, was measured and is listed in Table 15.4. The tarnish-resistant coating avoids oxidation of the electrotextile and a DC surface resistivity $R_s = 0.05 \, \Omega/sq.$ is specified by the manufacturer [30].

For each substrate, two inset-fed patch antennas, with a geometry depicted in Figure 15.4, one based on copper foil and one using an electrotextile, were designed. The conductive layers as well as the fabric layers were glued on to the substrates using an adhesive sheet. The ground plane size of the antennas was chosen large enough so that the effect of a finite size ground plane on the resonance frequency was negligible, that is, keeping at least a $\lambda/20$ distance between the patch edge and the substrate edge [31]. The resulting dimensions of the

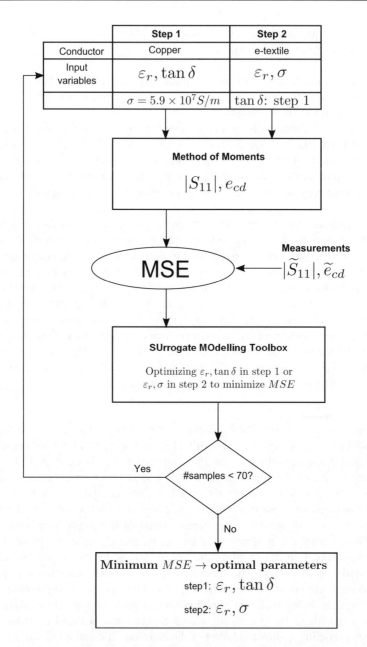

Figure 15.7   Schematic representation of the characterization process.

Table 15.4 Description antenna substrates

	Antenna materials composition	h (mm)	MR (%)
Fabric 1	4 layers, twill weave: 98% polyamide fibres + 2% antistatic carbon fibres	1.67	4.9
Fabric 2	Nonwoven synthetic fleece fabric: polyethylenteraftalate (PET)	2.20	0.2
Foam	Polyvinyl chloride + acrylonitrile butadiene rubber	3.40	1.9
Flectron$^{\circledR}$	Electrotextile copper-coated nylon fabric	$t = 80\,\mu m$	0.5

Table 15.5 Dimensions of the copper-based textile antennas in mm

Copper	L	W	x_f	L_f	s_i	w_f	GND
Fabric 3	43	55	8.5	25	2	6	75×80
Fabric 5	55	69	13	32	3	9	95×85
Foam 2	53	70	13	32	3	16	90×100

copper-based and electrotextile-based antennas are given in Tables 15.5 and 15.6, respectively.

15.2.4.2 Measurement Setup

The antenna prototypes were placed in a climate chamber (WK 350 from Weiss Technik) and conditioned for 24 hours at a specific relative humidity (RH) level at 23°C, assuming that an equilibrium state was reached between moisture absorbed by the material and moisture present in the climate chamber. The considered relative humidity levels are 10, 20, 30, 40, 50, 60, 70, 80, and 90%. After conditioning at each RH level, each antenna was removed from the climate chamber to perform a free-space reflection coefficient measurement in the frequency range from 2 to 3 GHz. For the efficiency measurements, the generalized Wheeler cap method was used [32]. This technique calculates antenna radiation efficiency from free-space input impedance measurements and measured input impedance when the antenna is placed inside a closed metallic cavity. A carefully designed circular cavity was used in order not to disturb the near-field distribution of the antenna or introduce cavity resonances that disturb the reflection coefficient measurement in the frequency band of interest.

Table 15.6 Dimensions of the electrotextile-based textile antennas in mm

Electrotextile	L	W	x_f	L_f	s_i	w_f	GND
Fabric 3	44	55	7	24	2	6	80×90
Fabric 5	55	70	13	33	3	10	95×90
Foam 2	55	70	11	36	3	16	100×105

15.2.4.3   Results and Validation

SUMO Toolbox Configuration
Version 6.2 of the surrogate modeling (SUMO) toolbox [33], implementing the surrogate-based optimization algorithm, is used to solve the inverse problem. An initial set of 24 data points is created using a maximum Latin hypercube design of 20 points together with four corner points. The standard expected improvement function, used for selecting the infill points, is optimized using the dividing rectangles (DIRECT) algorithm. The optimization is halted when the number of samples exceeds 70. The surrogate model, relying on kriging, is by default an interpolation technique. However, since the measurements are prone to errors, the cost functions will be noisy. Therefore, the kriging surrogate model was adapted to a regression technique. The substrate parameters as well as the conductivity are limited to a prescribed optimization range. These bounds depend on measured *MR* and the estimated electromagnetic property value used in the design of the textile patch antennas. For the characterization using copper-based antennas, based on the materials fabric 2 and foam 1 – which have an estimated permittivity close to one and a small *MR* – the bounds are $\varepsilon_r = [1\ 1.5]$ and $\tan\delta = [0.0001\ 0.05]$. For fabric 1 exhibiting a high *MR* and with an expected $\varepsilon_r \approx 1.8$, the bounds are $\varepsilon_r = [1.5\ 2.6]$ and $\tan\delta = [0.005\ 0.15]$. For the characterization process applied to the electrotextile-based antennas, that is, in step 2, where the loss tangent extracted from step one is reused, the permittivity optimization range was chosen to be the same as in step 1. In this second step, the bounds of the effective bulk conductivity are $\sigma = [5.0 \times 10^3\ \text{S/m}\ 5.0 \times 10^6\ \text{S/m}]$.

Results from the Copper-Based Antennas (Step 1)
The final kriging model resulting from the optimization applied to the copper-based antenna relying on fabric 1 at *RH* = 40% is depicted in Figure 15.8. The optimum values of the substrate's permittivity and loss tangent are clearly visible. Local optima as a function of loss tangent are also observed, due to multiple solutions of $(\varepsilon_r, \tan\delta)$ that yield a fit between

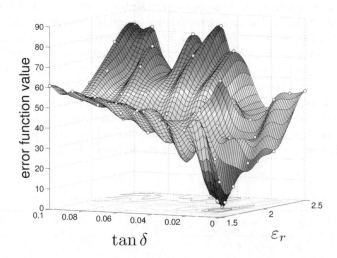

Figure 15.8 Final kriging surrogate model of the characterization process (step 1): material: fabric 3, copper, RH = 40%.

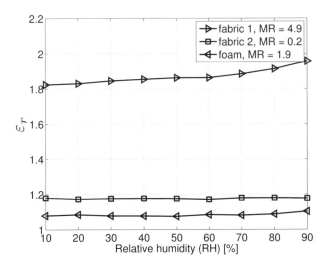

Figure 15.9 Estimated permittivity versus relative humidity.

simulated and measured efficiencies at their respective resonance frequencies. The permittivity and loss tangent results as a function of relative humidity are given in Figures 15.9 and 15.10, respectively. For the permittivity and loss tangent results, a clear relation between material MR and electromagnetic properties is observed. Figure 15.11 depicts the absolute error in resonance frequency $|\Delta f_r|$ for all optimization runs applied to the copper-based antennas. A comparison between simulated and measured antenna efficiencies e_{cd} at their respective resonance frequencies as a function of relative humidity is given in Figure 15.12. A good fit is found between measured and simulated efficiencies and the resonance frequency error does not exceed 8 MHz.

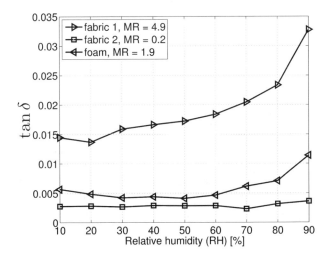

Figure 15.10 Estimated loss tangent versus relative humidity.

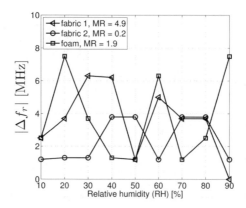

Figure 15.11   Resonance frequency error.

If measurement errors in $|S_{11}|$ and e_{cd} are small and a highly accurate simulation model is available, then a fit between measured and simulated reflection coefficients will automatically result in a good agreement between measured and simulated efficiency. If a discrepancy is present in the measurement results or in the simulation model, then the weights a_1 and a_2, determining the error on $|S_{11}|$ and e_{cd}, respectively, will determine the outcome of the characterization process. Discrepancies between the simulation model and the real-life antenna combined with measurement errors increase the uncertainty in the estimation of the material's electromagnetic properties.

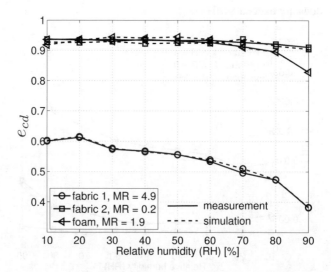

Figure 15.12   Comparison between measured and simulated antenna efficiency at their respective resonance frequency.

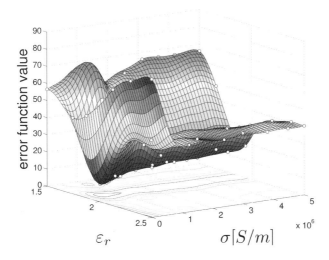

Figure 15.13   Final kriging surrogate model of the characterization process (step 2): material: fabric 3, e-textile, RH = 40%.

Results for the Flectron®-Based Antennas (Step 2)

The loss tangent extracted during step 1 is used in step 2 of the characterization process. This loss tangent is kept constant for each individual relative humidity level. Now, permittivity ε_r and effective bulk conductivity σ are optimized in order to minimize the error function and hence extract substrate permittivity and effective conductivity of the electrotextile. The final kriging model and the contour plot, resulting from the optimization applied to the electrotextile-based fabric 3 antenna at $RH = 40\%$, is depicted in Figure 15.13. For low conductivity values, the error function is more affected by the conductivity compared to the region where conductivity is high. This results from the fact that antenna efficiency drops rapidly when conductivity is relatively poor. For higher conductivity values, antenna efficiency is less affected by the conductivity. Here, the global optimum is less pronounced compared to the kriging model in step 1. The permittivity results for the materials are depicted in Figure 15.14. The extracted effective conductivity of the electrotextile and the surface resistivity R_s, calculated by $R_s = \sqrt{\pi f_{r,s} \mu_0 / \sigma}$, are depicted in Figure 15.15. The extracted conductivity lies in the range of 10^5 S/m, yielding a corresponding surface resistivity of 0.2 Ω/sq. Figure 15.16 depicts the absolute error in resonance frequency $|\Delta f_r|$ for all optimization runs applied to the electrotextile-based antennas and demonstrates a good estimation of the measured resonance frequencies. A comparison between simulated and measured antenna efficiencies e_{cd} at their respective resonance frequencies as a function of relative humidity is provided in Figure 15.17. A larger difference is observed compared to the results from step 1. The large variation observed in effective conductivity is a direct result from the deviations seen between simulated and measured antenna efficiency. In addition, the uncertainty in extracted tanδ during step 1 yields an additional uncertainty in the simulation model used in step 2 of the characterization process.

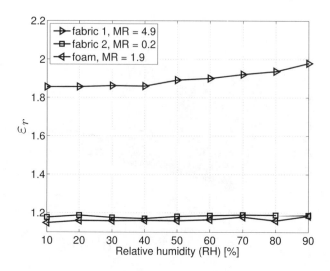

Figure 15.14   Estimated permittivity versus relative humidity.

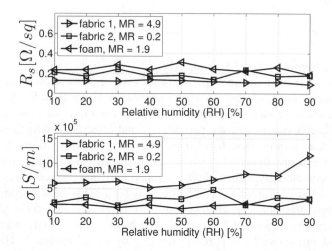

Figure 15.15   Estimated surface resistivity versus relative humidity.

15.3   Active Antenna Modules for Wearable Textile Systems

15.3.1   Active Wearable Antenna with Optimized Noise Characteristics

This section presents the design of a wearable active receive textile antenna operating in the 2.45 GHz ISM band, completely based on flexible materials. The structure of the active receive antenna with a low-noise amplifier directly located underneath the antenna ground plane is depicted in Figure 15.18. The ground plane shields the low-noise amplifier from the antenna radiation, thereby avoiding parasitic coupling that creates instabilities in the active circuitry [34]. The radiating element for the 2.4 − 2.485 GHz ISM band is a rectangular ring antenna. Since noise and gain performance are the most important characteristics in active

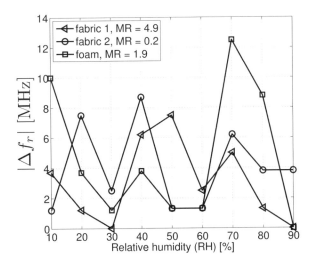

Figure 15.16 Resonance frequency error.

receive antenna designs, the LNA is directly fed by the antenna signal through a via without any matching network that transforms the antenna impedance into the optimal impedance for the minimum noise figure (NF). Instead, the antenna impedance is directly tuned to the optimal impedance for the minimum noise figure. This approach eliminates additional loss introduced by matching networks. When using textile substrates, transmission line losses can be considerable due to the high loss tangent of textile substrates. Hence long microstrip feed lines are avoided in the design, improving the overall noise performance. Additionally, small RF paths result in a small LNA footprint, minimizing EMI issues. The design of such

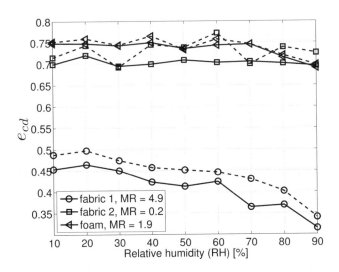

Figure 15.17 Comparison between measured and simulated antenna efficiency at their respective resonance frequency.

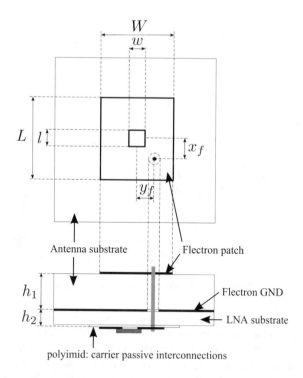

Figure 15.18   Geometry of the active wearable receive antenna.

an active antenna is quite challenging as it differs from the traditional $50\,\Omega$-based design methodologies. Since LNA and antenna can no longer be designed independently, an efficient design method based on EM/circuit co-optimization techniques is required [35]. Additionally, a dedicated multiplatform simulation approach between ADS-Momentum and CST Microwave Studio is used to account for all the losses induced by the use of conductive textile materials [12].

15.3.1.1   Active Receive Antenna Topology

The wearability of the active receive antenna is ensured by using mainly (breathable) textile materials for the antenna as well as for the LNA substrate, together with a small flexible polyimide layer on which the interconnections for the LNA are defined. The former substrate consists of flexible polyurethane foam of thickness $h_1 = 3.56$ mm. The conductive patch and ground plane are made of Flectron®, a copper-plated nylon fabric. The hybrid flexible LNA substrate is multilayered and consists of one layer of aramid textile fabric with a thickness $d = 400\,\mu$m and one polyimide layer with a thickness of $25\,\mu$m, resulting in an overall substrate height $h_2 = 425\,\mu$m. The passive copper interconnections of the active circuitry are patterned on the polyimide layer. An overview of the electromagnetic properties of the active antenna materials is given in Table 15.7. All layers were assembled by means of an adhesive sheet. The via connecting the radiating patch to the LNA and the vias of the LNA circuit consist of copper wires with a diameter of 1 and 0.5 mm, respectively.

Table 15.7 Electromagnetic properties of active antenna materials

	ε_r	$\tan \delta$	Materials
Antenna substrate	1.28	0.016	Polyurethane foam
LNA substrate	1.84	0.015	Aramid textile fabric + polyimide layer
	R_s	σ	
Flectron$^{®}$	0.45 Ω/sq.	4.8×10^4 S/m	

15.3.1.2 Low-Noise Amplifier Design

For the LNA circuit a grounded source topology using an ATF-54143 +e-PHEMT from Avago Technologies, as depicted in Figure 15.19, was used. The LNA was designed using the ADS circuit simulator in order to provide stability, sufficient gain, and low noise figure. The bias conditions of the single-voltage transistor accomplished by the voltage divider R_1, R_2 are $V_{ds} = 3\text{V}$, $I_d = 60\,\text{mA}$, and $V_{gs} = 0.56\text{V}$. A DC bias is provided at the drain of the transistor by means of the distributed bias network consisting of a high-impedance $\lambda/4$ microstrip line $(100\,\Omega)$ and capacitor C_4, providing a low-impedance point at the operating frequency to enhance stability of the design. Biasing the gate of the PHEMT is achieved by means of L_1. Resistive damping is provided by R_5 and the bypass capacitor C_2 improves the low-frequency stability without degrading the LNA performance. Additionally, resistive loading of the drain by R_6 further improves stability. Output matching is provided by L_2 and C_3 is used as a decoupling capacitor.

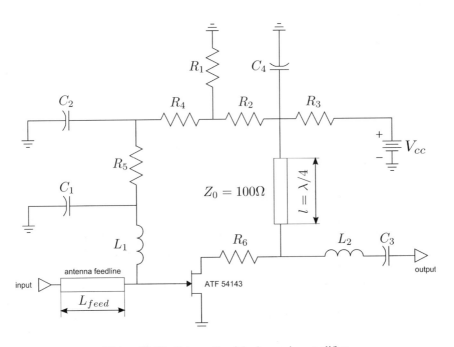

Figure 15.19 Schematic of the low-noise amplifier.

Table 15.8 Optimized geometrical dimensions in mm and optimized component values of the active receive antenna

L_{feed}	L_1	L_2	L	W	l_{gap}	w_{gap}	x_f	y_f
4.73	10 nH	1.2 nH	49.5	48	13	12	15.5	12

15.3.1.3 Full-Wave/Circuit Co-Design Strategy

Direct integration of the patch antenna and LNA into a compact design requires joint circuit/full-wave optimization, where the antenna and LNA are designed simultaneously to meet the desired specifications. The main idea of the design is to match the complex impedance of the antenna to the impedance for optimal noise performance of the LNA. Therefore, the antenna dimensions and several well-chosen parameters of the LNA were retained as degrees of freedom in the optimization process. Since a finite conductivity cannot be defined for an infinite ground plane embedded between two substrates in ADS Momentum, a multiplatform design approach is required to account for the losses induced by the Flectron® ground plane. The LNA was designed in a co-simulation setup, during which the passive interconnections were modeled by means of the full-wave simulator Momentum and the lumped and active components were simulated in the ADS-circuit simulator. The full-wave frequency-domain simulator of CST Microwave Studio was used to model the passive radiator. Interfacing the complex antenna impedance with the S-parameters of the LNA in a dynamic link between CST Microwave Studio and ADS-Momentum allowed us to optimize the antenna parameters, $L, W, l_{gap}, w_{gap}, x_f, y_f$, and the LNA parameters, L_1, L_2, and L_{feed}. The following design goals within the entire 2.45 GHz ISM band were put forward:

1. A noise figure $NF = NF_{min}$ (minimum achievable noise figure of the LNA).
2. Sufficient available amplifier gain, G_A.
3. Flat gain response, $(G_{A,max} - G_{A,min} < 0.5 \text{ dB})$.
4. Output matching, $|\Gamma_{out}| < -10 \text{ dB}$.

The optimized antenna dimensions and LNA parameters are given in Table 15.8. The back and front of the active wearable receive antenna are depicted in Figures 15.20 and 15.21, respectively. The total size of the active antenna is about 11 cm × 11 cm.

15.3.1.4 Simulation and Measurement Results

First, the measurements and simulations of the LNA in a 50 Ω reference system are discussed. In addition, an on-body measurement of the LNA was performed to investigate the feasibility of designing wearable RF electronics on flexible textile substrates. Second, the active antenna is completely characterized in terms of available gain and noise figure. Finally, on-body measurements that characterize gain performance and output matching of the active receiver antenna when integrated into a fire-fighter jacket (Figure 15.22) are discussed. The active antenna is located between the inner liner and the combined moisture and thermal barrier. Due to the curvature of the human body, the spacing between the body and the active antenna differs when integrating the antenna into the garment at two different positions, yielding different

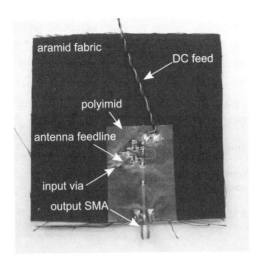

Figure 15.20 Backside of the active wearable antenna.

effects of the human body on active antenna performance. At positions 1 and 2, the spacing between the antenna and the body is about 1 and 5 cm, respectively.

Validating LNA Performance

Placing a thru-hole straight SMA connector at the input of the LNA allowed us to investigate the LNA as a two-port network in a 50 Ω environment. Two-port scattering parameter and noise figure measurements were performed in the range of 0.1–6 GHz. The on-body measurement was performed by integrating the LNA at position 1 as depicted in Figure 15.22. The measured and simulated gain, measured on-body gain and noise with respect to a 50 Ω reference impedance are depicted in Figure 15.23. The simulated $|S_{21}|$ and NF at 2.45 GHz are 11.79 dB and 0.87 dB, respectively. The measured $|S_{21}|$ and NF at 2.45 GHz are 11.18 dB

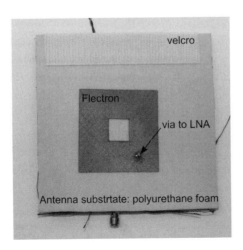

Figure 15.21 Front of the active wearable antenna.

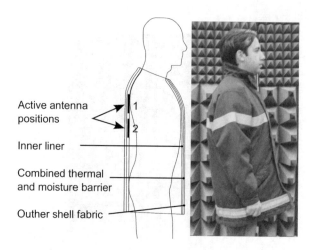

Figure 15.22 Antenna positions for the on-body measurements.

and 1.25 dB, respectively. A good agreement over the measured frequency range is observed between the measured and simulated noise and gain performance of the LNA. Note that the simulation model does not include the SMA connectors at the input and at the output of the LNA. The on-body gain measurement demonstrates that power loss in the LNA is negligible although the active circuitry is unshielded and directed towards the body. An LNA on-body gain of 10.87 dB at 2.45 GHz is obtained.

Active Antenna Free-Space Measurements and Simulations

As the amplifier and antenna are integrated in a single entity, traditional two-port measurements are no longer possible to characterize the transducer gain G_T, the available gain G_A, and the noise figure NF of the amplifier. Therefore, the measurement technique as proposed

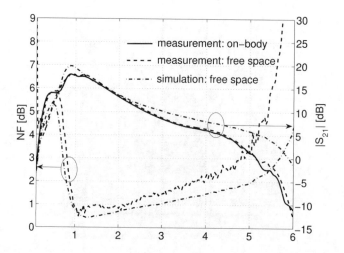

Figure 15.23 Gain and noise performance of the LNA in a 50 Ω reference system.

in Reference [36] was used. The technique measures the total gain G_{tot} of the active receive antenna and gain of a passive antenna G_p with identical dimensions and compares both in order to extract the transducer gain and available gain of the low-noise amplifier. The total gain $G_{tot}(\theta, \phi) = G_p(\theta, \phi)G_T$ of an active antenna is defined in a standard way if a single RF input port is available. Once G_T is known, the available gain G_A is defined by

$$G_A = \frac{G_T}{1 - |\Gamma_{out}|^2} \tag{15.10}$$

The noise in an active receive antenna can be attributed to the external noise caused by the environment, characterized by the effective noise temperature T_a and the internal noise of the active circuit, characterized by the noise factor F. The noise factor of the active circuitry is independent from its environment and solely determined by the antenna impedance and the active circuit. The noise figure measurement of the active receive antenna is based on the measurement of the active antenna's transducer gain G_T and absolute noise power density P_n referred to 290 K. The noise factor of the active antenna is given by

$$F = 1 + \frac{P_n}{G_T} - \frac{T_a}{290}. \tag{15.11}$$

The noise figure NF is then found as $10 \log(F)$.

The active antenna performance was measured inside the anechoic chamber in the frequency range from 2 to 3 GHz. The ambient room temperature T_a inside the anechoic chamber was 300 K. The simulated and measured available gain as well as the simulated absolute minimum noise figure, the simulated noise figure, and the measured noise figure in the frequency range from 2 to 3 GHz are given in Figure 15.24. A good agreement is shown between the measured and simulated active antenna performances. An available gain of about 12 dB, a gain ripple of 0.5 dB and a noise figure of about 1.3 dB are obtained within the ISM band.

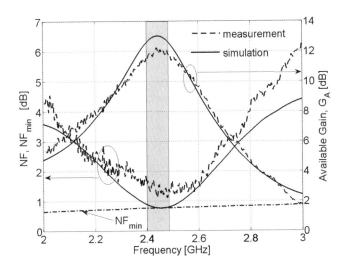

Figure 15.24 Simulated and measured noise figure and available gain of the active wearable antenna.

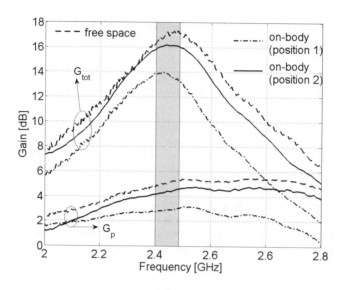

Figure 15.25   Measured total gain of the active receive antenna and passive antenna.

Active Antenna On-Body Measurements

In Figure 15.25 one observes the active antenna's free space total gain and on-body total gain in the frequency range from 2 to 3 GHz and this for the on-body positions 1 and 2. The total free-space gain of the active antenna is about 17 dBi. Comparing this with the on-body gains of the active antenna yields approximately 1 and 3 dB of power loss for on-body positions 2 and 1, respectively. As shown previously, the losses in the LNA are negligible after integration into the fire-fighter jacket, although the active circuitry is directed toward the body. As seen from the on-body gain and free-space measurements of the passive antenna in Figure 15.25,

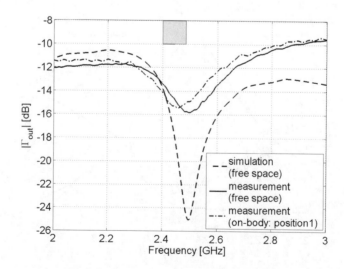

Figure 15.26   Simulated and measured output reflection coefficient of the active receive antenna.

the power loss in the passive gain for the on-body positions 1 and 2 is comparable to the power loss in the active antenna's on-body gain. Furthermore, a comparison between the gain for positions 1 and 2 shows that a smaller spacing between the body and the passive antenna results in significantly higher losses due to higher power absorption in the human body. Moreover, the change in antenna input impedance due to the presence of the human body will modify the active transducer gain, resulting in a change of the total active antenna gain. The textile layers of the garment covering the antenna will slightly decrease the passive gain of the radiating element. Hence, it is concluded that most of the power loss can be attributed to additional losses in the passive antenna which is caused by bulk power absorption in the human body when placing the active antenna close to the body. In addition, the maximum gain shifts towards the frequencies when placing the active antenna close to the body. Increasing the size of the passive radiator's ground plane will reduce both effects when placing the active antenna close to the body. Figure 15.26 shows the measured free space, the measured on-body, and the simulated reflection coefficient of the active receive antenna. A good agreement between measured and simulated reflection coefficient is observed, even though the output SMA connector was not modeled in the simulations. The measured on-body, measured free-space, and simulated $|\Gamma_{out}|$ are all below $-10\,\mathrm{dB}$ in the complete ISM band.

15.3.2 Solar Cell Integration with Wearable Textile Antennas

15.3.2.1 Shorted Solar Patch Antenna: Topology and Design

In this subsection we discuss the integration of solar cells onto a wearable aperture-coupled shorted patch antenna operating in the 902–928 MHz UHF band. This novel design uses the same area for RF radiation and for DC power generation, minimizing the overall size of the body-centric wireless communication system. Moreover, the shorting wall allows simple routing of the DC output connections from the solar cells to the energy management unit without disturbing antenna radiation. The solar cells can be placed on top or next to the radiating patch without disturbing its radiation function as long as the radiating edges of the patch remain uncovered [37, 38]. The geometry of the aperture-coupled shorted patch antenna is depicted in Figure 15.27. Aperture coupling avoids a fragile soldered probe feed connection and thereby improves the flexibility of the overall design. Given the large wavelength at the frequency of operation, a PIFA topology was selected to obtain a compact antenna suitable for wearable applications. Moreover, the presence of a ground plane minimizes bulk power absorption in the human body. An H-shaped coupling slot is introduced to improve coupling further while minimizing backward radiation toward the human body [39]. The antenna height was carefully chosen to obtain a low-profile design while providing a sufficiently large antenna volume, yielding a sufficiently large bandwidth to account for fabrication inaccuracies, variations in material parameters, and frequency detuning caused by the proximity of the human body.

Antenna Materials

The antenna substrate, consisting of flexible polyurethane protective foam, typically used in professional garments, provides a thickness $h_1 = 11$ mm, a permittivity $\varepsilon_r = 1.16$ and a loss tangent, $\tan \delta = 0.010$. The feed substrate is constructed by assembling two aramid textile layers typically used as the outer layer in fire-fighter jackets, resulting in a thickness h_2 of 0.95 mm, a permittivity of 1.97, and $\tan \delta = 0.020$. The conductive patch and ground plane

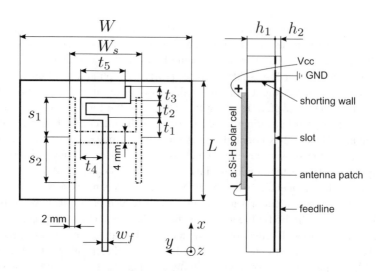

Figure 15.27 Geometry of the aperture coupled shorted solar wearable antenna.

are based on Flectron®, a copper-coated nylon fabric with a surface resistivity $R_s =$ 0.1 Ω/sq. The microstrip feed line is constructed using copper foil. The radiating patch with length L and width W is shorted to the ground plane by a conducting wall constructed using Flectron®. The shorting wall is created by folding the patch, inserting it through the antenna substrate, and connecting it to the ground plane using conductive tape. The H-shaped slot, with dimensions s_1, s_2, and W_s, coupling the electromagnetic energy from the 50 Ω microstrip line into the shorted patch, is centered underneath the patch. The footprint of the microstrip feed line is minimized by meandering the stub with dimensions t_1, t_2, t_3, t_4, and t_5.

Antenna Design

The antenna was designed using the time-domain solver of CST Microwave Studio. During the design process the conductive Flectron® textile layer with a thickness of 150 μm was modeled as a lossy metal with an effective bulk conductivity calculated by $\sigma = \pi f \mu_0 / R_s^2 = 361230$ S/m, with $f = 915$ MHz the frequency of operation and $\mu_0 = 4\pi \times 10^{-7}$ H/m the magnetic permeability. During the design process, the antenna dimensions L, W, s_1, s_2, W_s, and t_3 were optimized to meet the specified bandwidth requirement of $|S_{11}| \leq -10$ dB in the frequency range from 902 to 928 MHz. The resulting antenna dimensions are given in Table 15.9. The total ground plane size is about 120 $\times$ 120 mm. Note that the simulation model did not include the SMA connector.

Table 15.9 Antenna dimensions

Patch (mm)	Slot (mm)	Stub (mm)
L, W	s_1, s_2, W_s	t_1, t_2, t_3, t_4, t_5
62, 80	24, 28, 36	11, 12, 10, 12, 24
Feed line (mm)	w_f	3

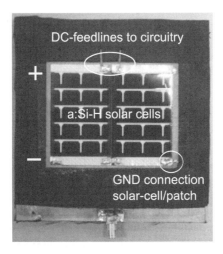

Figure 15.28 Picture of the antenna with two solar cells glued on top (parallel configuration).

15.3.2.2 Solar Cell Integration and DC Characteristics

Solar Cell Integration

The amorphous silicon (a:Si–H) solar cell is a multilayered structure, consisting of a flexible polyimide carrier, an aluminum layer, and a p-i-n silicon layer that is located between two transparent conductive zinc oxide (ZnO) layers. The total thickness of the solar cell is about 200 μm. The size of the solar cell aperture is 50 mm × 37 mm, yielding an active area of 18.5 cm^2 for one solar cell. Figure 15.28 depicts the antenna with two solar cells glued on top of the Flectron$^{®}$ patch and Figure 15.29 shows the meandered microstrip feed line. The solar cells are placed in such a manner that the edge opposite to the shorting wall of the

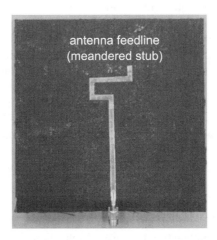

Figure 15.29 Picture of the meandered microstrip line feed.

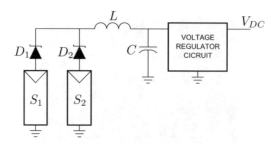

Figure 15.30   Schematic of two parallel connected solar cells and RF decoupling network.

shorted aperture-coupled patch antenna is not covered, since the fringing electric fields at this radiating edge are responsible for the antenna radiation. The DC contact of the solar cell is soldered on to the patch using a copper wire, while the DC+ contact wires are routed through the substrate at the side of the shorting wall. Since the side of the shorting wall is not a radiating edge, the positive voltage DC connections will not affect the radiation characteristics of the antenna. In this implementation the solar cells are connected in parallel. A schematic representation of two parallel connected solar cells S_1 and S_2 is depicted in Figure 15.30. Considering two differently illuminated solar cells, a diode D connected in series with each cell will prevent the flow of reverse current. The total open-circuit output voltage will be equal to the largest of both voltages and the short-circuit current will be equal to the sum of all short-circuit currents [40]. Schottky diodes are chosen since the voltage drop across the diodes is smaller compared to conventional diodes. The regulator circuit providing a constant voltage supply will be positioned below the ground plane to minimize the coupling between the circuitry and the patch antenna. Since the solar cells are connected to the patch and antenna ground plane, a series inductor and RF-decoupling capacitor are introduced.

Solar Cell DC Properties
The solar cell's DC I–V characteristic was measured using a solar simulator providing an illumination of $100\,mW/cm^2$, representing sunlight directly overhead and a turbidity-free sky. The measurement results are depicted in Figure 15.31. The measured open-circuit voltage V_{oc} and short-circuit current I_{sc} amount to 4.23 V and 26.25 mA, respectively. The maximum power point occurs at 3 V and 19.1 mA, resulting in a maximum deliverable power of 57.3 mW. Including the blocking diodes, a forward voltage of 500 mV reduces the maximum deliverable power to 48 mW per solar cell. Considering the parallel connection of the two solar cells as depicted in Figurer 15.30, a maximum DC output power of about 96 mW can be obtained. In a real-life application the solar cell will not be illuminated as strongly as with the solar simulator. Also the position of the antenna in the clothing will be as such that the sunlight will not be directly overhead, resulting in a lower DC power output.

15.3.2.3   Simulation and Measurement Results

A comparison between simulated and measured reflection coefficients of the antenna is shown in Figure 15.32. The simulated and measured bandwidth of the antenna without solar

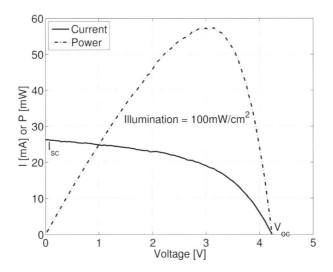

Figure 15.31 I–V characteristics of the solar cell.

cells is 48 MHz. With the solar cells placed on top of the patch, a slightly larger bandwidth is observed due to additional losses caused by the presence of the solar cell. For the on-body measurement, the antenna was positioned on the chest in order to minimize bending of the antenna. The measured on-body bandwidth has increased to 64 MHz due to the additional losses introduced by bringing the antenna in proximity to the human body. Gain measurements were performed inside an anechoic chamber and the radiation pattern in both the XZ plane and YZ plane were measured for the antenna with and without solar cells. The

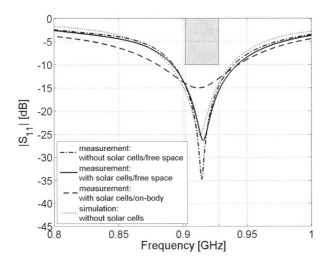

Figure 15.32 Measured and simulated reflection coefficients.

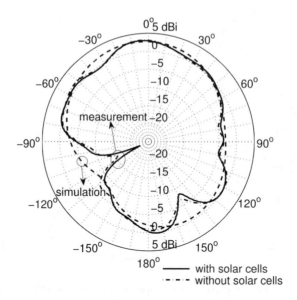

Figure 15.33   Measured and simulated gain in the *XZ* plane at 915 MHz.

measured and simulated free-space gain at 915 MHz in the *XZ* plane and *YZ* plane are depicted in Figures 15.33 and 15.34, respectively.

A very good agreement is observed between measured and simulated results. From these measurements it is concluded that the solar cells combined with the DC+ connection wires have a minor influence on antenna radiation. A maximum measured and simulated gain of

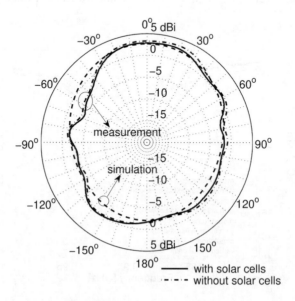

Figure 15.34   Measured and simulated gain in the *YZ* plane at 915 MHz.

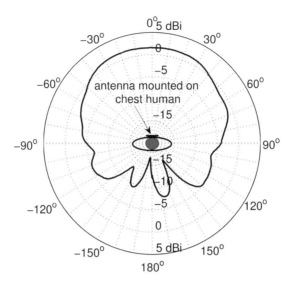

Figure 15.35 Measured on-body gain of the antenna with solar cells at 915 MHz.

about 3 dBi and 2.7 dBi are observed, respectively. The on-body gain measurement result in the *YZ* plane is depicted in Figure 15.35 and a maximum gain of 1.6 dBi is observed.

15.4 Conclusions

This chapter focused on the development and characterization of active wearable antenna modules constructed from fabrics and flexible materials in order to facilitate integration into wearable textile systems. Design aspects such as electromagnetic characterization of the materials used in a textile antenna design, computer-aided design, and characterization of active textile antenna modules were covered. In addition, solar energy scavenging by integrating solar cells with textile antennas is presented.

First, two electromagnetic characterization techniques dedicated to extract the constitutive parameters of fabrics and flexible foam materials used in a textile antenna design were treated. The matrix-pencil two-line method effectively removes the perturbations in the calculated propagation constant of the de-embedded transmission line, caused by geometry uncertainties and microstrip transmission line inhomogeneities. The surrogate-based inverse planar antenna characterization method requires only two reflection coefficient measurements, making the technique suitable for fast electromagnetic material characterization as a function of changing environmental conditions. A two-step characterization process allowed the effective conductivity of the electrotextile to be extracted.

Second, a novel active integrated wearable textile receive antenna for communication in the 2.45 GHz ISM band is presented. The antenna is constructed entirely from textile and flexible foam materials and the LNA substrate is constructed on a composite polyimide–aramid textile substrate. A dedicated full-wave electromagnetic/circuit optimization strategy in a novel multiplatform simulation setup was applied, allowing efficient design of the active receive textile antenna. The on-body gain performance of the active antenna was investigated and it was shown that gain performance is mainly affected by the spacing between the body and the passive radiator. The active circuitry is less influenced by the presence of the human body. Additionally, integration of flexible amorphous silicon solar cells on to a textile antenna was presented and it was shown that a novel interconnection scheme to route the DC routing wires along the textile antenna allows efficient interconnection without affecting the antenna characteristics.

References

1. Scanlon, W.G. and Evans, N.E. (2001) Numerical analysis of bodyworn UHF antenna systems. *Electronics and Communication Engineering Journal*, **13** (2), 53–64.
2. Salonen, P., Rahmat-Samii, Y. and Kivikoski, M. (2004) Wearable antennas in the vicinity of human body. Antennas and Propagation Society International Symposium, Montery, CA, June 20–25, 2004, vol. 1, IEEE, Piscataway, NJ, pp. 467–470.
3. Tronquo, A., Rogier, H., Hertleer, C. and Van Langenhove, L. (2006) Robust planar textile antenna for wireless body LANs operating in 2.45GHz ISM band. *IEE Electronics Letters*, **42** (3), 142–143.
4. Hertleer, C., Van Laere, A., Rogier, H. and Van Langenhove, L. (2010) Influence of relative humidity on textile antenna performance. *Textile Research Journal*, **80** (2), 177–183.
5. Lilja, J., Salonen, P. and de Maagt, P. (2009) Environmental characterization of industrial fabric for softwear antenna. IEEE International Symposium on Antennas and Propagation and USNC/URSI National Radio Science Meeting, Charleston, CA, June 1–5, 2009, IEEE, Piscataway, NJ.
6. Hertleer, C., Rogier, H., Vallozzi, L. and Van Langenhove, L. (2009) A textile antenna for off-body communication integrated into protective clothing for firefighters. *IEEE Transactions on Antennas and Propagation*, **57** (4), 919–925.
7. Salonen, P., Yang, F., Rahmat-Samii, Y. and Kivikoski, M. (2004) WEBGA – wearable electromagnetic bandgap antenna. IEEE Antennas and Propagation Society Symposium, Monterey, CA, June 20–25, 2004, IEEE, Piscataway, NJ, pp. 451–454.
8. Zhu, S. and Langley, R. (2009) Dual-band wearable textile antenna on an EBG substrate. *IEEE Transactions on Antennas and Propagation*, **57** (4), 926–935
9. Vallozzi, L., Rogier, H. and Hertleer, C. (2008) Dual polarized textile patch antenna for integration into protective garments. *IEEE Antennas Wireless and Propagation Letters*, **7**, 440–443.
10. Rogier, H., Hertleer, C., Vallozzi, L. *et al.* (2009) Indoor off-body communication based on a textile multi-antenna system integrated in clothing for rescue workers. Paper presented at 2nd IET Seminar on Antennas and Propagation for Body-Centric Wireless Communication, London, UK, April 20, 2009.
11. Kennedy, T.F., Fink, P.W., Chu, A.W. *et al.* (2009) Body-worn e-textile antennas: the good, the wow-mass, and the conformal. *IEEE Transactions on Antennas and Propagation*, **57** (4), 910–918.
12. Declercq, F. and Rogier, H. (2010) Active integrated wearable textile antenna with optimized noise characteristics. *IEEE Transactions on Antennas and Propagation*, **58** (9), 3050–3054.
13. Roller, D.P. and Slane, S. (1998) Wearable lithium-ion polymer batteries for military applications. Thirteenth Annual Battery Conference on Applications and Advances, Long Beach, CA, January 13–16, 1998, IEEE, Piscataway, NJ, pp. 71–74.
14. Hahn, R. and Reichl, H. (1999) Batteries and power supplies for wearable and ubiquitous computing. Third International Symposium on Wearable Computers, San Francisco, CA, October 18–19, 1999, IEEE, Piscataway, NJ, pp. 168–169.
15. Starner, T. (1996) Human-powered wearable computing. *IBM Systems Journal*, **35** (34), 618–629.
16. Schubert, M.B. and Werner, J.H. (2006) Flexible solar cells for clothing. *Materials Today*, **9** (6), 42–50.
17. Declercq, F., Georgiadis, A. and Rogier, H. (2011) Wearable aperture-coupled shorted solar patch antenna for remote tracking and monitoring applications. 5th European Conference on Antennas and Propagation (EuCAP), Rome, Italy, April 11–15, 2011, IEEE, Piscataway, NJ, pp. 2992–2996.
18. Lilja, J. and Salonen, P. (2009) Textile material characterization for softwear antennas. Proceedings of the 28th IEEE Conference on Military Communications, Boston, MA, October 18–21, 2009,, IEEE, Piscataway, NJ.

19. Cottet, D., Grzyb, J., Kirstein, T. and Troster, G. (2003) Electrical characterization of textile transmission lines. *IEEE Transactions on Advanced Packaging*, **26** (2), 182–190.

20. Declercq, F., Rogier, H. and Hertleer, C. (2008) Permittivity and loss tangent characterization for garment antennas based on a new matrix-pencil two-line method. *IEEE Transactions on Antennas and Propagation*, **56** (8), 2548–2554

21. Sankaralingam, S. and Gupta, B. (2010) Determination of dielectric constant of fabric materials and their use as substrates for design and development of antennas for wearable applications. *IEEE Transactions on Instrumentation Measurement*, **59** (12), 3122–3130.

22. Couckuyt, I., Declercq, F., Dhaene, T. *et al.* (2010) Surrogate-based infill optimization applied to electromagnetic problems. *International Journal of RF and Microwave Computer-Aided Engineering*, **20** (5), 492–501.

23. Declercq, F., Couckuyt, I., Rogier, H. and Dhaene, T. (2010) Complex permittivity characterization of textile materials by means of surrogate modelling. Antennas and Propagation Society International Symposium, Toronto, ON, Canada, July 11–17, 2010, IEEE, Piscataway, NJ, pp. 1–4.

24. Lee, M.Q. and Nam, S.W. (1996) An accurate broadband measurement of substrate dielectric constant. *IEEE Microwave and Guided Wave Letters*, **6** (4), 168–170.

25. DeGroot, D.C., Walker, D.K. and Marks, R.B. (1996) Impedance mismatch effects on propagation constant measurements. Electrical Performance of Electronic Packaging, 5th Topical Meeting, Napa, CA, 1996 October 28–301996, IEEE, Piscataway, NJ, pp. 141–143.

26. Agilent Technologies (2008) Linecalc [Internet], March 2008, available from http://cp.literature.agilent.com/litweb/pdf/ads2002/linecalc/index.html.

27. Edwards, T. (1992) *Foundations for Microstrip Circuit Design*, 2nd edn., John Wiley & Sons, Ltd, Chichester, West Sussex, England.

28. Hua, Y. and Sarkar, T.K. (1990) Matrix-pencil method for estimating parameters of exponentially damped/undamped sinusoids in noise. *IEEE Transactions on Acoustics Speech and Signal Processing*, **38** (5), 814–824.

29. Declercq, F. (2011) *Characterization and design strategies for active textile antennas. PhD Dissertation*, Ghent University, Ghent, Belgium.

30. Less, E.M.F. (2010) Pure copper polyester tafetta fabric [Internet], November 2010, available from http://www.lessemf.com/fabric.html.

31. Garg, R., Bhartia, P., Bahl, I. and Ittipiboon, A. (2001) *Microstrip Antenna Design Handbook*, Artech House, Inc., Norwood, MA.

32. Johnston, R.H. and McRory, J.G. (1998) An improved small antenna radiation-efficiency measurement method. *IEEE Antennas and Propagation Magazine*, **40** (5), 40–48.

33. Gorissen, D., Crombecq, K., Couckuyt, I. *et al.* (2010) A surrogate modeling and adaptive sampling toolbox for computer baised design. *JMRL*, **11**, 2051–2055.

34. De Mulder, B., Rogier, H., Vandewege, J. and De Zutter, D. (2003) Highly sensitive, co-optimised active receiver antenna: its use in Doppler radar in 2.4GHz ISM band. *IEE Electronics Letters*, **39** (18), 1299–1301.

35. Rogier, H., De Zutter, D., De Mulder, B. and Vandewege, J. (2004) Design of active planar antennas based on circuit/full-wave co-optimization. Antennas and Propagation Society International Symposium, Monterey, CA, June 20–25, 2004, vol. 3, IEEE, Piscataway, NJ, pp. 2329–2332.

36. An, H., Nauwelaers, B., Vandecapelle, A.R. and Bosisio, R.G. (1994) A novel measurement technique for amplifier-type active antennas. MTT-S International Microwave Symposium (ed. H.J.W.C.P. Kuno), vols 1–3, San Diego, CA, May 23–27, 1994, IEEE, Piscataway, NJ, pp. 1473–1476.

37. Tanaka, M., Suzuki, Y., Araki, K. and Suzuki, R. (1995) Microstirp antenna with solar-cells for microsatellites. *IEE Electronics Letters*, **31** (1), 5–6.

38. Vaccaro, S., Torres, P., Mosig, J.R. *et al.* (2000) Integrated solar panel antennas. *IEE Electronics Letters*, **36** (5), 390–391.

39. Rathi, V., Kumar, G. and Ray, K.P. (1996) Improved coupling for aperture coupled microstrip antennas. *IEEE Transactions on Antennas and Propagation*, **44** (8), 1196–1198.

40. Slonim, M.A. and Shavit, D.S. (1997) Linearization of the output characteristics of a solar cell and its application for the design and analysis of solar cell arrays. Proceedings of the 32nd Intersociety Energy Conversion Engineering Conference, vol. 3, Honolulu, HI, July 27–August 1, 1997, IEEE, Piscataway, NJ, pp. 1934–1938.

16

Novel Wearable Sensors for Body Area Network Applications

Chomora Mikeka and Hiroyuki Arai
Yokohama National University, Yokohama, Japan

16.1 Body Area Networks

In free space, electromagnetic waves emitted by isotropic radiators are transmitted as spherical waves. Spherical transmission is effective for multicasting services. On the other hand, to realize point-to-point links by radio in areas where cables were previously used, a precisely directed antenna transmitting a narrow beam is necessary to focus the energy emitted by a given base station to certain receive antenna.

It is also essential to reduce the output power of the transmitter and to limit interference with other systems in order to separate the coverage area of one radio system from another. Bluetooth or ZigBee have been developed to provide power control by protocol when using free-space transmission. In addition, low-power radios with a modest transmission capacity are available for unlicensed use.

Many wireless devices are currently used, so controlling the propagation of the electromagnetic waves so as to limit the cell covered by one device is expected to be the challenge in many applications. Body area networks (BANs) have to confine the coverage area to the human body, for example. As a result, the potential use of a subsidiary waveguide is being considered as a new method of focusing the transmitted energy, lowering transmission loss, and controlling the transmission power of wireless devices. The coverage area is limited to a region around the subsidiary waveguide, to which the wireless device is electromagnetically coupled. Although a leaky coaxial cable is not suitable for short-range wireless access, it is an example of a subsidiary waveguide [1–4]. However, when the coverage area is assumed to be in a plane 1–2 m in extent, it is desirable to realize a sheet-like waveguide to which the wireless device can be coupled at an arbitrary location.

Microwave and Millimeter Wave Circuits and Systems: Emerging Design, Technologies, and Applications,
First Edition. Edited by Apostolos Georgiadis, Hendrik Rogier, Luca Roselli, and Paolo Arcioni.
© 2013 John Wiley & Sons, Ltd. Published 2013 by John Wiley & Sons, Ltd.

A sheet-like waveguide proves to have smaller transmission loss than that experienced in free-space transmission and users can use their own waveguides independently so a high level of security could be achieved. In addition, because the channel is sheet-shaped, it can be fabricated into a wearable fabric, given that bending effects on the communication channel and signal transmission properties are thoroughly studied.

If communication by cable is described as one-dimensional (1-D) communication, and wireless access via antennas as three-dimensional (3-D) communication, then a sheet-like waveguide could be called a two-dimensional (2-D) communication, which suggests a new physical layer of communication. Radial waveguides and parallel planar waveguides can be considered as 2-D waveguides. They exhibit small transmission losses, but it is necessary to introduce an aperture or slit to excite or couple to the waveguide from the outside. Thus, a novel sheet-like waveguide in which signals travel freely between arbitrary points is desirable.

16.1.1   Potential Sheet-Shaped Communication Surface Configurations

Networks along a surface using a 2-D interface can combine the advantages of both wired and wireless networking in the aspects of reasonable connectivity, bandwidth, and independence of the system as well as power source availability. This subsection introduces some potential sheet-shaped communication surface configurations.

16.1.1.1   A Networked Surface

A networked surface [5] consists of an array of conducting plates as shown in Figure 16.1. For any position and orientation of the object on the surface, adequate links between conducting plates in different groups are guaranteed by choosing a suitable topology. Data or power-related functions are provided as busses on the surface. Information on the position

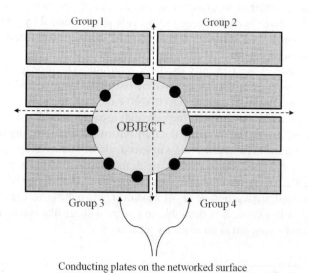

Conducting plates on the networked surface

Figure 16.1   Networked surface.

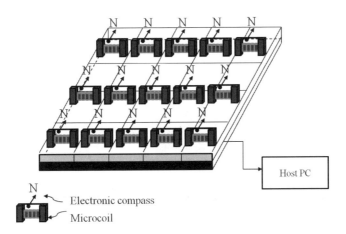

Figure 16.2 MAGIC Surface consisting of an array of microcoils and electronic compasses. The combination of microcoil and electronic compass form the 'magnetic communication device.'

and orientation of the object can also be obtained. This system has some disadvantages. A first problem occurs due to exposing the conducting parts to air. A second problem is related to sustaining the contacts. The corrosion of the bare metal parts also causes maintenance difficulties. The conductors may also produce unwanted short circuits. Furthermore, a practical method of power transmission that takes into account human safety issues should be provided.

16.1.1.2 A Magic Surface

A MAGIC Surface [6, 7] consists of an array of small magnetic communication units. Each unit is a combination of a microcoil and an electronic compass as shown in Figure 16.2. The microcoil produces a magnetic field, and the electronic compass detects the magnitude and direction of the field. The MAGIC Surface detects the position and orientation of objects placed on it. Power can be transferred by electromagnetic induction. Functions such as direction search, communication, power supply, position detection, and time synchronization are integrated on the intelligent plane of the MAGIC Surface.

16.1.1.3 A CarpetLAN

A CarpetLAN [8] combines the technologies of body-centric wireless networks and networks along a surface. It consists of carpet units and wearable devices as shown in Figure 16.3. The carpet unit consists of an electrode, floor transceiver, node unit, and carpet bus. One unit is about 1 m × 1 m. The portable device has two electrodes, one in contact with a human body (electrode 1) and the other radiating to the air (electrode 2). From electrode 1 to the carpet unit on which a person stands, the signal travels through intra-body communication, in which the human body is regarded as a wire. Between electrode 2 and the adjacent carpet unit, the signal is transmitted through free space. Together, the electrodes form a closed circuit. To detect the electric field, CarpetLAN uses the electro-optical sensors.

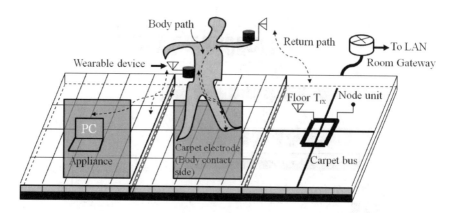

Figure 16.3 CarpetLAN.

16.1.1.4 A Wireless Power Transmission Sheet

A wireless power transmission sheet [9] consisting of a contactless position-sensing sheet and a power transmission sheet, as shown in Figure 16.4, is another intelligent surface. The position of electric devices on the sheet can be sensed contactless by electromagnetic coupling using a position-sensing coil array and an organic field-effect transistor (FET) active matrix. Contactless position sensing is achieved *by sensing changes in the impedance of the position-sensing units* when a receiver coil approaches the position-sensing coil.

Power is transferred to the objects by an electromagnetic field using a printed MEMS switching matrix and power transmission coil array. Electrical power is fed to the receiver coils wirelessly by electromagnetic induction. The bottom sides of all the power transmission electrodes in the plastic MEMS switch are connected to an AC power source operating at a frequency of 13.56 MHz. Once the position of the object is determined, one of the sender

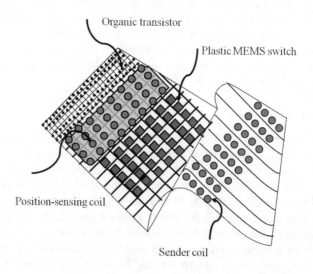

Figure 16.4 Power transmission sheet.

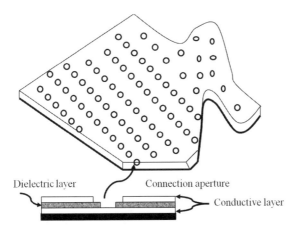

Figure 16.5 Two-dimensional transmission sheet.

coils on which the object is placed is selected by one of the MEMS switches and current starts flowing through the sender coil, generating a magnetic field. The magnetic field induces a current in the receiver coil. The efficiency of electromagnetic coupling is reasonably high at 81.4%. However, this high efficiency is available only if the transmitting and receiving coils are oriented properly. The coils can be fabricated by printed electronics technology and are about 1 mm thick.

16.1.1.5 A 2-D Transmission Sheet

A 2-D transmission sheet [10–15] consists of three layers, as shown in Figure 16.5. A dielectric layer is placed between two conductive layers. The upper conductive layer has an array of small apertures or a meshed structure. As in a *parallel plate capacitor*, electromagnetic waves exist only in the dielectric layer. The apertures or meshes allow slight leakage of electromagnetic waves around them. The 2-D transmission sheet traps signals around it. Wireless devices can receive signals at an arbitrary position on the sheet and the signals travel point to point. Exclusive resonant proximity connectors are also proposed to receive signals.

16.1.1.6 Summary

In summary, the networked schemes described in Sections 16.1.1.1 to 16.1.1.5 above are generally suitable for applications covering a short range of about 1–2 m because most of them consist of an array of units. However, the CarpetLAN is not, by design, suited for short-range networks such as BANs. It also lacks flexibility.

Each surface medium has a different method or way of exciting it. If contact between electrodes is used, difficulties arise in sustaining the contact and in maintenance, and it is difficult to avoid unexpected shorts. With magnetic coupling, the efficiency may drop for certain positions and/or orientations of the coils. In contrast, the transmission sheet and MAGIC Surface adopt an array of microcoils, but further advances are needed to lower the cost. It is desirable to integrate a power supply method as well as a communication facility.

Flexibility would greatly improve the user-friendliness of applications for a BAN. In this regard, a 2-D transmission sheet is perfect, but an external coupler is required to receive an adequate signal. The goal of this chapter is to provide a new option in which typical wireless sensing devices are freely accessible, particularly in a BAN application.

16.1.2   Wireless Body Area Network

The wireless body area network (WBAN) is a key technology to realize convenient continuous monitoring by removing cumbersome wires. The WBAN is also currently discussed in the IEEE 802.15.6 Task Group for WBAN standardization.

Recent achievements in the field of biomedical systems such as medical imaging, digital hearing aid, neuroprosthetic devices, and coclear implant, from sensing and stimulation integrated circuit (IC) technology, low-energy bio-signal processing, and wireless communication techniques, to E-textile and fabric circuit board technology in general, give opportunities to shift the healthcare paradigm toward applications in the field of patient-centric wearable healthcare to continuously manage chronic diseases. This chapter will present a 'flexible interface for body-centric wireless communications.'

16.1.3   Chapter Flow Summary

This chapter proposes a novel, simple 2-D sheet-shaped medium as a low-cost interface for BAN applications, though not limited to BANs. The proposed sheet-shaped medium can be coupled by common antennas (resonators). The proposed sheet-shaped medium allows free access to various wireless devices.

The medium offers short-range wireless communication that covers 1–2 m. It creates an intelligent surface acting as an interface for communication. In addition, it presents a solution for problems with interference and enables harmonious coexistence of multiple wireless applications. This solution is provided by limiting the directions of arrival of electromagnetic waves near the medium. When using E-textile or conductive fabric, the proposed sheet-shaped medium provides a degree of flexibility that is required for BAN application. Indisputably, this flexibility would ensure comfortable interaction between a human being and the network. The proposed sheet-shaped medium is called a 'free access mat,' the basic material for tailoring a smart suit, belt, or belly-band type wear. In the term free access mat, 'free access' means no special structure is needed to receive the signals and the receivers can couple to it at an arbitrary position, while 'mat' indicates the shape of the material.

As in other sheet-shaped media, the proposed medium successfully demonstrated the capability for wireless power transfer (WPT) at reasonable efficiency, but such a discussion is beyond the scope of this chapter. In brief, the goals and scope for this chapter are shown in Figure 16.6.

16.2   Design of a 2-D Array Free Access Mat

Based on a practical calculation, this section presents the transmission characteristics for a 2-D array with sheet-like geometry as depicted in Figure 16.7. The basic elements for the 2-D array include four patch resonators on the lower layer and one parasitic element centered over them on the upper layer, as shown in Figure 16.8(a). The length of the rectangular patch resonators on the lower layer is determined by the resonance length. In this 2-D array, square

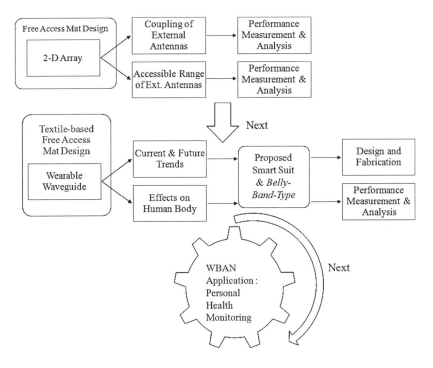

Figure 16.6   Chapter flow.

patch resonators are used in order to propagate electromagnetic waves in both the y and z directions while maintaining a symmetrical geometry.

Figure 16.8(a) shows the optimized parameters for the 2-D array. The size of the patch resonators on the lower layer yields a central frequency of around 5 GHz. The size and degree of overlap of the parasitic element on the upper layer were optimized to yield a small insertion loss. Each port is terminated by a 50 Ω load and is perfectly matched to the patch resonators on the lower layer. The output port (port 2) is on a diagonal line from the input port (port 1), in order to calculate the 2-D propagation characteristics between the two. In this model, the transmission loss is calculated without including antenna coupling in order to determine the basic characteristics of the free access mat. When the patch resonators are configured on the lower layer as in the following matrices, 1×2, 2×2, 3×3, and 4×4 arrays, the propagation lengths L (mm) are 52.4, 62.7, 103.9, and 159.2, respectively.

The transmission characteristics of each model are shown in Figure 16.8(b). Generally, an increase in the transmission loss is observed as the array matrix size n × n increases. The large difference in transmission loss between the 1×2 and 2×2 arrays is caused by cylindrical wave propagation inside the mat. The worst-case scenario results from the 4×4 array ($L = 159.2$ mm). Nevertheless, the resulting loss is still smaller than the free-space transmission loss. The transmission loss at 4.8 GHz of the 2×2, 3×3, and 4×4 arrays is shown in Figure 16.8(c). The average transmission loss per wavelength is 7.5 dB/λ, which is much smaller than the free-space transmission loss, 22 dB/λ. The losses in the mat including free-space transmission for 1 m are 19.54 dB and 45.59 dB, respectively. These results suggest

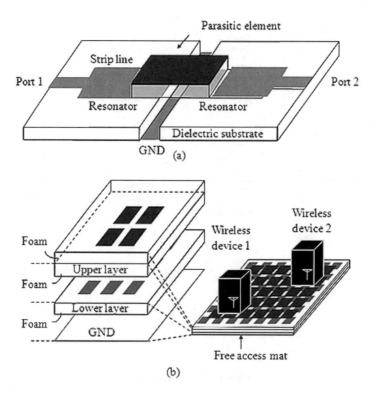

Figure 16.7 Free access mat: the sheet-like geometry and its basic elements. Configuration of (a) the ribbon-wire interconnect and (b) the free access mat.

that the proposed 2-D array, free access mat model provides sufficiently low loss characteristics.

In the discussion above, the output port was parallel to the input port. Next, a model that has an output port (port 3) orthogonal to the input port (port 1) for a 3×3 device is simulated, as shown in Figure 16.9. The calculated $|S_{21}|$ and $|S_{31}|$ characteristics are also shown in the figure. Evidently, $|S_{31}|$ is almost the same as $|S_{21}|$, which suggests that the direction of the output port is independent of that of the input port because the wave inside the mat propagates radially inside the mat owing to the symmetry of the upper and lower elements. The independence of the polarization is important for demonstrating the independence of the direction of an external antenna coupled to the mat at an arbitrary location.

16.2.1 Coupling of External Antennas

For practical use, the coupling between the mat and an external antenna is a key factor. However, it is not easy to include an antenna structure with a feed cable. In this section, the transmission characteristics of the mat including antenna coupling loss are calculated.

To apply this waveguide in a wireless access system, the coupling loss between an external antenna and the free access mat should be considered. In the above calculation (Section 16.2), ideal coupling was assumed both at the input and output ports. To find the antenna coupling effect for a 2-D array, two dipole antennas are placed above the center of the patch

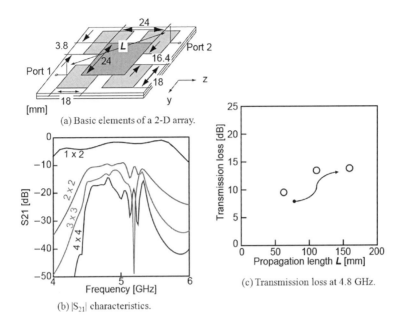

(a) Basic elements of a 2-D array.

(b) |S$_{21}$| characteristics.

(c) Transmission loss at 4.8 GHz.

Figure 16.8   2-D array and its simulated transmission characteristics.

resonators at the corners of a 4 × 4 device at a height of 5 mm, as shown in Figure 16.10. Each dipole has a resonant frequency around 5 GHz. The simulated |S$_{21}$| characteristics between two dipoles (I), one dipole and one perfectly matched port (II), two perfectly matched ports (III), and two dipoles without the free access mat (IV) are shown in the same

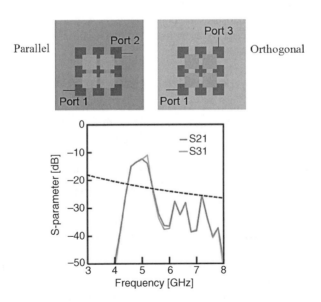

Figure 16.9   Simulated |S$_{21}$| (for parallel ports) and |S$_{31}$| (orthogonal ports) characteristics.

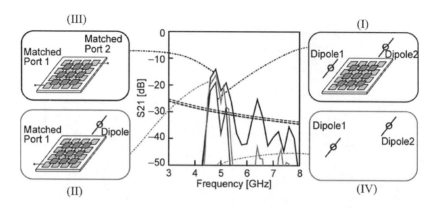

Figure 16.10 Simulated coupling loss of two external dipole antennas (I), one dipole antenna and one perfectly matched port (II), two perfectly matched ports (III), and two dipole antennas without the free access mat (IV).

figure. In case (I), waves from the dipoles couple to the free access mat and are transmitted through the mat with small loss. In case (II), waves impinge on to the free access mat by means of the perfectly coupled antenna, are then transmitted through the mat, and then couple to the dipole. In case (IV), the dipoles couple to each other directly.

The transmission losses for (I)–(III) is small, around 4.8 GHz, where direct coupling between the two dipoles is negligible, that is, smaller than −45 dB. The transmission losses for (I)–(III) are 18, 16, and 14 dB, respectively, at 4.8 GHz. The coupling loss of two antennas is given by the difference between the transmission losses of (I) and (III), being 4 ∼ 6 dB. In a practical application of the mat, the coupling of dipoles located above the corner of the mat is assumed to be the weakest, so the coupling of antennas on the mat would be stronger than presented in this case. These results show that the free access mat can easily be excited by a common antenna, and strong coupling of antennas to the mat is expected.

16.2.2 2-D Array Performance Characterization by Measurement

For a more realistic measurement, a 2-D 5×5 array device was fabricated, as shown in Figure 16.11(a). All the parameters are the same as those in Figure 16.8(a). Two dipole antennas oriented in the same direction in parallel were moved over the same device to measure the $|S_{21}|$ characteristics for 2×2, 3×3, 4×4, and 5×5 arrays. The antenna height is 5 mm, and the antenna coupling at a height of 25 mm is confirmed. S-parameter characteristics were measured using an Anritsu 37347C Vector Network Analyzer (40 MHz–20 GHz).

The measured $|S_{21}|$ characteristics of the device are shown in Figure 16.11(b). Direct coupling between two dipole antennas without the free access mat is also shown. The direct coupling is less than −30 dB, which is weak enough to be neglected. The transmission loss of the free access mat is small, around 4.54 GHz. The coupled antennas are so close to the mat that the resonant frequency has shifted to a lower frequency band. The fluctuations in the transmission characteristics are caused by noise.

The transmission losses at 4.54 GHz, which are smaller than 20 dB, are plotted in Figure 16.11(c). Cylindrical wave transmission is also confirmed by measurement. The transmission

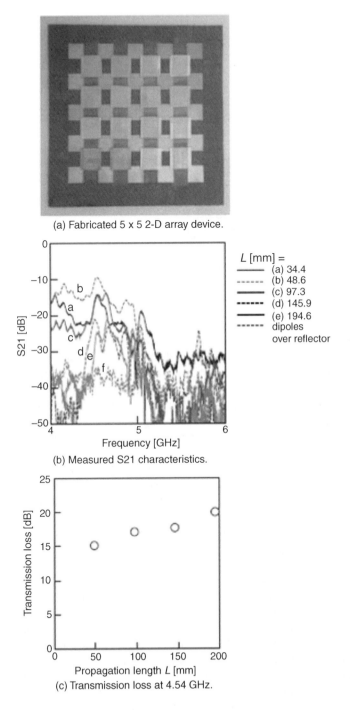

(a) Fabricated 5 x 5 2-D array device.

L [mm] =
—— (a) 34.4
----- (b) 48.6
—— (c) 97.3
----- (d) 145.9
—— (e) 194.6
----- dipoles
over reflector

(b) Measured S21 characteristics.

(c) Transmission loss at 4.54 GHz.

Figure 16.11 2-D 5 × 5 array and its measured transmission characteristics.

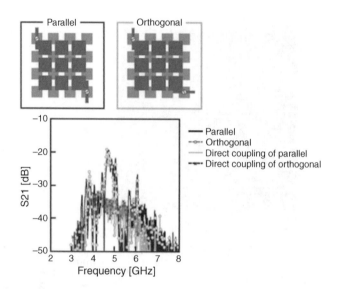

Figure 16.12   Measured $|S_{21}|$ characteristics of parallel and orthogonal ports.

loss per wavelength also remains low, at an average of 11.75 dB/λ. The transmission loss for 1 m would be 23.56 dB, which is much smaller than the free-space transmission loss, so the multipath effect caused by the mat can be neglected. The difference between the simulated and measured average loss is less than 4.25 dB and is the result of antenna coupling. Thus, the measured loss caused by coupling of the two antennas is 4.25 dB for the 2-D array, whereas the largest simulated coupling loss for the 2-D array is estimated to be 6 dB. These measured low-loss characteristics when the antenna coupling loss is considered suggest that the strong coupling between external antennas and the free access mat is confirmed and the free access mat is effective as a practical model.

The $|S_{21}|$ characteristics of parallel and orthogonal ports were also measured. Two dipole antennas were placed 5 mm above the 4 × 4 device. For the parallel ports, two dipoles were oriented in the same direction, whereas for the orthogonal ports, the dipoles were oriented in the y and z directions, respectively, as shown in Figure 16.12. The mutual coupling of dipoles in the two arrangements is assumed to be the same because of the structure's symmetry. The measured $|S_{21}|$ characteristics for the parallel and orthogonal ports are almost the same, as also shown in Figure 16.12, because cylindrical waves were transmitted inside the mat owing to the symmetry of the structure.

The direct coupling of the two antennas was measured without the mat when the antennas were located in the same position in each arrangement. Almost the same transmission characteristics were obtained in the limit that the direct coupling in the two arrangements is small enough to be neglected, that is, less than −30 dB. In addition, the transmission characteristics of the two arrangements when the dipoles are located above a 3 × 4 array were also measured. Although the difference in the losses at 4.54 GHz between the two arrangements is 0.99 dB, the transmission characteristics are almost the same. These results suggest radial propagation of electromagnetic waves inside the free access mat.

Coupling loss in the measurement antenna is not discussed because it is not easy to include a standard dipole antenna in the simulation. This is left as a future problem.

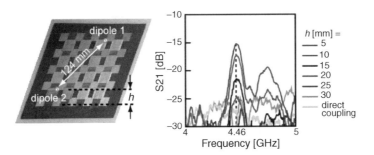

Figure 16.13 2-D array with external antennas and measured accessible range (coupling height) of the antennas.

16.2.3 Accessible Range of External Antennas on the 2-D Array

Because the free access mat is designed to limit the coverage area of a radio system, it is necessary to confirm the accessible range of external antennas. The range should be small in order to maintain security, limit the communication area, and reduce interference.

In these investigations, standard dipole antennas were placed over the 2-D array device as shown in Figure 16.11(a). The distance between the two antennas was 124 mm and the height h of the two antennas was varied simultaneously in the range 5–30 mm. The transmission characteristics for each height are shown in Figure 16.13, which also shows direct coupling of two dipoles over a reflector instead of over the free access mat.

The transmission loss increases with height. In addition, when the height is greater than 25 mm, the transmission characteristics are almost the same as those for the direct coupling of two antennas over the reflector because the antennas are too far away from the free access mat to couple to it and electromagnetic waves are reflected from the mat. Thus, the accessible range of the external antennas is limited to a height of 10 mm. The height, and hence access area, is small enough that only wireless devices on the mat could couple to the mat, which is desirable for maintaining a high level of security and limiting the coverage area, especially in BAN applications as proposed herein. More importantly, through investigations reported in this chapter, it has been confirmed that the free access mat can provide wireless devices with a contactless network along the surface as long as the devices are within a well-defined proximity or directly placed on the mat, that is, the sheet-shaped medium.

16.3 Textile-Based Free Access Mat: Flexible Interface for Body-Centric Wireless Communications

> Planar thin/thick film technology on a fabric, namely Planar-Fashionable Circuit Board (P-FCB), provides new possibility to directly integrate silicon chips and electronics into textiles. Since the substrate material is the textile fabric, a compact wearable electronics system with enhanced flexibility, durability, as well as body compatibility can be realized with a low cost.
>
> Hoi-Jun Yoo

The free access mat provides a concentrated network along a surface. If the mat can be made of flexible materials, it would have a very powerful advantage for short-range wireless

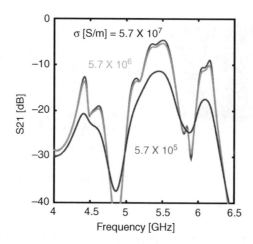

Figure 16.14   Effect of low conductivity.

networks. In particular, the flexibility would promise comfortable interactions between humans and the network in a BAN. The mats described above consisted of metallic elements and were fabricated using a dielectric substrate. To create a flexible hand-held medium, a light, thin, and flexible material is necessary. For this reason, a textile-based free access mat was fabricated using thin foamed polystyrene and conductive fabrics (ST2050 and SC8100; Teijin Fibers Ltd.). The conductive part consists of Cu–Ni, with densities of 80 (ST2050) and 112 (SC8100) g/m^2, respectively. For comparison, a free access mat using Cu tape was also fabricated. Cu, a perfect electrical conductor, has a density of about 312 g/m^2. It was necessary to check the effects of a low-density conductor, which can lead to low conductivity.

Figure 16.14 shows the effect of low conductivity simulated by a IE3D moment method-based electromagnetic (EM) simulation and optimization software. The free access mat was initially designed for a structure in which all the conductors are made of copper, which has a conductivity of $\sigma = 5.7 \times 10^7$ S/m. The conductivity was then substantially decreased and it is observed from the figure that the conductivity is not related to the resonant frequency but rather to the transmission loss. The transmission loss at the conductivity of 5.7×10^6 S/m is very similar to that of the initial model. The transmission loss increases markedly when the conductivity is reduced to 5.7×10^5 S/m and is attributable to conductivity losses in the patch elements. The simulated results show the low-conductivity limit for practical implementations of the free access mat using conductive textiles, which is estimated to be greater than 5.7×10^6 S/m.

Figure 16.15(a) shows samples of the free access mats fabricated with Cu, ST2050, and SC8100 materials. They were designed for a resonant frequency of 5 GHz. Two standard dipoles having the same resonant frequency were coupled to the samples at a height of 5 mm. The distance between the two dipoles was fixed at 176 mm, as shown in Figure 16.15(b). The $|S_{21}|$ characteristics, measured using a Vector Network Analyzer are shown in Figure 16.15(c). The variation in the resonant frequency of the Cu sample is attributed to fabrication errors. The measured transmission losses in dB/λ of the samples using ST2050, SC8100, and Cu were 4.5, 4.3, and 3.9, respectively. The variation is due to conductivity losses, but is less than 0.6 dB, and hence acceptable.

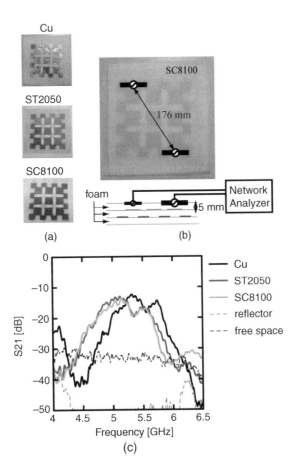

Figure 16.15 Free access mat using Cu and conductive textiles. Pictures of (a) fabricated devices, (b) measurement setup, and (c) measured $|S_{21}|$ characteristics.

Next, it was necessary to investigate the tolerance to bending or folding of the sample using SC8100. It was confirmed that the sample using ST2050 also showed the same characteristics. The SC8100 fabricated mat's $|S_{21}|$ characteristics were measured in four scenarios, as shown in Figure 16.16. Case (a) is the measured transmission loss for the unfolded, flat, free access mat. Case (b) is that for the mat when it was folded in half. The position of the two dipole antennas on the mat is fixed in these two cases. In case (c), a reflector blocks coupling of the two dipole antennas through free space. The antenna position in case (c) is the same as in case (a). Case (d) shows the free-space transmission loss when the distance between the two dipoles is the same as in case (a).

Around 5 GHz, low transmission losses were measured for cases (a), (b), and (c). They were much smaller than that for case (d), with a variation of more than 16 dB. The concentrated electromagnetic waves in the free access mat greatly reduce the transmission loss. The bandwidth in case (b) is narrower than that in case (a). In case (a), the transmission includes coupling between the two antennas through free space as well as through the mat. The

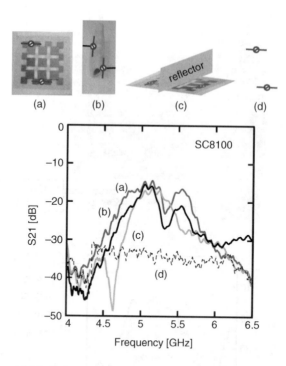

Figure 16.16   Tolerance to folding or bending of the free access mat.

variation in transmission loss at the resonant frequency between the cases (a) and (b) is negligibly small (less than 0.5 dB). This proves that transmission around the resonant frequency occurs through the free access mat. It was also found that bending or folding barely increases the transmission loss. The free access mat based on a light, soft, thin, and flexible conductive textile is scarcely degraded by the material properties of the textile or by bending or folding. It is thus attractive to think of it as comfortable and easy to wear for BAN applications, especially in ubiquitous personal health applications.

16.3.1   Wearable Waveguide

This subsection proposes a wearable textile-based free access mat. A novel concept waveguide, the smart suit, smart robe, or smart blanket, is proposed and its applicability for a BAN is demonstrated. The growing interest in BANs has motivated many studies in wearable wireless devices and their on-body or intra-body channel characteristics. This section focuses on improving the on-body channel characteristics, a major challenge for BANs. The problems are inherent to the presence of the human body. A wearable waveguide consisting of a flexible free access mat could provide an alternative approach for this purpose. The waveguide provides an independent path for electromagnetic waves that differs from the original on-body channel. This subsection presents the concept of the wearable waveguide, a prototype, and the measured improvement in the transmission characteristics. Several potential frequency bands are available for BANs. As described earlier, the central frequency of the free access mat can be controlled by changing the size of the resonators. As a matter of convenience, this section discusses a design for the 2.45 GHz ISM band.

16.3.1.1   In Sync with the Current and Future Trends: Research Interests in WBANs

People now carry many electronic devices, such as cellular phones, smartphones, personal stereos, personal digital assistants, laptops, and tablets. Networking allows these devices to share interfaces, storage, data, and computational resources as in 'cloud computing.'

Networking allows greater convenience and new services, and interest in body-centric wireless communications (Wireless BANs, or WBANs) has increased accordingly. A WBAN is a small-scale wireless communication network. Interest in WBANs has been growing because of their abundant applications: personal healthcare, smart home, personal entertainment, identification systems, space exploration, robots, and military usage, among others. A generic WBAN concept may include scenarios in which wireless sensor nodes are placed in or on the body. Because of the presence of the human body, on-body antennas and propagation have distinctive properties. For example, wearable antennas can suffer from efficiency redundancy, radiation pattern distortion, and variation in impedance at the feed [16–18]. Some issues with on-body radio channels include shadowing effects, dynamic variation in path loss, protection of anonymity, and delay elements. However, the proposed scenario in which robots or people wear a sheet-shaped waveguide, as shown in Figure 16.17(a),

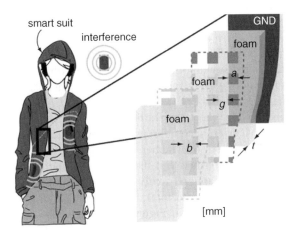

(a) Realization of a smart suit.

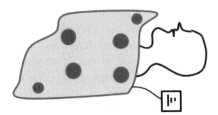

(b) Our conceptual smart blanket.

Figure 16.17   Smart suit and smart blanket, where (a) is the image and configuration of a smart suit where $a = 25.5\,\text{mm}$, $b = 33.9\,\text{mm}$, $g = 23.2\,\text{mm}$, and $t = 3\,\text{mm}$; cf. Section 16.3.1.2.

represents a new approach to these problems. The wireless sensor nodes contained in the worn waveguide communicate with each other through the waveguide. The electromagnetic waves are concentrated in the waveguide so the channel is scarcely distorted by the presence of the human body. This waveguide concept is named the 'smart suit.'

Interest in ubiquitous medical healthcare has also been growing. A generic concept of ubiquitous medical healthcare includes telemonitoring of vital data such as ECG and EEG results and telecontrol of medical devices and actuators. A wearable or implantable BAN can be applied to these types of telemetry and telecontrol. For example, vital information collected by *in vitro* or *in vivo sensors* is gathered by a master unit placed on the human body. The master unit or command/coordinator unit as described in References [19] and [20] decides an appropriate treatment method on the basis of the collected vital data and sends commands to action units that apply the treatment or medication. At the same time, the information sequence is sent to a healthcare management center such as a medical institute. These sensing, command, and action units are expected to be connected wirelessly for convenience and user-friendliness. Therefore, in this chapter a novel solution using the smart suit is proposed, which collects vital and healthcare data and applies proper treatment to objects. People simply wear the smart suit or blanket, which collects data as shown in Figure 16.17(b). The smart suit, robe, or blanket would contain within its fabric sensors, devices, and actuators or would simply collect vital data from wearable devices or would activate them. In either case, the smart suit, which acts as a receiving and transmitting (transceiver) element of the master unit, would be useful. Many challenges must be met to realize this concept of a wearable medium: the desired system offers a lightweight, flexible, sheet-shaped waveguide that enables wireless networking using various sensor units, which are placed at an arbitrary position on or near the human body.

A lightweight, thin, flexible, sheet-shaped device is suitable for a smart suit, robe, or blanket. Textile-based devices are also appropriate. A waveguide that sensor devices can access would be useful regardless of whether the smart suit itself includes sensor units. In addition, a wireless connection between the waveguide and the sensor units is desirable so that sensor units can be applied freely on the patient's body and the waveguide. Sensor units on the human body record vital data continuously or periodically using the smart suit or smart blanket as shown in Figure 16.17. It is also necessary to be free to choose the sensor units. Thus, a waveguide that can be accessed by any common antenna or coupler is strongly desired. In this case, the free access mat is a suitable waveguide for the smart suit and the smart blanket. Its simple configuration enables easy fabrication. Electromagnetic waves are concentrated in a certain frequency band. The central frequency can be controlled by changing the size of each element. In addition, shaping of the resonators can provide dual-band operations. The transmission losses of 2-D concentrated electromagnetic waves are much smaller than those in free-space transmission. Common antennas can couple to the free access mat at an arbitrary position at a height of several millimeters. The small accessible range of the radio rarely produces interference with other networks close to the mat.

Therefore, as interest in WBANs grows, a wide variety of consumer electronics and communication devices are expected to be wearable or built into clothing in the near future, including textile-based wearable devices. In this section, a basic study using a textile-based mat is presented. In Section 16.4, a belt type, wearable communication device with sensing capabilities for personal health monitoring in an ubiquitous fashion will be presented.

16.3.1.2   Flexible Free Access Mat and Effects of the Human Body

In a WBAN, the effect of the human body is distinctive. A major issue hindering the implementation of WBANs is the limited range of commonly used antennas. The dielectric mismatch between antennas and human body parts causes local reflections and further shortens the operating range. The smart suit, a wearable sheet-shaped waveguide, provides an independent path for the wireless signals and is likely to reduce distortions caused by the human body.

In this study, a vest-type smart suit using a free access mat based on conductive fabrics was fabricated. It was designed for operation at 2.45 GHz ISM band, and its parameters were as follows: square patch resonator length, $a = 55.7$ mm; parasitic square patch length, $b = 74.2$ mm; spacing between patch resonators, $g = 50.7$ mm; separation depth between parasitic patch and patch resonators, $t = 3$ mm. The transmission characteristics through the smart suit were measured using two commercial sleeve antennas at 2.45 GHz, with one sleeve antenna placed on the front of the human body while the other on the back. With this antenna positioning, however, the transmission characteristics varied significantly, as shown in Figure 16.18. Case (a) shows the transmission loss when the antennas were on the human body; in case (b), a person was wearing the smart suit with antennas on it; and case (c) shows the transmission loss through the smart suit with the same antenna position but without the human body. The smart suit greatly reduced the transmission loss, which varied by more than 57 dB. By comparing cases (a) and (b), it can be appreciated how much the free access mat improves the on-body channel characteristics. It can also be seen from cases (b) and (c) how much the channel characteristics in the mat are affected by the presence of the human body. The measured results in cases (b) and (c) are quite similar at 2.45 GHz. The transmission inside the mat is scarcely affected by the presence of the human body. In addition, the

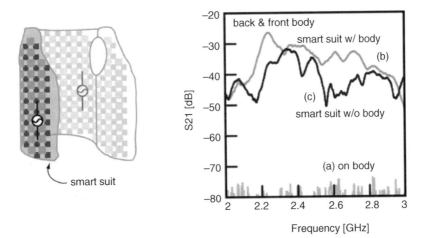

Figure 16.18   Measured $|S_{21}|$ characteristics of the *free access mat* smart suit with and without a wearer's body. One antenna is placed on the front of the body and the other on the back. The graph shows the measured $|S_{21}|$ characteristics for two antennas placed (a) on the human body, (b) on the smart suit with a person wearing it, and (c) on the smart suit without a human body. The transmission gain at 2.45 GHz exceeds 57 dB.

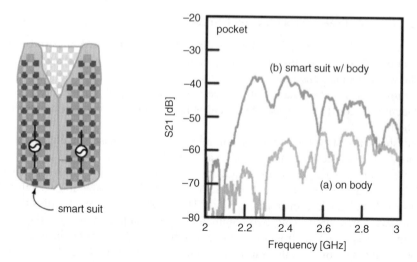

Figure 16.19 Measured $|S_{21}|$ characteristics of the *free access mat* smart suit with and without a wearer's body. Two antennas are placed in the front of the body. The graph shows the measured $|S_{21}|$ characteristics for two antennas placed (a) on the human body and (b) on the smart suit with a person wearing it. The transmission gain at 2.45 GHz exceeds 19 dB.

measured results for cases (a) and (b) indicate that the mat greatly decreases the transmission loss. This is because of the concentrated electromagnetic waves inside the free access mat and the small loss to coupling by external antennas.

Figure 16.19 shows the transmission characteristics when both antennas were placed on the front of the body. The transmission loss for case (a) is less than that shown in Figure 16.18(a). However, the transmission loss for case (b) is larger than that shown in Figure 16.18(b) because the transmission occurs by means of the smart suit and the distance between the two antennas is larger. Nevertheless, the smart suit greatly decreased the transmission loss, which varied by more than 19 dB. The smart suit greatly improved the transmission characteristics regardless of the antenna position on the human body. In addition, the transmission characteristics were scarcely distorted by the human body owing to the presence of the ground plane.

Wearable devices should also be breathable; however, the free access mat has a large ground plane. Although it is made of textile-based material, it is not breathable enough. It is therefore reasonable to use the smart suit if and only if needed. However, for a wider variety of applications, this chapter proposes using a meshed ground plane instead.

First, the chapter investigates what happens when there is no ground plane. A long free access mat is fabricated and wrapped around the human torso like a belly band, placing the antennas on the front and back of the body, as in the previous measurements shown in Figure 16.18. The mat was then moved around while the position at which the antennas couple to it were changed. Figure 16.20 shows the measured $|S_{21}|$ characteristics at different coupling positions. In Figure 16.20, the curve labeled 'w/GND' indicates the transmission loss through the initial free access mat with a perfect ground plane, whereas the curve labeled 'w/o GND' shows the loss through the mat without a ground plane. The transmission losses,

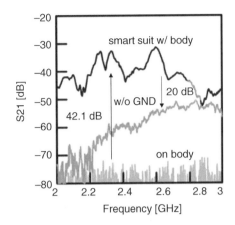

Figure 16.20 Measured $|S_{21}|$ characteristics of a worn belly-band-type free access mat with ground (w/GND) or without ground (w/o GND), compared to the 'on-body scenario.' In the graph, the black solid line is the transmission characteristic for the belly-band w/GND, while the gray solid line is the characteristic in the w/o GND case. Evidently, w/GND the gain exceeds 42.1 dB, while w/o GND a 20 dB drop in gain is observed.

which are around 10 dB, depend on the antenna coupling position, but are low enough and hence acceptable compared to those of on-body transmission. In all three cases, it can be seen that the transmission loss increases when the free access mat has no ground plane. The variation is about 13–20 dB. However, the mat without a ground plane still reduces the transmission loss compared to the case of on-body transmission; the variation is about 22–28 dB, which is large enough.

Next, the chapter investigates a meshed ground plane. Figure 16.21(a) shows a model simulated using the electromagnetic tool by Ansoft. The model is a rectangular cylinder, consisting of 3/4 muscle. The ground plane with many square holes in it is placed 0.8 mm above the muscle model, and the spacing between the patches and the ground plane is also 0.8 mm. The size of the holes p was changed from zero to $\lambda/4$, with w fixed. The simulated results are shown in Figure 16.21(b). The transmission loss increases with the hole size and the resonant frequency shifts to a lower frequency. When the hole size is $\lambda/10$, the transmission characteristics are quite similar to that of a model with a perfect ground plane. Thus, this meshed ground plane provides a more breathable device without changing the transmission characteristics.

16.3.2 Summary on the Proposed Wearable Waveguide

This section presented a textile-based free access mat. Foamed polystyrene and a conductive fabric, which were very light, thin, and flexible, were chosen and a textile based mat was made from them. The conductive fabric did not degrade the mat's transmission characteristics. In addition, it was highly robust to folding, indicating that the textile-based mat was suitable for a flexible sheet-shaped medium. The flexibility of the sheet-shaped medium greatly enhances comfort and user-friendliness, an advantage when the free access mat is used as a sheet-shaped medium for BAN applications.

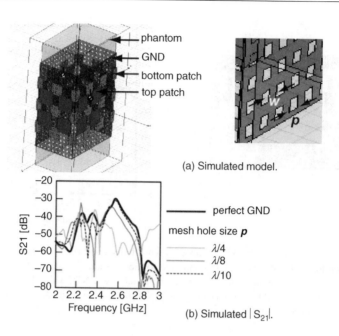

(a) Simulated model.

(b) Simulated $|S_{21}|$.

Figure 16.21 Model and the simulated $|S_{21}|$ characteristics of the belly-band-type free access mat with a meshed ground plane.

Based on the textile-based free access mat, a novel concept for a wearable waveguide has been described: the smart suit or blanket. Many frequency bands are candidates for use in WBANs, but, for convenience, a design for the 2.45 GHz ISM band was examined. A successful prototype fabrication for the wearable waveguide has been described and confirmed that it greatly improved the on-body channel characteristics by providing an independent path for the electromagnetic waves. The following section proposes a belt-type, wearable communication device with sensing capabilities for personal health monitoring in a ubiquitous fashion.

16.4 Proposed WBAN Application

16.4.1 Concept

The case of a person who is concerned about their blood pressure (BP) level (whether or not they have high or low BP) and body temperature condition (whether or not they have a fever) is considered. In BP diagnosis, $\leq 90/60 = $ low, $120/80 = $ normal, while $\geq 140/90 = $ high. In body temperature diagnosis, $\leq 35\,^{\circ}\text{C} = $ low, $36.9\,^{\circ}\text{C} = $ normal, while $\geq 39\,^{\circ}\text{C} = $ high. A person with low or high results becomes a patient and must be examined further for appropriate treatment.

In the proposed BAN application, potential patients are advised to wear a belt waveguide, proposed by the authors of this chapter, as shown in Figure 16.22. The *front* of the belt waveguide contains a sensing *mat radio*, operational at 433 MHz, a frequency recommended for in- and on-body medical applications [21]. The *rear* of the belt contains an *off-mat radio*.

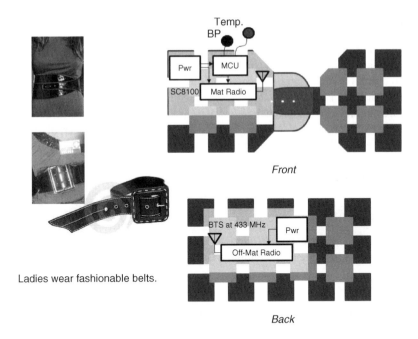

Figure 16.22 The proposed belt waveguide in the BAN application. In the figure, the belt's front integrates a blood pressure (BP) and human body temperature (TEMP) sensor radio for within waveguide low power operation, sensor data logging, and communication, while the rear (back) contains a high-power, long-range sensor for data transmission (flash operation).

The idea is to perform BP and temperature sensing at the front, log the data in the *mat radio*, then, via the belt waveguide fabric, send the data to the *off-mat radio* at the rear by means of low power propagation, ready for transmission to some data center, for example, the PC of a medical doctor (MD) or clinical officer (CO). When required, the data are flashed from the *off-mat radio* at high power transmission depending on the target range or throughput. The proposed belt waveguide configuration is shown in Figure 16.23.

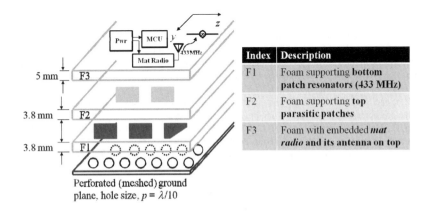

Figure 16.23   Belt waveguide configuration.

16.5   Summary

This chapter presented a novel sheet-shaped waveguide, the 2-D communication medium (free access mat). Using an SC8100 e-textile/fabric, the developed free access mat was integrated into a smart suit and later reduced to a belly-band for better performance analysis and ease of application to BAN. The smart suit exhibited little transmission loss because it acts as a waveguide, propagating electromagnetic waves along its surface.

A typical BAN application was proposed based on the belt-type wearable waveguide to measure a patient's BP and temperature, periodically. The belt waveguide has the potential for wireless power transfer (WPT) applications while the sensor *mat radio* could be powered from or by energy harvesting sources.

Acknowledgment

The authors of this chapter would like to thank the contributions by Dr. Kunsun Eom, of Samsung Advanced Institute of Technology, Bio & Health Lab in South Korea.

References

1. Hu, S.A. (1977) Leaky coaxial cable characteristics at 900 MHz. *IEEE Transactions on Vehicular Technology*, **VT-26** (4), 373–375.
2. Morgan, S.P. (1999) Prediction of indoor wireless coverage by leaky coaxial cable using ray tracing. *IEEE Transactions on Vehicular Technology*, **48** (6), 2005–2014.
3. Wang, J.H. and Mei, K.K. (2001) Theory and analysis of leaky coaxial cables with periodic slots. *IEEE Transactions on Antennas and Propagation*, **49** (12), 1723–1732.
4. Eom, H.J. and Kim, D.H. (2007) Radiation of leaky coaxial cable with narrow transverse slots. *IEEE Transactions on Antenna and Propagation*, **55** (1), 107–110.
5. Scott, J., Hoffmann, F., Addleseet, M. *et al.* (2002) Networked surfaces: a new concept in mobile networking. *ACM Mobile Networks and Applications*, **7** (5), 353–364.
6. Minami, M., Nishizawa, Y., Hirasawa, K. *et al.* (2005) MAGIC surfaces: magnetically interfaced surfaces for smart space applications. 3rd International Conference on Pervasive Computing, Munich, Germany, pp. 59–64.
7. Minami, M., Nishizawa, Y., Hirasawa, K. and Aoyama, T. (May 2007) Deployment Scalable Sensor Networks, ISSN 0913-5685.
8. Fukumoto, M. and Shinagawa, M. (2005) CarpetLAN: a novel indoor wireless(-like) networking and positioning system. 7th International Ubiquitous Computing, pp. 1–18.
9. Sekitani, T., Takamiya, M., Noguchi, Y. *et al.* (2007) A large-area wireless power-transmission sheet using printed organic transistors and plastic MEMS switches. *Nature Materials*, **6**, 413–417.
10. Makino, Y. and Shinoda, H. (2005) Selective stimulation to superficial mechanoreceptors by temporal control of suction pressure. 1st Joint Eurohaptics Conference and Symposium on Haptic Interfaces for Virtual Environment and Teleoperator Systems, 18–20 March 2005, World Haptics, pp. 229–234.
11. Makino, Y., Minamizawa, K. and Shinoda, H. (2005) Sensor networking using two-dimensional electromagnetic wave. IEEJ 22nd Sensor Symposium, pp. 83–88.
12. Makino, Y. and Shinoda, H. (2006) Wrist band type man–machine interface for two-dimensional measurement of mechanoreceptors. 23rd Sensing Forum, pp. 293–298.
13. Chigusa, H., Makino, Y. and Shinoda, H. (2007) Large area sensor skin based on two-dimensional signal transmission technology. World Haptics; Tsukuba, pp. 151–156.
14. Shinoda, H. (2007) High speed sensor networks along a surface (in Japanese). *Measurement and Control*, **46** (2), February, 98–103.
15. Shinoda, H., Makino, Y., Yamahira, N. and Itai, H. (2007) Surface sensor network using inductive signal transmission layer. 4th International Conference on Networked Sensing Systems (INSS07), Braunschweig, pp. 201–206.

16. Metev, S.M. and Veiko, V.P. (1998) *Laser Assisted Microtechnology* (ed. R.M. Osgood Jr.), Springer-Verlag, Berlin, Germany.
17. Breckling, J.E. (1989) *The Analysis of Directional Time Series: Applications to Wind Speed and Direction.* Lecture Notes in Statistics, Report 61, Berlin, Germany.
18. Zhang, S., Zhu, C., Sin, J.K.O. and Mok, P.K.T. (1999) A novel ultrathin elevated channel low-temperature poly-Si TFT. *IEEE Electron Device Letter*, **20**, 569–571.
19. Li, C., Li, H. and Kohno, R. (2009) Performance evaluation of IEEE 802.15.4 for wireless body area network (WBAN). ICC Workshops, pp. 1–5.
20. Li, H. and Kohno, R. (2008) Body area network and its standardization at IEEE 802.15.4 BAN. *Lecture Notes in Electrical Engineering*, 16, Part IV, 223–238.
21. Khan, J.M. and Yuce, M.R. Wireless Body Area Network (WBAN) for Medical Applications. New Developments in Biomedical Engineering, INTECH, pp. 591–628.

17

Wideband Antennas for Wireless Technologies: Trends and Applications

Bahattin Türetken, Umut Buluş, Erkul Başaran, Eren Akkaya, Koray Sürmeli and Hüseyin Aniktar
TÜBİTAK-UEKAE (National Research Institute of Electronics and Cryptology) Electromagnetic and Antenna Research Group, Istanbul, Turkey

17.1 Introduction

Electromagnetic waves are the waves including the electric and magnetic field components that carry energy/information by propagating within their environment. The earliest recorded recognition of the electric and magnetic phenomena is by Thales[1] in the sixth century BC, and this recognition has completed a process of almost 2500 years until today in scientific disciplines such as physics, mathematics and engineering. This stunning, mystical happening, which we use in every moment of our lives, is a divine gift not only for our World, but also for the entire Universe. Initially, the electric and magnetic forces had been considered independently of each other, but the observed phenomena have allowed us to find out a correlation between them. Taking into consideration the experiments and observations conducted between 1822 and 1845 as well as the equations proven by the static electric and magnetic phenomena, J.C. Maxwell[2] discovered that the electromagnetic phenomenon is actually a single phenomenon composed of two components, and described the basic rules of the classic electromagnetism by means of an elegant set of equations in his book, *A Dynamical Theory of the Electromagnetic Field*, published in 1864.

---

[1] 624–546 BC Miletus.

[2] James Clerk Maxwell, 1831 Edinburgh – 1879 Cambridge.

Microwave and Millimeter Wave Circuits and Systems: Emerging Design, Technologies, and Applications, First Edition. Edited by Apostolos Georgiadis, Hendrik Rogier, Luca Roselli, and Paolo Arcioni.
© 2013 John Wiley & Sons, Ltd. Published 2013 by John Wiley & Sons, Ltd.

Today these equations are the basic equations of electromagnetic and antenna theory. These equations imply the correlation between electrical and magnetic fields in any place in a location and while time dependent. The Maxwell equations are solved by considering the constitutive relations defining the universal field in which the event occurs independently from the field in which the event happens. The correlation between electric and magnetic fields occurs through both time dependent derivations and the $J_i = \sigma E$ term in J if the environment is conductive. Experimental showing of the electromagnetic wave radiation as a result of the Maxwell equations was realized by the German physicist R.H. Hertz[3] in 1886. Hertz developed a system to radiate the electromagnetic field generated by a resonant circuit out to the environment. This showed that electromagnetic waves radiating from this circuit stimulate another remote second circuit not connected to any conductor in between. In further experiments, Hertz also lay the foundations in geometrical optics issues by examining reflection, refraction and polarization of electromagnetic waves. In 1901, G. Marconi[4] realized communication over the Atlantic between Britain and America with an 820 kHz monopole antenna with 15 kW power. In 1907, J.A.W. Zenneck[5] suggested in his articles that a good antenna will not be solely effective in communication but the appropriate construction of the ground system will increase the efficiency of the antenna. Microwave antennas and radars intensified between the 1940s and 1945s, while starting from 1945 till 1949 VHF slot antennas, loop antennas, dipole antennas and dipole antenna arrays started to be used to a high degree [1].

This chapter starts with an overview of the antenna concept. A brief explanation of wideband antennas and applications are given in the second section. Wideband arrays are also in the scope of that section. The third part covers a short discussion of measurement techniques and rooms. In the last section, antenna trends including phased arrays, smart antennas, wearable antennas, capsule antennas for medical monitoring, RF hyperthermia, wireless energy transfer and implantable antennas are reviewed.

17.1.1 Antenna Concept

In its general definition, an antenna is the device or transducer transforming the alternating current and voltage into electromagnetic energy, receiving it from a system and radiating it to the environment (transmitting antenna), or transforming electromagnetic energy into alternating current and voltage, receiving from the environment and transmitting it to the system (receiving antenna). The word 'antenna' means 'scape of bug' in Latin and is known as 'antenna' in all languages.

Both transmitting and receiving antennas are connected to the related system via either a transmission line or pipe. Structures of transmitting and receiving antennas are very similar so one antenna can be used for both transmit and receive. An antenna problem is divided into three: finding appropriate radiation patterns for a selected geometry and feeding structure (analysis), designing geometry or feeding networks suitable for the desired parameters (synthesis) and connection of the antenna to a feeding unit (antenna matching network) [2–5].

---

[3] R.H. Hertz, 22 February 1857 Hamburg – 1 June 1894 Bonn.
[4] G. Marconi, 25 April 1874 Bologna – 20 July 1937 Roma.
[5] J.A.W. Zenneck, 15 April 1871 Ruppertshofen – 8 April 1959.

An antenna problem has a solution under boundary conditions of Maxwell equations. However, except for well-known geometries, this solution is not possible analytically. Therefore approximate solutions (numerical modelling) and practical configurations are suggested [6]. The formal and simplified approach is the 'equivalence principle'. In this principle the solution is realized by using equivalent electromagnetic sources placed on a boundary surface surrounding the antenna instead of placing them on the antenna itself. In this case, it is regarded as if there is no antenna material existing and relevant field expressions are calculated from the radiation of these equivalent sources [7].

There are some methods in the definition of equivalent sources. The most frequently used source is the one comprised of the combination of electric and magnetic surface flux densities. The best way to express these is to base them on the Huygens principle in which a field analysis of the volume can be expressed with the tangential field on the surrounding surface.

17.2   Wideband Antennas

Together with the technological development, the need for wideband antennas increases every day. Expansion of the communicable frequency band allows the transfer of more data. In recent years, the number of relevant studies in the literature has been increasing rapidly. The operating frequency band of an antenna is observed over two parameters: one of them is that the antenna's VSWR is less than 2 across the frequency band. The other requirement is that the antenna radiates across this frequency band, which can be measured as 'gain'. In some applications, it is also desired that the antenna has constant gain or that the 3 dB beamwidth (HPBW) remains the same across the frequency band.

The relation between the bandwidth and the Q-factor (quality factor) in the antennas indicates that if the bandwidth increases, the loss become larger:

$$Q = \frac{1}{BW} \tag{17.1}$$

Likewise, the radiation efficiency of the antenna with a high bandwidth is low.

This section discusses the principles utilized in the wideband antenna design and then shows the applications related to the log-periodic and spiral antennas.

17.2.1   Travelling Wave Antennas

In some kind of antenna types, such as the simple dipole, some part of the wave fed from the feed point is reflected from the end of the antenna arm and returns back to the feed point. Thus, the standing wave type current distribution occurs [8]. The travelling wave antenna, on the other hand, is the name given to the antennas with very small returning waves, as shown in Figure 17.1.

There are two methods used to transform standing wave antennas to travelling wave antennas. One of them is to increase the length of the antenna so that the wave going from the feed point to the open end of the antenna is weakened until it returns and, consequently, the standing wave decreases. The other method is to place a matched load on the antenna terminal. In both methods, the radiation loss of the antenna increases and the efficiency decreases.

The typical travelling wave antennas are the dielectric rod antenna, waveguide slot antenna and the tapered waveguide antennas like the horn and Vivaldi [9].

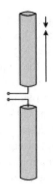

Figure 17.1 Travelling wave antenna.

17.2.2 Frequency Independent Antennas

When the dimensions of any antenna are changed a times, the wavelength is also changed equally, but the pattern and impedance remain the same. The following equation indicates that the dimensions ($h_{1,\,2}$) and the wavelengths ($\lambda_{1,2}$) are directly proportional:

$$\frac{h_1}{\lambda_1} = \frac{h_2}{\lambda_2} \tag{17.2}$$

This is a result of the principle of similitude.

In the article published in 1957, Rumsey [10] suggests that if the antenna dimensions are held constant, it is necessary to change the angle to obtain the same performance in a different frequency. In order to eliminate the wavelength dependency, the antenna dimensions must be infinite. In such a case, the only variable is the angle shown in Figure 17.2.

While the antennas are finite in practice, they have a wide bandwidth depending on the amount of truncation. If the surface of the antenna in a constant f_1 frequency is $r(\phi)$, it is

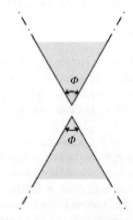

Figure 17.2 Frequency independent antenna.

multiplied by the coefficient a while being scaled to the f_2 frequency. If we consider an infinitely long metal, the same impact can be acquired by rotating it as much as ϕ_r. If $r(\phi)$ is the radius and ϕ is the angle at polar coordinates, the following equation shows the interdependency between the variables ϕ_r and a:

$$ar(\varphi) = r(\varphi + \varphi_r) \tag{17.3}$$

Any dimensional change is equivalent to a change in the angular value of the antenna.

The equiangular spiral, biconical antennas, and so on, are the frequency independent antennas based on this principle.

17.2.3 Self-Complementary Antennas

In 1948, Mushiake [11–14] introduced the constant input impedance characteristics of the self-complementary antennas.

When the Maxwell equations and the boundary conditions are written separately for each picture of Figure 17.3, according to the Babinet principle, a correlation is acquired between the impedances of the slot antenna in the infinite ground plane and the metal antenna in the free space. According to this correlation, if Z_1 is the slot antenna impedance, Z_2 is the metal antenna impedance and Z_0 is the air impedance (120π) for two complementary antennas;

$$Z_1 Z_2 = \left(\frac{Z_0}{2}\right)^2 \tag{17.4}$$

For an antenna structure with equivalent space and the metal parts, this expression turns into

$$Z_1 = Z_2 = Z = \frac{Z_0}{2} = 60\,\pi\,\Omega \cong 188\,\Omega \tag{17.5}$$

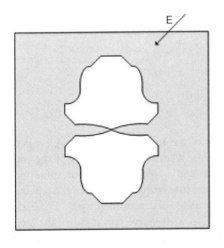
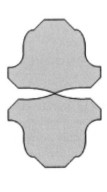

Figure 17.3 Self-complementary antenna.

Figure 17.4   Archimedian spiral antenna.

In this equation, it is important that the impedance is independent of the frequency, that is, in the ideal condition, the antenna impedance is equal in all frequencies and is 188 Ω. Since the finite structures are used in practice, it is observed that their impedances are less variable along a wideband. There are numerous self-complementary antennas in the literature: the turnstile antenna, spiral antenna and log periodic antenna. Figure 17.4 shows an Archimedian spiral antenna working in an ultra-wideband using the self-complementary principle.

17.2.4 Applications

17.2.4.1 Equiangular Spiral Antenna

The equiangular spiral antenna (Figure 17.5) was applied for the first time by Dyson [15] in 1959, providing the angle-dependence feature as specified by Rumsey [10]. Since the antenna cannot be infinitely long in practice, it is obvious that the bandwidth is not infinite. Its equations are as follows:

$$r = r_0 e^{a\varphi} \tag{17.6}$$

$$a = \frac{\ln(r_{i+1}/r_i)}{2\pi} \tag{17.7}$$

In Equation (17.6), r_0 is the beginning point of the radius, while the given a is the constant growth coefficient and r_{i+1} is the inner radius, where r_i is the radius of the previous turn. Therefore, the only correlation is between ϕ and r, and is an angular value mentioned by Rumsey. The positive limit value of ϕ gives the low frequency limit of the antenna. The high frequency limit of the antenna, on the other hand, is determined by r_0. For each frequency value, one region of the antenna becomes active. This active part is called the 'active region'. Since the active region is shifted towards the internal part of the antenna at high frequencies, the radiation occurs in the part close to the feed point.

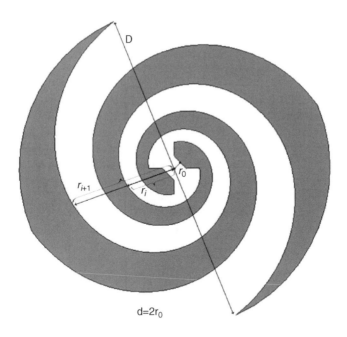

Figure 17.5 Equiangular spiral antenna.

According to the Mushiake [11–14] principle, the self-complementary equiangular spiral antenna must have almost 188 Ω impedance, but in practice this value appears to be 120–150 Ω. The first reason is the coaxial cable, which must be situated on the antenna, and the other reason is the antenna thickness. In order to achieve the symmetry, sometimes an extra coaxial cable might be used. In this case the outer conductor of the coaxial cable is soldered to the opposite arm, while the inner conductor is free.

Since the highest length D represents the lowest frequency and the shortest length d represents the highest frequency, the proportion between these two values gives the bandwidth.

For the antenna shown in Figure 17.6, the radiation pattern has the left-hand polarization inwards from the page plane and the right-hand polarization outwards from the page plane. It

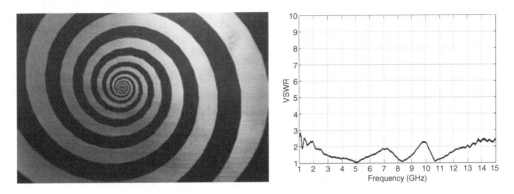

Figure 17.6 Equiangular spiral antenna and measured VSWR.

has a pattern called unidirectional, directed towards two directions. Therefore, its gain is about 5–6 dBi.

17.2.4.2 Log Periodic Dipole Array (LPDA) Antenna

In 1960, Dwight Isbell [16] created a wideband periodic structure by combining the narrow-band dipole antenna elements. In this structure, there are many dipole antennas with different lengths. A few of them become active for any frequency. Upon activation of these dipole antennas, an active region is created. This region contains the main dipole and a few elements in front of and behind it; as the frequency changes, the active region also shifts. The ratio, τ, of each dipole antenna's length (l_n) to the other is also the ratio of the antennas' distances (d_n) to each other:

$$\tau = \frac{l_{n+1}}{l_n} = \frac{d_{n+1}}{d_n} \qquad (17.8)$$

When the antenna begins from the zero point with the angle of α defined as

$$\alpha = \tan^{-1} \frac{l_n}{d_n} \qquad (17.9)$$

and extends logarithmically to infinite, it must theoretically have infinite bandwidth. Nevertheless, since this is practically impossible, it is used as a wideband antenna. With the feeding method shown in Figure 17.7(a) the phase of current on each element on the antenna is π higher than the previous one, and its impedance is 300 Ω. In such a case, the loss is high as the feeding structure is long. With the feeding structure in Figure 17.7(b), on the other hand, the current on each element has the same phase as the previous one, and it is possible to acquire 75 or 50 Ω impedances with this structure.

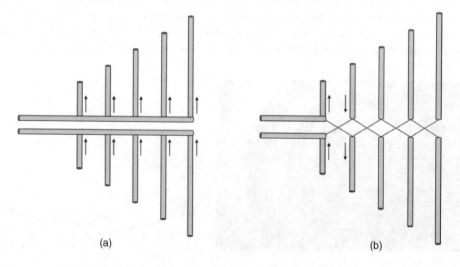

(a)   (b)

Figure 17.7   Log periodic antenna feeding structures.

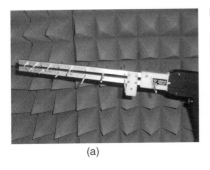

(a)

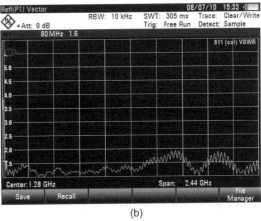

(b)

Figure 17.8   Log periodic antenna and VSWR.

The VSWR of the antenna in Figure 17.8(a) is shown in Figure 17.8(b). It has a rather wide band.

17.2.5   Ultra Wideband (UWB) Arrays: Vivaldi Antenna Arrays

The UWB antenna arrays have been contemplated to be a solution for meeting the wideband needs. Beam forming, high gain, low side lobe, and so on, are features that need to be acquired with the wideband array structure. In conventional methods, the UWB antenna array design procedure covers first the antenna element design and after that the array geometry.

Three basic problems encountered in the antenna array design are grating lobes, especially in the high frequencies depending on the array bandwidth, nulls in the radiation diagram because of surface wave modes and the array element impedance mismatch due to the interaction between the elements. They are the most important problems encountered by designers while designing a UWB array antenna.

In the conventional design of arrays, first the array element for the target frequency band is designed. In order to meet the lower end of the target frequency band, the size of the antenna element must be $\lambda_{low}/2$. In such a case, the radiation diagram of the array at high frequencies contains grating lobes because, at high frequencies, the distance between the antenna elements is more than $\lambda_{high}/2$.

Previous work indicates that the distance between antenna elements must not exceed half a wavelength of the highest frequency. In such a case, it can be thought that the array element impedance might not match at low frequencies. However, it is seen that this problem has been solved when the maximum dimension of the array is bigger than the lowest operating frequency wavelength in both axes [17]. It has been observed that increasing the interaction of the array elements under certain conditions has given successful results.

17.2.5.1   Vivaldi Antennas

Vivaldi antennas have shown up in the literature, with numerous different names given for its aperture geometry. The name 'Vivaldi', on the other hand, was used for the first time by Gibson in his article published in 1979 [18].

The Vivaldi antennas are travelling wave antennas having UWB impedance matching and endfire radiation characteristics. The biggest advantage provided by the microstrip structure is its ability to be connected directly with the TR (transmission-receiver) modules placed behind it. The antenna needs no additional component for impedance matching with its integrated balun structure.

Investigations on the geometrical structure of the Vivaldi baluns have produced different designs [19–21]. In early applications, Knorr's method has been used for the transition from microstrip to aperture [22]. The design example for a classical Vivaldi antenna is shown in Figure 17.9.

The dominant parameters in the single element design are L antenna radiation field length, W antenna aperture width, D_n cavity diameter, W_s slot width, L_{rs} stub length, θ stub angle and R_a the opening rate of the exponential curve forming the antenna aperture. The parameters L and W are dominantly effective on the antenna lower operating frequency limit. The D_n, W_s, L_{rs} and θ parameters are the balun region parameters affecting impedance matching of the antenna. High values of the R_a parameter provides a faster transition from the narrow to the wide part of the aperture. This facilitates the provision of impedance matching at the low end of the frequency band, but may also cause mismatching between the lower and upper frequency limits [23].

The length L shown in Figure 17.9 is generally 1.5 times the wavelength of the lowest frequency and the antenna aperture W is the about half the wavelength of the lowest frequency. The graph shown in Figure 17.10 is the VSWR result of the antenna in Figure 17.9. The antenna has the balun structure based on the microstrip–aperture coupling technique.

Another Vivaldi antenna type with the balun structure is the antipodal Vivaldi antenna. The antipodal Vivaldi antenna was introduced by Ehud Gazit [21] in 1988. In the stripline-fed classic Vivaldi structures, holes are needed to establish connection between the

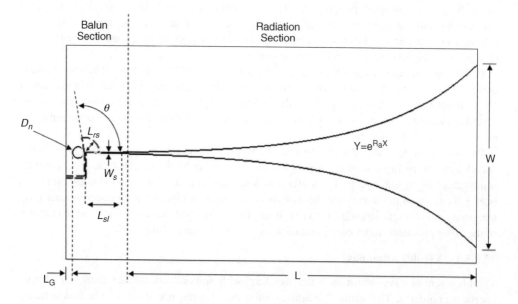

Figure 17.9 Vivaldi antenna.

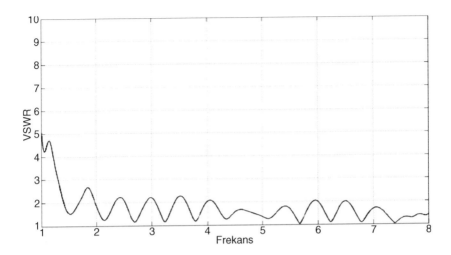

Figure 17.10   Classical Vivaldi antenna VSWR.

conductors on both sides of the dielectric. In the antipodal Vivaldi, however, there is no need for such holes. Balun structures based on microstrip–aperture coupling have the quarter wavelength lines. These lines limit the frequency band [21]. The balun technique used in the antipodal Vivaldi antennas brings a solution to such problems.

The balun structure of the antipodal Vivaldi antennas is shown in Figure 17.11. In this case, there are two transitions, one from the microstrip line to the parallel stripline and the other from the parallel stripline to the symmetrical double-sided slot line. The production of the antipodal Vivaldi antenna is easier than the classical Vivaldi antenna and there are fewer factors that can restrict its impedance matching. In order to avoid impedance mismatch, several studies have been conducted about narrowing the ground plane by different geometrical expressions [24, 25]. The balun geometry has been acquired by creating semicircle apertures on the ground plane and microstrip line in the antenna model shown in Figure 17.11.

Figure 17.12 shows the VSWR result of the antipodal Vivaldi antenna simulation study.

17.2.5.2   Vivaldi Antenna Arrays

The UWB antenna array section reviews the three basic problems of array design in an array created using the classic design methods, which have been stated in Section 17.2.5. The Vivaldi antenna element showed in Figure 17.9 has a 1:8 bandwidth. The antenna element is used to compose and simulate an eight-element linear array. In this study, the elements have been interspaced by disconnecting the direct contact between them to decrease the coupling between the antennas. In infinite Vivaldi arrays, when large spaces are left between the elements, the wave guide modes emerge, which leads to impedance mismatch in the frequency band. [26]. Figure 17.13 shows the array upper frequency limit radiation diagram. In order to decrease the simulation period, the upper frequency has been limited to 10 GHz.

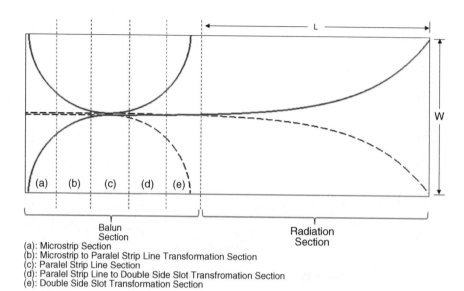

Figure 17.11   Antipodal Vivaldi antenna.

The grating lobes in the radiation diagram stems from the fact that the dimension of the array element is very large compared to the wavelength at high frequencies while the antenna aperture and interelement spacing is a half wave of the lowest frequency.

Figure 17.14 shows the radiation diagram of the 32 element antipodal vivaldi antenna array simulation at the upper frequency limit. Since the distance between the elements is $\lambda/2$ or less in the entire frequency band, no grating lobe has emerged in the radiation diagram.

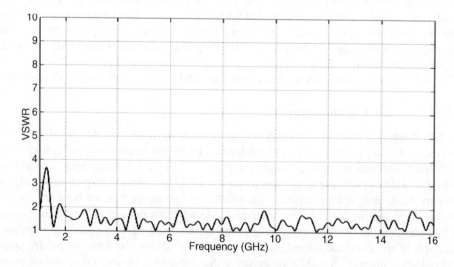

Figure 17.12   Antipodal Vivaldi antenna VSWR.

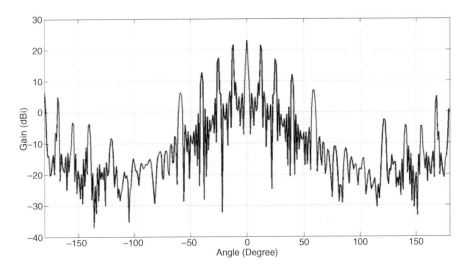

Figure 17.13   Radiation diagram of the Vivaldi array.

The designed 32-element array has been implemented and its VSWR response has been measured. Figure 17.15 shows the produced antenna. The array has been fed with the Wilkinson power divider that has a 1:10 bandwidth. Figure 17.16 shows the VSWR of the produced array. The impedance response of the array covers the target frequency band.

It has previously been specified that the element impedance performance at the lower frequency limit depends on whether the element is in the array. In such a case, since the arrays

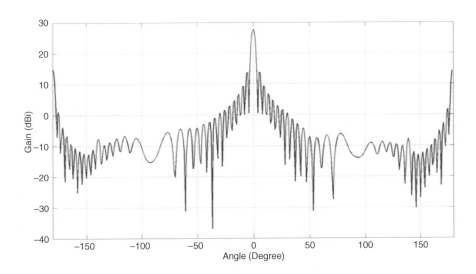

Figure 17.14   Radiation diagram of a 32 element Vivaldi array.

Figure 17.15   A 32 element array antenna.

implemented have finite array dimensions, the edge elements of the array are affected. In the previous works [27], it has been shown that the elements within a distance two times the wavelength of the study frequency, starting from the edge of the array, have been affected. The other elements, on the other hand, are not affected. The above mentioned situation is true for the first four elements of the upper frequency limit, while the number of the affected elements increases from the edge of the array to the centre as the lower frequencies are

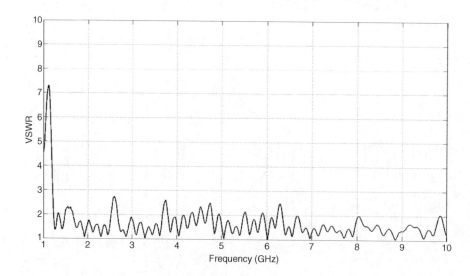

Figure 17.16   A 32 element array antenna VSWR.

approached. The negative effect observed appears in the form of impedance matching anomalies of several elements of the array in certain frequencies within the frequency band.

17.2.6   Wideband Microstrip Antennas: Stacked Patch Antennas

The microstrip antennas are used frequently in many areas for their advantages, such as low cost, small volume, ease of production and suitability for mounting on inclined surfaces. The most serious disadvantage of the microstrip antennas is that they are narrowband. With the classical microstrip antennas, it is difficult to acquire a bandwidth more than 8%.

In order to eliminate such disadvantages of the microstrip antennas, various methods have been introduced. The most frequently used one is the stacked microstrip antenna structure. In the stacked microstrip antenna structure, the first stack of the antenna is fed; this stack is called the driver stack. The second stack, on the other hand, is added to the first stack parasitically and coupled with the patch below electromagnetically. The patch antenna on the lower stack is fed with the microstrip line or coaxial probe. The coaxial probe-fed stacked patch antennas provide very good isolation between the radiation elements and the feeding network. Besides, the alignment problems are eliminated considerably as the feeding pin contacts the patch antenna.

In the stacked patch antenna, the bandwidth is restricted by the inductive structure of the feeding probe. This impact of the probe feeding was examined for the thick dielectric materials in 1987 [28]. In 1991, the full wave analyses of the probe-fed microstrip antennas were conducted with the spectral domain integral equation approach [29]. Again in 1991, the analyses were conducted to calculate the input impedance of the probe-fed stacked microstrip antennas [30]. In 1994, the full wave analyses of the probe-fed stacked patch antennas were conducted with the spectral domain integral equation approach [31]. In 1998, the aperture-coupled patch antenna, a kind of stacked patch antenna, was designed, and thus the antenna design with a rather wideband was introduced. However, since no full wave analysis was made in this study, it can be said that the bandwidth is lower in practice [32]. In the aperture-coupled patch antenna structure, the patch is allowed to be coupled electromagnetically through a slot opening on the ground plane placed on the feeding line. In order to adjust the amount of coupling made through the slot, various slot structures are used. In 1999, R.B. Waterhouse [33] examined the design of the probe-fed stacked patch antennas. The impacts of the dielectric materials used in the design, feed point and dimensions of the patches on the bandwidth have been examined and a 25% bandwidth has been acquired.

The selection of the dielectric material of the first stack is very important in the acquisition of a wide bandwidth in the stacked patch antennas. The current distribution on the patch on the lower stack has a very important impact on the antenna's bandwidth. The dielectric coefficient of the lower stack dielectric material should be bigger than the dielectric coefficient of the upper stack dielectric material. When this is achieved, the amplitude of the first degree mode on the patch antenna on the lower stack becomes bigger than the patch on the upper stack, and thus the wider bandwidths are acquired. Therefore, in order to acquire wider bandwidths, usually a material with a dielectric coefficient of 2.2 is selected for the first stack, while the foam material with a dielectric coefficient of 1.07 is selected for the second stack. Since the foam material does not have any metal surface, it is possible to add a third dielectric stack with a very low height and low dielectric coefficient. Thus, the production process is made easier and more professional.

The thickness of the selected dielectric materials also has a great impact on the bandwidth. The height of the second stack depends on the first stack, and generally the second stack is higher. In the studies conducted, it has been observed that it is appropriate to select the height of the first stack as 0.04 λ and the height of the second stack as 0.06 λ.

While making a design, first the analyses for the first stack are made. It is not desired that the first stack has a resonance at the centre frequency. Instead, it is desired that the first stack is capacitive as much as possible in the required frequency range. In order to achieve that, the feed point must be close to the edge of the first patch. Besides, it is desired that impedance is 250 Ω in the resonance frequency while adjusting the position of the feed point. At this point, the resonance frequency must be selected somewhat smaller than the lower point of the required frequency range.

After the analyses for the first stack have been completed, the second stack is added. Addition of the second stack shifts the capacitive impedance to almost match the impedance region. Thus, the high bandwidths are acquired.

After the impedance has approached the matching region, various adjustments can be made by changing the edge lengths of the patches. For example, when the short edge of the first stack patch antenna is lengthened, the real impedance decreases. When the short edge of the second stack patch antenna is increased, however, the real part at the centre of the impedance ring increases. Besides, the feed point can be used for fine tuning. When the value of the feed point is increased towards the edge, the real part at the centre of the impedance ring increases. However, this increase is never a large one.

A review of the microstrip-fed stacked patch antenna design was made in 1999, in which it has been suggested again that the dielectric coefficient of the first stack material must be higher than the dielectric coefficient of the second stack material [34].

As an example for the probe-fed stacked patch antenna, a design at the 7 GHz centre frequency has been made [2, 35]. The material of the first stack has been selected as Rogers 5880 and the materials of the second stack as Rohacell HF71. On the foam material, a 0.254 mm thick Rogers 5880 has been placed. The designed antenna's bandwidth is about 25%. The antenna and the return loss of the antenna are shown in Figure 17.17.

17.3 Antenna Measurements

There are three different methods used in the solution of the electromagnetics and antenna problems, which are analytic methods, numerical/computational and measurement methods. The solutions offered for problems impossible to be solved analytically and that can be solved (analyses) within the limits provided by simulation methods are supported by measurements. The measurements carried out with a well-defined method and with a mechanism boosted by accurate measurement systems give an approximate approach to reality. Therefore measurements are crucial in antenna problems impossible to be solved fully most of the time in an analytic way. Long phased and pretty costly, antenna measurements form an important part of the antenna designing process. In antenna measurements, conditions when the measurement systems are not completely ideal affect the measurement results negatively. If the distance between the measured antenna and source antenna is more than the far field, unwanted scattering can be received by the receiving antenna and when the far field limit is low, the depth of the zero points in antenna radiation patterns cannot be determined completely and these points close to the main lobe are seen as shoulder. Another risk

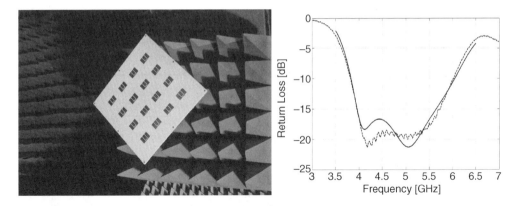

Figure 17.17 Stacked patch antenna and return loss graph (solid line is simulated, dashed line is measured result).

is that as the measurement is carried out in an uncontrolled environment, the effect of the environmental factors cannot be taken into consideration. The size of the antenna to be measured, measurement numbers, time and cost are also other difficulties of antenna measurement systems.

During the Second World War, various methods were developed to find out the characteristics of antennas. Using basic methods, parameters such as the antenna radiation diagram, input impedance, gain and routing were able to be measured and telecommunication and radar applications started to be used. In the 1960s, new systems and devices special for antenna measurements were in the course of development. These systems are generally antenna measurement environment, positioners, pattern recorders, signal generators, antenna measurement standards, automatic control systems, quick calculators making a 2D to 3D pattern transition and so on. The places where antennas are measured are called antenna ranges, which are divided into 'closed' (indoor) environments (Anechoic Chamber) and open (outdoor) environments (open area test fields, or OATSs). The antenna radiation diagram measurements are divided into two as 'far field' and 'near field' depending on the distance between the measurement point and the antenna [36]. Near field measurements are carried out by transforming amplitude, phase and polarization data measured in the near field of the antenna into far field data by using analytical or numerical techniques. These are the amplitude and phase measurements of tangential constituents of the electrical field with specific steps on a well-defined surface (plane, cylindrical, elliptic cylinder, cone, etc.). These data are used to calculate the angular spectrum of the plane, cylindrical or spherical surface. It is called 'modal expansion'. Far field measurement is to draw an amplitude diagram in an environment where the far field condition of the antenna is obtained. Moreover, the compact antenna test range, where approximate uniform plane waves are attained, is another environment for radiation diagrams of the antennas to be measured (Figures 17.18 and 17.19). In both environments, the radiation diagram is directly acquired without any need for transition from the near field to the far field [37].

Different measurement systems can be used depending on the sizes, frequency and radiation diagram of the antennas to be measured and critical antenna parameters necessary to be

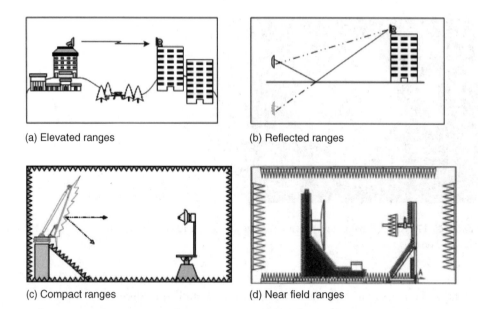

(a) Elevated ranges

(b) Reflected ranges

(c) Compact ranges

(d) Near field ranges

Figure 17.18   Antenna measurement systems: (a) elevated ranges; (b) reflected ranges; (c) compact ranges; (d) near field ranges.

measured. Categorized information about system frequency and sensitivity degree of measurement systems are given in Table 17.1.

17.4   Antenna Trends and Applications

The antennas are used almost everywhere in radio communication electronics. Their major application areas to date are as follows: radio, television, radar, radio-meteorology, radio-link, satellite communication, direction finding, mobile communication, medical use of electromagnetism and laser. Recently, researches and developments related to antennas are

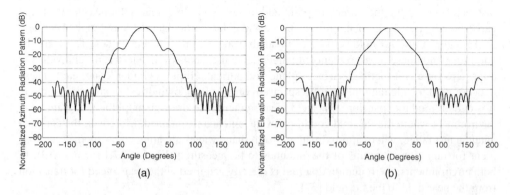

Figure 17.19   Antenna radiation pattern measurement results (a) azimuth, b) elevation, in spherical near field antenna systems in TUBITAK).

ongoing in rather different areas. Several popular antenna implementations and research subjects in recent years are as follows:

- Phase arrays and smart antennas
- Wearable antennas
- Capsule antennas for medical monitoring
- RF hyperthermia
- Antenna for wireless energy transfer
- Implantable antennas.

More detailed information on these applications is given herein under subheadings. In addition to the antenna implementations, the developments in the materials technologies are also important study fields, clearing the way for the antennas. The dimensions and electrical performances of the antennas are proportionate to the dielectric, conductivity and magnetic features of the materials used. Therefore, the innovations and developments in the material technologies have direct impacts on the antenna design and allow more compact and higher performance antennas to be designed.

17.4.1   Phase Arrays and Smart Antennas

Together with the developments in the microwave and software techniques today, it has become normal to expect the antennas used in radar applications to perform much more different tasks, because classic radars have begun to be ineffective against the threats. Therefore, it has become inevitable that radars now use the multifunctional and multitasking smart systems to perform multitasks such as engagement search, identification, tracking, volume scanning, guided-missile routing support, naval surface search, SAR and GMTI. Thus, the phase array structures have started to be used in modern radars to perform numerous functions [38, 39].

As known, the groups of identical antennas with suitable amplitude and phase relations designed to meet the required radiation features, which are arranged in various forms, are called antenna arrays [40]. The array antennas are used to narrow, form, steer the antenna beam and increase the gain. The most important radiation features of the array antennas are the main beam direction, side lobe levels and half power beamwidth. The phase array antennas constitute an array structure in which the amplitude and phase values of each one of the elements constituting the array can be controlled separately to create the required beam. The position of the beam is controlled electronically by adjusting the feeding phase values of the elements constituting the array. Thus, the main beam can be steered without moving the antenna physically.

The ability of the phase array antennas to perform fast and accurate beam steering at the microsecond level allows the systems to carry out numerous functions simultaneously. The radars that can perform beam steering electronically are also capable of tracking many targets and illuminating some of these targets with radiofrequency (RF) energy. Such a radar can also function as a communication system by steering its high gain beams to distant receivers and transmitters. The phase array antennas have enormous flexibility. Their scanning and tracking speeds can be adjusted so as to meet the special conditions in the best manner. For example, the data rate is increased in the case of uncertainty such as the

Table 17.1 Comparison of antenna pattern measurement systems

		High gain antenna	Low gain antenna	High frequency	Low Frequency	Gain measurement	Close side lobes	Far side lobes	Axial ratio
Near field	Planar	Excellent	Poor	Excellent	Poor	Excellent	Excellent	Adequate	Excellent
	Cylindrical	Good	Good	Excellent	Poor	Good	Excellent	Excellent	Excellent
	Spherical	Good	Good	Excellent	Good	Good	Excellent	Excellent	Excellent
Far field	Outdoor range	Adequate	Adequate	Poor	Good	Excellent	Good	Good	Good
	Anechoic Chamber	Adequate	Good	Poor	Fair	Good	Poor	Poor	Poor
	Compact Range	Excellent	Excellent	Excellent	Poor	Excellent	Good	Good	Good

manoeuvers of the targets. The antenna beamwidth can be changed electronically through phase changing. Thus, it is possible to cover certain areas much more rapidly, though with a lower gain. With the multilayer power generators placed along the aperture, it is possible to acquire very high power values. The power distribution can be controlled via computer in the scanning area. The electronically controlled phase array antennas are able to perform all of the various functions required to carry out a certain task in hand in the best manner. For example, they support the target classification functions such as the reverse synthetic aperture. The functions can be programmed rapidly with the digital beam steering computers. The phase array antennas can suppress the clutter more effectively [41].

The active phase array antennas are totally digital multifunctional arrays with programmable functionality. In the digital beam forming structure, only the high power and low noise amplifier units are analogue. In addition to beam forming, all amplitude and phase shifting functions are carried out digitally. The waveforms sent are generated directly by the digital synthesizers. The signals received are captured with very fast analogue-to-digital converters. In very long arrays, the correct time delays for beam steering can be realized digitally. As can be seen clearly, this architecture requires very fast data processing.

In general, the smart antennas are the antenna systems that are capable of processing digital signals and perform functions such as beam forming, beam steering and null point creating. They are composed of three main elements: the antenna unit, the RF unit and the DSP algorithms. Together with the rather widespread use of the wireless communication systems in recent years, several requirements have emerged, and meeting such requirements has also brought about a number of problems. These problems can be summarized as capacity, range, high data transfer rate, mobility, spectral efficiency and restriction of the biological effects. The way of overcoming such problems is by using a smart antenna system (SAS). With an SAS, in addition to routing the beam to the desired user, it is also possible to minimize the antenna gain in the direction of the unwanted region to ensure that the null points of the radiation diagram correspond to those regions for the purpose of decreasing the adverse effects (interference). These systems can also adapt the antenna beams in real time to minimize the multipath signals or improve the signal-to-noise ratio. The smart antenna systems have three types: switched, phase array and adaptive (active) phase array. In the switched beam structure, a constant beam can be turned to the desired point by being fed through shifting the suitable elements of the array with the help of switches. It has a simple structure and can be used in applications requiring low resolution. The smart antenna structure and examples of the beam forming and steering are shown in Figure 17.20.

The phase array structure is a passive structure and the process of steering a single beam with the phase shifters. This structure is a little bit more complex than the switched beam structure, and its biggest disadvantage is its inability to form the beams. The adaptive arrays, on the other hand, are active structures. They are more complex, more expensive and indispensable for applications requiring high resolution. The required forms of the beam, which must be created with these structures, are determined via various algorithms. The parameters, such as the required form, number, side lobe level, null points and direction of the beam, are acquired by adjusting the phase and amplitude values of the antenna elements using the digital beam forming algorithms.

Since it has high carrier mobility, the GaAs technology is used frequently in the high frequency applications. Its high voltage strength, low thermal resistance and high radiation immunity has enabled this technology to be used commonly in military applications as well.

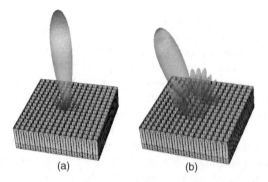

Figure 17.20   Beamforming with smart antenna: (a) equal phase feeding; (b) electronically steered.

However, the failure to find a solution for the low efficiency problem in the production processes of the GaAs technology results in the cost of the components produced using this
technology being very high. Since numerous transmitter/receiver units are used in the phase
array antennas, the contribution of the components to the total cost is high. Development of
alternative technologies not only to decrease the cost but also to design lighter and air-cooled
antennas, instead of being content with the current point reached with the GaAs technology,
is one of the current study issues.

   In the high frequency applications (cellular phones, WLAN, etc.), the LDMOS (laterally
diffused MOS) and SiGe HBT (heterojunction bipolar transistor) BiCMOS technologies are
used widely [42, 43]. Similarly, it is popular to use silicon (Si) technology in the transmit-
ter/receiver units of the phase array antennas as well. By starting to use Si technology in
military applications, it is expected that these systems will fall in price but rise in security.

   Until now, passive phase array antennas had been used in many phase array antenna applications because active phase antennas were yet to be finalized and they were very expensive
structures. Together with the introduction of the GaAs MMIC technology, more comprehensive
studies have been made on the active phase array antenna. The use of the MMIC technology
together with the automatic module assembly techniques have decreased the cost of the active
phase array antennas considerably, and thus they have started to be the preferred choice.

17.4.2   Wearable Antennas

Portable electronic devices have started to take an important place in the daily lives of people. In the near future, people will want to carry numerous devices and sensors on themselves
that communicate with each other. To achieve this, it is inevitable that wearable and implantable antenna technologies will be developed.

   The wearable antenna means using some part of the garment for communication purposes
such as tracking and navigation, mobile computing and public safety. Together with reduction in size of the wireless devices in recent years, the wearable or implantable antennas in
different frequencies have been designed and developed for various applications [44]. Since
it is desired that the wearable antennas are light, small, cheap, robust and easy to place into
accessories of any shape such as a button, belt, helmet, and so on, they are developed mostly
based on the microstrip technology. The cellular phone, GPS, WLAN and Hyper LAN are
the conspicuous applications.

The wearable antennas are monopole and dipole types in VHF and UHF frequencies [45–53]. In the microwave frequency, the planar inverted F antennas, rectangular/circular/triangular microstrip patch antennas, rectangular slot antennas, rectangular ring antennas, aperture coupled patch antennas, coplanar patch antennas and EBG textile antenna with two layers are used [54–77]. From the perspective of compatibility with the garments, the button, belt and helmet antennas are also some of the research subjects [78–85].

The trend is directed towards the miniature wearable antennas made using textile materials. Since the textile material has a low dielectric constant, the surface wave losses are less and it increases the impedance bandwidth of the antenna.

In each antenna design, it is necessary to consider the operating environment as well and also to analyse the performance of the wearable antennas on the human body. The human body is composed of multidielectric layers such as skin, fat, muscle and blood, all of which affect the antenna characteristics. In recent years, the close distance body and antenna interaction for the wearable antennas is one of the conspicuous research subjects. Since such antennas are placed on moving surfaces their performances in different positions must be analysed and the power absorbed by the human body must be examined. The SAR values of the power radiated by the antennas close to the body must be examined in accordance with the health criteria.

17.4.3   Capsule Antennas for Medical Monitoring

The capsule antenna is an electromagnetic radiating element patterned directly on to the surface of a hard shell capsule [86]. It can be used for medical monitoring like the endoscope. The application area of the capsule antenna is growing rapidly.

17.4.4   RF Hyperthermia

RF hyperthermia is one of the promising methods for cancer treatment with deep-body heating based on annular phased arrays of radiators, such as waveguides, coaxial TEM apertures and flat dipoles [87].

17.4.5   Wireless Energy Transfer

Wireless energy transfer is also an important issue to supply energy for implantable biomedical devices and RFID devices. The biggest challenge in the design of these devices is the maximization of the transferred energy, which increases the performance of the devices [88].

17.4.6   Implantable Antennas

The request for using antennas in or out of the human body for hi-tech medical and identity authentication applications is on the rise. The researches on this issue are concentrated on identity authentication, personal location data tracking, medical treatment and physiological data tracking, and hyperthermia generation for medical treatment and physiological data tracking.

The antennas are placed in or out of the patient's body to increase the temperature of the cancer tissues and are designed in different forms according to the place of use. The wideband implant antennas are the promising fields of study that have been receiving attention in recent years. The physiological and biological data tracking are some of the usage areas. The small-sized implant antennas are used in the industrial, scientific and medical (ISM) (2.4–2.48 GHz) band.

Technological advancements in healthcare provide a better life standard. Numerous studies are being made on remote patient monitoring. The usage area of the telemetry system is being improved to meet the need for communication between the patient and the base station.

In recent years, the ingestible antennas, in addition to the implantable antennas using the telemetry system, have also started to be included in the literature [41]. Wireless data telemetry applications such as in brain and cardiac pacemakers, in artificial eyes, in nerve stimulators, in implantable glucose sensors, in cochlear implants and in such similar areas are getting extensive interest in implantable antennas. Wireless data transmission at the implant between the patient and the base station that allows communication of the continuous monitoring of physiological parameters such as the glucose, blood pressure and temperature is very important for early diagnosis of diseases.

The implantable antenna design is difficult due to restrictions such as small size, low power requirement and biological compatibility. The problem space includes complex and frequency-dependent electrical features (dispersive permittivity, permeability, and conductivity and penetration depth) due to its layered tissue structure and heat. Since the antenna to be designed is inside the tissue, it is essential to make a design compliant with the electrical features of the tissue.

The most important factors affecting the implantable antenna system design are the position and the frequency range of the antenna. The antenna type must be appropriate for the desired target. The frequency knowledge is a measurement of the antenna size, penetration depth and the emission that will leave the body. The power absorption losses in the human body are frequency-dependent. Such dependencies are in fact characteristic of the tissue.

The SAR values of the power radiated by the antennas close to the body must be examined for implant antennas used in medical sensing, monitoring and curative applications.

Acknowledgements

The authors wish to thank Ömer Yılmaz for taking photos of antennas and drawing of figures. They also thank TÜBİTAK-BILGEM UEKAE and the Antenna Test and Research Centre (ATAM) for the manufacture of antennas, measurements and all technical infrastructure.

References

1. Akkaya, İ. (1982) Anten Teorisine Giriş Istanbul: ITU Maçka Elektrik Fakültesi Yayınları.
2. Surmeli, K. and Turetken, B. (2011) U-slot stacked patch antenna using high and low dielectric constant material combinations in S-band, URSI General Assembly and Scientific Symposium, Istanbul, 2011.

3. Ünal, İ., Turetken, B., Surmeli, K. and Canbay, C. (2011) An experimental microwave imaging system for breast tumor detection on layered phantom model. URSI General Assembly and Scientific Symposium, Istanbul.

4. Zengin, F., Akkaya, E. and Turetken, B. (2011) Design and realization of ultra wideband implant antenna for biotelemtry systems. URSI General Assembly and Scientific Symposium, Istanbul.

5. Altun, H., Korkmaz, E. and Turetken, B. (2011) Reconfigurable fractal tree antenna for multiband applications. URSI General Assembly and Scientific Symposium, Istanbul.

6. Turetken, B. (2006) Rigorous analysis of wide beam scalar feed horn with different impedance internal surfaces. *Electromagnetics*, **26** (5), 359–385.

7. Volakis, J.L. (2007) *Antenna Engineering Handbook*, 4th edn., McGraw-Hill, New York.

8. Stutzmann, W.L. and Thiele, G.A. (2012) *Antenna Theory and Design*, 2nd edn, John Wiley & Sons, Inc., New York.

9. Wiesbeck, W., Adamiuc, G. and Sturm, C. (2009) Basic principles and design principles of UWB antennas. *Proceedings of the IEEE*, **97** (2), 372–385.

10. Rumsey, V.H. (1957) Frequency independent antennas. *IRE National Convention Record*, **5** (1), 114–118.

11. Mushiake, Y. (1948) The input impedance of slit antenna. Joint Convention Record of Tohoku Sections of IEE and IECE of Japan, pp. 25–26.

12. Mushiake, Y. (1949) The Input Impedances of Slit Antennas. Technical Report, Tohoku University.

13. Mushiake, Y. (1959) Multiterminal constant impedance antenna. National Convention Record of IECE of Japan, p. 89.

14. Mushiake, Y. (1985) Self complementary antennas. Researches on Electronic Communication, Record of Electronic Communication Engineering Conversazione, Commemorative Issue, RIEC, Tohoku University, pp. 109–116.

15. Dyson, J. (1959) The equiangular spiral antenna. *IRE Transactions on Antennas and Propagation*, **7** (2), 181–187.

16. Isbell, D.E. (1960) Log periodic dipole arrays. *IRE Transactions on Antennas and Propagation*, **AP-8** (3), May, 260–267.

17. Balanis, C.A. (2008) *Modern Antenna Handbook Canada*, John Wiley & Sons, Inc., New York.

18. Gibson, P.J. (1979) The Vivaldi aerial. 9th European Microwave Conference, Brighton, pp. 101–105.

19. Abdel, A.Y., Hennawy, H.E., Nahrous, S. and Schunemann, K. (1984) Design of Vivaldi antenna for microwave integrated circuits applications. 14th European Microwave Conference, pp. 637–642.

20. Lewis, L., Fassett, M. and Hunt, J. (1974) A broadband stripline array element. Antennas and Propagation Society International Symposium, pp. 335–337.

21. Gazit, E. (1988) Improved design of the Vivaldi antenna. Microwaves, Antennas and Propagation. *IEEE Proceedings H*, **135** (2), 89–92.

22. Knorr, J.B. (1974) Slot-line transitions. *IEEE Transactions on Microwave Theory and Techniques*, **22**, 548–554.

23. Shin, J. and Schaubert, D.H. (1999) A parameter study of stripline-fed Vivaldi notch-antenna arrays. *Antennas and Propagation IEEE Transactions*, **47** (5), 879–886.

24. Lambrecht, A., Pauli, M., Ripka, B. and Zwick, T. (2009) A Vivaldi antenna for anti-electronics HPEM systems, Lambrecht. Antennas and Propagation Society International Symposium APSURSI '09, pp. 1–4.

25. Kim, S.G. and Chang, K. (2004) Ultra wideband exponentially-tapered antipodal Vivaldi antennas. Antennas and Propagation Society International Symposium, pp. 2273–2276.

26. Schaubert, D.H. (1996) A class of E-plane scan blindnesses in single-polarized arrays of tapered-slot antennas with a ground plane. *IEEE Transactions on Antennas and Propagation*, **44** (7)

27. Holter, H. and Steyskal, H. (2002) On the size requirement for finite phased-array models. *IEEE Transactions on Antennas and Propagation*, **50**, 836–840.

28. Hall, P.S. (1987) Probe compensation in thick microstrip patches. *Electronics Letters*, **23**, 606–607.

29. Aberle, J.T. and Pozar, D.M. (1991) Accurate and versatile solutions for probe fed microstrip patch antennas and arrays. *Electromagnetics*, **11**, January, 1–19.

30. Tulintsef, A.N., Ali, S.M. and Kong, J.A. (1991) Input impedance of a probe-fed stacked circular microstrip antenna. *IEEE Transactions on Antennas and Propagation*, **39**, March, 381–390.

31. Aberle, J.T., Pozar, D.M. and Manges, J. (1994) Phased arrays of probe-fed stacked microstrip patches. *IEEE Transactions on Antennas and Propagation*, **42**, July, 920–927.

32. Targonski, S.D., Waterhouse, R.B. and Pozar, D.M. (1998) Design of wideband aperture-stacked patch microstrip antennas. *IEEE Transactions on Antennas and Propagation*, **46**, September, 1245–1251.

33. Waterhouse, R.B. (1999) Design of probe-fed stacked patches. *IEEE Transactions on Antennas and Propagation*, **47**, December, 1780–1784.
34. Waterhouse, R.B. (1999) Stacked patches using high and low dielectric constant material combinations. *IEEE Transactions on Antennas and Propagation*, **47**, December, 1767–1771.
35. Turetken, B., Surmeli, K. and Basaran, E. (2011) Bandwidth enhancement for U-slot stacked patch antenna by using appropriate dielectric materials. Antenna Measurement Techniques Association, Colorado, pp. 131–134.
36. Nazlı, H., Bıçakçı, E., Turetken, B. and Sezgin, M. (2010) An improved design of planar elliptical dipole antenna for UWB applications. *IEEE Antennas and Wireless Propagation Letters*, **9**, 264–267.
37. Turetken, B., Buluş, U. and Yılmaz, Ö. (2011) Radar Anten Ölçümleri ve Hata Analizleri 1: Düzlemsel Yakın Alan Ölçüm Sistemi. UEKAE Dergis. Ocak-Nisan.
38. Türetken, B., Surmeli, K. and Çalışkan, A.U. (2011) Radar Antenleri – V: Faz Dizili Antenler – Besleme, Uygulama ve Gelişim Yönü. BİLGEM Dergisi, January–April 2011, pp. 111–119.
39. Parker, D. and Zimmermann, C. (2002) Phased arrays – Part II: implementations, applications and future trends. *IEEE Transactions on Microwave Theory and Techniques*, **50**, March.
40. Türetken, B. and Sürmeli, K. (2010) Radar antenleri – IV: faz dizili anten kuramına genel bakış. UEKAE Dergisi. Eylül-Aralık, pp. 118–125.
41. Türetken, B. and Sürmeli, K. (2010) Aktif faz dizili anten tasarımına sayısal modelleme tekniğiyle yeniden bakış. V. URSI-Türkiye'2010 Bilimsel Kongresi ve Ulusal Genel Kurul Toplantısı, Güzelyurt, pp. 452–455.
42. Gruner, D. *et al.* (2010) 17.43 A 1W Si-LDMOS power amplifier with 40% drain efficiency for 6GHz WLAN applications. IEEE MTT-S International Microwave Symposium, California, pp. 517–520.
43. Larson, L.E. (2004) SiGe HBT BiCMOS technology as an enabler for next generation communications systems. Proceedings of the 12th Gallium Arsenide Applications Symposium (GAAS® 2004), Amsterdam, pp. 251–254.
44. Gupta, B., Sankaralingam, S. and Dhar, S. (2010) Development of wearable and implantable antennas in the last decade. Technical Report, Department of Electronics and Tele-Communications Engineering, Jadavpur University, India.
45. Furuya, Y., Taira, Y., Iwasaki, H. *et al.* (2008) A wide band wearable antenna for DTV reception. International Symposium on Antennas and Propagation, pp. 1–4.
46. Peter, T. and Nilavalan, R. (2009) Study on the performance deterioration of flexible UWB antennas. Antennas and Propagation Conference, pp. 669–672.
47. Park, J.Y. and Woo, J.M. (2008) Miniaturization of microstrip line monopole antenna for the wearable applications. Asia-Pasific Microwave Conference, pp. 1–4.
48. Noury, N., Barralon, P. and Flammarion, D. (2005) Preliminary results on the study of smart wearable antennas. 27th Annual International Conference on Engineering in Medicine and Biology Society, pp. 3814–3817.
49. Kellomaki, T., Heikkinen, J. and Kivikovski, M. (2006) Wearable antennas for FM reception. First European Conference on Antennas and Propagation, pp. 1–6.
50. Kim, Y., Lee, K., Kim, Y. and Vhung, YC. (2007) Wearable UHF RFID tag antenna design using flexible electro-thread and textile. International Symposium on Antennas and Propagation, pp. 5487–5490.
51. Sychoudakis, D.P., Chen, C.C. and Volakis, J.L. (2007) Optimizing wearable UHF antennas for on-body operation. International Symposium on Antennas and Propagation, pp. 4184–4187.
52. Matthews, J.C.G. and Pettitt, G. (2009) Development of flexible, wearable antennas. 3rd European Conference on Antennas and Propagation, pp. 273–277.
53. Waterhouse, R. and Novak, D. (2006) Small uni-planar antenna suitable for body wearable applications. Military Communications Conference, pp. 1–6.
54. Salonen, P., Sydanheimo, L., Keskilammi, M. and Kivikoski, M. (1999) A small planar inverted-F antenna for wearable applications. Proceedings of the Third International Symposium on Wearable Computers, pp. 95–100.
55. Massey, P.J. (2001) Mobile phone fabric antennas integrated within clothing. Eleventh International Conference on Antennas and Propagation, pp. 344–347.
56. Salonen, P., Keskilammi, M., Rantanen, J. and Sydanheimo, L. (2001) A novel Bluetooth antenna on flexible substrate for smart clothing. IEEE International Conference, pp. 789–794.
57. Salonen, P. and Sydanheimo, L. (2002) Development of an S-band flexible antenna for smart clothing. International Symposium on Antennas and Propagation, pp. 6–9.
58. Salonen, P. and Humme, L. (2003) A novel fabric WLAN antenna for wearable applications. International Symposium on Antennas and Propagation, pp. 700–703.

59. Salonen, P., Rahmat-Samii, Y., Hurme, H. and Kivikoski, M. (2004) Effects of textile materials on wearable antenna performance: a case study of GPS antennas. International Symposium of the Antennas and Propagation Society, pp. 455–458.

60. Salonen, P., Rahmat-Samii, Y., Hurme, H. and Kivikoski, M. (2004) Effect of conductive material on wearable antenna performance: a case study of WLAN antennas. International Symposium on Antennas and Propagation, pp. 455–458.

61. Ouyang, Y., Karayianni, E. and Chappell, WJ. (2005) Effect of fabric patterns on electrotextile patch antennas. International Symposium on Antennas and Propagation, pp. 246–249.

62. Sankaralingam, S. and Gupta, B. (2010) Development of textile antennas for body wearable applications and investigations on their performance under bent conditions. *Progress in Electromagnetics Research B*, **22**.

63. Klemm, M., Locher, I. and Troster, G. (2004) A novel circularly polarized textile antenna for wearable applications. 34th European Microwave Conference, pp. 137–140.

64. Salonen, P., Rahmat-Sami, Y., Hurme, H. and Kivikoski, M. (2004) Dual-band wearable textile antenna. International Symposium on Antennas and Propagation, pp. 463–466.

65. Hsu, H.S. and Chang, K. (2006) Ultra thin CPW fed rectangular fed slot antenna for UWB application. International Symposium on Antennas and Propagation, pp. 2587–2590.

66. Tronquo, A., Hertleer, C. and Van Langenhove, L. (2006) Robust planar textile antenna for wireless body LANs operating in 2.45 GHz ISM band. *Electronics Letter*, **42** (3), February, 142–143.

67. Tronquo, A., Rogier, H., Hertleer, C. and Van Langenhove, L. (2006) Applying textile materials for the design of antennas for wireless body area networks. First European Conference on Antennas and Propagation, pp. 1–5.

68. Hertleer, C., Rogier, H. and Van Langenhove, L. (2007) A Textile Antenna for Protective Clothing. Antennas and Propagation for Body-Centric Wireless Communications, IET.

69. Hertleer, C., Tronquo, A., Rogier, H. *et al.* (2007) Aperture-coupled patch antenna for integration into wearable textile systems. *Antennas and Wireless Propagation Letters*, **6**.

70. Bai, Q. and Langley, R. (2009) Crumbled textile antennas. *Electronics Letters*, **45** (9), 436–438.

71. Bai, Q. and Langley, R. (2009) Wearable EBG antenna bending. 3rd European Conference on Antennas and Propagation, pp. 182–185.

72. Zhu, S. and Langley, R. (2007) Dual-band wearable antennas over EBG substrate. *Electronics Letters*, **43** (3), 141–142.

73. Salonen, P. and Keskilammi, M. (2008) Softwear antenna. Military Communications Conference, pp. 1–6.

74. Salonen, P. and Rahmat-Samii, Y. (2007) Textile antennas: effects of antenna bending on input matching and impedance bandwidth. *IEEE Aerospace and Electronic Systems Magazine*, **22** (12), 10–14.

75. Tanaka, M. and Jang, J.H. (2003) Wearable microstrip antenna for satellite communications. Proceedings of the 21st International Communications Satellite Systems Conference and Exhibit.

76. Thalmann, T., Popovic, Z., Notaros, B.M. and Mosing, J.R. (2009) Investigation and design of a multi-band wearable antenna. European Conference on Antennas and Propagation, pp. 462–465.

77. Sanz-Izquierdo, B., Huang, F., Batchelor, J.C. and Sobhy, M. (2006) Compact antenna for WLAN on body applications. Microwave Conference, pp. 815–818.

78. Sanz-Izquierdo, B., Huang, F. and Batchelor, J.C. (2006) Small size wearable button antenna. European Conference on Antennas and Propagation, pp. 1–4.

79. Sanz-Izquierdo, B., Batchelor, J.C. and Sobhy, M. (2006) UWB wearable button antenna. European Conference on Antennas and Propagation, pp. 1–4.

80. Sanz-Izquierdo, B., Batchelor, J.C. and Sobhy, M. (2007) Compact UWB wearable antenna. Antennas and Propagation Conference, Loughborough, pp. 121–124.

81. Sanz-Izquierdo, B. and Batchelor, J.C. (2007) WLAN jacket mounted antenna. Antenna Technology: Small and Smart Antennas Metamaterials and Applications, pp. 57–60.

82. Sanz-Izquierdo, B. and Batchelor, J.C. (2008) A dual band belt antenna. Proceedings of IWAT2008, Chiba, pp. 374–377.

83. Wang, J.J.H., Tillery, J.K., Bohannan, K.E. and Thompson, G.T. (1997) Helmet-mounted smart array antenna. International Symposium on Antennas and Propagation, pp. 410–413.

84. Wang, J.J.H. and Triplett, D.J. (2007) Multioctave broadband body-wearable helmet and vest antennas. International Symposium on Antennas and Propagation, IEEE, pp. 4172–4175.

85. Zengin, F., Akkaya, E., Türetken, B. and San, E. (2011) Design and realization of ultra wideband implant antenna for biotelemtry systems. URSI General Assembly and Scientific Symposium, Istanbul.

86. Bashirullah, R. and Euliano, N. (2009) Capsule antennas for medication compliance monitoring. IEEE Radio and Wireless Symposium (RWS'2009), pp. 123–126.
87. Wlodarczyk, W., Wust, P., Seebass, M. *et al.* (2002) RF hyperthermia: modeling and clinical systems. IEEE International Antenna and Propagation Society Symposium, pp. 827–830.
88. Gruosso, G., Mussetta, M. and Zich, R.E. (2011) Optimization of an antenna for wireless energy transfer. Proceedings of the 5th European Conference on Antennas and Propagation (EUCAP), pp. 1208–1209.

18

Concluding Remarks

Apostolos Georgiadis[1], Hendrik Rogier[2], Luca Roselli[3] and Paolo Arcioni[4]
[1]*Technological Telecommunications Centre of Catalonia (CTTC), Castelldefels, Spain*
[2]*Department of Information Technology (INTEC), Ghent University, Ghent, Belgium*
[3]*Department of Electronic and Information Engineering, University of Perugia, Perugia, Italy*
[4]*University of Pavia, Pavia, Italy*

Microwave and millimeter wave systems are commonly used in everyday life, from wireless communications to radar and wireless sensor networks. Nonetheless, technology and material advances combined with societal needs lead to novel emerging applications and challenges that such systems are further required to address. This book confronted the reader with a selected characteristic set of these challenges from different perspectives of simulation and optimization methodologies to circuit and system level design.

There is room to expand the presented material in the future along several ongoing research tracks. One of those is the special importance of flexible materials and the integration of electronics on flexible substrates. A fundamental category of flexible electronics addressed within the book was smart textiles and off and on body communication networks. Printed antennas and passive circuits with promising performance in microwave frequencies have already been reported combining dielectric textiles with conductive electrotextiles as well as large scale fabrication methods such as screen printing and inkjet printing, commonly used in the textile industry. One may expect further advances in printed active devices, which are presently limited to low frequency applications of up to few MHz. In addition, efforts are ongoing to improve reliability and efficiency, on the one hand, and to reduce cost of wearable systems, on the other hand. One way to go is to design complete wireless wearable modules that have all functionalities integrated on-board. In particular, these devices must be highly energy efficient and preferably autonomous. This calls for suitable energy scavenging/harvesting and power management circuits, implemented as an integral part of the wireless wearable system. Cooperative and cognitive networking might also help to achieve these goals.

Microwave and Millimeter Wave Circuits and Systems: Emerging Design, Technologies, and Applications,
First Edition. Edited by Apostolos Georgiadis, Hendrik Rogier, Luca Roselli, and Paolo Arcioni.
© 2013 John Wiley & Sons, Ltd. Published 2013 by John Wiley & Sons, Ltd.

The book also highlighted the importance of low cost and high performance millimeter wave technologies. Millimeter wave systems, utilizing high performance integrated devices, MEMS, as well as novel circuit topologies such as substrate integrated waveguides (SIW), realizing large operating bandwidths and low profile architectures, have led to the development of compact multiple input multiple output (MIMO) wireless systems. Commercial applications include 60 GHz wireless communications systems as well as radar systems including automotive radar, starting from 24, 77, and 94 GHz and moving to the 140 and 250 GHz frequency range. Application requirements for communication systems utilizing very large bandwidths extending to 40 and 100 Gbps, as well as for high resolution imaging systems, pave the way for the commercial application of THz systems operating above 300 GHz. Significant technological advances in these frequency ranges are expected to appear, with many challenges such as low cost and high yield fabrication, components, materials, instrumentation, and high power sources remaining to be addressed.

Finally, in addition to remarkable advances in microwave and millimeter wave technology, the book aimed to emphasize the importance of the integration of digital and analog electronics, the combined use of digital signal processing techniques, and the advantages of digital electronics in the design of microwave and millimeter wave systems, towards the implementation of highly reconfigurable and high performance radio transceivers, enabling the vision for a software defined radio.

Advances in microwave and millimeter wave systems and their further commercial deployment application will undoubtedly have an impact in enabling a number of applications for energy efficient systems, smart transportation systems, and services and toward longer and healthier lives, to name a few.

The structure of the book aimed to strike a balance between design and modeling trends, on the one hand, and applications, on the other, resulting in a book of wide utility.

Index

accessible range, 467
ACEPR, 15, 19–20, 23
acquisition, 279–87
  dual-band signals, 279–82, 284–7
  evenly spaced equal-bandwidth multiband
    signals, 282–3
  multiband signals, 281–2, 284–7
active balun, 312–15
active region 486, 488
adjacent channel error power ratio,
    see ACEPR
ADS, 296
AE, see Model average error
aliasing, 276–83, 285–7, 290–293
aliasing error, 13
AM-AM characteristic, 15, 23
amplifier, 302–3, 306–8, 312,
    327, 330, 337
  bi-directional, 327, 329, 333
    low-noise, 176, see also LNA
  buffer amplifier, 308
  differential amplifier, 306, 308, 312
  low noise amplifier (LNA), 302–3
  power, see PA
  reconfigurable, 329, 336–7, 341,
    344–8, 353
  time constant, 329
analogue-to-digital converter, 275–80,
    290, 293
anchor node, 138–40, 144–8, 152, 154–60, 165,
    167, 171–6, 178, 181
anechoic chamber, 425, 443, 449
angle of arrival (AOA), 143–4

antenna, 51–81, 325, 333, 369, 387–415, 417,
    482–3, see also SIW antennas
active wearable, 418, 436
antipodal Vivaldi antenna, 492
aperture-coupled patch antenna, 495
aperture-coupled shorted patch, 445
array pattern, 333
artificial magnetic conductor, 55–7
bow-tie, 57
broadband, 51–81
capsule antennas for medical monitoring, 503
corrugated horn, 374
diamond dipole, 65
directive, 51–81
EBG textile antenna, 503
efficiency, 427–30, 437
electronic beam-steering, 326, 336, 348
equiangular spiral antenna, 486
frequency independent antenna, 484
gain, 418, 445
Gaussian optics, 369
half-power beam width, 370
implantable antenna, 502–3
log periodic dipole array antenna, 488
low profile, 51–81
microstrip antennas, 495
noise temperature, 371–2, 380
normalized radiation pattern, 373
passive, 443
phase arrays and smart antennas, 499
realized gain, 57
rectangular ring, 436
reflector, 55, 59

Microwave and Millimeter Wave Circuits and Systems: Emerging Design, Technologies, and Applications,
First Edition. Edited by Apostolos Georgiadis, Hendrik Rogier, Luca Roselli, and Paolo Arcioni.
© 2013 John Wiley & Sons, Ltd. Published 2013 by John Wiley & Sons, Ltd.

antenna (*Continued*)
    self-complementary antenna, 485
    stacked microstrip antenna structure, 495
    steerable, 336
    textile, 418, 429, 436, 445
    travelling wave antenna, 483
    Vivaldi antenna arrays, 489–91
    wearable antennas, 418–9, 471, 502
antenna measurements, 496
    antenna measurement standards, 497
    closed (indoor) environments, 497
    far field, 496–7
    modal expansion, 497
    near field, 497
    open (outdoor) environments, 497
    pattern measurement systems, 500
    pattern recorders, 497
    sensitivity degree of measurement, 498
    shoulder, 497
    source antenna, 496
approximation
    numerical, 28–9
architecture, 275–8
    direct-sampling, 277–8
    low-IF, 276
    superheterodyne, 275–6
    zero-IF, 276
area-based localization, 139
array, 387–8, 406–14, *see also* SIW antennas
    antenna array for GNSS applications, 234
        controlled reception pattern antenna
            (CRPA), 237
        design, 234–5, 249, 255
        front-end, 239, 249, 259
        helical elements, 234
        patch elements, 235
        requirements, 234
        spiral elements, 237
        stacked patches, 236, 257
    array receiver for GNSS applications, 244
        automatic gain control, 232, 247
        calibration, of array, 250, 255
        clock domain crossing, FPGA design, 247
        clock skew, FPGA design, 247
        computational resources required, 244
        high-rate processing, 244
        low-rate processing, 245
        metastability, FPGA design, 247–8
        practical examples, 248, 253
    switched beam arrays, 313

artificial magnetic conductor, 52
    bandwidth, 53
artificial neural network, 28–35, 37, 45–9
    generalization, 29, 32–5, 43–7
    layer, 29, 34–5
    learning error, 31
        learning process, 28–35, 41–9
    network topology, 32
    neurons, 29, 33, 35
    testing set, 33, 35
    training set, 29–33, 37
atmosphere
    troposheric transmissivity, 365
    variability of, 365
Automatic Gain Control (AGC), 303
average error (AE), 8, 14–17, 22

Babinet's principle, 485
background subtraction, 212–13
    exponential averaging method, 213, 221
bandwidth, 418, 427, 445, 448
Bayesian approaches, 151–5
beam forming, 313
    for GNSS applications, 233, 239
        deterministic, 241
        eigenbeamformer, 243
        linearly constrained minimum variance
            (LCMV), 241
        minimum mean squared error (MMSE),
            242, 264
        narrowband, 240
        wideband, 240
    network, 333
        active, 333
beam scanning, 412–14, *see also* SIW antennas
behavioural prediction performance, 7, 15, 23
Belt waveguide, 476
Bessel functions, 7, 13, 24
Bessel-Fourier behavioural model, *see* PA model
BF model, *see* Bessel-Fourier behavioural
            model
BiCMOS, 342, 344
body area networks (BAN), 455–60
bow-tie antenna, 57
brightness temperature, 371–2
broadband transmission line technique, 419–20
Butler matrix, 313

CADENCE, 296
cascode, 307

cavity-backed antenna, 387–9, 395, 406, *see also*
    SIW antennas
  resonance frequency, *see* resonant mode
  resonant mode, 389–91, 393,
    399, 402
cavity perturbation technique, 419
CarpetLAN, 457
CCDF, 15, 18–19, 23
centroid, 138–40, 167–8
characteristic impedance, 420, 422
characterization of metamaterials, 52–55
chip, 175–6, 178–9, 185–6, 189, 191–2, 195
  test, 178–9, 185, 189, 191, 195
circuit, 175–7, 180–187, 189, 192, 194–7, 199–204
  analog and radio frequency, 175–6, 194
  digital, 175–6, 180, 182–3, 185, 192, 195–7,
    204
  mixed-signal, 184
climate chamber, 431
CMOS, 175, 185, 187, 189, 194–5, 205, 333–4,
    342
  inverter, 175, 187
  process, 175, 185, 187, 189, 194–5
coaxial probe, 495
coefficient spectrum, 12, 21
cognitive radio (CR), 3, 5, 10
common phase error, 113
  estimation, 119
  mitigation, 120
communication, 325
  body-centric, 417
  off-body wireless, 418
  point-to-point, 326, 342
  wireless, 336
complementary cumulative distribution functions,
    see CCDF
complex envelope transfer characteristic, 5
compression, 44–6
compression point, 1 db, 4–6, 11, 318
conductivity, 427, 429, 468
  effective, 419, 429, 435
conductor
  artificial magnetic, 52
  perfect electric, 55
  perfect magnetic, 55
constitutive parameters, 419, 427
  anisotropic, 419
  homogeneous, 419
  isotropic, 419
  inhomogeneous, 419

cooperative positioning, 135, 146, 151
corona, solar, 359
coronal mass ejections (CME), 358
coupled oscillator arrays, 406–14, *see also* SIW
    antennas
coupler, 315
coupling loss, 466
CPE interpolation, 112
Crámer–Rao bound (CRB), 158–9, 164
CST, 296
current mirror, 317
current steering, 317
cylindrical wave propagation, 461

delta dirac, 4–5
detection, 214–15
    CFAR detector, 215, 221–2
diamond dipole antenna, 65
dielectric constant, 420, 426
dipole, 463
direct-conversion architecture, 304
direct-conversion receiver, 109
direction finding, 498
direction of arrival (DOA), 143
DV-Hop, 138–40

ECMA, 297, 299
electrodes, 459
electromagnetic
  coupling, 422
  properties, 419, 427, 438
  property characterization, 419
electromagnetic bandgap, 52
electromagnetic waves, 482
electrotextile, 418–20, 429, 435,
EM envelope characteristic, 6–9, 15,
    18–20
end-fire radiation characteristics, 490
energy consumption, 136, 156–7,
    159–62, 174
equivalence principle, 483
equivalent memoryless envelope characteristic,
    see em envelope characteristic
error, mean absolute, 145, 156–7,
    164, 175–8
expected improvement function, 432
external antenna, 462, 474

factor graphs (FG), 151
fall time, 319

fast Fourier transform (FFT), 11
feed network, 333
  active, 333
  corporate, 333
feed point, 496
figures of merit, *see also* PA Model figures of
    merit
  error vector magnitude (EVM), 16
  normalized mean squared error (NMSE), 16
  peak-to-average power ratio (PAPR), 13
  signal-to-noise ratio (SNR), 5
filling factor, 372
filter, 276–8, 287–93, 327, 335
  active, 331
  band-pass, 327, 331
  frequency-agile, 326
  microstrip, 290–293
  multiband, 287–90, 326
    evenly spaced equal-bandwidth, 288–9
    stepped-impedance line, 289–90
  prototype, 290–293
    asymmetrical dual-band, 291–3
    quad-band, 290–291
  reconfigurable, 327
  recursive, 331
  tuneable, 326, 331–2, 353
flares, solar, 358
  accelerate particle populations, 362
  bremsstrahlung, 361
  flat or rising spectra, 363
  probability distribution, 364
  power-law density function, 365
  radiation mechanisms, 361
  radio-spectrum, 362
  rapid fluctuations, 363
  syncrotron emission, 361
flux, *see* Sun
foam material, 495
FPAA, 327
free access mat, 460
frequency detuning, 417
front-end, 325
  adaptive, 326, 337, 345, 353
  antenna, 337
  architectures, 326, 337, 353
  frequency-agile, 326
  multi-band, 326, 337, 345, 353
  reconfigurable, 326, 336–8
  wideband, 326, 337, 345, 353
Fourier series behavioural model, *see* PA model

GaAs, 326, 333, 336–41, 344–5, 353, 501
gain,
  available, 440
  free-space, 444, 450
  on-body, 441–2, 444, 451
  transducer, 442
gain enhancement, 64, 76
GaN, 326, 336–7, 341, 353
Gauss-Seidel, 168–70
Gaussian filter, 306
generalized wheeler cap method, 431
global navigation satellite system (GNSS), 228
  acquisition of signal, 230, 231
  COMPASS, 229, 231
  Galileo, 228, 231
  GLONASS, 228, 231
  interface control documents, 231
  navigation message, 229–30
  navigation solution, 230
  Navstar GPS, 228, 231
  pseudorange measurement, 230
  signal model, 230
  tracking of signal, 231
GPS, 136
gradient descent, 155–6, 167–70
grating lobes, 489, 492

harmonic balance, 388, 395–7, 399, 402–3, 410,
    412
HBT, 342, 502
HFSS, 296
high impedance surface, 51
homodyne, 302–3
homogeneous linear equation, 34
humidity
  relative, 418, 431

I/Q modulation, 110
  phase noise effect, 110
IBO, 4–5, 15, 17
IC, 182, 184, 194, 197
IC0803, *see* Action IC0803
ICI estimation, 120
IEEE 802.15.6, 460
IEEE 802.15.3c, 297
impedance matching, 490
  balun structure, 491
  impedance mismatch, 491
  microstrip-aperture coupling, 491
impulse radio architecture, 305

input backoff, *see* IBO
instantaneous voltage transfer characteristic, *see*
     IVT characteristic
intelligent microsystems, 336, 341
integrated circuit, 175–7, 181–2, 196
  mixed-signal, 181–2, 196
integration
  heterogeneous, 342
    hybrid, 336–7
  monolithic, 336–7, 341–2
intensity (of radiation), 359
inter-node measurements, 136–7, 140, 147, 150
intercarrier interference, 104, 111
interconnection, 419, 438
interferences, 55, 501
intermediate frequency (IF), 302–3
inverse fast fourier transform (FFT), 11
inverse model, 28–33, 35–8, 43–4, 47
  mapping, 28, 35
inverse planar antenna characterization
     method, 419
ionosphere
  Chasovitin, model of, 368
  collision frequency, 367
  electron density, 367
  F2 layer, 368
  stratified plasma, 367
  variability, 366
isolation, 316, 318, 329, 336, 338, 348, 495
IVT characteristic, 4, 7, 9–14, 17, 19–21, 23

Kalman filter, 151
Karhunen-Loeve transformation, *see* principal
     component analysis
KOSMA telescope on Gonergat (Swiss), 362
kriging, 432

laser-doppler-vibrometer, 342–3
lateration, 149–50
Latin hypercube design, 432
leaky coaxial cable, 455
least squares, 154, 158, 168, 174
LNA, 336
  frequency-agile, 326, 338, 340
  multi-band, 326
  reconfigurable, 327, 337–41, 344–8, 353
  tuneable, 326, 338
local oscillator (LO), 302, 307
localization, 217
  direct calculation method, 217, 223

localization algorithms
  centralized, 137, 146–9, 151–2, 173, 175
  distributed, 137, 139, 144, 146–53, 155–6,
     170, 173–4
location update phase, 145, 152, 162, 164
loss tangent, 419, 421, 432, 435, 437
low distance selection, 163–4, 170
low-noise amplifier, 436, 439
low path loss sleection, 137, 162
low-profile antenna concept, 387–9, *see also* SIW
     antennas
low-temperature co-fired ceramics (LTCC), 388
lumped element
  antenna's reflector, 61

magic surface, 457
matched impedance region, 496
matching networks, 338–40
  reconfigurable, 338–40
material, artificial, 51
matrix-pencil method, 420, 422
maximum likelihood estimation (MLE), 148,
     174–5
maximum power point, 448
Maxwell equations, 482
MDS, 147–8, 167, 174–5
MDS-MAP, 147
measurements (PA), 4, 7, 14–15, 19, 21
measurement phase, 152–3
medium access control (MAC), 299
memory to equivalent memoryless upper bound,
     see MEMUB
MEMS switch, 458
MEMUB, 7, 14, 18–19, 23
meshed ground plane, 475
metamaterial, 51
  phase diagram, 53
Metäshovi Radio Observatory (Finland), 359
microstrip, 420, 427, 439, 446
microwave filter, 27–49
  bandwidth, 37–8, 43
  center frequency, 37–8, 43
    cavities, 27, 30–31, 36, 40–43
    characteristic function, 28
  couplings, 27, 29–30, 36, 40–42
  cross-couplings, 29, 36, 40, 43
    detuning, 29
  order, 28–32, 35–7, 41–4, 49
  passband, 37–8, 43
  reflection characteristics, 29, 34–49

microwave filter (*Continued*)
　scattering characteristics, 37, 44
　　topology, 30, 36–43
　transmission characteristics, 34, 38–42, 49
　tuning elements, 27–43, 47–9
　tuning screws, 27–33, 36–7, 41, 47–8
Miller effect, 308
millimetre-wave, 325, 336
　applications, 337, 341–2
　circuits, 325, 338
　frequency, 338, 349, 353
　observations, 358
　radiometers, *see* radiometers
MIOS, 357
Mixer, 303, 309
MMIC, 326, 336–41, 344–7, 353
model, 173, 175–6, 182–7, 189–95, 204–5
　analytical, 182, 185–6, 195
modified Bessel-Fourier behavioural model, *see*
　　PA model
modified Saleh behavioural model, *see* PA
　　model
modulation schemes, 298
　binary phase shift keying (BPSK), 304–5
　phase shift keying (PSK), 298
　pulse-position modulation (PPM), 298
　quadrature amplitude modulation (QAM),
　　298
　quadrature phase shift keying (QPSK), 305
moisture regain, 429
monolithic microwave integrated circuit (MMIC),
　　296, 306
　BiCMOS, 296, 321
　CMOS, 296, 321
　GaAs, 296, 306
　HBT, 306
　InP, 306
　SiGe, 296, 306
multi-carrier, 10
multilateration, 138
multifunctional and multitasking smart systems,
　　499
　clutter, 501
　digital beam forming, 501
　guided-missile routing support, 499
　naval surface search, 499
　target classification, 501
　volume scanning, 499
multipath, 501
　propagation, 232–3

multiplatform design, 440
mushroom, 53
mutual shadowing, 222–3

NMSE, 14, 17–18, 23
Nobeyama Radioheliograph (Japan), 362
node selection, 136–7, 156–7, 159–65, 168,
　　170–171, 174, 176–7, 181
noise, 173, 175–87, 189, 191–6, 198–9, 201,
　　204–5
　coupling, 184–6, 193–4
　distribution, 422
　factor, 443
　figure, 304, 313, 315, 320, 437, 439
　power density, absolute, 443
　source, 193–4, 201
　substrate, 173, 175–81, 184, 192–6, 198–9,
　　204–5
　suppression, 195–6, 198
　switching, 175–8, 180, 182–7, 189, 191–6,
　　201, 204–5
　temperature, effective, 443
non-Bayesian, 150–68
non-cooperative, 137, 139, 145–6, 152,
　　157
non-located nodes, 139–40, 144–6, 149, 154–8,
　　161, 163, 166–7, 170–171
nonlinear analysis of oscillators, *see* harmonic
　　balance
nonlinear modeling, 5–8, 10, 13
normalized mean square error (NMSE), 17, *see
　　also* NMSE
null, 501
nulling interferometers, 366
Nyquist zone, 4–7, 10–11, 13, 16

OBO, 4
OFDM, 4, 24, 298, 302
OFDM capacity under phase noise, 114
　instantaneous capacity, average capacity,
　　114–18
OFDM system model, 113
OLPL-NS-LS, 168, 173–9
on-body transmission, 475
optimization
　forward, 427
　full-wave/circuit co-, 419
　surrogate based, 420, 429
orthogonal frequency division multiplex, *see*
　　OFDM

oscillator, 104–9, 327, 334–5
    feedback, 334
        free running, phase locked loop, 104–107
        generalized model, 107–9
        voltage controlled, 344–5, 353
oscillator phase noise, 104
output backoff, *see* OBO
oxygen absorption, 297

PA, *see* power amplifier
    frequency-agile, 326, 338
        multi-band, 326
        reconfigurable, 327, 337, 341
        tuneable, 326
PAPR, 5, 17, 20
parallel method, 31–3, 41–3
passive inter modulation (PIM), 30
patch resonators, 461
path loss, 135–7, 143, 153, 160, 162–5, 168–75,
    177–9
perfect electric conductor, 55
perfect magnetic conductor, 55
permittivity, 419, 427
    effective, 420–2
phase diagram, 53
    experimental determination, 54
    simulation, 53
phase-locked loop (PLL), 307
phase noise, 307, 334, 336
    suppression, 118
phase shifter, 326, 336–7, 501
    distributed MEMS transmission line, 348–50,
        352
    true-time delay line, 348–9, 352
    tuneable dielectric-block loaded-line, 348,
        350–2
phased arrays, 326, 348
physical layer (PHY), 298
    audio/visual (AV) PHY, 298
    high-rate PHY (HRP), 298, 301
    high-speed interface (HSI) PHY, 299
    low-rate PHY (LRP), 298, 301
    single carrier (SC) PHY, 298
PIFA, wearable, 419
PIN diode, 319
Plank's law, pag. 359
principal component analysis, 44–5
polyimide, 439, 447, 451
polynomials, 34
power amplifier (PA)

amplitude characteristic, 11, 16, 20
envelope characteristic, 5–7, 9–10, 13, 23
gain characteristic, 15–16
LDMOS PA, 4–9, 11, 13–15, 17, 19, 20
memoryless PA, 5, 23
model, 4, 18, 23
    amplitude model, 12
    approximation model, 12
    behavioural model, 3, 5, 22–4
    Bessel-Fourier model, 4, 7–10, 12–23
    BF model, *see* Bessel-Fourier model
    coefficients, 7, 16
    coefficient spectrum, 12, 21
    comparisons, 7, 17–22
    dynamic range ratio parameter, 7, 9–10, 13,
        23
    EM (equivalent memoryless) model, 18
    envelope model, 8–11, 14–15, 19–20
    error, 17
    extensibility, 15, 21
    extraction, 4, 13, 17, 19–20
    figures of merit, FOMs, 7, 14–15, 17–18, 23
    Fourier series (FS) model, 10, 12–14
    high order model, 14, 24
    IVTC, *see* instantaneous voltage transfer
        characteristic
    low order model, 3–4, 13–16, 19, 22–23
    margin of reliability, *see* margin of reliability
    margin of reliability, 15, 19–20, 23
    MBF, *see* Modified Bessel-Fourier
    mean absolute error, 7
    memoryless, 7, 23
    model accuracy, 3–4, 7–8, 10, 12, 14–17,
        22–3
    model average error (AE), 8, 14–16, 22
    model order, 7–8, 16
    Modified Bessel-Fourier (MBF), 4, 8–9,
        12–24
    modified Saleh model, 3, 16–18
    MS model, *see* modified Saleh model
    power series model, 16–18, 22
    Saleh model, 15–18, 22
    zero model, 7, 13, 15–16, 22
nonlinear PA, 3–4, 22–4
PA output, 5, 7, 17, 18
PA saturation, 14, 20
phase characteristic, 12, 21
power amplifier model, *see* PA model
RF power amplifier, 3
solid state power amplifier, 4

power combiner, 327, 333, 406–12
  active, 333
power harvesting, 418
power series behavioural model, *see* power
    amplifier (PA)
power spectrum, 181–3
power splitter, 327
preprocessing, 211–12
  time-zero setting, 211–12, 220–221
printed circuit board (PCB), 388
PROMFA, 326–35, 353
  amplifier, 329
  CMOS chip, 334
  sub-cells, 327–8
  tunable band pass filter, 331–2
  tunable oscillator, 334–8
propagation constant, 422

quality factor, 308, 483
Q-factor, *see* quality factor
quadrature-phase shift keying (QPSK), 13–17

radar, 277, 325, 335–6, 342, 348
  multifrequency, 277
radargram, 211, 220–222
radiation pattern, 417, 425, 449
radio interferometry, 144–5
radio frequency, 175–6, 183, 194, 196
  circuit, 175–6, 183, 194, 196
radio frequency interference, 232
  continuous wave, 232, 252
  intentional, 232
  pulsed, 232
  spoofing, 232
  swept, 232
  wideband, 232, 252, 265
radiometers, 335
  background, 373
  black-body, 379
  block diagram, 376
  calibration periodicity, 381
  calibration, 378
  detector sensitivity, 377
  detector, square-law, 376
  Dicke switch, 379
  directional coupler, 379
  gain stability, 373–4
  instrumentation amplifiers, 377
  noise source, 379
  noise-adding, 379

physical temperature, 381
  receiver noise temperature, 374
  resolution, 373
  sensitivity, 377
  sensors, 335
  systems, 325
radiometric background, *see* radiometers
range-based, 136–7, 140, 152
range-free, 136–7, 140, 144, 152
rational function, 34
raw radar signals, 211, 220
Rayleigh-Jeans approximation, 359
receiver architectures
  bandpass sampling receiver, 4–7, 10, 13, 16
  low-IF receiver, 4
  six-port interferometer receiver, 5
  super-heterodyne receiver, 3–4
  zero-IF receiver, 3–4
rectangular waveguide, 388
reflection coefficient, 420, 427–9
reflection phase method, 52
remote sensing, 325
resistivity, surface, 429, 435, 446
resonance frequency, 420, 425, 427
RF front-ends, 275–6
RF hyperthermia, 503
RF imperfections, 103
RF-MEMS, 326–7, 336–53
  cold-switching, 340–341
    hot-switching, 340–341
    phase shifters, 337, 348–52
  switches, 336–52
RFIC, 326, 336
rise time, 317
Robot, 36, 47–9
  IAFTT, 48
  SCARA, 48–9
round-trip time of arrival, 142
RSS-based 135–7, 143, 151–65, 167–8
RSSI, 142

Saleh behavioural model, *see* PA model
sampling, 277–83
  bandpass, 277–8
  direct, 277–8
  of frequency-sparse signals, 278
  minimum sub-Nyquist, 278–83
  multiband,, 278
  sub-Nyquist, 277–8
scientific and medical (ISM) band, 504

semidefinite programming, 148
sequential method, 30–31, 36–41
    subfilter, 30–31
shadowing, 143, 153
sheet-like waveguide, 456
short-range wireless communication, 460
shorting wall, 445
sievenpiper, 52
SiGe, 326, 336–7, 342–5, 353, 502
signal, 276–89
    acquisition, 278–83
    asymmetrical spectrum, 284
    dual-band, 279–82, 285–7
    interference, 287–9
    multiband, 277–8, 282–4
    multichannel, see signal, multiband
    spurious, 276
    symmetrical spectrum, 284
signal-to-interference-and-noise ratio,
        113–14
signal-to-noise ratio (SNR), 305, 307
simulation, 184–5, 189, 194–5, 198–9,
        201–4
    Matlab, 198
    SPICE, 184–5, 194–5, 201–3
simulator
    circuit, 439
    frequency domain, 440
    solar, 448
single-ended to differential converter (STD),
        307–8, 312
SIW antennas, 387–415
    array, 387–9, 402, 406–14
        beam scanning, 387–8, 406, 412–14
        beam steering, see beam scanning
        coupled oscillator, 387–8, 395, 401,
            406–14
        phased, 387, 406, see also coupled oscillator
        power combining, 406–12
    coupling between, 387, 406–9, 412, see also
            coupled oscillator
    electromagnetic analysis, see full-wave
            simulation
        feed line, 389–93, 392 (Fig.), 395–7,
            402–3 (Fig.)
    full-wave simulation, 388–9, 392, 395, 397,
            399, 402, 407, 412, see also S-parameters
    gain, 390, 393,
            396–7, 400–401, 404, 411–12, see also
            radiation pattern

input matching, see S-parameters
oscillator (antenna), 387–9, 395–405,
        406–14
    DC-to-RF conversion efficiency, 397, 401,
        404, 412
    effective radiated power (ERP), 396–7,
        398 (Fig.), 400–401, 404, 405 (Fig.),
        411–14
    frequency (of oscillation), 395–7, 400, 403,
        406, 410, 412–13
    frequency tuning in, 395, 397–406,
        412–14
    phase noise, 389, 397–8 (Fig.), 401, 402
        (Fig.), 404, 405 (Fig.), 410–411
    passive, 388–97, 399, 401, 404, 407–10, 412
    patch, 389–91, 392 (Fig.), 395–7, 398 (Fig.),
        401–5, 412–14
    radiation efficiency, 387, 389–90, 393
        radiation pattern, 390, 392 (Fig.), 393–4,
            397, 398 (Fig.), 400–401, 404–5,
            411–14
    return loss, see S-parameters
    S-parameters, 390, 392 (Fig.), 393, 394 (Fig.),
        395, 397, 402, 407–9, 412
    slot, 389, 391–4, 397–401, 402 (Fig.), 407
    synchronization, 406, 410, 412–13
SIW components, 388
smart suit, 472
software defined radio (SDR), 3, 5, 10, 275–8
solar cell, 419
    amorphous silicon, 418, 447
    integration, 445, 447
solar flux units (sfu), definition of, see Sun
space weather, 358
spectral domain integral equation approach, 495
spectral flux density, see Sun
standard IEEE temperature (for noise figure),
        373
statistical array processing, for GNSS
        applications, 243
    deflection coefficient, 253
    direction of arrival, estimation of, 264, 269
    GLRT colored version, signal acquisition, 251
    GLRT white version, signal acquisition, 251
    matched filter signal acquisition, 251–2
    maximum likelihood estimator
        (MLE), 244
    signal model, 243
stepped-impedance, 289–90
subsidiary waveguide, 455

substrate, 173, 175–84, 186, 191–6, 198–9, 204–5
  heavily doped, 175, 183–4, 205
  lightly doped, 175, 182, 184–5, 189, 195, 205
  textile, 418, 423, 427, 440
substrate integrated waveguide (SIW technology), 387–9, 395, 401, 406–7
Sun
  angular radius, 360
  quite sun properties, tab 13.1, 360
  solar flux units (sfu), definition of, 359
  spectral flux density, 359, 361, 363
super-heterodyne architecture, 302
surface current, 62, 64, 71–2, 74, 77, 79
switch
  capacitance ratio, 343, 349
    capacitive, 342–4, 349–50
    contact resistance, 338
    cycles, 344–5
    Dicke, 335
    matrices, 335
  high-speed switching, 317
  off-time, 342
    ohmic contact, 337
    on-time, 342
    redundancy, 335
    RF-MEMS, 336–52
    semiconductor, 335
    solid-state, 335
    SPDT, 345, 348
  pass transistor, 327
  power breakdown, 340–341
  power handling, 340–341, 350–351
  pull-down voltage, 337, 343, 349
  single-pole single-throw, see SPST
  SPST, 305, 317, 337, 339–41
  SPxT, 348–9
switches, see switch
system on substrate (SoS), 388

tail interpolation, 123
textile, 468
  system, wearable, 417, 436
third-order intermodulation product (IIP3), 304
time difference of arrival (TDOA), 141–2
time domain solver, 446
time of arrival (TOA), 140–2, 158
  association method, 216–17, 222–3
  estimation, 215–17

time-domain phase-noise mitigation
  Syrjälä, 126
tracking, 217–18
  multiple target tracking (MTT) system, 218, 223
  short-range, 209–10, 218–23
transformer, 315
transient analysis of oscillators, 395–6, 403, 410
transmission line, 421
  textile, 418, 422
transmission loss, 461, 474
transmission zero, 289, 291
trilateration, 138
tuning methods
  coupling matrix extraction, 27
  human operator, 27, 30, 35, 38, 43
  time domain tuning, 27
2-D transmission sheet, 459

ultra wide band, see UWB
unidirectional antenna, 51–81
unit cell, 52, 59
UWB, 173–8, 180–183, 195, 199, 204, 489
  circuit, 175, 204
  radar, 209
    handheld, 210
    M-sequence, 211, 218–20
    network, 223–4
    signal processing, 210–18, 223–4
  system, 173–4, 176, 195

V-band, 295, 297, 330
  standard, 297
Vapnik-Chervonenkis dimension, 32
Varactor, 307, 344–5
VCO, see oscillator; voltage controlled oscillator
vector network analyser, 30, 34, 48
Volterra series modeling, 7–8, 10–13
  least squares extraction, 12
  memory effects, 7–8, 10–13
  Volterra kernels, 8, 10–13
voltage controlled oscillator (VCO), 307

waveguide modes, 491
wavelet transform
  Daubechies D4 transform, 44–6
weak signal enhancement, 213–4
  advance normalization method, 213–4, 221

wearable antennas, *see* antenna
wearable devices, 474
WiFi, 297
WiGig, 301
Wilkinson power divider, 493
wireless body area network (WBAN), 460, 471
wireless energy transfer, 503
WirelessHD, 300

wireless power transmission, 458
wireless sensor networks (WSN), 135–9, 142,
        146, 149, 152, 156–7, 159–60, 162, 174
WPAN, 297

Zero-IF, 303, *see also* architecture; receiver
        architectures
zero model, *see* power amplifier (PA)